Macrocyclic Lactones in Antiparasitic Therapy

Macrocyclic Lactones in Antiparasitic Therapy

Edited by

J. VERCRUYSSE

Faculty of Veterinary Medicine
Department of Virology, Parasitology, Immunology
Ghent University
Belgium

and

R.S. REW

Pfizer Animal Health
Exton
Pennsylvania
USA

CABI *Publishing*

CABI *Publishing* is a division of CAB *International*

CABI Publishing
CAB International
Wallingford
Oxon OX10 8DE
UK

Tel: +44 (0)1491 832111
Fax: +44 (0)1491 833508
E-mail: info@cabi.org
Website: www.cabi-publishing.org

CABI Publishing
745 Atlantic Avenue
8th Floor
Boston, MA 02111
USA

Tel: +1 (617)682-9015

E-mail: cabi-nao@cabi.org

A catalogue record for this book is available from the British Library, London, UK.

The Library of Congress has cataloged the hardcover edition as follows:
Macrocyclic lactones in antiparasitic therapy / edited by J. Vercruysse and R.S. Rew
p. cm.
Includes bibliographical references (p.).
ISBN 0-85199-617-5 (alk. paper)
1. Avermectins. 2. Lactones. 3. Macrocyclic compounds.
4. Antiparasitic agents. I. Vercruysse, J. (Jozef) II. Rew, Robert S.
RM412 .M33 2002
616.9'6061--dc21 2002004075

ISBN 978 0 85199 617 2 (HB)

Typeset by AMA DataSet Ltd, UK
Printed and bound in the UK by CPI Group (UK) Ltd, Croydon, CR0 4YY

Contents

Contributors

M.R. Alvinerie, *INRA Laboratoire de Phamacologie, 180 Chemin de Tournefeuille, F-31931 Toulouse, France.*

J. Arends, *S&J Farms Animal Health, 2340 Sanders Road, Willow Springs, NC 27592, USA.*

W.J. Baker, *Pfizer Central Research, Eastern Point Road, Mailstop 8200-40, Groton, CT 06340, USA.*

I. Barger, *597 Rockvale Road, Armidale, NSW 2350, Australia.*

K.R. Brown, *8111 Winston Road, Philadelphia, PA 19118, USA.*

G. Buescher, *Novartis Animal Health Inc., CH-4002 Basel, Switzerland.*

J. Caracostantogolo, *Jose Paula Rodriguez Alvez 794, 1408 Ciudad de Buenos Aires, Argentina.*

G.A. Conder, *Pfizer Central Research, Eastern Point Road, Mailstop 8200-40, Groton, CT 06340, USA.*

R.L. Coop, *Moredun Research Institute, Pentlands Science Park, Bush Loan, Penicuik, Midlothian EH26 0PZ, UK.*

R.L. DeLay, *Fort Dodge Animal Health, Agricultural Research Center, PO Box 400, Princeton, NJ 08543-0400, USA.*

C. Eddi, *Alberti 664, 1714 Ituzalogo, Argentina. New address: Animal Production and Health Division, Room C-528, FAO, Vialle delle Terme di Caracalla-00100, Rome, Italy.*

T.G. Geary, *Pharmacia & Upjohn Co., 301 Henrietta Street, Kalamazoo, MI 49007-4940, USA.*

C. Genchi, *Dipartimento di Patologia Animale, Igiene e Sanita Pubblica Veterinaria, Sezione di Patologia generale e Parassitologia, Universitá degli Studi di Milano, Via Celoria 10, I-20122 Milan, Italy.*

M.J. Gliddon, *Fort Dodge Animal Health, Agricultural Research Center, PO Box 400, Princeton, NJ 08543-0400, USA.*

J.-F. Graf, *Novartis Animal Health Inc., CH-4002 Basel, Switzerland.*

C.B. Greenacre, *Department of Comparative Medicine, College of Veterinary Medicine, University of Tennessee, PO Box 1071, Knoxville, TN 37902-1071, USA.*

J. Guerrero, *Department of Pathobiology, School of Veterinary Medicine, University of Pennsylvania, Philadelphia, PA 19104, USA.*

D.R. Hennessy, *CSIRO Animal Production McMaster Laboratory, Clunies Ross St., Private Bag 1, Delivery Center Blacktown, Sydney, NSW 2148, Australia. New address: Veterinary Health Research Pty Ltd, 1 Rivett Rd, Riverside Corporate Park, North Ryde, NSW 2113, Australia.*

P.A. Holdsworth, *Director Scientific & Regulatory Affairs, Avcare, Locked Bag 916, Canberra, ACT 2601, Australia.*

F. Jackson, *Moredun Research Institute, Pentlands Science Park, Bush Loan, Penicuik, Midlothian EH26 0PZ, UK.*

M. Jung, *Novartis Centre de Recherche Santé Animal SA, CH-1566 St.-Aubin, Switzerland.*

R.M. Kaplan, *Department of Medical Microbiology and Parasitology, College of Veterinary Medicine, University of Georgia, Athens, GA 30602-7387, USA.*

T.R. Klei, *Department of Veterinary Microbiology and Parasitology, School of Veterinary Medicine, Louisiana State University, Baton Rouge, LA 70803, USA.*

S.E. Little, *Department of Medical Microbiology and Parasitology, College of Veterinary Medicine, University of Georgia, Athens, GA 30602-7387, USA.*

S.E. Marley, *Merial, 3239 Satellite Blvd, Duluth, GA 30096, USA.*

R.J. Martin, *Department of Biomedical Sciences, College of Veterinary Medicine, Iowa State University, Ames, IA 50011, USA.*

M. Maurer, *Novartis Centre de Recherche Santé Animal SA, CH-1566 St.-Aubin, Switzerland.*

J.W. McCall, *Department of Medicine, Microbiology and Parasitology, College of Veterinary Medicine, University of Georgia, Athens, GA 300602, USA.*

C.M. Monahan, *College of Veterinary Medicine, Ohio State University, 1900 Coffey Rd, Columbus, Ohio 43210, USA.*

A. Nari, *Via Odoardo Beccari 14 apt. 6, I-00154 Rome, Italy.*

R.K. Prichard, *Institute of Parasitology, McGill University, MacDonald Campus, 21111 Lakeshore Road, Ste-Anne-DeBellevue, Quebec H9X3V9, Canada.*

R.S. Rew, *formerly Pfizer Animal Health, 812 Springdale Drive, Exton, PA 19341, USA, now at Rewsearch Inc., 400 N Wawaset Road, West Chester, PA 19382, USA.*

D.W. Rock, *Fort Dodge Animal Health, Agricultural Research Center, PO Box 400, Princeton, NJ 08543-0400, USA.*

A.P. Robertson, *Department of Biomedical Sciences, College of Veterinary Medicine, Iowa State University, Ames, IA 50011, USA.*

A. Saito, *Sankyo Co. Ltd, Tokyo 104, Japan.*

W.L. Shoop, *Merck & Co., PO Box 2000, Rahway, NJ 07065, USA.*

M.D. Soll, *Merial 3239 Satellite Boulevard, Duluth, GA 30096, USA.*

J.W. Steel, *CSIRO Animal Production McMaster Laboratory, Clunies Ross St., Private Bag 1, Delivery Centre, Blacktown, Sydney, NSW 2148, Australia. New address: CSIRO Livestock Industries, 5 Julius Avenue (off Delhi Road), Riverside Corporate Park, North Ryde, NSW 1670, Australia.*

J. Vercruysse, *Faculty of Veterinary Medicine, Department of Virology, Parasitology, Immunology, Ghent University, Salisbury Laan 133, B-9820 Merelbeke, Belgium.*

K. Wardhaugh, *CSIRO Entomology, Black Mountain Laboratories, GPO Box 1700, Canberra, ACT 2602, Australia.*

A.J. Wolstenholme, *Department of Biology and Biochemistry, University of Bath, Bath BA2 7AY, UK.*

Preface

The first macrocyclic lactone (ML), ivermectin, was introduced as an antiparasitic drug in 1981, and its efficacy against nematodes and arthropods took parasite control to a new level. For the first time, a single product that was safe and efficacious against the majority of economically important internal and external parasites of all food-producing and companion animals was made available. The amount of the product required for activity was ten to 100 times less than that of previously used products. Ivermectin showed an unprecedented high efficacy – often up to 100% – against inhibited, larval and adult stages of the major nematodes and larval and adult arthropods. Because this product was highly lipophilic, it continued to remain in the treated animal and inhibit reinfection for extended periods of time. The extensive database on abamectin and ivermectin discovery, development and use was compiled into a book entitled *Abamectin and Ivermectin* edited by Campbell (1989).

Since that time, several new MLs including doramectin, eprinomectin, milbemycin A_3/A_4, moxidectin and selamectin have been developed for control of internal and external parasites. An extensive database has been generated for all MLs since *Abamectin and Ivermectin* was published. The overall objective of this book is to present the chemistry, pharmacology, mode of action, target animal safety, environmental impact, efficacy and resistance of all the MLs and to give the highlights of information on the use of MLs to control parasites in target animals, that is cattle, sheep/goats, horses, swine, dogs, cats, domesticated wild ruminants, man, mammalian pets and non-mammalians.

The authors selected to write the 14 chapters of this book are considered to be the most knowledgeable in the field for the particular subject they were asked to review. The first five chapters give a more general review on the MLs, while the following eight chapters review the specific use of MLs for a particular host. Authors of Chapter 1, covering the chemistry, pharmacology and safety of the products, were invited from the specific companies that discovered and developed the products. Pharmacokinetics are dealt with separately in Chapter 2 because they have such a direct impact on efficacy, persistent activity, safety, residues and even resistance that it deserved its own chapter. Chapter 3 provides the opportunity to review the new information on mode of action, giving glutamate-gated chloride ion channels the spotlight. The impact of the MLs on dung fauna is reviewed in Chapter 4. Resistance mechanisms and field resistance are reviewed in Chapter 5. The use section of the cattle chapter (Chapter 6) has been subdivided by geographic regions of the world, and then by management segments of cattle, since use of these products in cattle is so different from one geographic area to another and from one management segment to another. Chapter 7 reviews ML use in sheep and goats. Use of MLs in horses is reviewed in Chapter 8, with a specific focus on how we should evaluate resumption of egg appearance in faeces. Chapters 9 and 10 review publications on use of MLs in pigs and in dogs and cats. Chapters 11 and 12 attempt to cover use of MLs in a variety of mammals and non-mammals, for which many of the publications are anecdotal or on studies done with very few animals in poorly controlled tests, but since no labels are available for these minor species, these chapters may serve as starting points for further investigations and more extensive databases. Chapter 13 reviews the data on human use, still essentially only for ivermectin. The last chapter (Chapter 14) tries to answer the question of 'where do we go from here?' by examining scientific, social, political and economic issues that control the future of the MLs.

The target audience of this book is not only the basic researcher in antiparasitics and the field researcher involved in parasite control, but also the practising veterinarian. Until now, too often MLs have been misused, and we hope that the different chapters on the use of MLs in target animals will result in a more appropriate usage of MLs by veterinarians.

J. Vercruysse and R.S. Rew
December 2001

Chapter 1

Chemistry, Pharmacology and Safety of the Macrocyclic Lactones

Chapter 1.1

Ivermectin, Abamectin and Eprinomectin

W. Shoop and M. Soll

Introduction

The avermectins (e.g. ivermectin, abamectin and eprinomectin) are closely related 16-membered macrocyclic lactones derived from the soil micro-organism *Streptomyces*. Discovered in 1976, the first commercial use of these compounds came with the introduction of ivermectin for use in animals in 1981. Since then, the avermectins have been approved for use in a number of mammals, including sheep, horses, cattle, swine, dogs, cats and humans. Additional approved uses of ivermectin extend to goats, reindeer, camels, bison, rabbits, foxes and red deer, and the published literature contains reports of use to treat infections with more than 300 species of endo- and ectoparasites in a wide range of hosts.

Ivermectin

Ivermectin was the first macrocyclic lactone developed for use in animals and it revolutionized antiparasitic control in production animals, heartworm chemotherapy in companion animals, and antifilarial chemotherapy in humans. Ivermectin shares with abamectin, eprinomectin and all other avermectins/milbemycins a unique pharmacophore responsible for these activities (Fig. 1.1.1). The pharmacophore, consisting of a 16-membered macrocyclic backbone to which are fused both benzofuran and spiroketal functions, is a three-dimensional arrangement of structural and electronic molecular fragments which is recognized by specific chloride ion channel receptors. This pharmacophore is mechanistically responsible for the mode of action of ivermectin and its relatives which, in turn, defines the drug class.

It is the unique pharmacophore that accounts for the fact that ivermectin and all avermectins/milbemycins are structurally superimposable, that they bind to the same glutamate-gated chloride channel receptors, that they competitively displace one another at those receptors, that they are effective against the same spectrum of biologically diverse invertebrate parasites, that they kill these invertebrates through hyperpolarization and flaccid paralysis, that they are efficacious at similar dosages, that they elicit similar signs at toxic levels in mammals, and that they show cross-resistance to the same drug-resistant parasites (Shoop *et al.*, 1995).

Ivermectin belongs to the avermectin subclass within the avermectins/milbemycins. Although the pharmacophore of the avermectins and milbemycins is the same, these two subclasses differ in substituents at C-13, C-22,23 and at C-25. At C-13, the avermectins possess a sugar moiety known as a bisoleandrosyloxy, whereas in the milbemycins there is no substituent at that position. Therefore, one can think of avermectins as glycosylated milbemycins or, conversely, of the milbemycins as deglycosylated avermectins. Naturally occurring avermectins also possess single

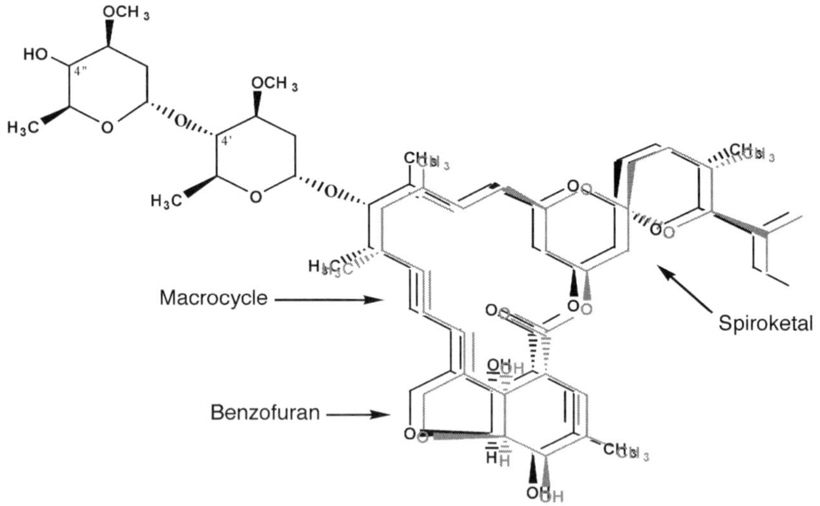

Fig. 1.1.1. Structure of ivermectin and a milbemycin (offset to right) showing the basic tri-partite pharmacophore.

or double bonds between C-22 and C-23, whereas the milbemycins have only a single bond at that position. Lastly, ivermectin, abamectin and eprinomectin have isopropyl and sec-butyl substituents at C-25, whereas milbemycins have simpler methyl or ethyl groups.

The avermectins/milbemycins are naturally produced by soil-dwelling actinomycetes from the genus *Streptomyces*. Strains of *Streptomyces* spp. which produce *milbemycin-type* compounds are found commonly in soil samples in screens for bioactivity. Because of their potency in bioassays, especially antiparasitic assays, tedious and expensive tests must continually be undertaken to isolate and identify the structures produced by these organisms before recognizing them to be known or previously described. Conversely, strains of *S. avermitilis* which produce *avermectin-type* compounds are rare. In fact, only two individual collections have ever been reported. The original culture from *S. avermitilis* produced a family of eight avermectins and, through various substrains, gave rise to ivermectin and every other commercialized avermectin. Since the discovery of this soil-dwelling species from Asia, it and its daughter strains have been kept in continuous culture. The second finding was a strain of *S. avermitilis* from Italy (US Patent 5,292,647). Unfortunately, no additional novel avermectins were isolated from this second strain.

The original *S. avermitilis* strain was collected in what has by now become legend in natural product discovery and development (Stapley and Woodruff, 1982). Through a collaborative agreement between Merck and Co., Inc. in the USA and Kitasato Institute in Japan, the latter was to collect naturally occurring microorganisms and the former was to test them for various biological activities. One of the culture broths from a Japanese golf course was found to be active in an *in vivo* parasite model consisting of mice experimentally infected with the gastrointestinal nematode, *Nematospiroides dubius*. The active broth was immediately assigned to isolation chemists to determine the active structures, which revealed the family of eight naturally occurring avermectins for the first time (Miller *et al.*, 1979; Albers-Schonberg *et al.*, 1981). It was estimated using high performance liquid chromatography (HPLC) that the original broth responsible for the anthelmintic activity contained only 9 $\mu g\ ml^{-1}$ of the avermectins. This modest yield was quickly increased tenfold through modification of the culture medium used to grow *S. avermitilis* and UV radiation yielded a high producing strain with a further fivefold improvement in avermectin metabolism (Burg *et al.*, 1979; Stapley and Woodruff, 1982). Production optimization continues unabated to this day.

The discovery of the avermectins resulted from a complex, high-risk screening strategy predicated on the knowledge that microorganisms compete with one another using bioactive chemicals. This screening strategy offers two significant advantages. First, it makes possible the discovery of complex molecules with biological activities that have already been optimized over millions of years of evolution. Incredibly, many of the threats to soil-dwelling organisms come from similar phylogenetic groups that are threats to livestock, companion animals and humans. Secondly, once these complex molecules are identified, the microorganism that produced them can be harnessed to ferment the target molecules on industrial scales. It is sobering to think that the discovery of the avermectins would only have been an academic exercise if their producer, *S. avermitilis*, had not been captured as well and developed to allow production by fermentation. For many animal health applications, a synthetic chemical process that requires five steps beyond starting material can potentially make development uneconomical, and total chemical synthesis of avermectin B_{1a}, ivermectin's starting material, requires more than 50 steps (White *et al.*, 1995). Manufacture on a commercial scale was therefore totally reliant on the ability to improve production of the organism that originally generated the compound.

The eight different avermectins produced by *S. avermitilis* are denoted A_{1a}, A_{1b}, A_{2a}, A_{2b}, B_{1a}, B_{1b}, B_{2a} and B_{2b}. The A-components possess a methoxyl group at C-5 where the B-group has a hydroxyl function; the 1-components have a double bond between C-22 and C-23 where the 2-components have a single bond with an hydroxyl group at C-23; and the a-components have a secondary butyl group at C-25 where the b-components have an isopropyl moiety.

It should be noted that separation of a- from b-components in large-scale fermentation is both impractical and unnecessary because these two homologues have virtually identical activities. Therefore, the avermectin literature most often refers only to A_1, A_2, B_1 and B_2 and it is usually inferred, if not stated explicitly, that each of these occurs as a mixture of a- and b-components; because the a-component is produced in greater proportion during fermentation, terminology such as ivermectin 'consists of not less that 90% a-component and not more than 10% b-component' is often used. These descriptions can lead to confusion because typically only the more abundant a-component of each mixture is shown in structural drawings.

Of the eight natural avermectins produced by *S. avermitilis*, only A_{2a}, B_{1a} and B_{2a} are produced in quantity during fermentation, making them desirable candidates for development. The B_1 homologues possess the highest potency and breadth of spectrum against nematodes, and are followed closely by the B_2 homologues. B_2, however, is safer to use; for example, the estimated oral LD_{50} in mice is approximately 15 mg kg^{-1} for B_1 and more than 50 mg kg^{-1} for B_2. It was data such as these that suggested to medicinal chemists that a semisynthetic analogue based on B_1 and B_2 components might provide a more optimized potency, spectrum, and safety profile than any of the other natural products. Consequently, much of the chemistry effort was directed toward members in those series.

Ivermectin, 22,23-dihydro-avermectin B_1, was the first avermectin/milbemycin to be developed for use in animals (Chabala *et al.*, 1980) and it was first made available commercially in 1981. In fact, because ivermectin is a mixture of B_{1a} and B_{1b}, it is more correct to say that it represented the first two avermectins to be commercialized for animal health. Ivermectin uses the B_1 mixture of natural components as the starting material and is synthesized by selective saturation of the *cis* 22,23 double bond, which gives it the same chair conformation found in the B_2 series. Structurally, ivermectin can be thought of as a hybrid between B_1 and B_2 (Fig. 1.1.2). It is virtually identical to B_2 except that it lacks the axial hydroxyl group at C-23 of the latter. Biologically, ivermectin maintains excellent potency and spectrum against nematode parasites, which is nearly as good as B_1, but it also has a greater safety factor (estimated LD_{50} in mice of approximately 30 mg kg^{-1}), which is more similar to the safety profile of avermectin B_2.

The broad spectrum of activity of ivermectin, which includes ecto-parasites, its excellent safety margins and new mode of action would have, on its own, produced a significant contribution to the world's anti-parasitic armamentarium. However, it was ivermectin's unprecedented potency that facilitated the formulation of a wide variety of oral, parenteral and topical dosage forms for cattle, sheep, goats, swine, horses, bison, camels, reindeer, dogs, cats and humans that has made it the largest selling antiparasitic drug in the world.

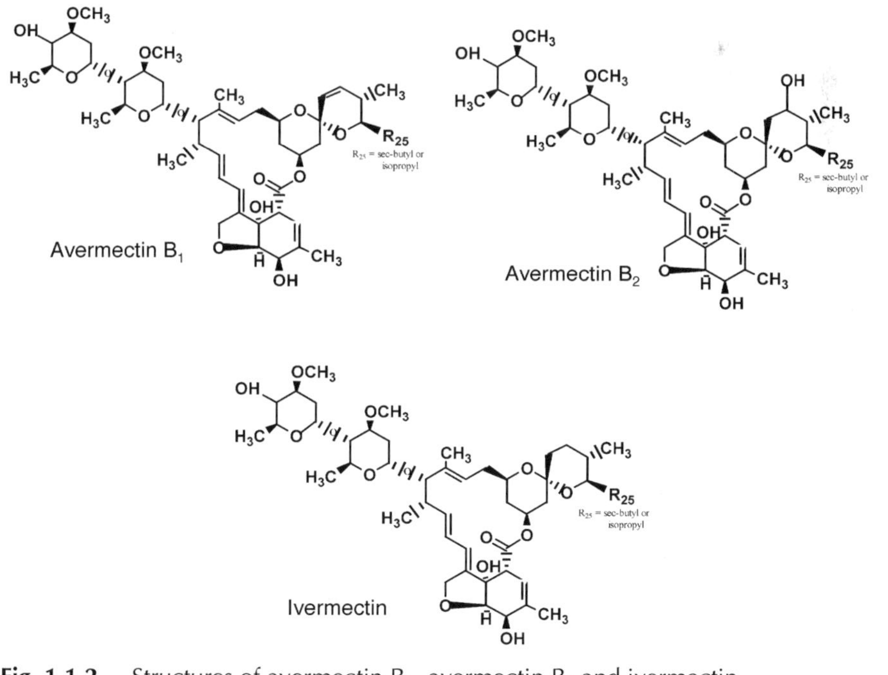

Fig. 1.1.2. Structures of avermectin B_1, avermectin B_2 and ivermectin.

Ivermectin pharmacology

Cattle

Egerton *et al.* (1981a) were the first to detail by titration the extraordinary potency of ivermectin against nematodes in cattle through oral and subcutaneous administrations. Treatment with 200 μg kg^{-1} of ivermectin either orally or subcutaneously eliminated >90% of immature and mature gastrointestinal nematodes such as *Haemonchus placei, Ostertagia ostertagi, Trichostrongylus axei, T. colubriformis, Cooperia oncophora, C. punctata* and *Oesophagostomum radiatum.* This study also showed elimination of the epidemiologically important hypobiotic stages of certain worms in the intestinal tract as well as extraintestinal activity against lungworms (*Dictyocaulus viviparus*).

Subsequently, a commercial dose of 200 μg kg^{-1} in a propylene glycol/glycerol formal vehicle (60:40) was adopted for subcutaneous administration to cattle. In this formulation, IVOMEC provides high levels of efficacy against all of the economically important gastrointestinal nematodes and lungworms (Campbell and Benz, 1984), as well as activity against other nematodes such as *Thelazia* and *Parafilaria* (Swan *et al.*, 1991; Soll *et al.*, 1992a). The product is also efficacious against a number of arthropod parasites including grubs (*Hypoderma bovis, H. lineatum, Dermatobia hominis*), sucking lice (*Haematopinus eurysternus, Linognathus vituli, Solenopotes capillatus*), mange mites (*Sarcoptes scabei, Psoroptes ovis*) and screw worms (*Chrysomya bezziana*).

Ivermectin treatment through either oral or subcutaneous administrations kills all three larval stages of *Hypoderma* spp. grubs with dosages as low as 0.2 μg kg^{-1} (Drummond, 1984). Ivermectin treatment through both administrations is also very effective against larval, nymphal and adult sucking lice, presumably through their ingestion of host blood. Activity of the injectable product against the surface-feeding biting louse (*Damalinia bovis*) may be more variable. The injectable product will also control *C. bezziana* and *D. hominis* infestations.

A single subcutaneous injection of ivermectin gives excellent control of *S. scabei* and *P. ovis*, but full efficacy against the surface-feeding *Chorioptes bovis* may require two treatments. Oral administration does not provide complete efficacy against mites. The injectable formulation also has activity against ticks, including *Boophilus microplus* and *B. decoloratus*, as well as the soft tick, *Ornithodoros.* These ticks show mortality and lessened engorgement, and those that do engorge produce fewer viable eggs after feeding on ivermectin-treated cattle.

Another important discovery was the persistent efficacy from subcutaneous injection of ivermectin against nematode genera such as *Cooperia, Ostertagia* and *Dictyocaulus* (Barth, 1983). Subsequent trials from several geographic regions have shown that ivermectin injection has extended persistent activity against a broad range of parasites including *T. axei, C. punctata, C. oncophora, H. placei, O. ostertagi, Oes. radiatum* and *D. viviparus*. Individual prophylactic periods coupled with epidemiological features from each nematode's life cycle contributed to strategic dosing programmes using ivermectin injections at specific intervals after turnout on to spring pasture in temperate areas (Ryan *et al.*, 1986).

A pour-on topical formulation of ivermectin in an isopropyl alcohol vehicle was developed for cattle at a dosage of 500 μg kg^{-1}. IVOMEC Pour-On is applied topically from the withers to the tailhead. Hotson *et al.* (1985) showed that it had activity against all of the economically important gastrointestinal and lung nematodes. In addition, it has extended persistent activity claims against a variety of gastrointestinal nematodes and lung worms and is also effective against the eyeworm *Thelazia*. This pour-on formulation of ivermectin is highly effective against arthropods controlled by the injectable formulation and is more completely effective against superficial-feeding mites (*C. bovis*) and biting lice (*D. bovis*). Additionally, the pour-on provides highly effective control of hornfly (*Haematobia irritans*) for up to 35 days following treatment (Foil *et al.*, 1998).

A sustained release bolus capable of ivermectin delivery for 135 days in the rumen of cattle was developed to provide worm control throughout an entire grazing season. Egerton *et al.* (1986) showed conceptually that it could kill ingested larvae and Baggott *et al.* (1986) showed that it would eliminate even established adult infections at 40 μg kg^{-1} day^{-1}. The commercial device is highly effective against all the important gastrointestinal nematodes (Rehbein *et al.*, 1997) and also has activity against ectoparasites, including lice, mange mites and grubs, as well as having an impact on a variety of tick species (Soll *et al.*, 1990). The IVOMEC SR Bolus delivers 12 mg of ivermectin per day designed to treat 300 kg animals at 40 mg kg^{-1} day^{-1} and to shut down promptly after the 135-day period to prevent underdosing.

A long-acting formulation of ivermectin (IVOMEC Gold) provides extended persistent activity against a range of endo- and ectoparasites, including 63 days of activity against lungworms, more than 75 days against ticks, and more than 140 days against grubs (*Dermatobia*) (Carvalho *et al.*, 1998; Alva *et al.*, 1999).

Sheep and goats

The tremendous potency of ivermectin against nematodes of sheep was first reported by Egerton *et al.* (1980). Therein it was disclosed that oral dosing of ivermectin to sheep at dosages almost 1000 times less than thiabendazole eliminated immature and adult stages of all of the major nematode species from the gastrointestinal tract including *Haemonchus contortus*, *Ostertagia circumcincta*, *T. axei*, *T. colubriformis*, *Cooperia curticei*, and *Oesophagostomum columbianum*.

Ivermectin is equipotent against most nematodes whether given orally or parenterally, but in general, efficacy against ectoparasites is better when treatment is given parenterally (Campbell, 1993). Consequently, both oral (micellar) and injectable (propylene glycol/glycerol formal vehicle (60:40)) formulations for sheep at a 200 $\mu g\ kg^{-1}$ dose remove virtually all of the important gastrointestinal parasites as well as itch mites (*Psorergates ovis*) and nasal bot (*Oestrus ovis*), and the injectable formulation provides highly effective control of sheep scab mites (*Psoroptes ovis*) (Soll *et al.*, 1992b).

Ivermectin has been developed for use in intraruminal controlled-release capsules providing the compound at the rate of 1.6 mg day^{-1} for 100 days. The controlled-release of ivermectin in sheep is very efficacious against established species of virtually all of the important lung and gastrointestinal nematodes and prevents reinfection with larval stages for the 100 days (Allerton *et al.*, 1998; Rehbein *et al.*, 1998). The capsule also provides control of established and new infestations for 100 days of itchmite (*Psorergates ovis*) and nasal bots (*Oestrus ovis*), and controls infestation of keds (*Melophagus ovinus*). It has been found to be a useful 'aid in control' for breech strike from blowfly (*Lucilia cuprina*), but provides only moderate reduction in the incidence of body strike (Rugg *et al.*, 1998).

Ivermectin is given to goats in the same oral formulation used in sheep and at the same dosage. It has a similar spectrum of claims as in sheep.

Horses

Egerton *et al.* (1981b) showed through titration in horses that a parenteral dose of 200 $\mu g\ kg^{-1}$ of ivermectin would eliminate the adult and immature stages of large (*Strongylus vulgaris*, *S. edentatus* and *S. equinus*) and small strongyles (*Cyathostomum pateratum*, *C. catinatum*, *Cylicocyclus nassatus*, *C. leptostomus*, *Cyliostephanus minutus*, *C. longibursatus* and *C. goldi*), as well as the immature stages of pinworm (*Oxyuris equi*), ascarid (*Parascaris equorum*), filariid (*Onchocerca cervicalis*), and gastrophilid bots (*Gasterophilus intestinalis* and *G. nasalis*). Notable was ivermectin's activity against immature stages of *S. vulgaris*, which during their migrations cause severe damage to the mesenteric artery of horses, and activity against microfilariae of *O. cervicalis*, which was to forecast activity against important filariids of dog and man.

Ivermectin was initially introduced as a product for intramuscular injection at 200 $\mu g\ kg^{-1}$ for horses, but was later replaced as EQVALAN in oral paste and liquid dosage forms.

Swine

Ivermectin is used in swine as a subcutaneous injection at 300 μg kg^{-1} in a propylene glycol/glycerol formal vehicle (60:40). At that dosage, IVOMEC removes all of the important gastrointestinal, lung and kidney nematodes. When given to sows 7–14 days prior to farrowing, ivermectin also controls prenatal transmission of somatic threadworm larvae (*Strongyloides ransomi*) to newborn pigs. In addition, it is highly efficacious against lice (*Haematopinus suis*) and mange mites (*S. scabiei*).

An in-feed, pre-mix ivermectin formulation (IVOMEC Premix) is designed to deliver a 100 μg kg^{-1} day^{-1} dosage to swine for 7 days, which is highly effective against major swine parasites.

Dogs and cats

Discovery of ivermectin's activity against developing heartworm (*Dirofilaria immitis*) was to revolutionize chemotherapy against that agent in dogs. Previous treatment required daily administration of diethylcarbamazine resulting in tedious compliance issues. Ivermectin dosages as low as 3 μg kg^{-1} interrupt the *D. immitis* life cycle by killing the L_3 and L_4 stage larvae. Transformation of the L_4 to the L_5 stage does not occur until about the third to fourth month of infection, which means the development of this species can be halted with ivermectin treatment within the first months of infection. Consequently, strategic dosing with either a tablet or beef-based chewable formulation of ivermectin (HEARTGARD) at a monthly dosage of 6 μg kg^{-1} provides highly effective control of heartworm in dogs. It has subsequently been shown to be similarly effective against developing heartworm infections and hookworms when administered at 24 μg kg^{-1} as HARTGARD-FX to cats.

Ivermectin is active against virtually all of the gastrointestinal nematodes of dogs at either an oral or subcutaneous dose of 200 μg kg^{-1}, but because of sensitivity of certain dogs of the collie breed to doses greater than 100 μg kg^{-1}, it has been marketed only for heartworm prophylaxis. Therefore, additional claims for nematodes have been acquired by adding pyrantel to the beef-based chewable formulation (HEARTGARD Plus).

Human

Donation of ivermectin for compassionate reasons to almost 30 countries in Africa and Central and South America where *Onchocerca volvulus* infections are endemic has been conducted since 1987. More than 25 million people are treated with MECTIZAN annually. Oral administration of ivermectin once a year at 150 μg kg^{-1} does not kill pre-adult or adult *O. volvulus*, but does destroy the developing embryos in the female worm's reproductive tract and the microfilariae in the skin. Clinically, destruction of these stages of the parasite's life cycle greatly reduces skin irritation and, more importantly, prevents the ocular lesions in the human that can lead to blindness. Epidemiologically, disruption of the parasite life cycle through once a year community-wide treatment has become the cornerstone of public health strategy to reduce the intensity and prevalence of this disease.

The donation programme for *O. volvulus* was expanded in 1999 to include lymphatic filariasis where both diseases were sympatric. Lymphatic filariasis, also known as elephantiasis, is caused by the filarial worms *Wuchereria bancrofti* and *Brugia malayi*. As with *O. volvulus*, ivermectin does not kill the adult worms which reside in the lymphatics, but is highly efficacious against the microfilariae. Since the microfilariae cause no clinical disease, treatment with ivermectin is used to reduce disease transmission.

Ivermectin has also been approved for intestinal strongyloidiasis caused by *Strongyloides stercoralis* at a single oral dose of 200 $\mu g\ kg^{-1}$ and for treatment of *Sarcoptes* infection in man.

Abamectin

Abamectin (Fig. 1.1.2) was developed for use as an injectable product for cattle. It is a naturally occurring avermectin approved for use in animal medicine and is the starting material for the production of ivermectin. As such, abamectin or avermectin B_1 differ from ivermectin only in the presence of a double bond at C-22,23. Abamectin has tremendous potency against most species of gastrointestinal nematodes through subcutaneous injection (Egerton *et al.*, 1979) and has a similar efficacy spectrum to ivermectin, although claims against ectoparasites are more limited.

Eprinomectin

Eprinomectin (Fig. 1.1.3) was approved as EPRINEX in 1997 for use in all cattle, including lactating dairy animals. Ivermectin, despite its excellent claim structure and safety record, cannot be used in lactating dairy cattle because of the levels of residue in milk. Over an 18-day period, approximately 5% of the total ivermectin dose given to dairy cows is found in the milk (Toutain *et al.*, 1988). Consequently, a medicinal chemistry programme was undertaken to identify a new avermectin/milbemycin that could be used without the requirement for any milk withdrawal following treatment.

Eprinomectin is the only avermectin/milbemycin available for animal health whose developmental programme not only included optimization against the multitude of endo- and ectoparasites of the host, but which

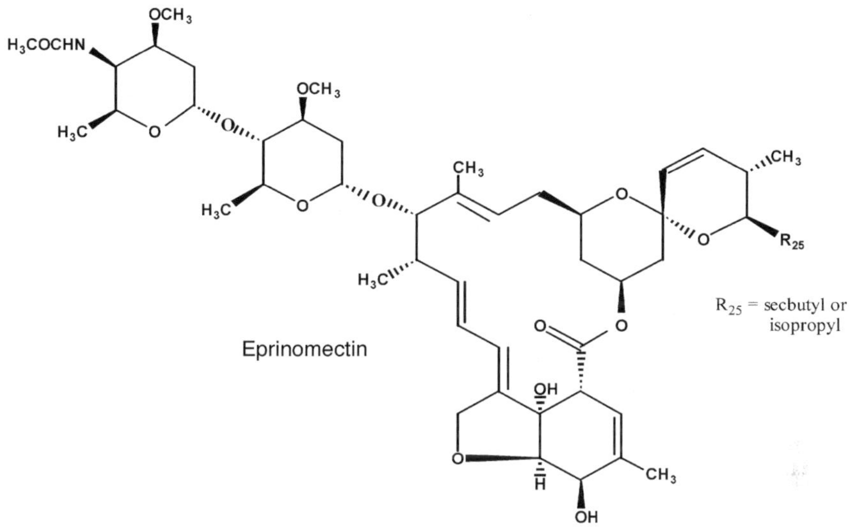

Fig. 1.1.3. Molecular structure of eprinomectin.

also sought to exclude it from specific physiological compartments within the production animal to enhance food safety. Parallel research programmes were instituted which were tasked with identifying, on the one hand, the potency of hundreds of avermectin/milbemycin analogues against gastrointestinal nematodes and, on the other, determining the concentrations of these analogues in the milk of lactating dairy cattle. Identifying the most potent analogues against the gastrointestinal parasites is an established procedure, but there was no reason at the time to believe that one could find any analogue from this highly lipophilic chemical class that would not distribute equally to all tissues, especially the mammary tissues of lactating animals.

Shoop *et al.* (1996a) were the first to show that the chemical structure of the avermectin/milbemycin molecule could be manipulated to change the milk partitioning coefficients in lactating dairy animals. They discovered a range of milk/plasma ratios among the molecules that first directed the search to those unsaturated at C-22,23, and then ultimately to those C-4″-epi-amino analogues unsaturated at C-22,23. It was this subgroup that showed one of the lowest proclivities to partition in the milk. The best from this series of compounds was 4″-epi-acetylamino-4″-deoxy avermectin B_1, which was given the name eprinomectin. Alvinerie *et al.* (1999) subsequently examined the pharmacokinetics of eprinomectin in lactating cattle and concluded that only 0.1% of the total dose was eliminated in the milk, which is 50-fold less than for either ivermectin or moxidectin.

Shoop *et al.* (1996b) showed titration data for eprinomectin in a topically applied experimental vehicle (the isopropyl alcohol vehicle used in the ivermectin pour-on product) on cattle against all of the major lung and gastrointestinal larval and adult stages of nematodes, as well as lice (*Linognathus vituli*), hornfly (*Haematobia irritans*) and mites (*Chorioptes bovis*). They calculated that 95% of all stages of helminths were eliminated at a dosage of 156 $\mu g\ kg^{-1}$, representing some threefold greater potency than ivermectin. This dosage was also very efficacious against lice and mites, but 500 $\mu g\ kg^{-1}$ was selected as the commercial dose in the final formulation to ensure control of all important ectoparasites.

Eprinomectin (EPRINEX) was subsequently approved for use at 500 $\mu g\ kg^{-1}$ in a pour-on for cattle and deer in a formulation consisting of natural oils. Pitt *et al.* (1997), Yazwinski *et al.* (1997) and Williams *et al.* (1997) reported results from world-wide trials showing the tremendous efficacy against all stages of all important helminths of cattle. The eprinomectin pour-on product also has significant persistent activity against a range of important nematodes (Cramer *et al.*, 2000). Additional trials demonstrated eprinomectin's potency against lice (*Linognathus vituli, Haematopinus eurysternus, Solenopotes capillatus* and *Damalinia bovis*) (Holste *et al.*, 1997), cattle grub (*Hypoderma* spp.) (Holste *et al.*, 1998) and mange mites (*C. bovis* and *Sarcoptes bovis*) (Barth *et al.*, 1997). Lastly, Gogolewski *et al.* (1997) showed that eprinomectin in its topically applied natural oil formulation was very efficacious against worms when administered to various hair coats on cattle and under a wide range of weather conditions.

The observed potency of eprinomectin against worms is partially explained by its greater bioavailability in cattle. Alvinerie *et al.* (1999) stated that it is generally accepted that the effect of a drug is closely related to its area under the curve (AUC) as determined pharmacokinetically. They calculated that following treatment of lactating cattle with commercial preparations, the AUC of eprinomectin was 239 $ng\ ml^{-1}\ day^{-1}$ compared with 115 $ng\ ml^{-1}\ day^{-1}$ for ivermectin. The twofold increase in levels of eprinomectin over ivermectin is similar to the threefold increase in potency against gastrointestinal worms.

Eprinomectin, like ivermectin, is a semi-synthetic compound derived from avermectin B_1. Despite their commonality of origin, the structural modifications provide each with dramatically different behaviours in cattle. When compared to ivermectin, eprinomectin not only penetrates the skin and doubles the concentration of drug in the blood, but it also partitions in the physiological compartments of the mammal in such a manner as to reduce excretion from the mammary glands to one-50th the amount of ivermectin.

Safety and Toxicology

Modes of action

It is likely that the entire family of avermectins and milbemycins shares a common mode of action, but most studies have been conducted with either avermectin B_{1a} or 22,23-dihydroavermectin B_{1a} (the major component of ivermectin). The mode of action of these molecules was reviewed by Turner and Schaeffer (1989) and further investigated by Arena *et al.* (1992, 1995) and by Cully *et al.* (1996).

In target organisms, the mode of action is receptor mediated, and ligand-gated chloride channels are the target proteins for this class of compounds. Avermectins potentiate and/or directly activate arthropod and nematode glutamate-gated chloride channels. There is a correlation between activation of glutamate-gated chloride channel current, membrane binding and nematocidal activity. Modulation of other ligand-gated chloride channels, such as those gated by the neurotransmitter γ-aminobutyric acid (GABA) may also be involved.

The consequence of the avermectin–receptor interaction is an increased membrane permeability to chloride ions. In nematodes and arthropods, avermectins potentiate the ability of neurotransmitters such as glutamate and GABA to stimulate an influx of chloride ions into nerve cells resulting in loss of cell function. This effect disrupts nerve impulses, resulting in paralysis and death in most affected invertebrates.

Several other actions have been proposed for avermectins in addition to their interaction with chloride channels but the significance of these is yet to be confirmed.

Secondary effects

At recommended therapeutic dose levels, ivermectin, abamectin and eprinomectin do not have any secondary effects on the normal host animal. Although reports have been published describing various effects of the avermectin subfamily in vertebrates (Turner and Schaeffer, 1989), these *in vitro* studies have generally been conducted at drug concentrations far in excess of those that could be obtained under practical conditions. Additionally, the effects described in vertebrates in these studies cannot necessarily be related to the chloride ion channel-modulated mode of action identified in invertebrate target species.

Idiosyncratic reactions were observed in some Murray Grey cattle after treatment with abamectin at 200 $\mu g\ kg^{-1}$ (Seaman *et al.*, 1987). The signs, which included ataxia, muscle fasciculation, lingual paralysis, apparent blindness and recumbency, were similar to those seen in some collie dogs after treatment with ivermectin at 200 $\mu g\ kg^{-1}$ (Pulliam *et al.*,

1985; Paul *et al.*, 1987). GABA is a known neurotransmitter in nematodes and arthropods, and GABA-ergic cell bodies and terminals are found in the central nervous system (CNS) of mammals. Concentrations of avermectins in the mammalian CNS following treatment are usually negligible, but elevated levels of drug were detected in brain tissue from affected cattle and collies. The signs observed in the affected animals indicate CNS dysfunction and are consistent with enhancement of GABA activity. It is postulated that P-glycoprotein deficiency in these animals allows avermectins to penetrate and accumulate in the CNS more readily than would normally be expected, causing unusual signs at dose levels considerably below those required to produce toxicity in normal animals.

The role of P-glycoprotein in the toxicity of avermectins

The toxicity of the avermectins is dependent in part upon the activity of P-glycoprotein. P-glycoprotein is a transmembrane protein located in a number of tissues, including the blood–brain barrier, the mucosal lining of the intestinal and hepatobiliary tract and the placenta. P-glycoprotein acts as a transport protein that carries certain drugs from the inside to the outside of the cell. Of importance to the toxicity of the avermectins, P-glycoprotein limits the entry of avermectins into potentially sensitive tissues. Thus, its presence serves to reduce tissue distribution and oral bioavailability, and enhance the elimination of the avermectins, all of which function to reduce the risk of avermectin-induced toxicity.

In the CNS, P-glycoprotein is found in the capillary-endothelial cells that form the blood–brain barrier. Once bound, the avermectins are transported by P-glycoprotein from the inside to the outside of the endothelial cell back into the lumen of the capillary, thus preventing further diffusion into the CNS. Hence, the presence of P-glycoprotein in the capillary-endothelial cells of the brain affects the tissue level and, ultimately, the susceptibility to the acute neurological effects caused by the avermectins. In the absence of P-glycoprotein the avermectins are capable of diffusing freely into the CNS and accumulating to higher tissue concentrations than in the presence of P-glycoprotein. Indeed, a subpopulation of the CF-1 mouse strain deficient in P-glycoprotein (Umbenhauer *et al.*, 1997), as well as mice genetically engineered to be deficient in P-glycoprotein (also referred to as knockout mice), are unusually sensitive to the adverse effects of ivermectin. In fact, the CF-1 deficient mice are about 100 times more sensitive than the fully P-glycoprotein competent animals, i.e. LD_{50} dosage of 0.3 mg kg^{-1} versus 30 mg kg^{-1}, respectively (Umbenhauer *et al.*, 1997). The toxicity of ivermectin and related compounds to CF-1 mice is related to a specific mutation in the P-glycoprotein *Mdr 1a* gene. The homozygous (*Mdr 1a*(–/–)*1b*(–/–)) mice in this strain lack P-glycoprotein in the blood–brain barrier and in the intestinal epithelium (Lankas *et al.*, 1997), and in the placental barrier (Lankas *et al.*, 1998). The heterozygous (*Mdr 1a*(+/−)*1b*(+/−)) mouse is deficient in P-glycoprotein in these tissues, but it is not completely lacking. The increased sensitivity to ivermectin correlates with the increased accumulation of ivermectin in the CNS. Twenty-four hours after a 0.2 mg kg^{-1} dose of ivermectin in the genetically engineered P-glycoprotein deficient mice, brain concentrations of 131 ± 165 ng g^{-1} were observed versus 1.5 ± 1.2 ng g^{-1} in normal mice – an 87-fold difference (Schinkel *et al.*, 1994). Thus, the failure to express P-glycoprotein results in the accumulation of 87-fold increase in brain concentrations in the deficient mice correlating closely with the 100-fold increase in sensitivity to acute neurological effects.

P-glycoprotein expression in the mucosal lining of the intestinal and hepatobiliary tract is another important factor that can increase susceptibility to the effects of ivermectin. Animals deficient in P-glycoprotein expression in the intestine absorb more ivermectin following oral administration and thus develop higher blood levels and an enhanced potential for acute neurotoxicity. For example, Kwei *et al.* (1999) demonstrated higher blood ivermectin concentration in P-glycoprotein deficient CF-1 mice. Likewise, Lankas *et al.* (1997) reported that CF-1 mice deficient in P-glycoprotein in their intestinal tracts showed 4-h post-treatment ivermectin blood concentrations of 22 ± 1.6 ng ml^{-1} versus 15 ± 1.8 ng ml^{-1} in normal CF-1 mice (1.5 times greater in the deficient CF-1 mouse). After 24 h, the blood ivermectin concentration was 2.5 times higher in the deficient vs. the normal mouse, 20 ± 2.6 vs. 8.1 ± 0.8 ng g^{-1}, respectively. The same results have been reported for the genetically engineered mice (Schinkel *et al.*, 1994). In general, the blood levels of ivermectin following oral administration are about three times higher in CF-1 mice lacking intestinal P-glycoprotein vs. CF-1 mice expressing this protein (Kwei *et al.*, 1999).

Deficient P-glycoprotein expression in the hepatobiliary tract exerts a slight effect on blood ivermectin concentrations by reducing the elimination of ivermectin via the bile. Thus, not only is more ivermectin absorbed in the P-glycoprotein-deficient mice but there is also a reduction in the amount of ivermectin that is actively excreted back into the intestinal lumen. Overall, higher blood ivermectin concentrations provide a greater gradient for blood–brain barrier diffusion, which would add to the fact that the deficient mice have no means to exclude ivermectin from the CNS.

In summary, the presence of P-glycoprotein in the intestines, blood–brain barrier and placenta serves as an important protective biological barrier to any adverse health effects of avermectins. Animals lacking P-glycoprotein absorb more ivermectin following oral administration, develop higher blood ivermectin levels, accumulate far greater amounts of ivermectin in the CNS, and appear to be more sensitive to the adverse health effects caused by these compounds than animals with a normal complement of P-glycoprotein.

Safety in laboratory animals

Safety has been evaluated in a broad range of formulations in target species (livestock, rats and humans) and the compounds have been widely tested in laboratory animals to meet regulatory requirements and to help define the safety profile for human exposure.

Acute toxicity

Clinical signs of acute toxicity for ivermectin in laboratory animals include: mydriasis (pupillary dilation) in dogs, emesis in monkeys, and ataxia, convulsions and/or tremors and coma at higher doses in most species. Although the exact mechanism of action remains to be elucidated, these adverse effects are likely mediated via an interaction with GABA receptors or other ligand-gated chloride channels in the CNS (Lankas and Gordon, 1989; Burkhart, 2000). Based on *in vitro* assays, high levels of avermectins may also activate ryanodine receptors in muscle and reduce calcium ion release in the sarcoplasmic reticulum, which may explain some of its toxic signs, particularly hyperthermia (Ahern *et al.*, 1999).

Oral LD_{50} values in rats range from 2–3 mg kg^{-1} in pups to 50 mg kg^{-1} in adults. The LD_{50} in mice is approximately 30 mg kg^{-1}, although certain strains of mice show greater sensitivity to ivermectin and related compounds. P-glycoprotein-deficient CF-1 mice show effects at doses 100-fold lower than doses causing toxicity in other species or strains (Lankas *et al.*, 1997).

The low observed effect level (LOEL) for clinical signs in primates (emesis) after an acute oral dose of either ivermectin or abamectin was 2 mg kg^{-1} (Lankas and Gordon, 1989). A female rhesus monkey inadvertently dosed intramuscularly with four doses of approximately 1.9 mg kg^{-1} (39× the therapeutic dose) on 2 consecutive days showed transient ataxia and attitudinal abnormalities. No mydriasis, emesis or tremors were observed. Clinical pathology findings included mild increases in liver enzyme levels (Iliff-Sizemore *et al.*, 1990).

Dermal LD_{50} values for ivermectin are high, indicating that avermectins are not readily absorbed by the dermal route; a dermal penetration study in the rhesus monkey with abamectin confirmed penetration of 0.5% or less of a dermally applied dose (Lankas and Gordon, 1989).

Subchronic toxicity

Rats showed no clinical signs of toxicity or changes in clinical pathology parameters from exposures up to 1.6 mg kg^{-1} day^{-1} ivermectin for 3 months. Microscopic evaluation of tissues showed enlargement of the spleen at 0.8 mg kg^{-1} day^{-1} and above. The no-observed adverse effect level (NOAEL) was 0.4 mg kg^{-1} day^{-1}. Dogs dosed daily for 3 months showed clinical signs including tremors, ataxia and anorexia at 2.0 mg kg^{-1} day^{-1}; the NOAEL was 0.5 mg kg^{-1} day^{-1}. In a 2-week study with rhesus monkeys, there were no treatment-related findings at the highest dose treated which was 1.2 mg kg^{-1} day^{-1} (Lankas and Gordon, 1989).

Ivermectin caused no increase or alteration in seizure incidence (induced by bicuculline) of seizure-prone or seizure-resistant mice dosed with 600 μg kg^{-1} day^{-1} in drinking water every other week for 6 weeks. Additionally, no effect was observed at this dose on the benzodiazepine-binding site on the GABA–chloride channel complex in mouse brain homogenates (labelled with [^{3}H]-flunitrazepam) (Diggs *et al.*, 1990).

Chronic toxicity

Results of chronic toxicity studies on abamectin showed a chronic NOAEL for abamectin given in the diet to dogs of 0.25 mg kg^{-1} day^{-1}. The NOAEL in a chronic (53-week) dietary rat study with abamectin was 1.5 mg kg^{-1} day^{-1}. The same population (with the high dose reduced to 2.0 mg kg^{-1} day^{-1}) was followed to 105 weeks to assess potential carcinogenicity. No treatment-related tumours were found in the rat chronic study, or in a 94-week dietary mouse study at doses up to 8 mg kg^{-1} day^{-1}. Decreased weight gain and tremors were seen in the high dose group; the NOAEL for chronic toxicity in mice was 4 mg kg^{-1} day^{-1} (Lankas and Gordon, 1989).

Conclusion

The observed LD_{50} values in experimental animals, measured in units of milligrams per kilogram of body weight, are well above the microgram per kilogram dosages used in humans and against target species for antiparasitic activity (Burkhart, 2000). This, together with the low affinity for mammalian ligand-gated chloride channels (affinity for binding sites in rat brain is 100-fold less than that in *Caenorhabditis elegans*) and the minimal accumulation of ivermectin in the CNS of mammalian species confers a wide margin of safety to the avermectins. Moreover, the studies above clearly demonstrate specific no-effect levels, indicating the toxicity of the avermectins to be dose dependent, and the dosages used therapeutically, in conjunction with pharmacokinetic and pharmacodynamic properties, are well below the dosages necessary to cause harm.

Safety in target animals

In target animal species (e.g. horses, swine, cattle, sheep, dogs and cats) avermectins are used commercially for the broad-spectrum control of nematode and arthropod parasites. Their widespread use is due to their potency against these endoparasitic and ectoparasitic organisms at low dose levels, coupled with wide margins of safety in the mammal due to the pharmacokinetic and pharmacodynamic features of the compounds and formulations used. Ivermectin distributes poorly into the brain of mammalian species and the affinity of avermectin for specific binding sites in rat brain is much lower (100-fold) than that in *C. elegans*. Additionally, glutamate-gated chloride channels have not been reported in mammals (but are found in nematodes) providing another reason for the selectivity and safety of the avermectins in target species at the dosages used (McKellar and Benchaoui, 1996).

At very high doses, toxic effects may occur and the acute toxic effect in mammals is manifested in CNS signs, and this may be related to their effect on GABA in the mammalian brain and spinal cord (Campbell, 1993; Schinkel *et al.*, 1994; McKellar and Benchaoui, 1996). Signs of acute toxicity include depression, ataxia, tremors, salivation, mydriasis and, in severe cases, coma and death (Campbell, 1993).

The safety of commercially available formulations has been exhaustively tested and extensive field use emphasizes the wide therapeutic index of these products when used according to label directions.

Cattle

The therapeutic dose of ivermectin for subcutaneous injection or oral administration to cattle is 200 $\mu g\ kg^{-1}$ (Campbell and Benz, 1984; Hsu *et al.*, 1989; Campbell, 1993). Campbell and Benz (1984) reported that single subcutaneous doses of 6 $mg\ kg^{-1}$ (30× the recommended use level), a single oral dose of 2 $mg\ kg^{-1}$ (10× the recommended use level), or three daily oral (paste) applications of 1.2 $mg\ kg^{-1}$ resulted in no clinical signs of toxicity. Drench doses of 4 $mg\ kg^{-1}$ (20×), as well as subcutaneous doses of 8 $mg\ kg^{-1}$ (40×), did produce signs of CNS depression (listlessness, ataxia and mydriasis) in some animals.

No effect on breeding performance, semen quality, pregnancy or on calves was observed when bulls or cows were given ivermectin at 0.4 $mg\ kg^{-1}$ (Campbell and Benz, 1984). In these studies, cows were treated repeatedly during the period of organogenesis to 56 days after insemination with no effects on pregnancy attributable to treatment and no teratogenic effects in calves. Similarly, no adverse effects were observed and normal calves were born to cows treated repeatedly in the second and third trimesters of pregnancy (Pulliam and Preston, 1989).

Controlled field trials involving many thousands of cattle of various breeds and types under a wide range of husbandry and climatic conditions were also conducted in the development of various formulations of ivermectin. Results demonstrated that dosing cattle at twice the use level did not result in an increased incidence of health problems. Further evidence of the remarkable safety profile of ivermectin is demonstrated by field use experience where more than 5 billion doses of ivermectin products are estimated to have been efficiently and safely applied to cattle worldwide since its introduction.

Eprinomectin applied topically to cattle at 1, 3 and 5× the recommended dose (0.5 mg kg^{-1}) at 7 day intervals for 3 weeks produced no adverse effects. Among calves treated at 5 mg kg^{-1} (10× the recommended use level) transient mydriasis was observed in a single animal, but no other adverse or unexpected systemic effect was observed in the eprinomectin-treated animals.

Extensive field trials and commercial use have shown eprinomectin to be safe for use in cattle of all breeds and ages, including lactating dairy cattle. Application of eprinomectin at 1.5 mg kg^{-1} (3× the recommended use level) to breeding cows prior to mating, or from mating to parturition had no effect on conception, organogenesis, fetal survival or parturition. Repeated treatment at this level similarly had no effect on the breeding soundness of bulls.

An injectable form of avermectin B_1 (abamectin) has been extensively evaluated for use in cattle. As discussed earlier in this chapter, avermectin B_1 has a different safety profile to ivermectin. Acute toxicity studies demonstrated signs of toxicosis in cattle treated subcutaneously with abamectin at 1.0 mg kg^{-1}, and at levels of 2.0–8.0 mg kg^{-1} and above animals showed more severe signs, including ataxia, recumbency, decreased lip and tongue tone, drooling, mydriasis, coma and death. Product labelling warns against use in calves under 4 months of age (Pulliam and Preston, 1989). Idiosyncratic toxic reactions have also been reported in a herd of Murray Grey cattle treated with abamectin in Australia. These animals were found to have higher levels of abamectin in the CNS than would normally be expected (Seaman *et al.*, 1987).

Abamectin was shown to be safe for use in breeding bulls and cows during all stages of breeding and pregnancy.

Sheep and goats

The usual therapeutic dosage of ivermectin given to sheep and goats (200 μg kg^{-1}) is well below levels that cause adverse health effects. Campbell and Benz (1984) reported that sheep given ivermectin at 4 mg kg^{-1} (20×) in a micelle formulation by stomach tube showed no ill effects. Sheep given 4–8 mg kg^{-1} of ivermectin in propylene glycol orally had ataxia lasting for 3 days; however, this effect was also observed in vehicle controls, making it attributable to the solvent as opposed to ivermectin. No reproductive effects were observed in rams and ewes given repeated oral ivermectin dosages of 0.4 mg kg^{-1} or repeated subcutaneous dosages of 0.6 mg kg^{-1} (3× the recommended use level). Extensive field testing in many countries under various conditions of management has shown ivermectin to be safe for use in sheep and goats of all breeds and ages when administered orally. Similar studies support the safe application of an injectable form with the only observation being a low incidence of transient pain reactions immediately after treatment.

Horses

The normal therapeutic dosage given to horses is also 200 μg kg^{-1} (Campbell and Benz, 1984; Hsu *et al.*, 1989; Campbell, 1993). By comparison, Campbell and Benz (1984) reported that an acute toxicity syndrome consisting of depression and ataxia was observed in horses injected with 12 mg kg^{-1} of ivermectin (60× the recommended use level). Campbell and Benz (1984) also reported that intramuscular injection of 3 and 6 mg kg^{-1} led to mydriasis. Repeated treatment of foals orally at doses of 0.6 (3×), 1.0 (5×), or 1.2 (6×) mg kg^{-1} elicited no signs of toxicosis, but foals treated at nine times the use level (1.8 mg kg^{-1}) displayed a slow pupillary light response and decreased menace reflex after repeated treatment (Pulliam and Preston, 1989).

Transient allergic ventral subcutaneous oedema has been reported following treatment of horses infected with *Onchocerca cervicalis* (Herd and Donham, 1983), but these swellings were attributable to the death of *O. cervicalis* microfilariae.

No effects on breeding performance or on foals were observed in mares given repeated oral or intramuscular dosages of 0.6 mg kg^{-1}. Likewise, no effect on breeding performance has been observed in stallions given a single intramuscular injection of 0.6 mg kg^{-1} (Campbell and Benz, 1984).

Swine

Campbell and Benz (1984) reported that four pigs treated at 30 mg kg^{-1} (100× the recommended use level) via injection became lethargic and ataxic within a day of treatment. Pigs treated at lower levels up to 15 mg kg^{-1} (50× the recommended use level) did not show signs of toxicity, and the wide margin of safety of ivermectin given by subcutaneous injection to pigs has been demonstrated in a number of field trials and through extensive commercial use.

Treatment of pigs with an in-feed formulation designed to provide 100 μg kg^{-1} per day for 7 days has been shown to be safe when levels up to 10 ppm (5× the recommended use level) are provided for 21 days (3× the use period). Subcutaneous treatment of boars at 0.6 mg kg^{-1} (2× use level) and repeated treatment of sows at this level during the initial 30 days of pregnancy or in the second and third trimesters demonstrated no significant adverse effects on breeding performance or on offspring (Pulliam and Preston, 1989). Similarly, treatment of breeding pigs fed a normal ration providing ivermectin at 300 μg kg^{-1} body weight daily (3× the recommended use level) has been shown to be safe when administered to boars and when administered repeatedly for 7-day periods to cover all stages of breeding and gestation in sows.

Dogs

Dogs, in general, have been shown to be relatively refractory to the toxic effects of avermectins. Treatment of beagles orally at 2.0 mg kg^{-1} (more than 300× the dose required to prevent heartworm disease) once did not elicit any signs, while single oral doses of 2.5 mg kg^{-1} are reported to cause mydriasis. Doses of 5.0 mg kg^{-1} caused mydriasis and tremors, while more severe signs including depression and ataxia were reported at 10 mg kg^{-1}. Repeated daily dosing for 14 weeks at 0.5 mg kg^{-1} did not result in signs of ivermectin toxicosis (Pulliam and Preston, 1989).

As a result of extra-label treatment of dogs with ivermectin products formulated for use in other species, it became apparent that certain dogs of the collie breed were more sensitive to the effects of ivermectin administered at dosages several times higher than those recommended for prevention of heartworm infection (6 μg kg^{-1}). The results of several clinical studies indicated a range of sensitivity to the effects of ivermectin in this breed with certain collies showing signs of toxicity at doses as low as 0.1 mg kg^{-1} while others were refractory to treatment at rates as high as 2.5 mg kg^{-1} (Paul *et al.*, 1987). These results confirm the existence of a subpopulation of collie dogs that is more sensitive to the effects of ivermectin administration at lower dose levels (>100 μg kg^{-1}) than those required to elicit toxicity in dogs of other breeds (>2000 μg kg^{-1}) or the remainder of the collie population. The higher ivermectin brain concentrations reported in avermectin-sensitive collies vs. non-sensitive individuals may be a function of limited P-glycoprotein expression (Rose *et al.*, 1998).

Due to the low dosage of ivermectin included in products developed for heartworm prophylaxis, toxic reactions are not seen in collie dogs treated at this level. Further, collies known to be sensitive to the effects of treatment with ivermectin at 150 μg kg^{-1} have not shown adverse reactions when treated repeatedly at doses of 60 μg kg^{-1} (10× the recommended use level for heartworm prophylaxis) (Fassler *et al.*, 1991; Paul *et al.*, 1991) and signs typical of ivermectin toxicity have not been reported in collies treated at levels less than 100 μg kg^{-1} – equivalent to 16× the recommended target dose level for prevention of heartworm disease.

Results of extensive field trials with commercially available dosage forms (Soll *et al.*, 1991) and extensive commercial use support the safety of dosage forms (HEARTGARD, HEARTGARD Plus, CARDOMEC, CARDOMEC SL) designed and registered specifically for dogs

Repeated treatment of breeding bitches at doses of 600 μg kg^{-1} had no negative effects on reproductive status as measured by numbers of implants, resorptions, and live or dead puppies, and continuation of treatment after whelping had no effects in puppies (Pulliam and Preston, 1989). No adverse effect on reproductive status was observed in dogs treated at 600 μg kg^{-1} monthly for 8 months and bred to untreated bitches (Daurio *et al.*, 1987)

A combination product containing ivermectin and pyrantel has also been shown to be safe when administered at 3× the recommended dose repeatedly in breeding males and females.

Cats

Ivermectin is approved for use in cats for prevention of heartworm disease and treatment and control of certain gastrointestinal nematodes at a dose of 24 μg kg^{-1}. A no-effect level of 750 μg kg^{-1} (30× safety margin over the target dose) was established in an acute toxicity study and the safety of HEARTGARD FX was demonstrated at 1, 3 and 5× the recommended use level administered repeatedly to kittens and cats. HEARTGARD FX was also shown to be safe when administered repeatedly to breeding queens and toms at 3× the recommended use level.

Humans

Ivermectin has been used or tested for use in a variety of parasitic diseases in humans, including onchocerciasis and other tropical filarial diseases and scabies (De Sole *et al.*, 1989; Shenoy *et al.*, 1992; Elgart, 1996; Barkwell and Shields, 1997; Dunyo *et al.*, 2000). In fact, ivermectin is the drug of choice for treatment of the nematode *O. volvulus*, which is the major cause of blindness in inhabitants of some tropical areas (Goa *et al.*, 1991; Pacque *et al.*, 1991; Campbell, 1993). The use of ivermectin in humans was the result of strategic planning, and initial doses were derived from NOAEL in rodents (Brown and Neu, 1990). Based on the results of early clinical trials, with support from the basic toxicology data, ivermectin was judged to be safe for large-scale treatment programmes for the control of onchocerciasis at dosages of 100–200 μg kg^{-1} (De Sole *et al.*, 1989; Brown and Neu, 1990; Goa *et al.*, 1991; Campbell, 1993). Large groups of people, from children to the elderly, have been treated in community-based programmes, with the largest study examining the efficacy and safety of ivermectin in 50,929 individuals (De Sole *et al.*, 1989; Campbell, 1993). It has been estimated that more than 25 million people worldwide are currently treated with ivermectin each year.

As with other mammals, the toxicity of the avermectins in humans is likely mediated via an interaction with GABA receptors or other ligand-gated chloride channels in the CNS (Burkhart, 2000). However, at the dosage used in humans (generally 150 μg kg^{-1} orally every 6–12 months), almost all of the reported adverse health effects are not due to the intrinsic toxicity of these chemicals. Rather, the adverse health effects reported during ivermectin treatment, as well as their severity, are a result of the patient's immune response to dead microfilariae (De Sole *et al.*, 1989; Rothova *et al.*, 1989; Goa *et al.*, 1991; Campbell, 1993; Dunyo *et al.*, 2000). The most common adverse effects include myalgia, rash, node tenderness, swelling of nodes, joints, limbs or face, itching, headaches, fever and chills, and in some cases postural hypotension. These adverse health effects generally appear within 3 days of treatment, are usually of mild to moderate severity, resolve within days of their occurrence, and respond to analgesics or antihistamines.

Conclusions

Ivermectin has proven to be both efficacious and safe for use in a broad range of production and companion animals and humans for the treatment of a number of parasitic diseases. The dosages employed therapeutically in both target species and humans are well below dosages that cause harm, even in the case of more sensitive populations such as some members of the collie breed.

The margin of safety of compounds of this class is further attributable to the fact that mammals do not have glutamate-gated chloride channels, the macrocyclic lactones have a low affinity for other mammalian ligand-gated chloride channels and they do not readily cross the blood–brain barrier. The potency of the active, together with innovative formulation and the wide therapeutic index of formulations containing ivermectin and eprinomectin continue to make them very important assets in the ongoing battle against parasites.

References

Ahern, G.P., Junankar, P.R., Pace, S.M., Curtis, S., Mould, J.A. and Dulhunty, A.F. (1999) Effects of ivermectin and midecamcin on ryanodine receptors and the Ca^{2+}-ATPase in sarcoplasmic reticulum of rabbit and rat skeletal muscle. *Journal of Physiology* 514, 313–326.

Albers-Schonberg, G., Arison, B.H., Chabala, J.C., Douglas, A.W., Eskola, P., Fisher, M.H., Lusi, A., Mrozik, H., Smith, J.L. and Tolman, R.L. (1981) Avermectins. Structure determination. *Journal of the American Chemical Society* 103, 4216–4221.

Allerton, G.R., Gogolewski, R.P., Rugg, D., Plue, R.E., Barrick, R.A. and Eagleson, J.S. (1998) Field trials evaluating ivermectin controlled-release capsules for weaner sheep and for breeding ewes. *Australian Veterinary Journal* 76, 39–43.

Alva, R., Cramer, L.G., Carvalho, L.A., Bridi, A.A., Cox, J.L. and Soll, M.D. (1999) The efficacy of ivermectin long-acting injection (LAI) against ectoparasites of cattle. *Proceedings of the IV Seminario Internacional de Parasitologia Animal, Puerto Vallarta, Mexico*, pp. 171–177.

Alvinerie, M., Sutra, J.F., Galtier, P. and Mage, C. (1999) Pharmacokinetics of eprinomectin in plasma and milk following topical administration to lactating dairy cattle. *Research in Veterinary Science* 67, 229–232.

Arena, J.P., Liu, K.K., Parress, P.S., Schaeffer, J.M. and Cully, D.F. (1992) Expression of a glutamate-activated chloride current in *Xenopus oocytes* injected with *Caenorhabditis elegans* RNA: evidence for modulation with avermectin. *Molecular Brain Research* 15, 339–348.

Arena, J.P., Liu, K.K., Paress, P.S., Frazier, E.G., Cully, D.F., Mrozik, J. and Schaeffer, J.M. (1995) The mechanism of action of avermectins in *Caenorhabditis elegans*: correlation between glutamate-sensitive chloride current, membrane binding, and biological activity. *Parasitology* 81, 286–294.

Baggott, D.G., Batty, A.F. and Ross, D.B. (1986) The control of mature nematode infection in cattle by sustained delivery of ivermectin. *Proceedings of the 14th World Congress on Diseases of Cattle*, Vol. 1. Dublin, Ireland, pp. 160–165.

Barkwell, R. and Shields, S. (1997) Deaths associated with ivermectin treatment of scabies. *Lancet* 349, 1144–1145.

Barth, D. (1983) Persistent anthelmintic effect of ivermectin in cattle. *Veterinary Record* 113, 300.

Barth, D., Hair, J.A., Kunkle, B.N., Langholff, W.K., Lowenstein, M., Rehbein, S., Smith, L.L., Eagleson, J.S. and Kutzer, E. (1997) Efficacy of eprinomectin against mange mites in cattle. *American Journal of Veterinary Research* 58, 1257–1259.

Brown, K.R. (1998) Changes in the use profile of Mectizan: 1987–1997. *Annals of Tropical Medicine and Parasitology* 92 (Suppl. 1), S61-S64.

Brown, K.R. and Neu, D.C. (1990) Ivermectin – clinical trials and treatment schedules in onchocerciasis. *Acta Leidensia* 59, 169–175.

Burg, R.W., Miller, B.M., Baker, E.E., Birnbaum, J., Currie, S.A., Hartman, R., Kong, Y.L., Monaghan, R.L., Olson, G., Putter, I., Tunac, J.B., Wallick, H., Stapley, E.O., Oiwa, R. and Omura, S. (1979) Avermectins, new family of potent anthelmintic agents: producing organism and fermentation. *Antimicrobial Agents and Chemotherapy* 15, 361–367.

Burkhart, C.N. (2000) Ivermectin: an assessment of its pharmacology, microbiology and safety. *Veterinary and Human Toxicology* 42, 30–35.

Campbell, W.C. (1993) Ivermectin, an antiparasitic agent. *Medicinal Research Review* 13, 61–79.

Campbell, W.C. and Benz, G.W. (1984) Ivermectin: a review of efficacy and safety. *Journal of Veterinary Pharmacology and Therapeutics* 7, 1–16.

Carvalho, L.A., Bianchin, I., Bridi, A.A., Maciel, A.E., Santos, A.C., Malacco, M.A., Cruz, J.B., Barrick, R.A. and Cox, J. (1998) Controle Antiparasitário em Gado de Corte com Ação Prolongada, em Comparação com Produto Convencional. *A Hora Veterinária Ano* 18, 53–58.

Chabala, J.C., Mrozik, H., Tolman, R.L., Eskola, P., Lusi, A., Peterson, L.H., Woods, M.F. and Fisher, M.H. (1980) Ivermectin, a new broad-spectrum antiparasitic agent. *Journal of Medicinal Chemistry* 23, 1134–1136.

Cramer, L.G., Pitt, S.R., Rehbein, S., Gogolewski, R.P., Kunkle, B.N., Langholff, W.K., Bond, K.G. and Maciel, A.E. (2000) Persistent efficacy of topical eprinomectin against nematode parasites of cattle. *Parasitology Research* 86, 944–946.

Cully, D.F., Paress, P.S., Liu, K.K., Schaeffer, J.M. and Arena, J.P. (1996) Identification of a *Drosophila melanogaster* glutamate-gated chloride channel sensitive to the antiparasitic agent avermectin. *Journal of Biological Chemistry* 271, 20187–20191.

Daurio, C.P., Gilman, M.R., Pulliam, J.D. and Seward, R.L. (1987) Reproductive evaluation of male Beagles and the safety of ivermectin. *American Journal of Veterinary Research* 48, 1755–1760.

De Sole, G., Remme, J., Awadzi, K., Accorsi, S., Alley, E.S., Ba, O., Dadzie, K.Y., Giese, J., Karam, M. and Keita, F.M. (1989) Adverse reactions after large-scale treatment of onchocerciasis with ivermectin: combined results from eight community trials. *Bulletin of the World Health Organization* 67, 707–719.

Diggs, H.E., Feller, D.J., Crabbe, J.C., Merrill, C. and Farrell, E. (1990) Effect of chronic ivermectin treatment on GABA receptor function in ethanol withdrawal-seizure prone and resistant mice. *Laboratory Animal Science* 40, 68–71.

Drummond, R.O. (1984) Control of larvae of the common cattle grub (Diptera: Oestridae) with animal systemic insecticides. *Journal of Economic Entomology* 77, 402–406.

Dunyo, S.K., Nkrumah, F.K. and Simonsen, P.E. (2000) A randomized double-blind placebo-controlled field trial of ivermectin and albendazole alone and in combination for the treatment of lymphatic filariasis in Ghana. *Transactions of the Royal Society of Tropical Medicine and Hygiene* 94, 205–211.

Egerton, J.R., Ostlind, D.A., Blair, L.S., Eary, C.H., Suhayda, D., Cifelli, S., Riek, R.F. and Campbell, W.C. (1979) Avermectins, new family of potent anthelmintic agents: efficacy of the B1a component. *Antimicrobial Agents and Chemotherapy* 15, 372–378.

Egerton, J.R., Birnbaum, J., Blair, L.S., Chabala, J.C., Conroy, J., Fisher, M.H., Mrozik, H., Ostlind, D.A., Wilkins, C.A. and Campbell, W.C. (1980) 22,23-Dihydroavermectin B1, a new broad-spectrum antiparasitic agent. *British Veterinary Journal* 136, 88–97.

Egerton, J.R., Eary, C.H. and Suhayda, D. (1981a) The anthelmintic efficacy of ivermectin in experimentally infected cattle. *Veterinary Parasitology* 8, 59–70.

Egerton, J.R., Brokken, E.S., Suhayda, D., Eary, C.H., Wooden, J.W. and Kilgore, R.L. (1981b) The antiparasitic activity of ivermectin in horses. *Veterinary Parasitology* 8, 83–88.

Egerton, J.R., Suhayda, D. and Eary, C.H. (1986) Prophylaxis of nematode infection in cattle with an indwelling rumino-reticular ivermectin sustained release bolus. *Veterinary Parasitology* 22, 67–75.

Elgart, M.L. (1996) A risk–benefit assessment of agents used in the treatment of scabies. *Drug Safety* 14, 386–393.

Fassler, P.E., Tranquilli, W.J., Paul, A.J., Soll, M.D., DiPietro, J.A. and Todd, K.S. (1991) Evaluation of the safety of ivermectin administered in a beef-based chewable formulation to ivermectin-sensitive collies. *Journal of the American Veterinary Medical Association* 199, 457–460.

Foil, L.D., Strother, G.R., Hawkins, J.A., Gross, S.J., Coombs, D.F., Rerouen, S.M., Wyatt, W.E., Kuykendall, L.K. and Spears, B.G. Jr (1998) The use of Ivomec

Pour-on and permethrin ear tags for hornfly control. *Southwestern Entomologist* 23(4), 317–323.
Goa, K.L., McTavish, D. and Clissold, S.P. (1991) Ivermectin. A review of its antifilarial activity, pharmacokinetic properties and clinical efficacy in onchocerciasis. *Drugs* 42, 640–658.
Godber, L.M., Derksen, F.J., Williams, J.F. and Mahmoud, B. (1995) Ivermectin toxicosis in a neonatal foal. *Australian Veterinary Journal* 72, 191–192.
Gogolewski, R.P., Allerton, G.R., Pitt, S.R., Thompson, D.R., Langholff, W.K., Hair, J.A., Fulton, R.K. and Eagleson, J.S. (1997) Effect of simulated rain, coat length, and exposure to natural climatic conditions on the efficacy of a topical formulation of eprinomectin against endoparasites of cattle. *Veterinary Parasitology* 69, 95–102.
Herd, R.P. and Donham, J.L. (1983) Efficacy of ivermectin against *Onchocerca cervicalis* microfilariae in horses. *American Journal of Veterinary Research* 44, 1102–1105.
Holste, J.E., Smith, L.L., Hair, J.A., Lancaster, J.L., Lloyd, J.E., Langholff, W.K., Barrick, R.A. and Eagleson, J.S. (1997) Eprinomectin: a novel avermectin for control of lice in all classes of cattle. *Veterinary Parasitology* 73, 153–161.
Holste, J.E., Colwell, D.D., Kumar, R., Lloyd, J.E., Pinkall, N.P., Sierra, M.A., Waggoner, J.W., Langholff, W.K., Barrick, R.A. and Eagleson, J.S. (1998) Efficacy of eprinomectin against *Hypoderma* spp. in cattle. *American Journal of Veterinary Research* 59, 56–58.
Hotson, I.K., Bliss, W.J., Cox, J.L., Roncalli, R.A. and Sutherland, I.H. (1985) Efficacy of topically administered ivermectin against cattle parasites. *Proceedings 11th Conference of the World Association for Advancement of Veterinary Parasitology*, Rio de Janeiro, Abstract 112.
Hsu, W.H., Wellborn, S.G. and Schaffer, C.B. (1989) The safety of ivermectin. *Compendium on Continuing Education for the Practicing Veterinarian* 5, 584–589.
Iliff-Sizemore, S.A., Partlow, M.R. and Kelley, S.T. (1990) Ivermectin toxicology in a Rhesus macaque. *Veterinary and Human Toxicology* 32, 530–532.
Jackson, T.A., Boivin, G.P., Hall, J.E., Suo, W. and Stedelin, J.R. (1997) Ivermectin toxicosis in mice of multiple transgenic lines. *Contemporary Topics in Laboratory Animal Science* 36, 77.
Kwei, G.Y., Alvaro, R.F., Chen, Q., Jenkins, H.J., Hop, C.E.A.C., Keohane, C.A., Ly, V.T., Strauss, J.R., Wang, R.W., Wang, Z., Pippert, T.R. and Umbenhauer, D.R. (1999) Disposition of ivermectin and cyclosporin A in CF-1 mice deficient in MDR1A P-glycoprotein. *Drug Metabolism and Disposition* 27, 581–587.
Lankas, G.R. and Gordon, L.R. (1989) Toxicology. In: Campbell, W.C. (ed.) *Ivermectin and Abamectin*. Springer-Verlag, New York, pp. 89–112.
Lankas, G.R., Minsker, D.H. and Robertson, R.T. (1989) Effects of ivermectin on reproduction and neonatal toxicity in rats. *Food and Chemical Toxicology* 27, 523–529.
Lankas, G.R., Cartwright, M.E. and Umbenhauer, D. (1997) P-glycoprotein deficiency in a subpopulation of CF-1 mice enhances avermectin-induced neurotoxicity. *Toxicology and Applied Pharmacology* 143, 357–365.
Lankas, G.R., Wise, L.D., Cartwright, M.E., Pippert, T. and Umbenhauer, D.R. (1998) Placental P-glycoprotein deficiency enhances susceptibility to chemically induced birth defects in mice. *Reproductive Toxicology* 12, 457–463.

McKellar, Q.A. and Benchaoui, H.A. (1996) Avermectins and milbemycins. *Journal of Veterinary Pharmacology and Therapeutics* 19, 311–351.

Miller, T.W., Chaiet, L., Cole, D.J., Cole, L.J., Flor, J.E., Goegelman, R.T., Gullo, V.P., Joshua, H., Kempf, A.J., Krellwitz, W.R., Monaghan, R.L., Ormaond, R.E., Wilson, K.E., Albers-Schonberg, G. and Putter, I. (1979) Avermectins, a new family of potent anthelmintic agents; isolation and chromatographic properties. *Antimicrobial Agents and Chemotherapy* 15, 368–371.

Pacque, M., Munoz, B., Greene, B.M. and Taylor, H.R. (1991) Community-based treatment of onchocerciasis with ivermectin: safety, efficacy, and acceptability of yearly treatment. *Journal of Infectious Diseases* 163, 381–385.

Paul, A.J., Tranquilli, W.J., Seward, R.L., Todd, K.S. and DiPietro, J.A. (1987) Clinical observations in collies given ivermectin orally. *American Journal of Veterinary Research* 48, 684–685.

Paul, A.J., Tranquilli, W.J., Todd, K.S., Wallace, D.H. and Soll, M.D. (1991) Evaluating the safety of administering high doses of a chewable ivermectin tablet to collies. *Veterinary Medicine* 86, 623–625.

Pitt, S.R., Langholff, W.K., Eagleson, J.S. and Rehbein, S. (1997) The efficacy of eprinomectin against induced infection of immature (fourth larval stage) and adult nematode parasites in cattle. *Veterinary Parasitology* 73, 119–128.

Pulliam, G.W. and Preston, J.M. (1989) Safety of ivermectin in target animals. In: Campbell, W.C. (ed.) *Ivermectin and Abamectin*. Springer-Verlag, New York, p. 153.

Pulliam, J.D., Seward, R.L., Henry, R.T. and Steinberg, S.A. (1985) Investigating ivermectin toxicity in collies. *Veterinary Medicine* 80, 36–40.

Rehbein, S., Batty, A.F., Barth, D., Visser, M., Timms, B.J., Barrick, R.A. and Eagleson, J.S. (1998) Efficacy of an ivermectin controlled-release capsule against nematode and arthropod endoparasites in sheep. *Veterinary Record* 142, 331–334.

Rehbein, S., Pitt, S.R., Langholff, W.K., Barth, D. and Eagleson, J.S. (1997) Therapeutic and prophylactic efficacy of the Ivomec SR Bolus against nematodes and *Psoroptes ovis* in cattle weighing more than 300 kg at the time of treatment. *Parasitology Research* 83, 722–726.

Rose, J.M., Peckham, S.L., Scism, J.L. and Audus, K.L. (1998) Evaluation of the role of P-glycoprotein in ivermectin uptake by primary cultures of bovine brain microvessel endothelial cells. *Neurochemical Research* 23, 203–209.

Rothova, A., vander Lelij, A., Stilma, J.S., Wilson, W.R. and Barbe, R.F. (1989) Side-effects of ivermectin in treatment of onchocerciasis. *Lancet* (June 24), 1439–1441.

Rugg, D., Thompson, D., Gogolewski, R.P., Allerton, G.R., Barrick, R.A. and Eagleson, J.S. (1998) Efficacy of ivermectin in a controlled-release capsule for the control of breech strike in sheep. *Australian Veterinary Journal* 76, 350–354.

Ryan, W.G., Armour, J., Bairden, K., Fox, M.T. and Jacobs, D.E. (1986) Early season use of ivermectin to control parasitic gastroenteritis and bronchitis in calves. *Proceedings of the 14th World Congress of Diseases of Cattle*, Vol. 1. Dublin, Ireland, pp. 185–190.

Schinkel, A.H., Smit, J.J.M., van Tellingen, O., Beijnen, J.H., Wagenaar, E., van Deemter, L., Mol, C.A.A.M., van der Valk, M.A., Robanus-Maandag, E.C., de Riele, H.P.J., Berns, A.J.M. and Borst, P. (1994) Disruption of the mouse mdr1a

P-glycoprotein gene leads to a deficiency in the blood–brain barrier and to increased sensitivity to drugs. *Cell* 77, 491–502.

Schinkel, A.H., Wagenaar, E., van Deemter, L., Mol, C.A.A.M. and Borst, P. (1995) Absence of the mdr1a P-glycoprotein in mice affects tissue distribution and pharmacokinetics of dexamethasone, digoxin, and cyclosporin A. *Journal of Clinical Investigation* 96, 1698–1705.

Schinkel, A.H., Wagenaar, E., Mol, C.A.A.M. and van Deemter, L. (1996) P-glycoprotein in the blood–brain barrier of mice influences the brain penetration and pharmacological activity of many drugs. *Journal of Clinical Investigation* 97, 2517–2524.

Seaman, J.T., Eagleson, J.S., Corrigan, M.J. and Webb, R.F. (1987) Avermectin B_1 toxicity in a herd of Murray Grey cattle. *Australian Veterinary Journal* 64, 284–285.

Shenoy, R.K., Kumaraswami, V., Rajan, K., Thankom, S. and Jalajakumari (1992) Ivermectin for the treatment of periodic malayan filariasis: a study of efficacy and side effects following a single oral dose and retreatment at six months. *Annals of Tropical Medicine and Parasitology* 86, 271–278.

Shoop, W.L., Mrozik, H. and Fisher, M.H. (1995) Structure and activity of avermectins and milbemycins in animal health. *Veterinary Parasitology* 59, 139–156.

Shoop, W.L., Demontigny, P., Fink, D.W., Williams, J.B., Egerton, J.R., Mrozik, H., Fisher, M.H., Skelly, B.J. and Turner, M.J. (1996a) Efficacy in sheep and pharmacokinetics in cattle that led to the selection of eprinomectin as a topical endectocide for cattle. *International Journal for Parasitology* 26, 1227–1235.

Shoop, W.L., Egerton, J.R., Eary, C.H., Haines, H.W., Michael, B.F., Mrozik, H., Eskola, P., Fisher, M.H., Slayton, L., Ostlind, D.A., Skelly, B.J., Fulton, R.K., Barth, D., Costa, S., Gregory, L.M., Campbell, W.C., Seward, R.L. and Turner, M.J. (1996b) Eprinomectin: a novel avermectin for use as a topical endectocide for cattle. *International Journal for Parasitology* 26, 1237–1242.

Soll, M.D., Benz, G.W., Carmichael, I.H. and Gross, S.J. (1990) Efficacy of ivermectin delivered from an intraruminal sustained release bolus against natural infections of five tick species in cattle. *Veterinary Parasitology* 37, 285–296.

Soll, M.D., Plue, R.E., Alva, R.A., Fulton, R.K., Seward, R.L., Daurio, C.P. and Cifelli, C.S. (1991) Field safety, efficacy and acceptability of ivermectin in a chewable form for dogs. *Canine Practice* 16, 5–8.

Soll, M.D., Carmichael, I.H. and Scherer, H. (1992a) The efficacy of ivermectin against *Thelazia rhodesii* in the eyes of cattle. *Veterinary Parasitology* 42, 67–71

Soll, M.D., Carmichael, I.H., Swan, G.E. and Abrey, A. (1992b) Treatment and control of sheep scab (*Psoroptes ovis*) with ivermectin under field conditions. *Veterinary Record* 130, 572–574.

Stapley, E.O. and Woodruff, H.B. (1982) Avermectin, antiparasitic lactones produced by *Streptomyces avermitilis* isolated from a soil in Japan. In: Umezawa, H., Demain, A.L., Hata, T. and Hutchinson, C.R. (eds) *Trends in Antibiotic Research*. Japan Antibiotics Research Association, Tokyo, pp. 154–170.

Sutherland, I.H. and Campbell, W.C. (1990) Development, pharmacokinetics and mode of action of ivermectin. *Acta Leidensia* 59, 161–168.

Swan, G.E. and Gross, S.J. (1985) Efficacy of ivermectin against induced gastrointestinal nematode infections of goats. *Veterinary Record* 117, 147–149.

Swan, G.E., Soll, M.D. and Gross, S.J. (1991) Efficacy of ivermectin against *Parafilaria bovicola* and lesion resolution in cattle. *Veterinary Parasitology* 40, 267–272.

Toutain, P.L., Campan, M., Galtier, P. and Alvinerie, M. (1988) Kinetic and insecticidal properties of ivermectin residues in milk of dairy cows. *Journal of Veterinary Pharmacological Therapy* 11, 288–291.

Turner, M.J. and Schaeffer, J.M. (1989) Mode of action of ivermectin. In: Campbell, W.C. (ed.) *Ivermectin and Abamectin*. Springer-Verlag, New York, p. 73–88.

Umbenhauer, D.R., Lankas, G.R., Pippert, T.R., Wise, L.D., Cartwright, M.E., Hall, S.J. and Beare, C.M. (1997) Identification of a P-glycoprotein-deficient subpopulation in the CF-1 mouse strain using a restriction fragment length polymorphism. *Toxicology and Applied Pharmacology* 146, 88–94.

White, J.D., Bolton, G.L., Dantanarayana, A.P., Fox, C.M.J., Hiner, R.N., Jackson, R.W., Sakuma, D. and Warrier, U.S. (1995) Total synthesis of the antiparasitic agent avermectin B1a. *Journal of the American Chemical Society* 115, 1908–1939.

Williams, J.C., Stuedemann, J.A., Bairden, K., Kerboeuf, D., Ciordia, H., Hubert, J., Broussard, S.D., Plue, R.E., Alva-Valdes, R., Baggott, D.G., Pinkall, N. and Eagleson, J.S. (1997) Efficacy of a pour-on formulation of eprinomectin (MK-397) against nematode parasites of cattle, with emphasis on inhibited early fourth-stage larvae of *Ostertagia* spp. *American Journal of Veterinary Research* 58, 379–383.

Wise, L.D., Lankas, G.R., Umbenhauer, D.R., Pippert, T.R. and Cartwright, M.E. (1997) CF-1 mouse sensitivity to abamectin induced cleft palate correlates with fetal/placental P-glycoprotein genotype. *The Teratology Society Thirty-Seventh Annual Meetings*, June 21–26, Abstract no. 30.

Yazwinski, T.A., Johnson, E.G., Thompson, D.R., Drag, M.D., Zimmerman, G.L., Langholff, W.K., Holste, J.E. and Eagleson, J.S. (1997) Nematocidal efficacy of eprinomectin, delivered topically, in naturally infected cattle. *American Journal of Veterinary Research* 58, 612–614.

Chapter 1.2

Chemistry, Pharmacology and Safety: Doramectin and Selamectin

G.A. Conder and W.J. Baker

Doramectin

Chemistry

Doramectin is the product of a mutational biosynthesis programme where a mutant strain of *Streptomyces avermitilis* was utilized to produce avermectins with substituents at the C-25 position that differed from those of avermectins produced by conventional strains of the bacterium. Based on antinematodal activity *in vitro* (Dutton *et al.*, 1991) and in laboratory models and in cattle (Goudie *et al.*, 1993), doramectin was selected for development as a livestock endectocide. Doramectin is a white-to-tan powder and its structure is provided in Fig. 1.2.1. The chemical name for

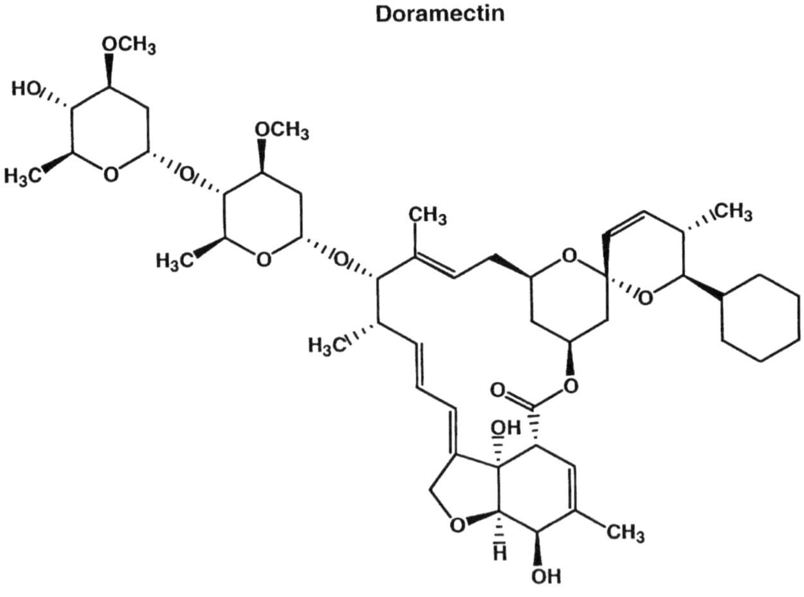

Fig. 1.2.1. Chemical structure of doramectin.

doramectin is 25-cyclohexyl-5-*O*-demethyl-25-de(1-methylpropyl) avermectin A_{1a}, its molecular formula is $C_{50}H_{74}O_{14}$, and its molecular weight is 899.14. The commercial name globally is Dectomax®. Dectomax® injectable is a 1% solution of doramectin in a non-aqueous vehicle of sesame oil and ethyl oleate. Dectomax® pour-on is a 0.5% solution of doramectin in a vehicle of cetearyl octanoate and isopropanol.

Pharmacology

Injectable formulations

Injectable formulations (1% solutions) of doramectin are licensed for use in cattle, swine, and sheep in various parts of the world. Goudie *et al.* (1993) demonstrated that the intrinsic, potent therapeutic and persistent activity in cattle of the doramectin molecule, parenterally administered in an aqueous micelle formulation, was consistent with its pharmacokinetic profile. Based on information generated for ivermectin (Campbell and Benz, 1984; Lo *et al.*, 1985), it was anticipated that doramectin's pharmacokinetics and efficacy would be affected by formulation. As lipophilic molecules, avermectins exhibit limited aqueous solubility. Although it was generally recognized at the time of doramectin's development that avermectins were soluble in oils, as is the case for most lipophilic compounds, the value of oil vehicles in optimizing the pharmacokinetics and efficacy of avermectins had not been reported. Wicks *et al.* (1993) conducted a series of studies to assess the effects of various pharmaceutically acceptable, synthetic and semisynthetic oil vehicles in modulating the pharmacokinetics and efficacy for doramectin when administered subcutaneously in cattle. They concluded that plasma levels of doramectin could be sustained using oil-based formulations, probably through a modulation of the rate of absorption from the subcutaneous space, and that the oil used determined the rate of absorption or drug release from the injection depot and, hence, bioavailability. Further, the optimum formulation of sesame oil/ethyl oleate (90:10 (v/v)) provided excellent therapeutic efficacy and was generally beneficial with respect to persistent efficacy, convenience of administration and toleration at the injection site. Nowakowski *et al.* (1995) demonstrated bioequivalence for doramectin administered subcutaneously and intramuscularly.

Clearly, each macrocyclic lactone in its respective formulation will exhibit a unique pharmacokinetic profile, with C_{max} driving efficacy and spectrum and the depletion characteristics determining persistence.

For example, a study (Toutain *et al.*, 1997) comparing the pharmacokinetic profiles of the commercial formulations of doramectin and ivermectin, administered subcutaneously at use level (200 $\mu g\ kg^{-1}$), demonstrated similar C_{max} values of about 32 $ng\ ml^{-1}$ achieved in 5.31 ± 0.35

or 3.98 ± 0.28 days, respectively. These data reflect the similar efficacy and spectrum seen with these compounds. The area under the curve (AUC) value for doramectin was 511 ± 16 (ng ml^{-1}) × days compared to 361 ± 17 (ng ml^{-1}) × days for ivermectin, with the difference being significant ($P < 0.0001$). These authors concluded that the difference in AUC was indicative of a greater doramectin availability for a longer time, which is consistent with the longer periods of persistence generally reported for doramectin relative to ivermectin. This conclusion is supported further by the work of Lifschitz and colleagues (2000) where they showed that ivermectin and doramectin concentrations remained above 1 ng g^{-1} of target tissues (skin, lung and mucosal tissues) for 18 or 38 days, respectively. The difference in pharmacokinetic profiles is probably a function of both formulation (sesame oil/ethyl oleate versus glycerol formal/formaldehyde, respectively) and intrinsic molecular traits (e.g. greater non-polarity of doramectin due to its cyclohexyl group at the C-25 position).

Pharmacokinetic data from swine and sheep given the injectable formulation at a dose of 300 μg kg^{-1} are provided in Table 1.2.1.

Pour-on formulation

A pour-on formulation is available for use in cattle in many countries. A study comparing the pharmacokinetic profiles of the commercial injectable (200 μg kg^{-1}) and pour-on (500 μg kg^{-1}) formulations of doramectin in cattle was conducted following label recommendations for each formulation. Figure 1.2.2 shows the plasma concentration profiles obtained, and Table 1.2.2 shows the values determined for the pharmacokinetic variables. The values obtained for the injectable formulation were similar to those reported by Toutain *et al.* (1997) for this route of administration, as noted above. In contrast to the injectable formulation, the pour-on formulation, despite its higher dose, produced dramatically reduced AUC and C_{max} values but increased T_{max} and mean (harmonic) half-life. The reduced C_{max} for the pour-on is sufficient to allow efficacy against the

Table 1.2.1. Pharmacokinetic values (mean AUC ((ng ml^{-1}) × days), C_{max} (ng ml^{-1}) and T_{max} (days) ± SD) in plasma from swine treated intramuscularly with doramectin or from sheep given doramectin subcutaneously or intramuscularly at 300 μg kg^{-1}.

	Swine	Sheep	
Variable	Intramuscular	Subcutaneous	Intramuscular
AUC	228 ± 75	264 ± 58	238 ± 46
C_{max}	22.9 ± 6.2	26.3 ± 4.4	25.4 ± 4.7
T_{max}	2.6 ± 1.3	1.8 ± 0.5	2.3 ± 1.0

same spectrum of nematodes and arthropods as for the injectable, although activity against the dose-limiting species, *Cooperia oncophora*, is marginal. In addition, the topical application provides greater bioavailability at the skin surface and, hence, allows efficacy against the biting louse, *Damalinia bovis*. The protracted depletion of doramectin in the pour-on formulation generally provides enhanced duration of persistent activity relative to the injectable, as outlined in Table 1.2.3. As for the injectable formulation, comparison of pharmacokinetic profiles for ivermectin and doramectin commercial pour-on formulations in cattle (Gayrard *et al.*, 1999) demonstrated similar C_{max} and T_{max} values for both drugs, but significantly greater AUC and mean residence time for doramectin, reflecting the relative efficacy and persistence profiles for the two drugs.

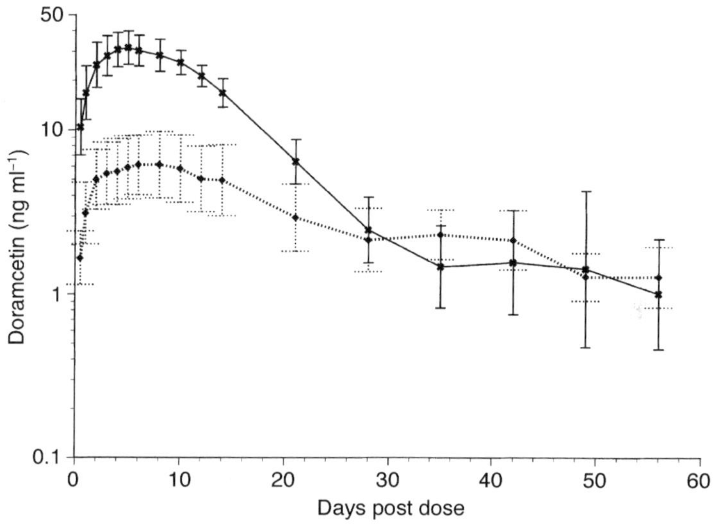

Fig. 1.2.2. Plasma concentration profiles for doramectin from cattle treated subcutaneously at 200 μg kg^{-1} (x) or topically at 500 μg kg^{-1} (◆). Error bars show 99.9% confidence intervals.

Table 1.2.2. Pharmacokinetic values (mean AUC ((ng ml^{-1}) × days), C_{max} (ng ml^{-1}) and T_{max} (days) ± SD) in plasma from cattle treated subcutaneously (200 μg kg^{-1}) or topically (500 μg kg^{-1}) with doramectin.

Variable	Subcutaneous	Topical
AUC	520 ± 30	210 ± 30
C_{max}	35 ± 2	9 ± 2
T_{max}	6 ± 2	9 ± 2
$t_{1/2}$	5.8	10

Table 1.2.3. Persistence period (days)[a] against selected parasite species in cattle for doramectin administered subcutaneously (200 μg kg^{-1}) or topically (500 μg kg^{-1}).

Parasite	Subcutaneous	Topical
Dictyocaulus viviparus	28	21
Ostertagia ostertagi	21	28
Haemonchus placei	14	35[b] (28)
Cooperia oncophora	14	21
Cooperia punctata	28	28
Oesophagostomum radiatum	28	28
Haematobia irritans	ND	7

[a]As registered in the USA.
[b]This value is based on geometric means; all others are based on arithmetic means. The number in parentheses is based on arithmetic means.

Target animal safety

Drug tolerance

CATTLE. Two studies, one examining doramectin at 2 mg kg^{-1} and the other at 5 mg kg^{-1} (10× and 25× the commercial dose), were conducted in cattle using the injectable formulation administered subcutaneously. No adverse effects were attributable to doramectin with respect to clinical condition or weight gain and no pathologically significant changes occurred in any of the haematology and clinical chemistry values at either overdose level. Based on the reduced bioavailability of the pour-on formulation of doramectin in cattle relative to the injectable formulation, as indicated by the comparative pharmacokinetic study detailed on pages 32–33, it was determined that a toleration study was not required for the pour-on formulation.

SWINE. For swine, two overdose studies were conducted, one each at 3 or 7.5 mg kg^{-1} (10× or 25× the commercial dose of 300 μg kg^{-1}) in an attempt to induce toxicity. Doramectin was administered intramuscularly. Clinical signs of ataxia and depression were observed in two of four animals at 25× the dose; none were observed at 10×. No significant treatment-related changes occurred for any haematological or clinical chemistry value. There was no adverse effect on weight gains.

SHEEP. Two studies evaluating tolerance were conducted in sheep. In both studies, doramectin was used at 3 mg kg^{-1} (15× the commercial dose of 200 μg kg^{-1}); in one study, drug was administered intramuscularly and in the other it was given subcutaneously. No adverse effects were noted in clinical signs, weight gain, haematological or clinical chemistry, or on macroscopic post-mortem examination.

Injection site toleration

Cattle dosed intramuscularly or subcutaneously with doramectin at 200 μg kg^{-1} in an injection site toleration study exhibited no pain reaction. Regardless of route, most doramectin injection sites, on post-mortem examination, showed pale discoloration at 4 days post-inoculation. However, by 30 days post-inoculation, there was no significant difference in frequency of discoloration between sites injected with doramectin versus saline. Similar results were obtained for doramectin at use level in swine (300 μg kg^{-1}). At 3 mg kg^{-1} (15× use level), there were no changes or reactions to treatment in sheep at the injection site.

Margin of safety

CATTLE. In one study each, injectable (subcutaneous) and pour-on formulations of doramectin were evaluated at 1, 3 and 5× the use dose (the use doses were 200 or 500 μg kg^{-1}, respectively) given on 3 consecutive days in cattle to assess margin of safety. No significant haematological, clinical chemistry or histopathological abnormalities were observed for either formulation. The only clinical observation for the injectable formulation was transient post-treatment salivation, particularly at the 5× dose. In the case of the pour-on, at approximately 2 weeks post-treatment, superficial skin flaking at the site of administration was observed clinically, which correlated with microscopic observations of mild acanthosis and hyperkeratosis of the epidermis. This skin flaking was not highly dose-correlated.

SWINE. A study using the same 1, 3 and 5× (use dose of 300 μg kg^{-1}) for 3 consecutive days was conducted in swine. No drug-related clinical, haematological, clinical chemistry, gross pathological or histopathological abnormalities were observed.

SHEEP. For sheep, two studies were conducted in which sheep were dosed at 300, 900 or 1500 μg kg^{-1} (1.5, 4.5 or 7.5× the use dose of 200 μg kg^{-1}). In one study, sheep were dosed intramuscularly, and in the other, dosing was by the subcutaneous route. No clinical, haematological, clinical chemistry or gross pathological abnormalities were observed in either study.

Reproductive safety

CATTLE. Reproductive safety in female cattle was determined for the injectable formulation of doramectin in two studies, one covering segment I and the other segments II/III. In the segment I study, synchronized and artificially inseminated heifers were treated subcutaneously with doramectin at 600 μg kg^{-1} (3× use dose) on post-oestrus day 3 or 11, or 18 or 10 days post-insemination (~31 days post-oestrus). No clinical effects were noted and doramectin had no adverse effects on reproductive performance, duration of the oestrous cycle or pregnancy rates. In the segment II/III study, synchronized and artificially inseminated heifers were treated subcutaneously with doramectin twice at 600 μg kg^{-1}. Each animal received one treatment on a given, randomly assigned day from day 12 to 55 post-insemination (during organogenesis) and all animals received the second treatment at day 224 postinsemination (third trimester). The only clinical sign noted in the study was a few instances of transitory hypersalivation. Data are presented in Table 1.2.4 on the outcome of pregnancy. Doramectin appeared to have no effect on pregnancy. There was no evidence of abortion and no significant differences relative to saline-treated controls in duration of gestation, length of parturition, incidence of dystocia or agalactia, or post-partum health or lactation for heifers treated with doramectin. Viability of neonatal calves (ability to suckle, stand and walk) and survival of calves for 7 days did not differ between the doramectin- and saline-treated groups. It can be concluded that doramectin treatment during organogenesis and again during the third trimester of pregnancy had no adverse effects on embryo development, maintenance of pregnancy, parturition, or neonatal calf viability or survival. Safety in bulls was evaluated for the injectable doramectin formulation in mature, reproductively sound Angus bulls experienced in semen donation. Bulls treated subcutaneously with 600 μg kg^{-1} (3× use dose) of doramectin exhibited no significant differences in semen quality or reproductive organs/structures relative to saline-treated controls.

The comparative bioavailability study detailed on pages 32–33 demonstrated significantly lower systemic exposure to doramectin in the pour-on than in the injectable formulation in cattle. Based on this information and that from the studies on reproductive safety done with the injectable formulation described above, it was concluded that the doramectin pour-on formulation would be safe in breeding cattle.

Table 1.2.4. Pregnancy rates and live calf numbers born to animals treated twice with doramectin (subcutaneously at 600 μg kg^{-1}) or saline (in a volume equal to the doramectin), once during organogenesis (day 12–55 post-insemination; at least one animal treated on each day) and a second time in the third trimester of pregnancy (day 224 post-insemination).

Treatment	Number inseminated	Number not returned to heat and treated	Number pregnant on day 220	Number calved	Percentage survival
Saline	50	29	23	20	95
Doramectin	50	32	28	28	100

SWINE. Safety of doramectin in breeding female swine was examined in three studies in which the injectable formulation was administered subcutaneously at 900 μg kg^{-1} (3× use dose) to gilts at various stages of the reproductive cycle. In the first study, gilts were treated on post-oestrus day 6 or 11, or 18 (metoestrus, dioestrus or pro-oestrus) or 10 days post-insemination (implantation). Gilts remained clinically normal throughout the study, cycled normally, exhibited good visual and standing heats, and responded well to vasectomized boars. Treatment with doramectin at 6 or 11 days post-oestrus or 10 days post-insemination had no effect on pregnancy rate, while treatment at 18 days post-oestrus resulted in a significant reduction in the percentage of gilts becoming pregnant compared with saline-treated controls. There were no significant differences in the mean number of viable fetuses between the doramectin- and saline-treated groups. The second study re-examined the effects of doramectin during pro-oestrus using more animals (~2×). In this study, gilts were treated on post-oestrus day 18. No significant effects of doramectin treatment were noted in oestrus, duration of oestrus, number of gilts inseminated, pregnancy rate, mean number of viable fetuses or general health of gilts relative to saline-treated control gilts. Together, these two studies demonstrated that doramectin had no adverse effects on oestrus, conception or embryo implantation. The third study evaluated the effects of doramectin on organogenesis and during the last third of pregnancy. In this study, each gilt was treated twice, once on a day from 12 to 45 days post-insemination and again on post-insemination day 90. No adverse effects were observed on embryo development, maintenance of pregnancy, parturition or neonatal viability or survival in the doramectin-treated group compared with the saline-treated group. Safety in boars was determined in mature, reproductively sound animals experienced in semen donation. Boars were administered doramectin at 900 μg kg^{-1} (3× use dose) or an equal volume of saline by the intramuscular route. No clinical signs related to treatment were noted during the study and there were no significant differences between the two treatment groups in semen volume, sperm motility, sperm concentration, total sperm output, major or minor sperm defects or any reproductive organs/structures.

SHEEP. One study each was conducted to assess reproductive safety in breeding ewes or rams. All doramectin-treated animals received multiple, subcutaneous injections of doramectin at 600 μg kg^{-1} (3× use level) over the reproductive period. In the ewe study, three groups of doramectin-treated animals were used, with each group receiving two treatments prior to mating (36 and 15, 29 and 8, or 22 and 1 day(s) prior) and at 3-weekly intervals following introduction of male sheep (1, 22, 43, etc., 8, 29, 50, etc., or 15, 36, 57, etc., days following introduction) until lambing. No adverse effects on fertility, gestation, fecundity or lamb viability and survivability were observed in doramectin-treated animals relative to saline-treated controls (the saline-control group was treated identically to one of the doramectin-treated groups). In the ram study, animals were treated five times with doramectin or saline at approximately 2-week intervals to cover the full cycle of the seminiferous epithelium. No adverse effects were noted in semen quality or on physical examination of the reproductive system in doramectin-treated relative to saline-treated rams.

Neonatal safety

CATTLE. Neonatal safety in cattle was evaluated for the injectable formulation of doramectin using clinically normal cow–calf pairs. Three groups of six calves were treated subcutaneously with doramectin at 200 μg kg^{-1} (use level), saline in the same volume as the preceding group or doramectin at 600 μg kg^{-1} (3× use dose) within 1–12 h of birth. All 18 dams, regardless of calf treatment, were treated simultaneously with the calves, using a 200 μg kg^{-1} dose of doramectin administered subcutaneously. Calves and cows were observed for 14 days following treatment. No clinical abnormalities related to treatment were observed in any calf or cow, and no macroscopic abnormalities were found at necropsy in any calf. From a systemic drug availability standpoint, as detailed on pages 32–33, the pour-on formulation has a dramatically reduced bioavailability relative to the injectable formulation and, hence, the neonatal safety study with the injectable formulation supports the safety of the pour-on formulation in this class of animal.

SWINE. One study evaluated neonatal safety of doramectin injectable solution in 3- to 4-day-old piglets. Piglets were treated intramuscularly with doramectin at 900 μg kg^{-1} (3× recommended dose) or an equivalent volume of saline and were observed for 7 days post-treatment. No adverse effects were noted clinically, in weight gain or at necropsy that were considered to be related to doramectin treatment.

SHEEP. Safety in neonatal sheep has not been evaluated for doramectin.

Field efficacy

Field efficacy studies were run globally under field-use conditions in cattle (injectable and pour-on), swine (injectable) and sheep (injectable), and these studies demonstrated that the recommended dose could be used safely under diverse field conditions. For example, in North America, doramectin (pour-on) was examined for safety and efficacy against nematodes at seven sites representative of climatic conditions and management systems in a total of 336 cattle. No adverse reaction to treatment was observed in any doramectin-treated animal, and the reduction in faecal egg counts across all studies was 99.7% (Conder *et al.*, 1998).

Human food safety

Toxicology

GENOTOXICITY

1. *Ames test.* Doramectin dissolved in dimethylsulphoxide (DMSO) was tested at levels of 0.005–10 mg per plate for induction of reverse mutation in *Salmonella typhimurium* strains TA 1535, TA 1537, TA 98 and TA 100. No indication of mutagenic activity was observed. Cells from the same strains of *S. typhimurium* were exposed to urine collected from mice treated intraperitoneally once with 0.2, 2 or 4 mg kg^{-1}. These studies provided no evidence of any mutagenic excretory product.
2. *Mouse lymphoma.* At non-cytotoxic test concentrations (8–50 μg ml^{-1}) providing 20–80% total relative growth, doramectin dissolved in DMSO produced no substantial dose-related increase in mutant frequency in L5178Y cells exposed for 3 h.
3. *Unscheduled DNA synthesis.* The ability of doramectin (DMSO solvent) to produce unscheduled DNA synthesis in primary cultures of rat hepatocytes was assessed after 18 h of exposure at concentrations of 1.7, 5, 7.5 and 10 μg ml^{-1}. Higher concentrations were cytotoxic. No significant increase in unscheduled DNA synthesis was detected under these conditions.
4. In vivo *micronucleus assay.* Doramectin administered *per os* at 500, 1000 or 2000 mg kg^{-1} day^{-1} was examined in male and female mice for its ability to induce micronuclei in bone marrow. The drug was dissolved in distilled water containing 0.5% methylcellulose. No evidence of micronucleus induction was observed. At ≥ 1000 mg kg^{-1} day^{-1}, target organ toxicity was indicated by a reduced ratio of polychromatic erythrocytes to normochromatic erythrocytes.

TOXICITY

1. *Three-month oral study in rats with* in utero *exposure.* Long-Evans rats of both sexes were given doramectin (sesame oil solvent) *per os* at 0, 0.5, 2 or 8 mg kg^{-1} day^{-1} for 90 days. Rats were F_1 pups produced in a two-generation study and were, therefore, exposed to drug *in utero* and during lactation. Dosing was started approximately 3–4 weeks post-weaning. Females treated with 8 mg kg^{-1} had significantly increased absolute and relative liver weights. The NOEL was 2 mg kg^{-1} day^{-1}.
2. *Three-month oral study in beagle dogs.* Two studies were conducted. In the first study, dogs of both sexes (four per sex per dose) were treated *per os* with doramectin (sesame oil solvent) at 0, 0.5, 1 or 2 mg kg^{-1} day^{-1} for 91 days. A NOEL could not be assigned from this study. Adverse effects were dose-related, transient, reversible mydriasis in one, two and five animals in the low, moderate and high dose groups, respectively. A second study was conducted where the dogs (three per sex per dose) were

treated at 0, 0.1 or 0.3 mg kg^{-1} day^{-1} of doramectin for 92 days. Transient, reversible mydriasis was seen in one dog given the 0.3 mg kg^{-1} day^{-1} dose. The NOEL was 0.1 mg kg^{-1} day^{-1}.

3. *Fetotoxicity in rats (oral route).* Pregnant albino rats (Crl: COBS-VAF-CD(SD)BR) were given doramectin (sesame oil solvent) *per os* at 0, 1.5, 3 or 6 mg kg^{-1} day^{-1} on days 6–15 of gestation. A pharmacokinetic study was conducted in parallel, in which inseminated rats were given doramectin *per os* at 6.0 mg kg^{-1} day^{-1} on days 6–15 of gestation, and, on the last day, blood, amniotic fluid and fetuses were harvested for assay. Adverse signs were limited to a threshold increase in embryo mortality in the 6 mg kg^{-1} day^{-1} group. The NOEL was 3 mg kg^{-1} day^{-1}.

4. *Fetotoxicity in mice (oral route).* Pregnant albino mice (Crl: COBS-VAF-CD1(ICR)BR) received doramectin (sesame oil solvent) *per os* at 1.5, 3 or 6 mg kg^{-1} day^{-1} on days 6–13 post-insemination. A pharmacokinetic study was conducted in parallel, in which inseminated mice were given doramectin at 6 mg kg^{-1} day^{-1}, and, on the last day, blood, amniotic fluid and fetuses were harvested for assay. There were no clinical signs or deaths, no effects on maternal body weights and reproductive variables, and no evidence of teratogenicity. The embryo mortality rate was higher, but not significantly so, in the mice that received 6 mg kg^{-1} day^{-1} of doramectin. The NOEL for embryotoxicity was 3 mg kg^{-1} day^{-1}.

5. *Fetotoxicity in rabbits (oral route).* Artificially inseminated New Zealand White rabbits were treated *per os* with doramectin (sesame oil solvent) at 0.75, 1.5 or 3 mg kg^{-1} day^{-1} on days 7–18 post-insemination. A pharmacokinetic study run in parallel used inseminated rabbits treated *per os* with doramectin at 3 mg kg^{-1} day^{-1} on days 7–18 post-insemination, and, on the last day, blood, amniotic fluid and fetuses were harvested for assay. Direct teratogenic effects were limited to some delay in fetal bone ossification at 1.5 and 3 mg kg^{-1} day^{-1}. Cleft palate was seen at 3 mg kg^{-1} day^{-1}, a dose that also produced maternal toxicity and, therefore, this effect was considered to be an indirect effect on the fetuses. The NOEL was 0.75 mg kg^{-1} day^{-1}.

6. *Two-generation oral study in rats.* Long-Evans rats were treated for two generations (F_0 and F_1) with doramectin (sesame oil solvent) at 0, 0.1, 0.3 or 1 mg kg^{-1} day^{-1}. Body weight gain during lactation was adversely affected in F_1 females receiving 1 mg kg^{-1} day^{-1} of the drug. Mean weights for F_2a and F_2b pups from the 1 mg kg^{-1} day^{-1} treated group were lower than controls on day 21 of lactation. Since doramectin is excreted in milk at high concentrations, this was considered a direct effect of the drug. The NOEL for reproductive effects was 0.3 mg kg^{-1} day^{-1}.

RESIDUES

1. *Cattle.* For the injectable formulation of doramectin, radiotracer and doramectin residue depletion studies were conducted (both sexes) in cattle. In the radiotracer depletion study, samples of injection site, liver, kidney, muscle and perirenal fat were harvested for radioassay at 7, 14, 21, 28, 35 and 42 days post-treatment from animals treated intramuscularly with [^{3}H]-doramectin at 200 μg kg^{-1}. Residue levels detected in this study are shown in Table 1.2.5. This study established that, of the residue-monitoring target tissues (liver, kidney, muscle and fat), liver contains the highest levels of total drug-related doramectin residues and that this is the tissue in cattle from which residues are the last to deplete to the safe concentration. Therefore, liver is the target tissue, and parent doramectin is the marker residue.

Two marker residue studies were conducted to establish the depletion of doramectin residues from tissues of cattle treated at 200 μg kg^{-1}; in one study, doramectin was given by subcutaneous and in the other by intramuscular injection. Samples of injection site, liver, muscle, kidney and perirenal fat were collected for residue analysis at 14, 21, 28 and 35 days post-treatment. The results are summarized in Table 1.2.6. Statistical analysis of the marker residue studies were used to establish the withdrawal time, which varies geographically based on local regulatory agency requirements.

Table 1.2.5. Mean total tissue residues (ppb ± SD) in cattle treated intramuscularly with [^{3}H]-doramectin at 200 μg kg^{-1}.

Day post-treatment	Injection site	Liver	Kidney	Muscle	Perirenal fat
7	2540 ± 1800	470 ± 110	108 ± 15	40 ± 5	551 ± 42
14	672 ± 790	415 ± 180	60 ± 4	20 ± 6	265 ± 27
21	421 ± 400	257 ± 69	35 ± 8	13 ± 3	180 ± 30
28	571 ± 630	120 ± 24	22 ± 5	10 ± 4	115 ± 26
35	<24 ± 38	42 ± 28	7 ± 3	<3 ± 1	<36 ± 17
42	18 ± 10	24 ± 1	4 ± 1	<3	23 ± 6

Table 1.2.6. Mean total tissue residues (ppb ± SD) in cattle treated parenterally with doramectin at 200 μg kg^{-1}.

Route of administration	Day post-treatment	Injection site	Liver	Kidney	Muscle	Perirenal fat
Subcutaneous	14	7300 ± 6200	88 ± 14	23 ± 2	13 ± 2	288 ± 36
	21	1900 ± 1300	44 ± 16	11 ± 5	<7 ± 4	182 ± 104
	28	380 ± 300	25 ± 11	9 ± 4	<4 ± 2	94 ± 34
	35	930 ± 910	14 ± 6	<5 ± 2	<3 ± 2	57 ± 22
Intramuscular	14	838 ± 1653	89 ± 26	24 ± 4	12 ± 3	182 ± 36
	21	1033 ± 1198	39 ± 9	12 ± 2	7 ± 2	97 ± 25
	28	162 ± 317	13 ± 7	4 ± 2	3 ± 1	48 ± 28
	35	177 ± 309	10 ± 5	3 ± 1	<2 ± 1	37 ± 19

For the pour-on formulation of doramectin, a marker residue study was conducted in cattle (both sexes). Samples of liver, kidney, perirenal fat and muscle (*semimembranosus* and *longissimus dorsi*) were harvested for assay at 14, 21, 28, 42 and 56 days post-treatment from animals treated topically with doramectin at 625 μg kg^{-1}. Residue levels detected in this study are shown in Table 1.2.7.

2. *Swine.* For the injectable formulation of doramectin, radiotracer and doramectin residue depletion studies were conducted (both sexes) in swine. In the radiotracer depletion study, samples of injection site, liver, muscle, kidney and perirenal fat were harvested for radioassay at 7, 14, 21 and 28 days post-treatment from animals treated intramuscularly with [^{3}H]-doramectin at 300 μg kg^{-1}. Residue levels detected in this study are shown in Table 1.2.8. This study established that liver and fat contain the highest levels of total doramectin residues. Liver was selected as the target tissue, in part because this was the target tissue selected for cattle and it facilitates regulatory agency monitoring of carcasses for doramectin residues using a single assay. Therefore, liver is the target tissue, and parent doramectin is the marker residue.

Table 1.2.7. Mean total tissue residues (ng g^{-1} ± SD) in cattle treated topically with doramectin at 625 μg kg^{-1}.

Day post-treatment	Liver	Kidney	Perirenal fat	Muscle	
				Semimembranosus	*Longissimus dorsi*
14	90 ± 40	20 ± 8	130 ± 60	7 ± 3	9 ± 4
21	70 ± 30	17 ± 5	90 ± 20	7 ± 2	6 ± 2
28	60 ± 30	13 ± 6	70 ± 30	<4	<4
42	28 ± 7	6 ± 2	38 ± 11	<3	<3
56	15 ± 8	<4	18 ± 9	[a]	[a]

[a]Below the lower limit of assay quantitation.

Table 1.2.8. Mean total tissue residues (ppb ± SD) in swine treated intramuscularly with [^{3}H]-doramectin at 300 μg kg^{-1}.

Day post-treatment	Injection site[a]	Liver	Kidney	Muscle	Perirenal fat
7	5132 ± 4431	186 ± 47	79 ± 19	35 ± 7	412 ± 54
14	2512 ± 1802	111 ± 20	46 ± 7	19 ± 6	255 ± 42
21	1078 ± 721	46 ± 10	17 ± 4	6 ± 1	90 ± 18
28	118 ± 38	37 ± 20	8 ± 3	4 ± 3	58 ± 33

[a]Injection site data were collected in a separate study from the data for the other tissues.

A marker residue study was conducted to establish the depletion of doramectin residues from tissues of swine treated by intramuscular injection with doramectin. Samples of injection site, liver, muscle, kidney and perirenal fat were collected for residue analysis at 7, 14, 21, 28 and 35 days post-treatment. The results are summarized in Table 1.2.9. Statistical analysis of the marker residue studies was used to establish the withdrawal time, which varies geographically based on local regulatory agency requirements.

METABOLISM.

1. *Rats/dogs*. Liver and faeces from Sprague–Dawley rats and beagles treated *per os* with a single dose of [^{3}H]-doramectin at 5 or 3.5 mg kg^{-1}, respectively, were profiled for doramectin metabolites. Livers were harvested 48 h post-treatment. The principal metabolite profile is shown in Table 1.2.10.
2. *Cattle*. Liver and fat tissues from cattle treated intramuscularly with [^{3}H]-doramectin at 200 μg kg^{-1} were harvested 3 and 21 days

Table 1.2.9. Mean total tissue residues (ppb ± SD) in swine treated intramuscularly with doramectin at 300 μg kg^{-1}.

Day post-treatment	Injection site	Liver	Kidney	Muscle	Perirenal fat
7	7000 ± 4000	160 ± 30	80 ± 20	40 ± 9	470 ± 120
14	5000 ± 3000	83 ± 8	43 ± 7	24 ± 8	290 ± 40
21	900 ± 500	40 ± 20	18 ± 7	11 ± 5	130 ± 50
28	700 ± 500	23 ± 13	<13	<7	80 ± 50
35	160 ± 150	18 ± 8	<7	<6	50 ± 20

Table 1.2.10. Main metabolite profile (percentage of sample activity recovered) of doramectin in Sprague–Dawley rats and beagles treated *per os* with [^{3}H]-doramectin at 5 or 3.5 mg kg^{-1}, respectively.

	Rat		Dog	
Metabolite	Liver	Faeces	Liver	Faeces
Doramectin	18	22	28	6
3″-*O*-desmethyldoramectin	12	19	12	8
24-hydroxymethyldoramectin	3	14	ND[a]	5
24-hydroxymethyl-3″-*O*-desmethyldoramectin	2	16	ND	4

[a]ND = not detected.

post-treatment to profile doramectin metabolites. At 3 days post-treatment, 95% of the radioactivity was extractable from the liver, indicating that bound residues constitute only a small fraction of the total residue. Over the period of study, unchanged doramectin was the major metabolite (58–70% of extracted radioactivity) found in liver. Three minor metabolites also were observed: 3″-*O*-desmethyldoramectin (7–9%), 24-hydroxymethyldoramectin (0–4%) and 24-hydroxymethyl-3″-*O*-desmethyldoramectin (7–8%). This profile is similar to that observed in rats and dogs as shown in Table 1.2.10. In fat, unchanged doramectin was again the major metabolite (91%); a minor metabolite (7%), identified by mass spectrometry as an epimer of doramectin, was also found. Although *epi*-doramectin was not observed in rats or dogs, it is a very close structural analogue of doramectin and, hence, no further toxicological evaluation of this metabolite was warranted.

3. *Swine.* The profiling of doramectin metabolites in swine was conducted with liver tissue from swine treated with 300 μg kg^{-1} [^{3}H]-doramectin and sacrificed 7 days later. The metabolite work-up revealed that approximately 90% of the radioactivity was extractable, indicating that bound residues constitute, at most, a small fraction of the total residue. Unchanged doramectin was the major metabolite of doramectin in liver.

WITHDRAWAL PERIOD

1. *Cattle injectable.* Withdrawal times vary by country. Meat withdrawal times are as follow: USA, 35 days; Europe, 32–75 days; Latin America, 35–50 days; Middle East/Africa, 35 or 42 days; Asia, 28–70 days; Australia/New Zealand, 42/49 days. In general, this product is restricted from use in lactating animals. In Europe, some countries have determined a 60-day pre-calving withdrawal time, while in some Latin American countries a 35-day milk withdrawal time has been set.

2. *Cattle pour-on.* Meat withdrawal times are as follows: USA, 45 days; Europe, 35 days; Australia/New Zealand, 42/49 days. In general, this product is restricted from use in lactating animals. In Europe, some countries have determined a 60-day pre-calving withdrawal time.

3. *Swine injectable.* Meat withdrawal times are as follows: US, 24 days; Europe, 28–77 days; Latin America, 28–50 days; Kenya, 28 days; Asia, 28 or 60 days.

4. *Sheep injectable.* Meat withdrawal times are as follows: Europe, 28–70 days; Latin America, 35 or 50 days; Middle East/Africa, 35 or 42 days; Asia, 35 days.

Selamectin

Chemistry

Selamectin is a semisynthetic monosaccharide oxime derivative of doramectin that was identified from a targeted effort to identify a macrocyclic lactone that retained efficacy against heartworms consistent with existing products from the class, while providing utility against fleas at a dose safe for use in dogs and cats (Banks *et al.*, 2000; Bishop *et al.*, 2000). Monosaccharide derivatives and, most notably, their C-5 oximes showed flea activity, with selamectin emerging as the best compound from the targeted screening effort. Selamectin is a white powder and its structure is provided in Fig. 1.2.3. The chemical name for selamectin is 25-cyclohexyl-25-de(1-methylpropyl)-5-deoxy-22,23-dihydro-5-(hydroxyimino)-avermectin B_{1a}, its molecular formula is $C_{43}H_{63}NO_{11}$ and its molecular weight is 770. Revolution™ is the commercial name everywhere, except for Europe, where the name is Stronghold™. The commercial product comes as a ready-to-use, topical 6 or 12% solution of selamectin in an isopropyl alcohol/dipropylene glycol methyl-ether vehicle. Based on commercial tube size and animal weight, the dose for selamectin is 6–12 mg kg^{-1}.

Selamectin

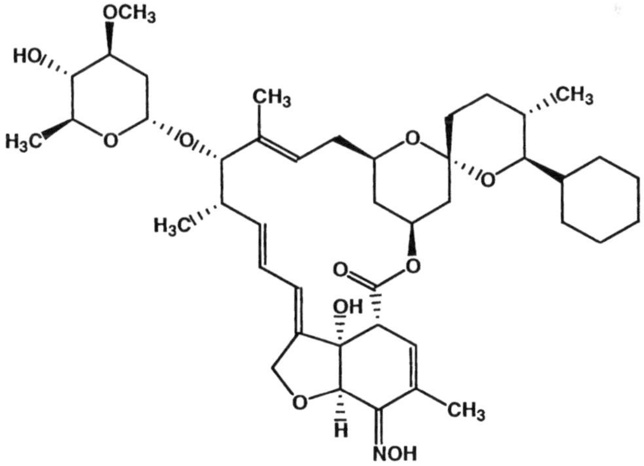

Fig. 1.2.3. Chemical structure of selamectin.

Pharmacology

The topical formulation of selamectin is licensed for use in dogs and cats of 6 weeks or older in various countries. Selamectin administered as a single topical dose at 24 mg kg^{-1} (four times the minimum commercial dose) to dogs or cats in an isopropyl alcohol/butylated hydroxytoluene + dipropylene glycol monomethyl-ether vehicle provided a C_{max} of 86.5 ± 34.0 ng ml^{-1} at 72 ± 48 h post-treatment and an AUC of 15,229 ± 4078 (ng ml^{-1}) × h in dogs and a C_{max} of 5513 ± 2173 ng ml^{-1} at 15 ± 12 h post-treatment and an AUC of 743,349 ± 443,430 (ng ml^{-1}) × h in cats. In contrast, oral administration of selamectin at a dose of 24 mg kg^{-1} in a sesame-seed oil vehicle achieved a C_{max} of 7630 ± 3140 ng ml^{-1} at 8 ± 5 h post-treatment with an AUC of 227,901 ± 121,866 (ng ml^{-1}) × h in dogs, and a C_{max} of 11,929 ± 5922 ng ml^{-1} at 7 ± 6 h post-treatment with an AUC of 1,109,933 ± 726,616 (ng ml^{-1}) × h. There was a mean systemic availability of selamectin in dogs from the topical and oral doses of 4% (range 3–7%) and 62% (range 17–96%), respectively. The mean relative bioavailability for selamectin from the topical dose compared with the oral dose in dogs was 7% (range 4–25%). For cats, mean systemic availability of selamectin from the topical and oral doses was 74% (range 41–127%) and 109% (13–174%), respectively. The mean relative bioavailability for selamectin from the topical dose compared with the oral dose in cats was 68% (range 24–385%). Sex-related differences in both dogs and cats for C_{max} and AUC were not significant for either topically or orally administered selamectin, although the 90% confidence intervals were not contained within ±20% of the overall means for the topical application in either species. Single, 30-min intravenous infusions of 0.05, 0.1 or 0.2 mg kg^{-1} of selamectin to dogs or the same doses given as a single intravenous bolus to cats in a glycerol formal/benzyl alcohol/solutol/glucose/saline vehicle demonstrated that mean maximum plasma concentration at the end of dosing and AUC were linearly related to dose. Plasma concentrations declined biphasically in dogs and polyexponentially in cats. Systemic clearance, mean residence time and terminal half-life were independent of intravenous dose level at the doses evaluated for both dogs and cats. There were no drug-related clinical signs or any effect on weight in either the dog or cat study.

A study was conducted in dogs to evaluate the skin distribution and localization of selamectin following topical application at the base of the neck with 120 mg of [^{3}H]-selamectin in a 12% prototype formulation (Pillai *et al.*, Connecticut, 2000, personal communication). Based on biopsy samples subjected to tissue oxidation and scintillation counting, selamectin (18–71 ng) was present at all skin locations (back to belly) examined, although levels were low relative to the applied dose. Autoradiography demonstrated that the drug was localized in the sebaceous glands and basal layer of the epithelium, which suggested that the sebaceous gland may play a role in the distribution of selamectin on the skin and in the extended duration of activity for this compound.

Target animal safety

Drug tolerance/margin of safety

For both dogs (Novotny *et al.*, 2000) and cats (Krautmann *et al.*, 2000), drug tolerance and margin of safety were evaluated together in a single study. For each species, selamectin in the commercial formulation was applied topically at the base of the neck. Unit doses of 0, 1, 3, 5, or 10× (0, ≥6, ≥18, ≥30 or ≥60 mg kg^{-1}, respectively) the recommended dose were administered once every 28 days for seven treatments to 6-week-old puppies or kittens. No adverse effects were observed clinically or histopathologically in either study. For cats (6-week-old), a second study (Krautmann *et al.*, 2000) was conducted using exact doses of 0, 16, 48 or 80 mg kg^{-1} of selamectin in a prototype 8% formulation applied topically to the base of the neck weekly for four treatments. No adverse effects were seen in the selamectin-treated animals.

Oral safety

Oral safety was evaluated in both dogs (Novotny *et al.*, 2000) and cats (Krautmann *et al.*, 2000). In each case, animals ≥5 months of age were treated *per os* with saline or the recommended unit dose (≥6 mg kg^{-1}) of selamectin in the commercial formulation. No adverse effects were observed in selamectin-treated dogs. In cats, salivation was observed in three of the six selamectin-treated cats within 4 h after treatment, and vomiting was seen in two of the selamectin-treated cats within 24 h following treatment and again on day 2 post-treatment.

Safety in heartworm-positive animals

Safety of selamectin in heartworm-positive dogs (Novotny *et al.*, 2000) and cats (Krautmann *et al.*, 2000) was assessed using animals experimentally infected with adult heartworms as outlined by Rawlings and McCall (1985). Heartworm infection was verified prior to treatment by a positive Knott's test and a positive adult heartworm antigen test in dogs and a positive Knott's test in cats. For both dogs and cats, animals (6–10 months of age) were treated (~28 or 21 days post-infection, respectively) topically at the base of the neck with 4× the recommended minimum dose (24 mg kg^{-1}) of selamectin in the commercial formulation (dog) or a prototype 8% formulation (cat) at 28-day intervals for a total of three treatments in dogs and seven treatments in cats. No adverse effects of treatment were observed in either dogs or cats.

Reproductive safety

For dogs (Novotny *et al.*, 2000) and cats (Krautmann *et al.*, 2000), one study each was conducted to examine the safety of selamectin in reproductively active females or males. For females of both species, animals with proven reproductive status were treated topically at the base of the neck with saline or selamectin in the commercial formulation at 3× (≥18 mg kg^{-1}) the recommended unit dose every 28 days until mating (minimum of two doses prior to mating). After mating, treatments at the threefold level were adjusted such that the next treatment occurred on either the 1st or 15th day following mating, and subsequent treatments were given every 28 days until pups or kittens were weaned at 6 weeks of age. No adverse effects on health or any phase of the reproductive cycle, including fertilization, implantation, organogenesis, fetal growth and development, birth, lactation or weaning, were detected in female dogs or cats. In the male reproductive studies, males with proven reproductive status received saline or selamectin in the commercial formulation topically at the base of the neck at 3× (≥18 mg kg^{-1}) the recommended unit dose every 14 days until each male had mated with two oestrous females (16 or 17 treatments/animal). No adverse effects on health, libido or any reproductive variable were observed in males of either species.

Collie safety

Due to the idiosyncratic toxicty seen for macrocyclic lactones in collie breeds, two studies were conducted to evaluate safety of topically applied selamectin in macrocyclic lactone-sensitive collies (Novotny *et al.*, 2000). In the first study, dogs were treated topically at the base of the neck with saline or selamectin at 40 mg kg^{-1} in a prototype 16% formulation. After approximately 60 days, the dogs were administered the opposite treatment in a cross-over design. No abnormalities were observed. In the second study, dogs were treated topically at the base of the neck with saline or selamectin in the commercial formulation at 1, 3, or 5× (≥6, ≥18, or ≥30 mg kg^{-1}) the recommended unit dose at 28-day intervals for three treatments. Mild and sporadic salivation was noted in all treatment groups, including the saline-treated control group.

A third study examined the safety of selamectin in macrocyclic lactone-sensitive collies when the drug was administered orally in sesame oil. Each dog received 2.5, 5, 10 and 15 mg kg^{-1} of selamectin in an escalating manner on days 0, 7, 14 and 22, respectively. One dog exhibited slight ataxia 8 h after treatment with 5 mg kg^{-1} of selamectin and this sign continued into the next day. On subsequent treatment at 10 and 15 mg kg^{-1}, no clinical signs were observed in this dog. No adverse effects were noted at any dose in any other dog.

Acknowledgements

All those within Pfizer and outside of Pfizer who contributed to the work on doramectin and selamectin outlined in this chapter are gratefully noted, particularly the efforts of members of Pfizer's Drug Safety Evaluation, Pharmaceutical Sciences, and Veterinary Medicine Clinical Development, Discovery, Regulatory Affairs, and Safety and Metabolism groups.

References

Banks, B.J., Bishop, B.F., Evans, N.A., Gibson, S.P., Goudie, A.C., Gration, K.A.F., Pacey, M.S., Perry, D.A. and Witty, M.J. (2000) Avermectins and flea control: structure activity relationships and the selection of selamectin for development as an endectocide for companion animals. *Bioorganic and Medicinal Chemistry* 8, 2017–2025.

Bishop, B.F., Bruce, C.I., Evans, N.A., Goudie, A.C., Gration, K.A.F., Gibson, S.P., Pacey, M.S., Perry, D.A., Walshe, N.D.A. and Witty, M.J. (2000) Selamectin: a novel broad-spectrum endectocide for dogs and cats. *Veterinary Parasitology* 91, 163–176.

Campbell, W.C. and Benz, G.W. (1984) Ivermectin: a review of efficacy and safety. *Journal of Veterinary Pharmacology and Therapeutics* 7, 1–16.

Conder, G.A., Rooney, K.A., Illyes, E.F., Keller, D.S., Meinert, T.R. and Logan, N.B. (1998) Field efficacy of doramectin pour-on against naturally-acquired, gastrointestinal nematodes of cattle in North America. *Veterinary Parasitology* 77, 259–265.

Dutton, C.J., Gibson, S.P., Goudie, A.C., Holdom, K.S., Pacey, M.S., Ruddock, J.C., Bu'Luck, J.D. and Richards, M.K. (1991) Novel avermectins produced by mutational biosynthesis. *Journal of Antibiotics* 44, 357–365.

Gayrard, V., Alvinerie, M. and Toutain, P.L. (1999) Comparison of pharmacokinetic profiles of doramectin and ivermectin pour-on formulations in cattle. *Veterinary Parasitology* 81, 47–55.

Goudie, A.C., Evans, N.A., Gration, K.A.F., Bishop, B.F., Gibson, S.P., Holdom, K.S., Kaye, B., Wicks, S.R., Lewis, D., Weatherley, A.J., Bruce, C.I., Herbert, A. and Seymour, D.J. (1993) Doramectin – a potent novel endectocide. *Veterinary Parasitology* 49, 5–15.

Krautmann, M.J., Novotny, M.J., De Keulenaer, K., Godin, C.S., Evans, E.I., McCall, J.W., Wang, C., Rowan, T.G. and Jernigan, A.D. (2000) Safety of selamectin in cats. *Veterinary Parasitology* 91, 393–403.

Lifschitz, A., Virkel, G., Sallovitz, J., Sutra, J.F., Galtier, P., Alvinerie, M. and Lanusse, C. (2000) Comparative distribution of ivermectin and doramectin to parasite location tissues in cattle. *Veterinary Parasitology* 87, 327–338.

Lo, P.K.-A., Fink, D.W., Williams, J.B. and Blodinger, J. (1985) Pharmacokinetic studies of ivermectin: effects of formulation. *Veterinary Research Communications* 9, 251–268.

Novotny, M.J., Krautmann, M.J., Ehrhart, J.C., Godin, C.S., Evans, E.I., McCall, J.W., Sun, F., Rowan, T.G. and Jernigan, A.D. (2000) Safety of selamectin in dogs. *Veterinary Parasitology* 91, 377–391.

Nowakowski, M.A., Lynch, M.J., Smith, D.G., Logan, N.B., Mouzin, D.E., Lukaszewicz, J., Ryan, N.I., Hunter, R.P. and Jones, R.M. (1995) Pharmacokinetics and bioequivalence of parenterally administered doramectin in cattle. *Journal of Veterinary Pharmacology and Therapeutics* 18, 290–298.

Rawlings, C.A. and McCall, J.W. (1985) Surgical transplantation of adult *Dirofilaria immitis* to study heartworm infection and disease in dogs. *American Journal of Veterinary Research* 46, 221–224.

Toutain, P.L., Upson, D.W., Terhune, T.N. and McKenzie, M.E. (1997) Comparative pharmacokinetics of doramectin and ivermectin in cattle. *Veterinary Parasitology* 72, 3–8.

Wicks, S.R., Kaye, B., Weatherley, A.J., Lewis, D., Davison, E., Gibson, S.P. and Smith, D.G. (1993) Effect of formulation on the pharmacokinetics and efficacy of doramectin. *Veterinary Parasitology* 49, 17–26.

Chapter 1.3

Chemistry, Pharmacology and Safety: Milbemycin Oxime

M. Jung, A. Saito, G. Buescher, M. Maurer and J.-F. Graf

Introduction

The discovery of milbemycins dates back to 1967 when A. Aoki at Hokkai Sankyo, found that the metabolite B-41 (actually a mixture of several structurally similar compounds) from the fermentation of an actinomycete showed dramatic acaricidal activity against plant mites and their eggs. J. Ide and co-workers, from Sankyo's Medicinal Chemistry Research Laboratories, in collaboration with Hokkai Sankyo, succeeded in improving the productivity of the strain, and, in 1972, H. Mishima was able to elucidate the structure of one of the metabolites of B-41, using X-ray crystallographic analysis of the *p*-bromophenylurethane derivative, mass spectrometry and ^{1}H and ^{13}C NMR. B-41 was named 'milbemycin' after the German word for mites and in combination with the conventional suffix for streptomycete antibiotics, although milbemycins have no antimicrobial activity (Ide *et al.*, 1993)

Basic Chemistry and Fermentation

Structure of naturally occurring milbemycins

The milbemycin structures are closely related 16-membered macrocyclic lactones with a spiroketal ring system consisting of two six-membered rings and cyclohexenediol or phenol. Unlike the avermectins, isolated subsequently in 1977 from *Streptomyces avermitilis*, milbemycins are not substituted at the C-13 position. The major metabolites produced by natural fermentation are milbemycin A_4, milbemycin A_3 and milbemycin D, differing from each other by the substitution at the C-25 position (Fig. 1.3.1.).

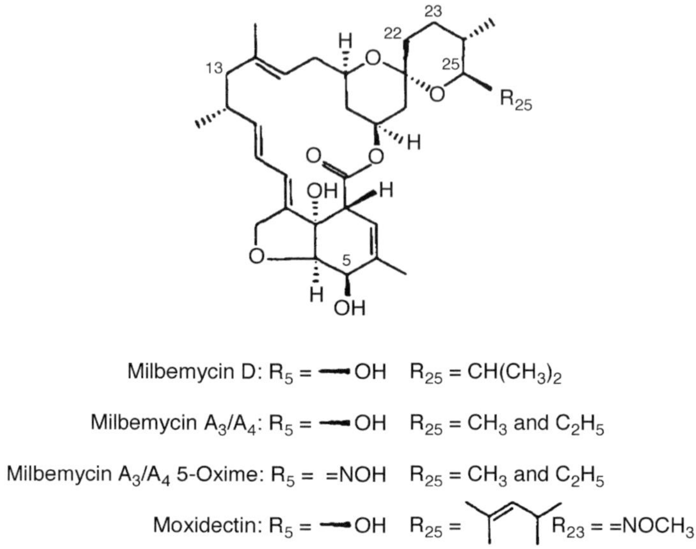

Fig. 1.3.1. Structure of important milbemycins.

Producing strains

Strain SANK 60576 (=B-41–146) was isolated by A. Aoki, Hokkai Sankyo Co. Ltd, from soil collected at Kuttian-cho in the Abuta District of Hokkaido, Japan. Based on its morphological, cultural and physiological characteristics, the strain was assigned to the genus *Streptomyces* (Waksman and Henrici, 1943). Among the known species of the genus, *S. hygroscopicus* most closely resembles strain SANK 60576. However, SANK 60576 differs from typical *S. hygroscopicus* for example by its growth colour (dark yellowish vs. colourless to pale yellowish brown) and the formation of golden yellow drops of exudate on aerial mycelium. It was therefore proposed that SANK 60576 represent a new subspecies of *S. hygroscopicus, S. hygroscopicus* subsp. *aureolacrimosus* (Ide *et al.*, 1993).

In order to obtain strains producing new families of mibemycins or high-yielding strains, several spontaneous or artificial mutants of SANK 60576 were isolated. Thus, strains producing high yields of milbemycin D and milbemycin A_3/A_4 could be identified.

Biosynthesis

The biosynthetic origin of milbemycins was studied by feeding of culture broth with [1–^{13}C]acetate, [1–^{13}C]propionate, [3–^{13}C]propionate, [1–^{13}C] isobutyrate, DL-[2–^{13}C]valine or L-[methyl-^{13}C]methionine and by the subsequent analysis of the distribution of the radiolabel. The ^{13}C NMR spectra revealed that the basic skeleton of milbemycin D was formed from seven acetate and five propionate units. The C-25 and associated isopropyl group was derived from isobutyrate. Feeding of either [1–^{13}C]isobutyrate or DL-[2–^{13}C]valine caused enrichment specifically at C-25, suggesting that DL-valine is converted to isobutyryl-CoA, which is directly incorporated into C-25 and its isopropyl group (Ide *et al.*, 1993).

Chemistry

The first milbemycins to be developed as commercial products were a mixture of milbemycin A_3 and A_4 as an acaricide for crop use and milbemycin D for the control of canine heartworm disease caused by *Dirofilaria immitis*.

The latter was marketed in Japan for several years following its launch in 1986. However, as the fermentation yield of milbemycin D remained low, further investigations were undertaken to come up with safer, more effective products. In this context, a chemical synthesis programme was initiated aiming at improving the potency and other characteristics of natural milbemycins such as milbemycin A_3, A_4 and D.

Modifications of the C-5 hydroxyl group were to reveal the most promising antiparasitic activity in biological tests for *Nippostrongylus brasiliensis* motility *in vitro* as well as in dogs naturally infected with microfilariae of *D. immitis*. While 5-amido, 5-oxo and 5-hydrazone derivatives were only marginally active in the various antiparasitic screens, the 5-oxime derivatives, synthesized from milbemycin A_3 and A_4 offered high efficacy against the microfilariae of *D. immitis*, combined with a broad anthelmintic spectrum, suitable stability and improved target animal safety.

Thus, a mixture of the 5-oxime derivatives of A_4 and A_3 (ratio A_4:A_3 = 80:20) was developed jointly by Sankyo and Ciba-Geigy (today Novartis) for the prevention of heartworm disease and control of other endoparasites in dogs. The resulting product was launched as Interceptor® in the USA and Canada in 1990, followed by many other countries.

Basic Pharmacology

Introduction

Milbemycins are 16-membered macrocyclic lactones and are very closely related to avermectins (Figs 1.3.1 and 1.3.2). Both classes are produced by fermentation and subsequent chemical modification, and have a similar range of biological activity.

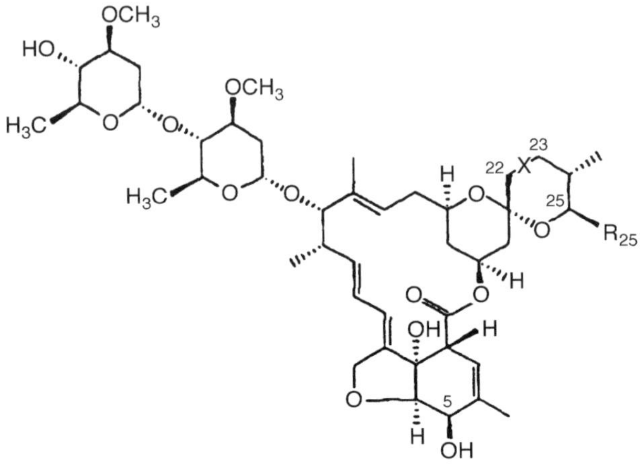

Ivermectin: X = $-CH_2CH_2-$ R_{25} = $CH(CH_3)CH_2CH_3$ and $CH(CH_3)_2$
Abamectin: X = $-CH{=}CH-$ R_{25} = $CH(CH_3)CH_2CH_3$ and $CH(CH_3)_2$
Doramectin: X = $-CH{=}CH-$ R_{25} = Cyclohexyl

Fig. 1.3.2. Structure of important avermectins, to show the similarity to the milbemycins.

Chemically, milbemycins are unglycosylated avermectins, lacking the bisoleandrosyl moiety in the 13-position. Apart from the 13-position, the various investigated milbemycins and avermectins all have very similar structures. Otherwise, the commercialized active ingredients differ from each other only by their substituents in the 5-position (e.g. hydroxyl, hydroxylimino) and in the 25-position (usually a mixture of >80% ethyl and <20% methyl for the milbemycins but, for example, secondary butyl or isopropyl for the avermectins). In addition, some avermectins, such as abamectin and doramectin, have a double bond between the 22- and 23-positions that milbemycins lack.

The use of milbemycins in the animal health field, including numerous data on their efficacy, safety, tolerability, toxicity, metabolism and resistance potential, has been described in many publications and reviewed in detail (Ide *et al.*, 1993; Shoop *et al.*, 1995). Apart from the commercially available products, a large number of other milbemycin derivatives has been synthesized and tested in animals.

Some pharmacokinetic studies have been published on avermectins in livestock animals, mostly after pour-on or subcutaneous administration (Ali and Hennessy, 1996; Oukessou *et al.*, 1996, 1999; Shoop *et al.*, 1996; Toutain *et al.*, 1997; Gayrard *et al.*, 1999; Lifschitz *et al.*, 1999; Perez *et al.*, 1999; Atta and Abo-Shihada, 2000). However, concerning milbemycins, other than the work performed at Sankyo Co. which has been published in Ide *et al.* (1993), little information has become available about their pharmacokinetics and metabolism in dogs and cats (after oral or i.v. administration). On the following pages, the published work will be reviewed briefly and supplemented by pharmacokinetic data from Novartis.

Distribution, metabolism and elimination

Study with radiolabelled milbemycin A_3/A_4 in rats

Studies with radiolabelled compounds are considered by far the best source of information about the distribution and elimination of the sum of a given parent substance and all its metabolites, which enables elucidation of the metabolic pathway as well as metabolic balance calculations. Whereas such information from dogs is unavailable, a study has been performed by Sankyo scientists in rats with milbemycin A_3/A_4 ('milbemectin', as obtained by fermentation, >80% A_4 and <20% A_3) during the development of an acaricide for crop protection (Ide *et al.*, 1993). They administered milbemectin at 2.5–25 mg kg^{-1} *per os* to rats. Similar results were obtained for the two homologues milbemycin A_3 and milbemycin A_4. It can be assumed that the following results are more or less representative also for other, similar milbemycin derivatives.

The milbemycin molecule is metabolized extensively by hydroxylation. It initially is hydroxylated in the 13-position, then a large number of different dihydroxy- and trihydroxy-milbemycins are formed (Tables 1.3.1 and 1.3.2). 13-Hydroxy-milbemycin A_4 and A_3 are also obtained by

Table 1.3.1. Urinary metabolites of milbemycin[a] in rats (Ide *et al.*, 1993).

		% of dose	
		Male	Female
U-1	Unknown metabolite	0.1	0.1
U-2	Milbemycin A_4	0.1	0.1
U-3	13-Hydroxy A_4	0.1	0.2
U-4	Unknown metabolite	0.2	0.2
U-5	13,29-Dihydroxy A_4	0.2	0.4
U-6	13,23-Dihydroxy A_4	4.1	1.5
U-7	13,30-Dihydroxy A_4	1.3	1.8
U-8	Unknown metabolite	0.3	0.4
U-9	13,26-Dihydroxy A_4	0.7	0.5
U-10	13,28-Dihydroxy A_4	0.4	0.2
U-11	Unknown metabolite	0.3	0.4
	Others[b]	0.5	0.4
	Total	8.3	6.2

[a]The sample excreted 2 days after single oral administration of [^{14}C]-milbemycin at 25 mg kg^{-1}.
[b]Radioactivity except for U-1~11.

Table 1.3.2. Faecal metabolites of milbemycin[a] in rats (Ide *et al.*, 1993).

		% of dose	
		Male	Female
Neutral fraction		46.2	44.7
F-1	Unknown metabolite	0.4	0.3
F-2	Milbemycin A_4	6.4	5.3
F-3	Unknown metabolite	0.3	0.3
F-4	13-Hydroxy A_4	2.4	1.6
F-5	Unknown metabolite	0.5	0.3
F-6	13,29-Dihydroxy A_4	3.0	3.6
F-7	13,23-Dihydroxy A_4	3.8	1.1
F-8	13,30-Dihydroxy A_4	1.7	2.0
F-9	Unknown metabolite	0.7	0.7
F-10	13,26-Dihydroxy A_4	3.5	4.3
F-11	13,28-Dihydroxy A_4	2.1	2.0
F-12	13,23,29-Trihydroxy A_4	1.0	0.2
F-13	13,26,29-Trihydroxy A_4	2.2	2.6
F-14	13,26,30-Trihydroxy A_4	0.7	0.7
F-15	13,23,26-Trihydroxy A_4	1.0	0.4
	Others[b]	7.7	7.0
	Origin of TLC	8.8	12.3
Acidic fraction		7.1	7.4
Aqueous fraction		16.8	18.5
Unextracted ^{14}C		15.9	16.2
	Total	86.0	86.8

[a]The sample excreted 2 days after single oral administration of [^{14}C]-milbemycin A_4 at 25 mg kg^{-1}.
[b]Radioactivity except for F-1~15 and the origin of TLC.

in vitro experiments using rat liver microsomes. The same is also true for the corresponding oximes, i.e. milbemycin A_3 and A_4 5-oxime, where the 13-hydroxylated product was identified and isolated as a major metabolite. Its structure is shown Fig. 1.3.3.

More than 98% of the radioactivity is excreted within 7 days (Fig. 1.3.4), mainly in the faeces, but some also in the urine. The main path of excretion is via the bile, as 42% of the administered radioactivity could be collected within the first 24 h from the cannulated bile duct. The level of radioactivity in blood reaches a maximum after 3 h and then decreases with a half-life of approximately 7–8 h. Also in tissues, the radioactivity decreases rapidly and becomes undetectable after 7 days, except for the liver and kidney. In the blood and liver at 6 h after administration, 13-hydroxy-milbemycin A_4 accounts for over 50% of the radioactivity. Apart from the gastrointestinal duct, radioactivity was detected mainly in the liver and fat.

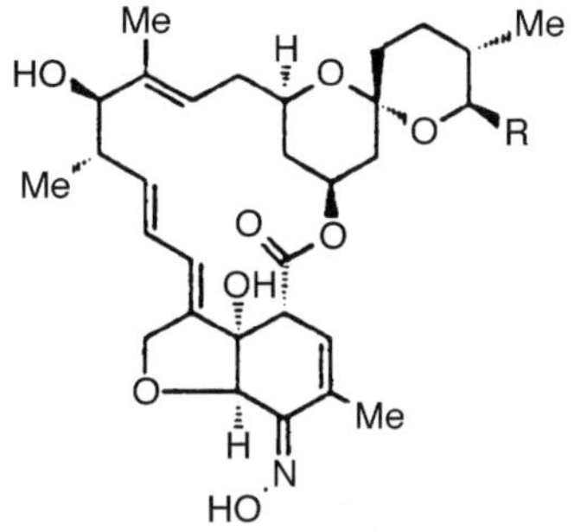

Fig. 1.3.3. Structure of 13β-hydroxymilbemycin A_4 5-oxime and A_3 5-oxime, identified as the first metabolite of milbemycin A_3A_4 5-oxime (Ide *et al.*, 1993).

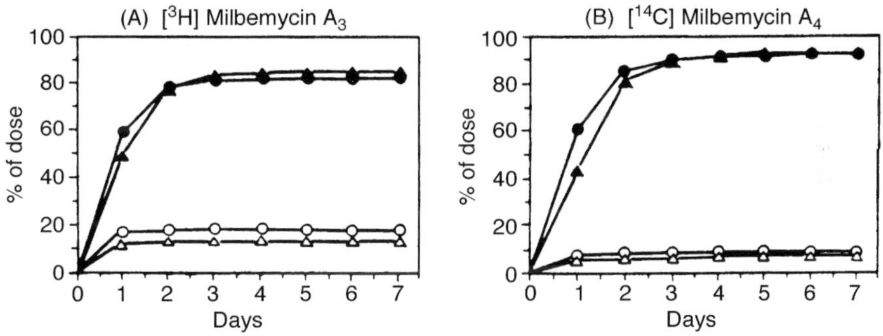

Fig. 1.3.4. Excretion of radiolabelled milbemycin in rats (Ide *et al.*, 1993) after a single oral administration ([^{3}H]milbemycin A_3/[^{14}C]milbemycin A_4 = 3/7) 25 mg kg^{-1}. —●—, faeces in male rats; —○—, urine in male rats; —▲—, faeces in female rats; —△—, urine in female rats. Milbemycin A_3 and A_4 were administered orally at 7.5 and 17.5 mg kg^{-1}, respectively. Animals were housed in a metabolic cage individually and the urine and faeces were collected every 24 h for 144 h after administration.

Tissue distribution of milbemycin D and milbemycin A_3A_4 5-oxime in rats

The tissue distribution of milbemycin D and milbemycin A_3A_4 5-oxime in rats was investigated using 'cold' (non-radiolabelled) substance (Ide *et al.*, 1993). This type of study revealed the tissue distribution of the parent substance only, not considering the metabolites.

After oral administration of milbemycin D at 1, 5 and 10 mg kg^{-1} to rats, the lung acts as a temporary depot organ, together with the kidney, fat, liver and heart, as determined 3 h after oral administration (Table 1.3.3). The results indicated that the substance was deposited initially in the lungs and then transferred to and accumulated in the adipose tissues.

After oral administration of milbemycin A_3A_4 5-oxime (the oxime of 'milbemectin', mixture of homologues, >80% A_4 and <20% A_3) at 1 and 2.5 mg kg^{-1} to rats, the main depot tissues were the fat and liver, whereas the lungs seemed less important. The corresponding tissue profiles are displayed in Table 1.3.4. Maximum tissue concentrations were observed

Table 1.3.3. Tissue concentration of milbemycin D in rats 3 h after oral administration (Ide *et al.*, 1993).

	Average tissue concentration (ng ml^{-1} or ng g^{-1})						
Dose	Plasma	Liver	Kidney	Heart	Lung	Muscle	Fat
1 mg kg^{-1}	2.2	16.1	20.3	6.1	39.4	1.7	17.1
5 mg kg^{-1}	109.5	675.6	448.0	455.9	485.2	57.7	354.7
10 mg kg^{-1}	45.3	188.4	233.8	263.2	236.3	48.7	315.2

at time points between 1 and 8 h post-administration, with C_{max} levels in the order fat > liver > kidneys > lungs > plasma > muscle.

For both milbemycin D and milbemycin A_3A_4 5-oxime, excretion was less than 2% of the total dose after 2 days. This is not surprising, since this figure represents unmetabolized parent substance only, whereas the above-mentioned results obtained with radiolabelled milbemycin A_3A_4 revealed the high degree of metabolic degradation prior to excretion.

Pharmacokinetics

Pharmacokinetics of milbemycin D and milbemycin A_3A_4 5-oxime after single oral or intravenous administration to dogs

Sankyo scientists (Ide *et al.*, 1993) performed a pharmacokinetic study where they administered milbemycin D orally as a 1% powder formulation to six beagle dogs. The results were compared with an intravenous administration to three beagle dogs. Analysis was performed using a highly sensitive and specific fluorescent high-performance liquid chromatography (HPLC) method similar to the one originally developed by Merck scientists for ivermectin (De Montagny *et al.*, 1990; Alvinerie *et al.*, 1995). The results are shown in Table 1.3.5. The absolute bioavailability of milbemycin D as a powder formulation was estimated at 10.0 ± 1.4%.

The Sankyo scientists also performed a pharmacokinetic study in three beagle dogs with milbemycin A_3A_4 5-oxime tablets. Since the 5-position is blocked by the oxime group, the fluorescent HPLC method could not be applied. An enzyme-linked immunoassay, which is not capable of distinguishing between the homologues (>80% A_4 and <20% A_3), was used instead. The results are displayed in Table 1.3.6 and Fig. 1.3.5. Additional results were obtained at Novartis (Tables 1.3.7 and 1.3.8, Fig. 1.3.6) after either oral administration of Interceptor® tablets or another novel tablet formulation or a corresponding intravenous administration of milbemycin A_3A_4 5-oxime to groups of eight beagle dogs (Internal Report). The latter study used an HPLC method with UV detection, where the two homologues A_3 and A_4 were resolved as separate

Table 1.3.4. Tissue distribution of milbemycin 5-oxime in rats (Ide *et al.*, 1993).

Tissue/ organ	Dose ($mg\ kg^{-1}$)	Concentration (mean ± SE: $ng\ ml^{-1}$ or $ng\ g^{-1}$)						
		0.5 h	1 h	2 h	4 h	8 h	24 h	48 h
Plasma	1	16.5 ± 1.9	105.7 ± 13.6	142.2 ± 23.2	187.4 ± 9.7	123.8 ± 18.2	9.3 ± 2.1	2.6 ± 1.3
	2.5	75.5 ± 14.5	177.6 ± 14.2	220.6 ± 21.0	245.7 ± 25.9	192.1 ± 2.7	29.5 ± 2.7	12.0 ± 4.5
Liver	1	64.1 ± 17.2	306.1 ± 62.9	387.7 ± 78.5	344.5 ± 48.9	155.6 ± 30.7	3.0 ± 3.0	0.0
	2.5	435.3 ± 110.5	1397.8 ± 121.3	1221.6 ± 172.1	1065.1 ± 120.8	672.9 ± 71.6	62.2 ± 20.1	17.2 ± 8.0
Kidney	1	13.5 ± 8.7	170.6 ± 50.7	242.4 ± 53.5	305.6 ± 32.9	199.2 ± 26.4	28.4 ± 9.3	0.0
	2.5	117.4 ± 21.8	613.8 ± 115.5	693.2 ± 114.1	775.5 ± 80.4	444.1 ± 86.2	65.9 ± 14.7	15.9 ± 2.2
Heart	1	40.5 ± 10.1	148.9 ± 24.2	206.1 ± 35.6	220.3 ± 25.2	138.5 ± 13.6	5.9 ± 3.8	0.0
	2.5	144.9 ± 29.2	507.0 ± 66.0	580.1 ± 72.0	586.3 ± 54.2	513.5 ± 94.0	27.5 ± 6.5	11.7 ± 11.7
Lung	1	39.5 ± 18.5	136.1 ± 17.5	221.4 ± 53.8	225.8 ± 18.7	115.4 ± 15.8	5.2 ± 3.2	0.0
	2.5	131.7 ± 21.1	426.4 ± 58.3	412.7 ± 55.8	456.4 ± 37.9	370.3 ± 42.8	125.8 ± 36.9	70.7 ± 45.2
Muscle	1	2.3 ± 2.3	59.7 ± 9.7	109.2 ± 18.9	173.3 ± 19.0	116.2 ± 16.6	2.8 ± 2.8	0.0
	2.5	14.8 ± 4.6	95.9 ± 13.3	185.7 ± 28.1	230.0 ± 20.9	188.2 ± 29.5	19.5 ± 2.1	11.4 ± 7.1
Fat	1	13.9 ± 5.8	260.4 ± 67.2	746.9 ± 149.4	1295.4 ± 271.9	1243.3 ± 229.6	373.8 ± 63.9	129.5 ± 46.7
	2.5	54.2 ± 16.5	146.3 ± 29.9	818.2 ± 65.7	1610.4 ± 156.7	2344.3 ± 232.6	583.4 ± 136.8	175.9 ± 64.6

Table 1.3.5. Pharmacokinetic parameters (± SE, SE = SD/√*n*) of milbemycin D in beagle dogs (9.9–13.2 kg) (data from Ide *et al.*, 1993).

	Route	
	Oral	Intravenous
Dose rate (mg kg^{-1})	1.0	0.5
C_{max} (ng ml^{-1})	55 ± 6	3701 ± 573
T_{max} (h)	1.5	—
$t_{1/2}$ (h)	3.0 ± 0.5	2.9 ± 0.6
AUC (0–inf) ((ng ml^{-1}) × h)	292 ± 39	1478 ± 105

Table 1.3.6. Pharmacokinetic parameters (± SE, SE = SD/√*n*, *n* = 3) of milbemycin A_3A_4 5-oxime (mixture of homologues, >80% A_4 and <20% A_3) in beagle dogs (~11 kg) (data from Ide *et al.*, 1993).

	Route		
	Oral	Oral	Intravenous
Dose (mg)	1.25	2.5	5
C_{max} (ng ml^{-1})	46 ± 61	102 ± 10	204 ± 22
T_{max} (h)	2.3 ± 0.9	3.3 ± 0.7	3.7 ± 1.5
$t_{1/2}$ (h)	14.2 ± 3.3	14.8 ± 1.7	15.1 ± 1.6
$AUC_{(0-inf)}$ ((ng ml^{-1}) × h)	754 ± 133	1692 ± 165	3843 ± 829

peaks. The dose rate in this study (11.5 mg tablets, i.e. the upper end of the recommended dose rate of Interceptor®) was higher than that used in the Sankyo study (1.25, 2.5 or 5 mg tablets).

An interesting feature is the different elimination half-lives ($t_{1/2}$) observed in the above two studies. This difference is attributed mainly to the higher dose rates used at Novartis, which resulted in enhanced profiles for a longer period of time (7 days) compared with the Sankyo study (2 days). Hence, a protracted slower phase of elimination with very good linearity could be measured at Novartis, while the plasma levels in the Sankyo study were obviously below the limit of quantitation.

Milbemycin A_3 5-oxime was eliminated more rapidly than milbemycin A_4 5-oxime.

Comparison of plasma levels following oral and intravenous administration (Tables 1.3.7 and 1.3.8) yielded absolute bioavailabilities of 40–86% for the A_3 and A_4 oximes and showed, furthermore, that the bioavailability of both homologues from tablets was statistically significantly higher (~ 15%) in fed than in fasted dogs ($n = 8$).

On the basis of the Sankyo study and yet another study performed at Novartis (Internal Report), in which one, three and five tablets containing 12.5 mg milbemycin A_3A_4 5-oxime were administered to beagle dogs

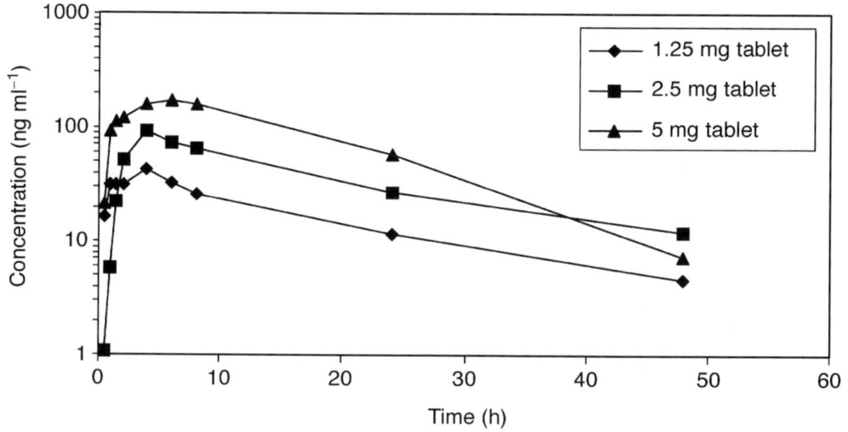

Fig. 1.3.5. Blood profiles of milbemycin A_3A_4 5-oxime (mixture of homologues, >80% A_4 and <20% A_3) in beagle dogs (~11 kg) (data from Ide *et al.*, 1993).

Table 1.3.7. Pharmacokinetic parameters (±SE, SE = SD/√*n*, *n* = 8) of milbemycin A_4 5-oxime (the *major* homologue of milbemycin A_3A_4 5-oxime, >80%) in beagle dogs (11.7–16.0 kg).

	Route			
	Oral, fasted	Oral, fasted	Oral, fed	I.V., fasted
Dose (mg)	11.5	12.5	12.5	12.5
Treatment	Interceptor tablet	Milbemax tablet	Milbemax tablet	Injectable
Dose rate of A_4 5-oxime (mg kg^{-1})	0.676 ± 0.021	0.734 ± 0.025	0.740 ± 0.036	0.745 ± 0.031
C_{max} (ng ml^{-1})	130 ± 11	132 ± 14	241 ± 10	679 ± 50
T_{max} (h)	1.9	2.4	1.0	0
$t_{1/2}$ (h)	92.1 ± 13.1	94.0 ± 12.6	108.8 ± 12.6	95.5 ± 12.4
MRT[a] (h)	122.3 ± 17.5	126.1 ± 16.5	141.3 ± 16.6	123.3 ± 15.9
$AUC_{(0-inf)}$ ((ng ml^{-1}) × h)	7072 ± 960	8248 ± 1173	13,156 ± 1649	15,422 ± 1450
Bioavailability (%)	45 ± 3	54 ± 6	77 ± 2	100 —
Clearance (l kg^{-1} day^{-1})	— —	— —	— —	1.23 ± 0.13
Volume of distribution in elimination phase (l)	— —	— —	— —	6.6 ± 0.6

[a]MRT, mean residence time. MRT = area under the moment curve (AUMC) divided by the area under the curve (AUC).

(Tables 1.3.9 and 1.3.10, Fig. 1.3.7), dose linearity of the kinetic disposition could not be statistically proven, but there was no evidence for any systematic deviation from linearity between the administered dose and area under the curve (AUC) or C_{max}.

Table 1.3.8. Pharmacokinetic parameters (± SE, SE = SD/√*n*, *n* = 8) of milbemycin A_3 5-oxime (the *minor* homologue of milbemycin A_3A_4 5-oxime, <20%) in beagle dogs (11.7–16.0 kg).

	Route			
	Oral, fasted	Oral, fasted	Oral, fed	I.V., fasted
Dose (mg)	11.5	12.5	12.5	12.5
Treatment	Interceptor tablet	Milbemax tablet	Milbemax tablet	Injectable
Dose rate of A_3 5-oxime (mg kg^{-1})	0.138 ± 0.004	0.150 ± 0.005	0.151 ± 0.005	0.153 ± 0.006
C_{max} (ng ml^{-1})	26 ± 2	25 ± 2	39 ± 2	85 ± 5
T_{max} (h)	1.8	2.5	1.5	0
$t_{1/2}$ (h)	23.5 ± 2.8	39.2 ± 6.8	39.1 ± 5.2	40.7 ± 4.8
MRT (h)	33.1 ± 4.0	55.3 ± 9.2	53.5 ± 6.9	55.9 ± 6.7
$AUC_{(0–inf)}$ ((ng ml^{-1}) × h)	577 ± 90	929 ± 185	1230 ± 122	1505 ± 169
Bioavailability (%)	40 ± 6	61 ± 7	86 ± 8	100 —
Clearance (l kg^{-1} day^{-1})	— —	— —	— —	2.69 ± 0.38
Volume of distribution in elimination phase (l)	— —	— —	— —	6.01 ± 0.41

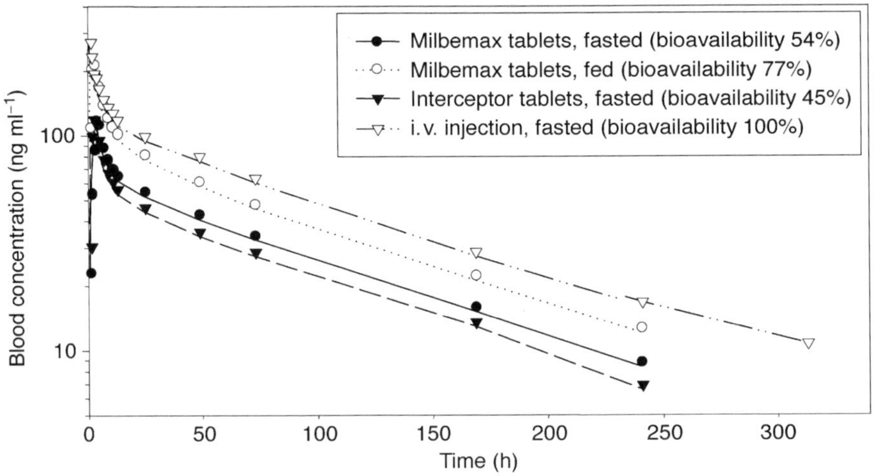

Fig. 1.3.6. Blood profiles of milbemycin A_4 5-oxime (the *major* homologue of milbemycin A_3A_4 5-oxime, >80%) in beagle dogs (11.7–16.0 kg).

Pharmacokinetics of milbemycin A_3A_4 5-oxime after repeated oral administration to dogs

In collaboration with Novartis, Toutain *et al.* (unpublished report) performed a pharmacokinetic study involving repeated daily or weekly

Table 1.3.9. Dose linearity of pharmacokinetic parameters (± SE, SE = SD/√*n*, *n* = 8) for milbemycin A_4 5-oxime (the *major* homologue of milbemycin A_3A_4 5-oxime, >80%) in young beagle dogs (4.0–7.6 kg).

	Route		
	Oral	Oral	Oral
Dose (mg)	12.5	37.5	62.5
Dose rate of A_4 5-oxime (mg kg^{-1})	1.76 ± 0.12	5.49 ± 0.45	8.91 ± 0.66
C_{max} (ng ml^{-1})	423 ± 80	1185 ± 119	1604 ± 126
$AUC_{(0-inf)}$ ((ng ml^{-1}) × h)	13,589 ± 1924	45,499 ± 2140	72,363 ± 5408

Table 1.3.10. Dose linearity of pharmacokinetic parameters (± SE, SE = SD/√*n*, *n* = 8) for milbemycin A_3 5-oxime (the *minor* homologue of milbemycin A_3A_4 5-oxime, <20%) in young beagle dogs (4.0–7.6 kg).

	Route		
	Oral	Oral	Oral
Dose (mg)	12.5	37.5	62.5
Dose rate of A_3 5-oxime (mg kg^{-1})	0.359 ± 0.025	1.125 ± 0.092	1.825 ± 0.135
C_{max} (ng ml^{-1})	69 ± 13	196 ± 18	263 ± 20
$AUC_{(0-inf)}$ ((ng ml^{-1}) × h)	1426 ± 231	4984 ± 217	7893 ± 495

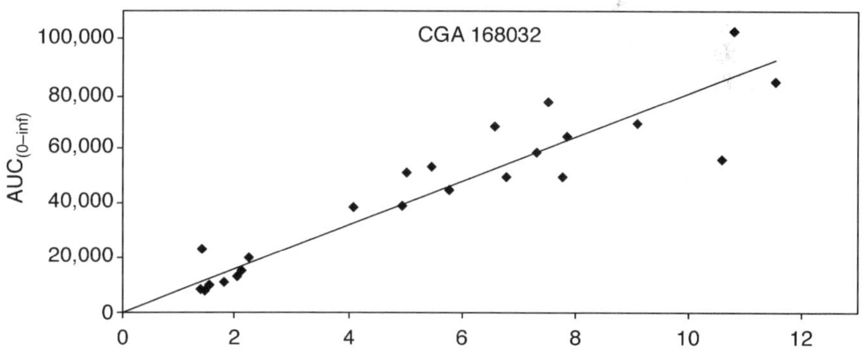

Fig. 1.3.7. Dose linearity of AUC for milbemycin A_4 5-oxime (the *major* homologue of milbemycin A_3A_4 5-oxime, >80%) in young beagle dogs (4.0–7.6 kg).

oral administration of milbemycin A_3A_4 5-oxime (Interceptor® tablets) to beagle dogs. Two dogs received the recommended daily dose of 0.5 mg kg^{-1} day^{-1}, whereas two other dogs received a weekly treatment at 3 mg kg^{-1} $week^{-1}$. The purpose was to document the most appropriate dosage regimen for long-term treatment at high dose rates, as required, for example for the control of *Demodex*. The results (one dog per group) are displayed in Figs 1.3.8 and 1.3.9. The measured plasma concentrations of milbemycin A_4 5-oxime were interpreted using a two-compartment model, and the respective curves were calculated for best fit with the experimental data points. The results displayed dose linearity and kinetic stability (absence of induction, inhibition, etc.). The delay to reach steady

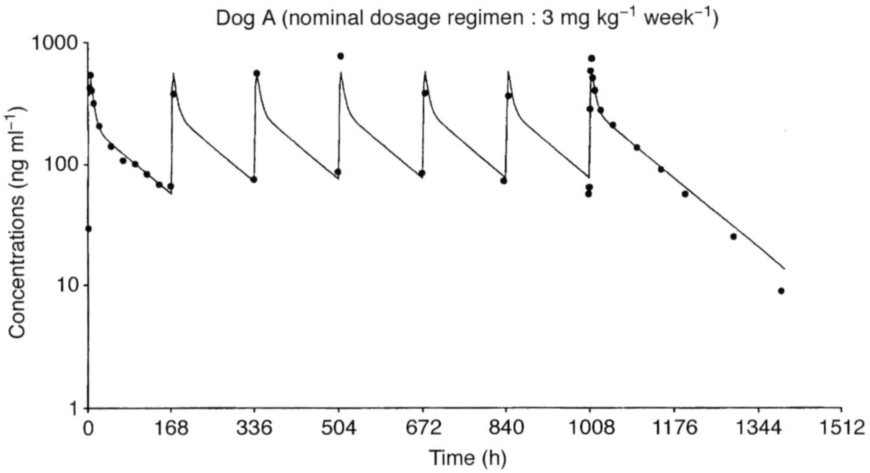

Fig. 1.3.8. Blood profiles of milbemycin A_4 5-oxime in one beagle dog during repeated weekly treatment with milbemycin A_3A_4 5-oxime (Interceptor) at 3 mg kg^{-1} $week^{-1}$.

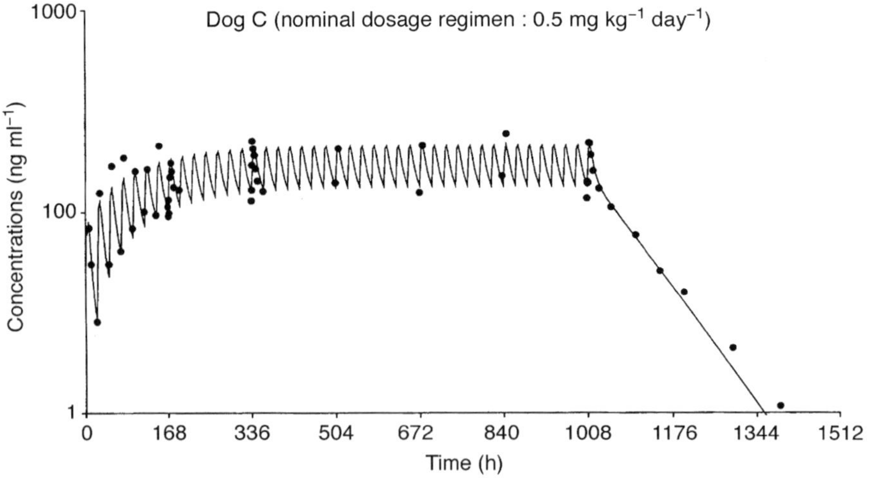

Fig. 1.3.9. Blood profiles of milbemycin A_4 5-oxime in one beagle dog during repeated daily treatment with milbemycin A_3A_4 5-oxime (Interceptor) at 0.5 mg kg^{-1} day^{-1}.

state was close to 2 weeks. For the weekly dosage regimen, there was no accumulation, that is plasma profiles after the first administration were close to steady-state conditions. For the daily dosage regimen, the accumulation ratio was approximately 4, and a delay of 1–2 weeks (i.e. approximately the time to reach steady-state conditions) was needed to reach plasma concentrations higher than 100 ng ml^{-1}. Hence, as the higher peak concentrations are well tolerated, the weekly dosage regimen may be an option for long-term treatment against *Demodex*, and time to reach steady state may be shortened by a suitable loading dose.

Pharmacokinetics of milbemycin A_3A_4 5-oxime after single oral administration to cats

A pharmacokinetic study of milbemycin A_3A_4 5-oxime with eight cats was performed (Internal Report) in order to compare the results with those previously obtained with dogs. Depending on its weight, each cat received either an entire or half a tablet containing 16 mg milbemycin A_3A_4 5-oxime. The results are shown in Table 1.3.11. In summary, as compared with dogs, the AUC and C_{max} in cats were rather low and highly variable, indicating lower bioavailability. In addition, the elimination half-life was much shorter, which is probably due to the 'real' elimination phase being below the limit of quantitation, as mentioned above also for dogs at low dose rates.

Toxicology and Safety in Target Animals

Toxicology

Results from acute, subacute and special toxicity for milbemycin 5-oxime are given in Table 1.3.12.

Table 1.3.11. Pharmacokinetic parameters (± SE, SE = SD/$\sqrt{n}$, $n = 8$) of milbemycin A_4 5-oxime (the *major* homologue of milbemycin A_3A_4 5-oxime, >80%) in cats (2.9–4.8 kg).

	Oral, fasted
Dose (mg)	8 or 16
Dose rate of A_4 5-oxime (mg kg^{-1})	2.520 ± 0.155
C_{max} (ng ml^{-1})	86 ± 37
T_{max} (h)	1.9 ± 0.2
$t_{1/2}$ (h)	17.6 ± 3.7
MRT (h)	23.2 ± 4.3
$AUC_{(0-inf)}$ ((ng ml^{-1}) × h)	1215 ± 290

Table 1.3.12. Results from acute, subacute and special toxicity for milbemycin 5-oxime (data from Ide *et al.*, 1993).

Test/species	Results
LD_{50} *per os* rats (mg kg^{-1})	863 (M)/532 (F)
LD_{50} *per os* mice (mg kg^{-1})	946 (M)/722 (F)
LD_{50} s.c. rats (mg kg^{-1})	>3000 (M)/>3000 (F)
LD_{50} s.c. mice (mg kg^{-1})	>3000 (M)/>3000 (F)
LD_{50} i.p. rats (mg kg^{-1})	454 (M)/318 (F)
LD_{50} i.p. mice (mg kg^{-1})	138 (M)/120 (F)
4 weeks feeding study rats	No effect level 10 mg kg^{-1} day^{-1}
3 months feeding study rats	No effect level 3 mg kg^{-1} day^{-1}
Reproductive toxicity in rats (daily oral application from day 7 to day 17 of pregnancy)	No teratogenic toxicity at 30 mg kg^{-1} day^{-1}
Reproductive toxicity in rabbits (daily oral application from day 6 to day 24 of pregnancy)	No teratogenic toxicity at 30 mg kg^{-1} day^{-1}
Mutagenicity (bacterial reverse mutagenicity test)	No induction of revertant colonies regardless of the presence of metabolic activator
Mutagenicity (chromosomal aberration test)	No significant increase in the number of cells with chromosomal abnormalities

Tolerability of milbemycin oxime in the target animal

General tolerability in dogs

In a first trial (Anonymous, 1990; Ide *et al.*, 1993), five groups of beagles aged 8, 10 or 12 weeks received on each of 3 consecutive days cocktails of tablets containing milbemycin oxime equivalent to minimum daily dose rates of 0.5, 2.5, 7.5 or 12.5 mg kg^{-1}. One group was left untreated. The dogs were observed for clinical signs for 2 weeks, they were weighed before and 4, 7 and 14 days after the first administration and their eyes were examined before and 2 weeks after treatment. There was a clear dose dependency and a slight age dependency of incidence, severity and duration of clinical signs. No effects were seen after giving 3× 0.5 mg kg^{-1}. At higher dose rates, symptoms appeared only after the second dose and were more pronounced after the third dose which may suggest accumulation. Six of the 12 dogs were affected after 3× 2.5 mg kg^{-1} and all after higher doses. At the lower dose rates, the clinical signs (ataxia, trembling) were mild. At higher dose rates, they were more severe, and prostration and salivation were seen in addition. In all cases, symptoms were transient and, in general, all dogs had recovered within 1–2 days without treatment. Other treatment-related deviations were not detected.

Table 1.3.13. Daily dose rates of milbemycin oxime administered to beagle dogs for the first 2 months.

Month	Tablets	Dose per dog (mg)	Dose rate (mg kg^{-1})		
			Minimum	Mean	Maximum
1	1	5.68	1.9	2.8	3.8
	3	17.04	6.2	8.2	12.3
	5	28.4	11.1	14.4	19.2
2	1	5.68	1.1	1.5	2.4
	3	17.04	3.3	4.7	8.2
	5	28.4	6.1	7.6	9.0

In a second trial (Anonymous, 1990; Stansfield and Hepler, 1991; Blagburn, 1993; Ide *et al*., 1993), four groups of beagles (≥ 1.4 kg) aged 8 weeks received placebo tablets or one, three or five tablets containing milbemycin oxime daily for 3 consecutive days each month for 10 months. This resulted in the daily dose rates for the first 2 months shown in Table 1.3.13.

The dose rates of later administrations were lower as the dogs grew. The animals were observed for clinical signs. They received physical examinations. Weight and food consumption were determined. Before and 2 days after the last administration each month, blood was taken for haematology (12 parameters) and clinical chemistry (18 parameters). Urine was analysed (13 parameters) monthly 2 days after administration beginning on day 95. The eyes were examined before and at the end of the trial, at which time the dogs were necropsied for gross pathology, determination of absolute and relative organ weights (ten organs) and histology (40 organs or tissues). No reaction was observed after giving one tablet per day. Transient trembling and ataxia were observed after giving three (five of 16 dogs) or five (13 of 15 dogs) tablets per day, but only during the first 3 days. The reaction was very slight at the lower dose. Other treatment-related deviations were not detected.

In a third trial (Anonymous, 1995), four groups of beagles aged 2 weeks received placebo tablets or one, three or five tablets containing milbemycin oxime once every 2 weeks for 6 consecutive weeks. This resulted in the dose rates for the four administrations shown in Table 1.3.14.

The dogs were observed for clinical signs during the administration period and for one more day. They were weighed. Blood was taken for haematology (15 parameters) and clinical chemistry (18 parameters) as well as urine for analysis (seven parameters) after the last administration. One placebo puppy showed trembling 8 days after treatment. Two puppies in the one-tablet group showed diarrhoea 1–3 h after the first administration. No reaction to the following administrations was

Table 1.3.14. Dose rates for four administrations of milbemycin oxime to beagle dogs.

Administration	Tablets	Dose per dog (mg)	Dose rate (mg kg^{-1})		
			Minimum	Mean	Maximum
1	1	2.3	2.2	3.1	4.4
	3	6.9	7.4	10.0	12.7
	5	11.5	10.6	15.6	20.9
2	1	2.3	1.4	2.0	2.4
	3	6.9	4.9	6.3	8.8
	5	11.5	6.4	10.3	14.8
3	1	2.3	0.9	1.2	1.6
	3	6.9	2.8	4.0	5.8
	5	11.5	4.4	6.6	10.5
4	1	2.3	0.7	0.9	1.1
	3	6.9	1.9	2.8	3.5
	5	11.5	3.2	4.8	7.8

detected. Six puppies in the three-tablet group were affected, showing one or more of the following transient symptoms: trembling, diarrhoea or vocalizing; they recovered generally within 6 h without treatment. No reaction to the following administrations was detected. Seven puppies in the five-tablet group were affected after the first administration, demonstrating transient symptoms such as trembling, reduced activity, unsteady gait and vocalizing; all puppies had recovered within 2 days without treatment. Three of the puppies reacted slightly to the second dose and recovered from the same symptoms within 1 day without treatment. Four of the puppies reacted very slightly to the third administration with the same transient symptoms, and recovered within 6 h. No reaction to the last administration was detected. The γ-glutamyl transferase was slightly increased in the medium and high dose group but not in the low dose group. Other treatment-related deviations were not detected.

CONCLUSION. Milbemycin oxime is well tolerated in dogs of all ages, but very young puppies may be slightly more susceptible to overdoses. Clinical signs generally are only seen after overdosing ten times or more. More specific symptoms commonly observed after overdosing more than ten times are reduced activity, unsteady gait, incoordination and, after very high doses, also prostration. The incidence, severity and duration are dose dependent. Trembling commonly seen after overdosing need not be treatment related in all cases, as untreated dogs may also show it. Diarrhoea, also observed occasionally, is probably not treatment related as it is too sporadic and not dose dependent. Affected dogs recover normally within 1 day without treatment. There may be accumulation with daily administration of overdoses as symptoms are often seen only after the second overdose.

Tolerability during reproduction of dogs

This trial (Anonymous, 1990; Stansfield and Hepler, 1991; Blagburn, 1993; Ide *et al.*, 1993) was conducted in two parts.

Part 1. Male and female beagles, older than 2 years and with at least two successful litters, received three milbemycin tablets appropriate for their weight daily before and during mating, and the bitches continued up to 1 week before the anticipated whelping day. Thus, the animals received daily a threefold overdose, the males for ≥18 weeks and the females for ≥30 weeks. A similar group of animals received placebo. An ejaculate was examined before and 2 and 3 months after the start of the treatment and 1 month after mating, that is at the end of the observation period for males. Puppies were examined by X-rays when 3 days old. Puppies were weaned when 6 weeks old, that is at the end of the observation period for females and puppies.

Part 2. Seven bitches, older than 2 years and with at least two successful litters, received 4.5 milbemycin tablets, that is a threefold overdose, once 6 to 3 days before whelping, another three similar bitches 1 day before whelping, five more similar bitches each on the day of whelping or 1 or 2 days after whelping. Puppies were weaned when 6 weeks old, that is at the end of the observation period for females and puppies.

Parents and litters were observed for clinical signs and received a physical examination each week. Food consumption of parents and weights of all animals were determined.

There was no treatment-related effect on the health of parents or puppies nor on any reproductive parameter (e.g. volume, colour and pH of ejaculate; quantity, motility, speed of progression and morphology of sperm; length of oestrous cycle; mating behaviour; duration of pregnancy; ease of whelping; number, anatomy, viability and weight development of puppies).

CONCLUSION. Milbemycin oxime does not interfere with reproduction even when given at a threefold overdose and daily instead of monthly. There is no effect on suckling puppies after a single threefold overdosing of the bitch.

Tolerability in collies

In a first trial (Blagburn *et al.*, 1989; Anonymous, 1990; Stansfield and Hepler, 1991; Ide *et al.*, 1993), heartworm-free adult collies received 0.5 or 2.5 mg milbemycin oxime per kg as 1% powder twice at an interval of 1 week. After another week, the dogs received the other dose rate, again twice at an interval of 1 week. Finally, they received 200 μg kg^{-1}

ivermectin orally. The dogs were observed for clinical signs and their blood (six haematology parameters, 19 clinical chemistry parameters) and urine (ten parameters) were analysed before and 4 h after dosing. In addition, the eyes were examined. None of the collies reacted to milbemycin, whereas three of the ten animals reacted to ivermectin, showing the typical symptoms: ataxia, recumbence, mydriasis, tremors, emesis and panting.

In a second trial (Anonymous, 1990; Stansfield and Hepler, 1991; Blagburn, 1993; Ide *et al.*, 1993) adult collies, 11 of them microfilaraemic, received single doses of 2.5, 5, 10 and 12.5 mg kg^{-1} milbemycin oxime, at 2-week intervals. Forty days after the last dose, they received 150 μg kg^{-1} ivermectin. The dogs were observed for clinical signs and their blood was examined for haematology and clinical chemistry before and 3 h after each milbemycin administration. Only one dog reacted after the highest dose rate with marked ataxia, recumbence, elevated temperature and salivation. At necropsy, adult heartworms and heartworm disease alterations were detected. No other reaction was detected in any other dog, whereas three collies reacted severely to ivermectin with the typical clinical signs.

In a third trial (Sasaki *et al.*, 1990), rough-coated heartworm-free collies 10–18 weeks old and similar shiba dogs received 0.25, 0.5, 1.0 and 2.5 mg kg^{-1} milbemycin oxime as 0.25% powder daily for 10 consecutive days at 2-week intervals. Afterwards, the dogs were given a single dose of 12.5 or 25 mg kg^{-1}. The dogs were observed for clinical signs. Before and 3 and 24 h after the first and tenth administration, blood was taken for haematology (two parameters) and clinical chemistry (five parameters), the arterial blood pressure was measured and the body temperature was taken. All dogs were necropsied 1 day after the last dose. One collie staggered slightly 1.5 h after receiving 2.5 mg kg^{-1}. This animal did not show any clinical sign after higher dose rates. No reactions were detected in any other dog even after the high dose rates.

In a fourth trial (Tranquilli *et al.*, 1991), heartworm-free adult collies known to be sensitive to ivermectin, as they react to the oral administration of 120 μg kg^{-1}, were left untreated or received 5 or 10 mg kg^{-1} milbemycin oxime, using intact commercial tablets and fragments. The animals were observed for clinical signs, paying particular attention to depression, ataxia, mydriasis and salivation. No reaction was observed in the untreated dogs. Two dogs showed clinical signs (depression, ataxia and salivation) after the 5 mg kg^{-1} dose. They recovered within 1 day without treatment. After the higher dose, all dogs were affected, showing the same symptoms and mydriasis. They recovered within 2 days without treatment. Other treatment-related deviations were not detected.

CONCLUSION. Milbemycin oxime is well tolerated by collies. Tenfold overdose may lead to mild transient clinical signs such as reduced activity, ataxia and salivation.

Tolerability in dogs infected with heartworms

In a first trial (Anonymous, 1990; Ide *et al.*, 1993), three groups of adult mongrel dogs infected with heartworm microfilariae were left untreated or received one or three commercial tablets appropriate for their weight in order to administer the recommended dose or a threefold overdose three times at monthly intervals. The dogs were observed for clinical signs during the administration period and for another 2 weeks. Weight and food consumption were determined. Blood was taken before and 3 and 24 h after each administration for haematology (seven parameters), clinical chemistry (13 parameters) and estimations of microfilarial numbers. Urine (nine parameters) and eyes were examined before the first and after the second (urine only) and third administrations. All dogs were necropsied for counts of adult heartworms and histology. A slight to moderate laboured respiration, pale mucous membranes and reduced activity were observed in the majority of treated dogs after the first administration irrespective of the dose, but all dogs with high microfilaraemia were affected. The affected dogs recovered within 1–2 days without treatment. After the second and third dosing, fewer dogs reacted and the reaction was less pronounced. The microfilaraemia of treated dogs dropped after the first treatment to zero or very low levels, and little further change was observed after the other two treatments. Other treatment-related deviations were not detected. All dogs harboured adult heartworms.

A second very similar trial (Anonymous, 1990; Ide *et al.*, 1993) in which the high dose group received five tablets instead of three resulted in similar observations.

In a third trial (Ide *et al.*, 1993), 70 microfilaraemic mongrel dogs received 0.25 mg kg^{-1} milbemycin oxime and 42 microfilaraemic mongrels received 0.5 mg kg^{-1}. They were observed for clinical signs. Most of the dogs exhibited several clinical signs such as pale mucous membranes, vomiting and reduced activity after both dose rates. No dog died.

CONCLUSION. Clinical signs such as laboured respiration, pale mucous membranes, coughing, vomiting, salivation and reduced activity may be observed after treating microfilaraemic dogs. The clinical signs were slight to moderate, and transient. Dogs recovered without treatment. The incidence and severity of the clinical signs appear to depend on the parasitaemia and not on the dose rate. As the parasitaemia drops after treatment, it is suggested that the clinical signs are due to dying microfilariae rather than direct toxicity.

General tolerability in cats

In a first trial (Anonymous, 1998), three male and three female domestic short hair cats aged 9–11 months received once cocktails of tablets containing 2.3, 5.75 or 11.5 mg milbemycin oxime such that a minimum dose rate of 20 mg kg^{-1} was given. A similar group received placebo tablets. The cats were observed for clinical signs during 2 weeks, weighed at the start and end of the trial and their food consumption was determined. Treatment-related deviations were not detected.

In a second trial (Anonymous, 1998), three male and three female domestic short hair kittens aged 8 weeks received once cocktails of tablets containing 2.3 or 5.75 mg milbemycin oxime such that a minimum dose rate of 20 mg kg^{-1} was given. A similar group received placebo tablets. The kittens were observed for clinical signs during 2 weeks, weighed at the start and end of the trial and their food consumption was determined. Treatment-related deviations were not detected.

In a third trial (Anonymous, 1998), three male and three female domestic short hair kittens aged 2–3 weeks received seven times at fortnightly intervals cocktails of intact and half tablets containing 2.3, 5.75, 11.5 or 23 mg milbemycin oxime such that a minimum dose rate of 2, 6 or 10 mg kg^{-1} was given. The actual dose rates given at the first dosing were 3.7–6.3, 6.1–12.0 and 11.0–12.7 mg kg^{-1}, respectively. A similar group received placebo tablets. The kittens were observed for clinical signs during the administration period and an additional week. They received a weekly physical examination and their weight and food consumption were determined weekly after weaning. The eyes were examined just before weaning and at the end of the trial. Blood was taken at termination for haematology (13 parameters) and clinical chemistry (18 parameters) and urine for analysis (12 parameters). All cats were necropsied for gross pathology, determination of absolute and relative organ weights (ten organs) and for histology (41 organs or tissues). Treatment-related deviations were not detected.

In a fourth trial (Anonymous, 1998), three male and three female domestic short hair cats aged 3–4 months received seven times at fortnightly intervals cocktails of intact and half tablets containing 2.3, 5.75, 11.5 or 23 mg milbemycin oxime such that a minimum dose rate of 2, 6 or 10 mg kg^{-1} was given. A similar group received placebo tablets. The kittens were observed for clinical signs during the administration period and for an additional week. They received a weekly physical examination and their weight and food consumption were also determined weekly. The eyes were examined at the start and end of the trial. Blood was taken before the first dosing, in week 6 and at termination for haematology (13 parameters) and clinical chemistry (18 parameters) and urine for analysis (12 parameters). All cats were necropsied for gross pathology, determination of absolute and relative organ weights (ten organs) and histology (41 organs or tissues). Treatment-related deviations were not detected.

CONCLUSION. Milbemycin oxime is well tolerated by cats. Tenfold overdose did not lead to reactions, even in kittens.

Acknowledgements

The authors would like to thank Dr R. Steiger, Novartis Animal Health Inc., for his critical review of the manuscript and his constructive remarks.

References

Ali, D.N. and Hennessy, D.R. (1996) The effect of level of feed intake on the pharmacokinetic disposition and efficacy of ivermectin in sheep. *Journal of Veterinary Pharmacology and Therapeutics* 19, 89–94

Alvinerie, M., Sutra, J.F., Badr, M. and Galtier, P. (1995) Determination of moxidectin in plasma by high performance liquid chromatography with automated solid phase extraction and fluorescent detection. *Journal of Chromatography B* 674, 119–124.

Anonymous (1990) Freedom of Information Summary. NADA 140–915 original approval http://www.fda.gov/cvm/efoi/section2/foiabst2.html

Anonymous (1995) Freedom of Information Summary. NADA 140–915 supplemental approval http://www.fda.gov/cvm/efoi/section2/foiabst2.html

Anonymous (1998) Freedom of Information Summary. NADA 140–915 supplemental approval http://www.fda.gov/cvm/efoi/section2/foiabst2.html

Atta, A.H. and Abo-Shihada, M.N. (2000) Comparative pharmacokinetics of doramectin and ivermectin in sheep. *Journal of Veterinary Pharmacology and Therapeutics* 23, 49–52.

Blagburn, B.L. (1993) Milbemicina ossima: un nuovo farmaco per la profilassi della filariosi. *Veterinaria* 7 (Suppl.), 33–36.

Blagburn, B.L., Hendrix, C.M., Lindsay, D.S., Vaughan, J.L., Mysinger, R.H. and Hepler, D.I. (1989) Milbemycin: efficacy and toxicity in Beagle and Collie dogs. In: Otto, G.F. (ed.) *Proceedings of the Heartworm Symposium.* American Heartworm Society, Washington, DC, pp. 109–113.

De Montagny, P., Shim, J.-S.K. and Pivnichny, J.V. (1990) Liquid chromatography determination of ivermectin in animal plasma with trifluoroacetic anhydride and *N*-methylimidazole as derivatization agent. *Journal of Pharmaceutical and Biomedical Analysis* 8, 507–511

Gayrard, V., Alvinerie, M. and Toutain, P.L. (1999) Comparison of pharmacokinetic profiles of doramectin and ivermectin pour-on formulations in cattle. *Veterinary Parasitology* 81, 47–55.

Ide, J., Okazaki, T., Ono, M., Saito, A., Nakagawa, K., Naito, S., Sato, K., Tanaka, K., Yoshikawa, H., Ando, M., Katsumi, S., Matsumoto, K., Toyama, T., Shibano, M. and Abe, M. (1993) Milbemycin: discovery and development. *Annual Report Sankyo Research Laboratories* (Sankyo Kenkyusho Nenpo) 45, 1–98.

Lifschitz, A., Virkel, G., Pis, A., Imperiale, F., Sanchez, S., Alvarez, L., Kujanek, R. and Lanusse, C. (1999) Ivermectin disposition kinetics after subcutaneous and intramuscular administration of an oil-based formulation to cattle. *Veterinary Parasitology* 86, 203–215.

Oukessou, M., Badri, M., Sutra, J.F., Galtier, P. and Alvinerie, M. (1996) Pharmacokinetics of ivermectin in the camel (*Camelus dromedarius*). *Veterinary Record* 139, 424–425

Oukessou, M., Berrag, B. and Alvinerie, M. (1999) A comparative kinetic study of ivermectin and moxidectin in lactating camels (*Camelus dromedarius*). *Veterinary Parasitology* 83, 151–159.

Perez, R., Cabezas, I., Garcia, M., Rubilar, L., Sutra, J.F., Galtier, P. and Alvinerie, M. (1999) Comparison of the pharmacokinetics of moxidectin (Equest(TM)) and ivermectin (Eqvalan(TM)) in horses. *Journal of Veterinary Pharmacology and Therapeutics*, 22, 174–180.

Sasaki, Y., Kitagawa, H., Murase, S. and Ishihara, K. (1990) Susceptibility of rough-coated collies to milbemycin oxime. *Japanese Journal of Veterinary Science* 52, 1269–1271.

Shoop, W.L., Mrozik, H. and Fisher, M.H. (1995) Structure and activity of avermectins and milbemycins in animal health. *Veterinary Parasitology* 59, 139–156.

Shoop, W.L., Demontigny, P., Fink, D.W., Williams, J.B., Egerton, J.R., Mrozik, H., Fisher, M.H., Skelly, B.J. and Turner, M.J. (1996) Efficacy in sheep and pharmacokinetics in cattle that led to the selection of eprinomectin as a topical endectocide for cattle. *International Journal for Parasitology* 26, 1227–1235.

Stansfield, D.G. and Hepler, D.I. (1991) Safety and efficacy of milbemycin oxime for parasite control. *Canine Practice* 16, 11–16.

Toutain, P.L., Upson, D.W., Terhune, T.N. and McKenzie, M.E. (1997) Comparative pharmacokinetics of doramectin and ivermectin in cattle. *Veterinary Parasitology* 72, 3–8

Tranquilli, W.J., Paul, A.J. and Todd, K.S. (1991) Assessment of toxicosis induced by high-dose administration of milbemycin oxime in collies. *American Journal of Veterinary Research* 52, 1170–1172.

Chapter 1.4

Chemistry, Pharmacology and Safety: Moxidectin

D.W. Rock, R.L. DeLay and M.J. Gliddon

Basic Chemistry

Moxidectin is a potent broad-spectrum endectocide (endo- and ectoparasiticide) that is a semisynthetic methoxime derivative of LL F-29249α (F-alpha) (Fig. 1.4.1). F-alpha (nemadectin) is a fermentation product of *Streptomyces cyaneogriseus* subsp. *noncyanogenus*. This organism was isolated in 1983 from a sample of red sand from Victoria, Australia.

Moxidectin is a 16-member, second-generation pentacyclic lactone of the milbemycin class of compounds. It is approved globally for use in companion and food-producing animals. It is similar in structure to ivermectin, a member of the avermectins, a class of compounds previously developed with similar uses. Moxidectin and ivermectin differ only in the presence of a disaccharide at C-13 found in ivermectin but not moxidectin, a substituted olefinic side chain at C-25 in moxidectin and the characteristic methoxime moiety at C-23, which is unique to moxidectin.

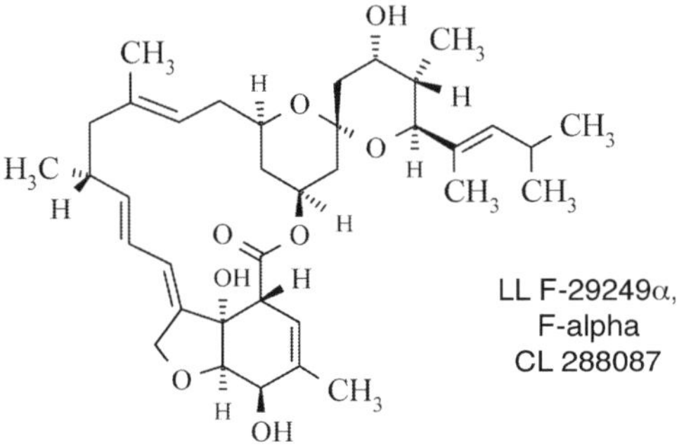

Fig. 1.4.1. Structure of LL F-29249α (F-alpha).

A brief summary of moxidectin structure and properties is given below:

Chemical name: (2a*E*,4*E*,5′*R*,6*R*,6′*S*,8*E*,11*R*,13*S*,15*S*,17a*R*,20*R*, 20a*R*,20b*S*)-6′-[(*E*)-1,2-dimethyl-1-butenyl]-5′,

	6,6′,7,10,11,14,15,17a,20,20a,20b-dodecahydro-20,20b-dihydroxy-5′6,8,19-tetra-methylspiro [11,15-methano-2*H*,13*H*,17*H*-furo[4,3,2-pq][2,6] benzodioxacyclooctadecin-13,2′-[2*H*]pyrano]-4′,17(3′*H*)-dione,4′-(*E*)-(*O*-methyloxime)
CAS Registry number:	113507–06–5
Physical form:	White to pale yellow powder
Structural formula:	

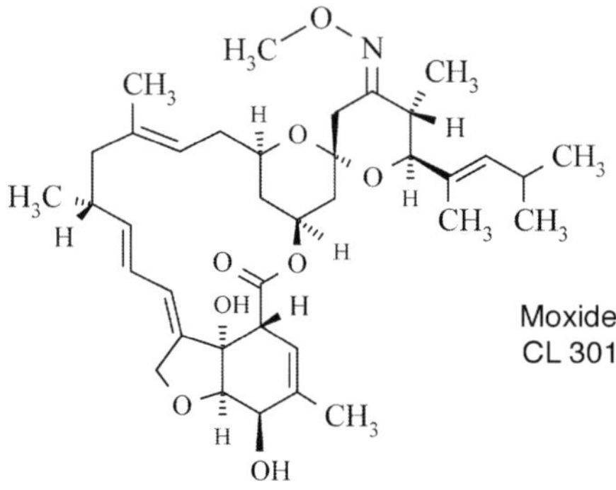

Moxidectin, CL 301423

Molecular formula:	$C_{37}H_{53}NO_8$
Relative molecular mass:	639.84
Chirality:	Degrees of rotation +104.1°
Melting point:	145–154°C
Octanol/water partition coefficient:	58,300

Moxidectin is slightly soluble in water, 0.51 mg l^{-1}, with solubility not affected by change in pH. It is readily soluble in organic solvents such as methylene chloride, diethyl ether, ethanol, acetonitrile, ethyl acetate, propylene glycol and benzyl alcohol.

F-alpha was discovered and patented independently by Glaxo Group Ltd, UK (Bain *et al.*, 1987; Sutherland *et al.*, 1987) and American Cyanamid, USA (Asato and France, 1988; Rudd and Ramsay, 1989). Negotiations led to the worldwide transfer of Glaxo's rights to American Cyanamid in 1988, and F-alpha fermentation development and chemical process development for the molecule commenced. The first product marketed was an injectable for cattle introduced in Argentina in 1990. Subsequent formulations introduced worldwide have included a moxidectin dog tablet, a sheep drench, horse gel, a cattle pour-on, a sheep injectable plus 6 in 1 vaccine for sheep, a swine injectable and a canine sustained-release injectable.

Basic Pharmacology

The means by which macrocyclic lactones have their effect on parasites continues to be researched. A series of physiological function tests were used to demonstrate the activity of moxidectin (D. Ingle and I.B. Wood,

Princeton, New Jersey, 1990, Company Report). These studies indicated that moxidectin had activity at the γ-aminobutyric acid (GABA)-A complex. Activity of milbemycin compounds has also been associated with glutamate-gated chloride channels (Fisher, 1997). In addition, P-glycoprotein has also been investigated as a possible mechanism of action for moxidectin as well as other macrocyclic lactones (Blackhall *et al.*, 1998). P-glycoprotein has also been implicated in the mechanism of resistance development to macrocyclic lactones (Sangster *et al.*, 1999). Cole and Casida (1992) demonstrated a different binding activity for a GABA-gated chloride ion channel between avermectin and moxidectin. Paiement *et al.* (1999) demonstrated a difference in the activity of avermectins and milbemycins in experiments involving insulin.

Metabolism and Tissue Residues

Total residue studies in target species

Radiolabelled moxidectin has been used to determine the tissue distribution, biotransformation and elimination of moxidectin in several animal species. Studies have been carried out in cattle dosed subcutaneously or topically, and sheep and horses dosed orally. The metabolic profile in milk taken from cows treated topically with moxidectin has also been studied. These studies were carried out using [^{14}C]-moxidectin prepared by chemically modifying its precursor nemadectin. Nemadectin was isolated from a fermentation using a mixture of ^{14}C-labelled substrates (Ahmed *et al.*, 1993). In most studies, deuterium-labelled moxidectin was included in the test material to facilitate metabolite characterization. A qualitatively similar metabolic pattern has been seen in all animal species, with unaltered moxidectin being the major residue in the tissues at all time points studied.

In one such study, three steers were administered a single subcutaneous dose of a mixture of ^{14}C- and deuterium-labelled moxidectin providing 0.2 mg kg^{-1} moxidectin (Zulalian *et al.*, 1994). Total excreta were collected following treatment until the animals were sacrificed at 7, 14 and 28 days after treatment. Samples of 24 organs, tissues and body fluids were collected at each sacrifice for the determination of absorption, distribution and elimination of radioactivity. Tissue residue levels are shown in Table 1.4.1. Among the tissues and fluids analysed, the lowest residue levels were found in the brain. Of the edible tissues, fat had the highest moxidectin residue levels and muscle the lowest levels at all time points. There was a steady decline in residue levels over time, demonstrating that there was no bioaccumulation of moxidectin in the tissues. The depletion half-lives were 14 and 12 days for back fat and omental fat, respectively, 11 days for liver and kidney, and 9 days for muscle.

In other studies using radiolabelled moxidectin, sheep were dosed with an oral drench providing 0.2 mg kg^{-1} moxidectin (Afzal *et al.*, 1997), horses were dosed with an oral gel providing 0.4 mg kg^{-1} moxidectin (J. Afzal, Princeton, New Jersey, 1996, Company Report, MET 96-003) and cattle were dosed topically with a pour-on formulation providing 0.5 mg kg^{-1} moxidectin (S.-S. Wu, Princeton, New Jersey, 1996, Company Report, PD-M 29-43). The distribution of moxidectin-related residues among the major tissues followed a similar pattern in all species and the different routes of administration. Residues were always highest in the fat and lowest in the muscle, with liver and kidney intermediate (Table 1.4.2). The major route of excretion of the administered radioactive dose was via the faeces in all species. For cattle dosed subcutaneously, 32% of the dose was excreted in the faeces after 7 days, 41% after 14 days and 58% after 28 days, with only 3% of the radioactivity recovered in the urine through

Table 1.4.1. Total ^{14}C residues (ppb) in tissues from steers dosed subcutaneously with ^{14}C- and ^{2}H-labelled moxidectin.[a]

	Days post-treatment		
Tissue	7	14	28
Abdominal fat	898	636	275
Back fat	495	424	186
Kidney	42	38	13
Liver	109	77	31
Muscle	21	10	4
Adrenals	88	120	29
Bile	159	82	42
Bladder	33	61	21
Blood	10	7	3
Brain	7	3	<2
Carcass	115	67	42
Oesophagus	80	67	61
GI contents	22	17	2
Heart	94	91	37
Injection site	383	192	96
Intestines, large	139	167	65
Intestines, small	111	100	19
Lungs	32	161	12
Pancreas	83	69	32
Stomach compartments	74	42	30
Spleen	8	26	10
Thyroid	403	90	127
Thymus	197	83	48
Tongue	120	90	61

[a]One animal per sacrifice time.

28 days post-treatment. For sheep and horses dosed orally, 43 and 77% of the administered radioactivity was recovered in the faeces by 7 days post-treatment, respectively, with <1% excreted in the urine.

The partitioning of moxidectin into milk was determined using radiolabelled moxidectin administered topically to cows (J. Afzal, Princeton, New Jersey, 1998, Company Report). Six lactating Holstein cows were dosed individually with a pour-on formulation of moxidectin at 1.5 ml kg^{-1}, providing 0.75 mg kg^{-1} moxidectin or 150% of the approved use level. Milk samples collected at approximately 12-h intervals for 10 days after treatment were assayed for total radioresidues by direct scintillation counting. Residue levels for consecutive 12-h periods were combined proportionally according to milk production to determine pooled daily levels of moxidectin-related residues in the milk. Residues were generally below the limit of detection of the assay (4 ppb) until day 5 after dosing, when measurable residues were present for five of the six cows (Table 1.4.3). Total moxidectin residues in milk from cows treated with the pour-on formulation at 1.5 times the approved use level peaked at 10.9 ± 6.4 ppb at 8 days post-treatment.

Table 1.4.2. Total ^{14}C residues (ppb) in tissues from cattle, sheep and horses dosed with ^{14}C- and ^{2}H-labelled moxidectin.[a]

Species	Cattle	Cattle	Sheep	Horse
Route	Subcutaneous	Topical	Oral (drench)	Oral (gel)
Dose (mg kg^{-1})	0.2	0.5	0.2	0.4
Tissue				
Abdominal fat	636	113	322	884
Back fat	424	55	287	864
Liver	77	12	50	184
Kidney	38	8	22	51
Muscle	10	<3	12	21

[a]14 days post-dosing for cattle, 7 days post-dosing for sheep and horses; mean of three animals except one animal for cattle dosed by subcutaneous injection.

Table 1.4.3. Pooled daily residues (ppb moxidectin equivalents) in milk following the topical treatment of lactating cows with 0.75 mg kg^{-1} moxidectin.[a]

	Days post-treatment									
	1	2	3	4	5	6	7	8	9	10
Mean	<4	<4	<4	<4	5.6	8.5	10.0	10.9	10.2	9.6
SD	—	—	—	—	7.8	8.4	6.7	6.4	4.8	5.1
Range	<4	<4–4	<4–8	<4–18	<4–21	<4–25	<4–19	<4–19	<4–16	<4–17

[a]Mean of six animals.

Biotransformation

Since moxidectin is widely used in food-producing animals, toxicological considerations require that the structure of the major components of the residues in the edible tissues (parent compound and/or metabolites) be determined. Therefore, organic extracts were prepared from the primary edible tissues (muscle, liver, kidney, fat and milk) and faeces from the animal studies just described. Because total residue levels frequently were very low in the tissues, faecal extracts were used as the primary source of metabolites for identification. Once identified, the distribution of the metabolites in the tissues could be confirmed. Aliquots of the extracts and post-extracted solids were analysed for radioactivity by liquid scintillation counting and combustion. High-pressure liquid chromatography (HPLC) and thin-layer chromatography were used to separate and purify the metabolites. The metabolites were characterized structurally using thermospray liquid chromatography/mass spectrometry (LC/MS) and thermospray liquid chromatography/tandem mass spectrometry (LC/MS/MS; Stout *et al.*, 1994). In the horse study, chromatographic retention times of the metabolites relative to parent moxidectin (identified by LC/MS) were used to establish their identity (J. Afzal, Princeton, New Jersey, 1996, Company Report).

Moxidectin residues were extracted readily from edible tissues, with extraction efficiencies generally greater than 90%. In fat, the primary site of moxidectin residues, over 95% of the radioactivity was extracted. This indicates that bound moxidectin residues are of no concern for the safety of consumers. The primary residue in all tissues is parent moxidectin, accounting for approximately 35–60% of the residue in liver, 55–80% in kidney, 40–90% in muscle and 75–95% in fat, the target tissue. Only two metabolites were present in the tissues at levels routinely above 2% of the total radioactivity. These were identified (Stout *et al.*, 1994) as monohydroxylated derivatives of moxidectin with hydroxylation at C-29 and on the methyl group at C-14 (Fig. 1.4.2). The relative distribution of the parent moxidectin and these two metabolites in the fat and liver of cattle, sheep and horses is shown in Table 1.4.4. Up to five other minor metabolites were identified. These were mostly mono- or di-hydroxylated derivatives on the parent molecule. One metabolite, found primarily in liver, was shown to be the 23-keto derivative of moxidectin (Fig. 1.4.2).

The composition of the radioresidues in the milk from lactating dairy cows treated topically with radiolabelled moxidectin has also been studied (J. Afzal, Princeton, New Jersey, 1998, Company Report). Individual milk samples with the highest radioresidues from the study described in the previous section were extracted and partitioned using techniques similar to those described for the tissue samples. The metabolic profile in milk was very similar to that identified in fat tissue from

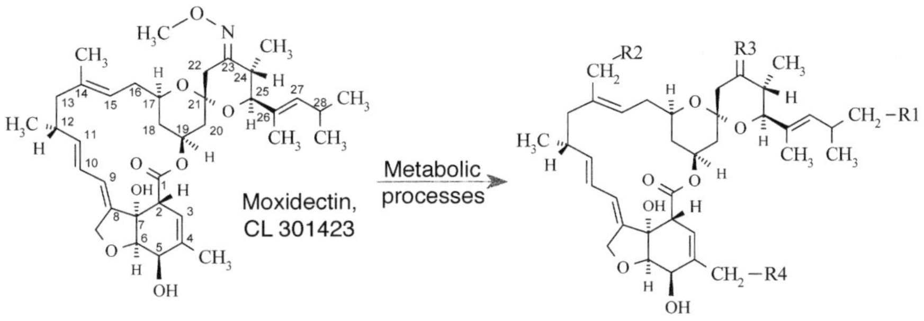

	Found in				Structure			
Component	Cattle	Sheep	Horse	Rat	R1	R2	R3	R4
Moxidectin	X	X	X	X	H	H	N-OCH_3	H
C-29 hydroxymethyl	X	X	X		OH	H	N-OCH_3	H
C-14 hydroxymethyl	X	X	X	X	H	OH	N-OCH_3	H
23-keto		X		X	H	H	O	H
C-4 hydroxymethyl				X	H	H	N-OCH_3	OH

Fig. 1.4.2. The metabolic transformation of moxidectin.

cattle. Parent moxidectin accounted for 76% of the total radioactivity, with C-29 hydroxymethyl moxidectin and C-14 hydroxymethyl moxidectin accounting for 3.8 and 2.7%, respectively.

Metabolism studies in rats

Comparative metabolism studies were carried out in rats to establish the relevance of rodent toxicology studies to human safety. In a preliminary mass balance study, male and female rats were administered 1.5 mg kg^{-1} ^{14}C-labelled moxidectin as a single oral dose in corn oil (S.-S. Wu, Princeton, New Jersey, 1991, Company Report). Total faeces, urine and respired carbon dioxide were collected through 72 h post-dosing. Animals were sacrificed and the cages rinsed at the end of the study, and all samples were assayed for total ^{14}C content. Total recovery of the administered radioactivity ranged from 92 to 95%. Faeces was the primary route of excretion of radioactivity, accounting for 88–96% of the administered dose. Urine accounted for less than 1%. No radioactivity was detected in the respired air, demonstrating that carbon dioxide is not involved in the metabolism of moxidectin.

In a second study, male and female rats were administered ^{14}C- and deuterium-labelled moxidectin as a single oral dose (1.5 or 12 mg kg^{-1}

Table 1.4.4. Primary components of the moxidectin radioresidue in the fat and liver of cattle, sheep and horses.[a]

Species	Cattle	Cattle	Sheep	Horse
Route	Subcutaneous	Topical	Oral (drench)	Oral (gel)
Dose (mg kg^{-1})	0.2	0.5	0.2	0.4
Fat (abdominal)				
Moxidectin	88	79	95	89
C-29 hydroxymethyl	1	2	1	1
C-14 hydroxymethyl	3	2	1	ND
Liver				
Moxidectin	40	39	51	60
C-29 hydroxymethyl	9	11	12	9
C-14 hydroxymethyl	12	17	6	11

[a]14 days post-dosing for cattle, 7 days post-dosing for sheep and horses; mean of three animals except one animal for cattle dosed by subcutaneous injection and for the liver data for cattle dosed topically. ND, not detected.

body weight) or a daily oral dose (1.5 mg kg^{-1} body weight) for 7 days (Wu *et al.*, 1993). Total excreta and tissues were collected at intervals up to 7 days post-dosing to determine the nature and depletion of the radioresidues. In both the high and low dose groups, faeces was the primary route of excretion, accounting for 60–91% of the radioactivity for all rats after 7 days. Less than 2% of the dose was eliminated in the urine. The residue levels in the tissues (muscle, liver, kidney and fat) were dose dependent and approximately 20 times higher in the fat than in the other tissues. There was no bioaccumulation of total ^{14}C residues in the tissues. Depletion half-lives averaged 11.5 days in fat, 3.9 days in muscle and 2.4 days in kidney and liver. Moxidectin was the major component of the radioresidue in the tissues and faeces. Six metabolites were isolated from liver and faeces samples, none of which accounted for more than 10% of the radioactivity in tissue samples collected from animals 7 days after dosing. One of the primary metabolites was identified as the C-14 hydroxymethyl derivative of moxidectin, designated as the same as found in the target animal studies. The other major metabolite in rats was shown to be C-4 hydroxymethyl moxidectin, a compound not found in the target animal metabolism studies. The 23-keto derivative of moxidectin, also seen in the target animal studies, was a minor component of the radioresidue in rats. The other minor metabolites identified were also hydroxylated derivatives of moxidectin, indicating that hydroxylation is the principal route of metabolism of moxidectin in the rat. Since the chromatographic metabolite profile is qualitatively similar between rats and food-producing animals, the rat is a suitable animal for toxicological studies.

In vitro *studies*

The low levels of many of the metabolites found in tissues in the *in vivo* studies made identification difficult. Therefore, *in vitro* techniques were used to facilitate structural characterization. As a complement to the cattle (Zulalian *et al.*, 1994), sheep (Afzal *et al.*, 1994) and rat (Wu *et al.*, 1993) *in vivo* studies, liver microsomal pellets for the study species were prepared according to the procedure of Miwa *et al.* (1982). The microsomal preparations were incubated with [^{14}C]moxidectin under aerobic conditions. The samples were extracted with diethyl ether and the radioactive components separated and purified using chromatographic techniques. Structural characterization of the metabolites was carried out using liquid chromatography in conjunction with mass spectrometry (Stout *et al.*, 1994). The major metabolites found in the tissues *in vivo* were recovered from the microsomal incubations. The *in vitro* studies greatly facilitated the structural elucidation of the metabolites of moxidectin and confirmed that hydroxylation was the principal route of biotransformation for moxidectin in animals.

Liver microsomal preparations have also been used to compare directly the metabolism of moxidectin among animal species (Alvinerie and Galtier, France, 1998, unpublished data). Microsomal pellets were prepared from freshly collected rat, cattle, sheep, goat and deer livers. After incubation with ^{14}C-labelled moxidectin, the metabolites were detected and evaluated by HPLC using on-line radioactive detection. The radioactivity was quantitated by liquid scintillation counting. No attempt was made to define the metabolite structures. The results indicate that the biotransformation of moxidectin by liver microsomes is qualitatively similar among the species (Table 1.4.5). Moxidectin was the major compound recovered for all species, accounting for 96% of the radioactivity in the rat and 67–86% in the other species. Two components were predominant in most of the incubates (components 3 and 9), except in the sheep. These data illustrate the similarity of the metabolism of moxidectin among animal species.

Target animal residue depletion studies

Since the radiotracer studies demonstrated that the parent compound, moxidectin, was the primary component of the residues in tissues, a practical analytical method has been developed to monitor the depletion of moxidectin from animal tissues following treatment (Khunachak *et al.*, 1993). Moxidectin is extracted from tissues with acetonitrile. The extract is then partitioned with hexane and concentrated. Additional cleanup is achieved using a Florisil Sep-Pak. The compound is reacted with acetic anhydride and l-methylimidazole in dimethylformamide to produce a

Table 1.4.5. Comparative percentages of different compounds after incubation of [^{14}C] moxidectin with liver microsomal preparations from five animal species.[a]

Component[b]	Rat Area %	Cattle Area %	Sheep Area %	Goat Area %	Deer Area %
1	—	0.98 ± 0.33	2.10 ± 0.28	2.22 ± 0.69	1.23 ± 0.42
2.0	0.06 ± 0.05	1.46 ± 0.45	2.67 ± 0.55	1.03 ± 0.42	1.37 ± 0.39
2.1	0.09 ± 0.06	—	—	—	—
3	0.95 ± 0.23	9.21 ± 1.80	19.26 ± 2.46	4.84 ± 0.72	2.32 ± 0.24
4	0.45 ± 0.08	0.91 ± 9.19	0.68 ± 0.05	1.51 ± 0.65	1.32 ± 0.18
5	0.36 ± 0.14	1.98 ± 0.54	2.68 ± 0.14	1.20 ± 0.18	0.77 ± 0.16
6	0.27 ± 9.19	1.98 ± 0.42	0.64 ± 0.08	0.38 ± 0.08	0.52 ± 0.11
7	0.12 ± 0.06	0.67 ± 0.18	0.25 ± 0.09	0.68 ± 0.36	0.79 ± 0.09
8	0.08 ± 0.05	0.21 ± 0.08	0.35 ± 0.07	0.64 ± 0.22	0.97 ± 0.20
9	1.32 ± 1.78	4.38 ± 1.90	3.86 ± 1.33	1.02 ± 0.30	7.39 ± 1.67
Moxidectin	95.84 ± 1.93	78.21 ± 3.75	67.44 ± 1.64	86.28 ± 2.83	83.32 ± 1.88
10	0.16 ± 0.10	—	—	—	—
11	0.23 ± 0.18	—	—	—	—
12	—	0.03 ± 0.02	0.07 ± 0.06	0.19 ± 0.10	0.10 ± 0.10

[a]Values are mean ± SD of eight determinations (four animals in duplicate).
[b]Listed in order of increasing chromatographic retention time.

conjugated dehydration product, which is fluorescent. Quantitation of moxidectin is accomplished by liquid chromatography with fluorometric detection and the external standard technique. With minor modifications for specific tissues, this method is applicable to the tissues from various animal species (cattle, sheep, deer and horses) as well as blood (serum). For whole milk, the samples are extracted with methylene chloride/acetone followed by partitioning with hexane/acetonitrile. The validated sensitivity of the method is 10 ppb with a typical limit of detection of 1–2 ppb. A similar method utilizing 1-methyl imidazole derivation with fluorescence detection was reported for moxidectin in cattle plasma (Alvinerie *et al.*, 1995). This method had a limit of quantitation (LOQ) of 0.1 ng ml^{-1}. Numerous other methods have been reported in the literature including a cattle tissue method with an LOQ of 1 ppb (Alvinerie *et al.*, 1996) and multiresidue methods with a similar LOQ for moxidectin (Ishii *et al.*, 1998; Roudaut, 1998; Schenck and Lagman, 1999). Thermospray LC/MS has been used for the confirmation of moxidectin in cattle tissues (Khunachak *et al.*, 1993). Confirmation of macrocyclic lactone residues, including moxidectin, in food matrices and milk has also been reported (Turnipseed *et al.*, 1999; Ali *et al.*, 2000).

Using the method described by Khunachak *et al.* (1993) residue depletion studies have been completed for cattle, sheep deer and horses using parenteral, topical and oral routes of administration. The moxidectin residues in fat, the target tissue, and muscle, the most important tissue from a human safety standpoint, are summarized in Table 1.4.6.

Metabolism and residue summary

In vivo and *in vitro* metabolic studies with radiolabelled moxidectin have shown similar results across a variety of animal species. Moxidectin was excreted primarily in the faeces, with very little radioactivity recovered in the urine. The compound was highly lipophilic as evidenced by high levels of residues in fat relative to other edible tissues. Residue levels depleted with time after treatment, confirming a lack of bioaccumulation of moxidectin in edible tissues. The metabolic profile was qualitatively similar across tissues, including milk, and across animal species. The primary component of the residue was the parent compound, accounting for 80–95% of the radioresidue in fat. There were two primary metabolites in the tissues and milk of the target animals, both hydroxylated derivatives of the parent molecule. Neither of these metabolites occurred at levels to be of toxicological significance. These same two metabolites, along with a third hydroxylated metabolite, were present in the rat, establishing the rat as a suitable model for toxicological studies relating to human safety. Residue depletion studies measuring parent moxidectin confirmed that residues deplete with time after treatment. The highest residues levels were in fat, with little or no measurable residue in muscle.

Toxicology

The toxicological potential of moxidectin has been evaluated during the numerous regulatory processes globally since the first registration of a moxidectin-based product. The most extensive toxicological review of moxidectin was during the 45th meeting of the Joint FAO/WHO Expert Committee on Food Additives (JECFA) (Woodward, 1996). A tabular summary of that review is listed in Table 1.4.7 and is the basis of this summary.

Genotoxicity

The genotoxicity of moxidectin has been evaluated in a variety of tests. Negative results were obtained when moxidectin was tested in the Ames test, in reverse and forward mutation assay with *Escherichia coli*, or in a test using mammalian cells for chromosome aberrations in rat bone marrow cells, and moxidectin did not induce unscheduled DNA synthesis in primary rat hepatocytes.

Table 1.4.6. Moxidectin residues (ppb) in fat and muscle of animals following single treatment with various commercially available moxidectin formulations at the approved dose.

Species	Route	Dose (mg kg^{-1})	Days	Fat	Muscle
Cattle	Subcutaneous[a] (aqueous formulation)	0.2	14	275 ± 56	<10
			21	243 ±86	<10
			28	225 ± 45	<10
			35	153 ± 29	<10
			42	77 ± 41	<10
			49	141 ± 50	<10
	Subcutaneous[b] (non-aqueous formulation)	0.2	14	263 ± 76	<10–11.2
			21	199 ± 42	<10
			28	145 ± 46	<10
			35	110 ±71	<10
			42	58 ± 37	<10
		0.5	3	90 ± 76	<10
	Topical[c]		7	71 ± 27	<10
			10	65 ± 69	<10
			14	49 ± 26	<10
			21	31 ± 17	<10
Sheep	Oral[d]	0.2	14	43 ± 13	<10
			21	<10–23	<10
			28	<10–26	<10
			35	<10	<10
			42	<10	<10
	Subcutaneous[e] (aqueous formulation)	0.2	10	324 ± 83	29 ± 6
			20	234 ± 41	18 ± 9
			30	139 ± 42	11 ± 2
			40	164 ± 69	<10–22
Deer	Topical[f]	0.5	7	126 ± 17	<10
			14	155 ± 44	ND
			21	57 ± 32	ND
			28	31 ± 16	ND
Horse	Oral[g]	0.4	28	258 ± 96	<10–13.2
			35	216 ± 41	<10–13.1
			42	126 ± 52	<10
			49	115 ± 52	<10

[a]K.A. Rooney, Princeton, New Jersey, 1991, Company Report.
[b]L.S. de Montigny, Princeton, New Jersey, 1996, Company Report.
[c]C.A. Hirschlein, Princeton, New Jersey, 1997, Company Report.
[d]H. Berger, Princeton, New Jersey, 1991, Company Report.
[e]L.D. Parker, Princeton, New Jersey, 1995, Company Report. Animals were treated twice with a 10-day interval. Results are reported as days after the second treatment.
[f]J.C. Turner, New Zealand, 1992, unpublished data.
[g]R.L. DeLay, Princeton, New Jersey, 1996, Company Report.
ND, not determined.

Table 1.4.7. Toxicity studies with moxidectin.

Species, study and doses tested	NOEL mg kg^{-1}
Mouse (CD-1)	
28-day – diet (0, 6.9, 17.7, 23.2, 24.1, 32.2)	6.9
2-year – diet (0, 2.49, 5.10, 11.87/7.89)	5
Rat (Crl:CD)	
28-day – diet (0, 12.2, 22.8, 26.4, 31.2)	<12
90-day – diet (0, 1.95, 3.90, 7.99, 12.23)	4
2-year – diet (0, 0.8, 3.2, 9.8/5.1)	6
Teratology – gavage (0, 2.5, 5, 10, 12)	5
	5
Three-generation – diet (0, 0.07, 0.15, 0.41, 0.83)	0.4
	0.4
Rabbit	
Teratology – gavage (0, 1, 5, 10)	1
	10
Dog	
28-day – diet (0, 0.5, 2, 4/1.25)	0.5
91-day – diet (0, 0.3, 0.9, 1.6)	0.3
52-week – diet (0, 0.25, 0.49, 1.12)	1.1

Acute toxicity

Moxidectin was considered to be moderately toxic after oral and interperitoneal administration to rats and mice in acute toxicity testing (Table 1.4.8). The main clinical sign in mice administered a toxic dose of moxidectin was decreased activity. Animals observed with decreased activity had recovered by 4 days after treatment. Post-mortem evaluation of animals that had died or were sacrificed after 14 days showed no gross abnormalities. In rats, toxic doses resulted in decreased activity, prostration, tremors, chromodacryorrhoea, decreased respiration, diarrhoea, hypersensitivity to touch and sound, and epistaxis. In post-mortem examination, congestion of the liver, kidneys and lungs was observed in animals that died during the test, but for animals that survived to sacrifice at 14 days, no abnormalities were observed.

Dermal application of moxidectin to rabbits had no overt toxicity.

Subchronic toxicity

In mice, the subchronic toxicity of moxidectin was evaluated in a 28-day study. Mice were given 0, 6.9, 18, 23, 24 or 32 mg moxidectin per kg of body weight per day. High mortality (80–100%) was observed in the three highest dose groups. Only one animal died in the 18 mg kg^{-1} group, with

Table 1.4.8. Acute toxicity studies with moxidectin.

Species	Sex	Route	LD_{50} (mg kg^{-1})
Mouse	M and F	Oral	84
	F	Oral	42
	F	Oral	50
	M and F	i.p.	86
	M and F	s.c.	263
Rat	M and F	Oral	106
	M and F	i.p.	394
	M and F	s.c.	>640
	M and F	Inhalation	3.28 mg l^{-1} (5 h LC_{50})

no mortality observed in the remaining groups. Other signs of toxicity included tremors, hypersensitivity to touch and urine-stained fur in the 18, 23 and 24 mg kg^{-1} groups. No effects were observed on haematology, relative or absolute organ weights, or in gross or microscopic evaluation of the tissues. The NOEL in this study was 6.9 mg kg^{-1} day^{-1}.

In rats, subchronic toxicity was evaluated at 28 days (0, 12, 23, 26 or 31 mg moxidectin kg^{-1} day^{-1}) and 13 weeks (0, 1.9, 3.9, 7.9 or 12 mg moxidectin kg^{-1} day^{-1}). Clinical observation, mortality, organ evaluation, feed intake, haematology, clinical chemistry and urinalysis were used to evaluate toxicity. In the 28-day study, mortality was seen in the two highest dose groups (100%). Two female rats died in the 23 mg moxidectin kg^{-1} groups. There was no mortality in the lowest dose group. No NOEL was assigned in this study due to hypersensitivity to touch which was observed in the lowest dose group. In the 13-week study, three females in the highest dose rate died or were sacrificed in moribund condition. Other signs of toxicity observed in some groups included decreased food intake, ataxia, tremors, salivation, piloerection, hypersensitivity to touch and diuresis. The assigned NOEL was 3.9 mg moxidectin kg^{-1} day^{-1}.

The toxicity of moxidectin in dogs was evaluated in a 28-day (0, 0.5, 2 or 4 mg moxidectin kg^{-1} day^{-1}), a 91-day (0. 0.3, 0.9 or 1.6 mg moxidectin kg^{-1} day^{-1}) and a 52-week study (0, 0.26, 0.52 or 1.15 mg moxidectin kg^{-1} day^{-1}). In the 28-day study, dogs developed anorexia, ataxia, prostration and diarrhoea at the highest dose level and were changed to a dose of 1.25 mg moxidectin kg^{-1} day^{-1} at day 5. Toxicity signs were observed at the highest doses in the 28-day and the 91-day study, with clinical signs, tissue effects and effects on food consumption which were similar to those seen in mice and rats. The NOELs assigned to these studies were 0.5, 0.3 and 1.15 mg moxidectin kg^{-1} day^{-1} for the 28-day, 91-day and 52-week studies, respectively.

Developmental and reproductive toxicity

The reproductive and developmental toxicity of moxidectin has been evaluated in rats and rabbits. In a preliminary single generation test, rats were fed 0, 1.8, 3.9 or 9.8 mg moxidectin kg^{-1} day^{-1} for a 9-week period prior to mating and through gestation and lactation to produce F_{1a} litters. All pups died at the highest treatment level, and parental animals had reduced weight gains, a decreased number of live pups at birth and an increased number of dead pups at birth. No parental adverse reactions were observed at the two mid-dose treatments. However, all pups died during lactation. Subsequently, dietary levels were reduced to 0, 0.4, 0.8 or 1.1 mg moxidectin kg^{-1} day^{-1}. In the F_{1b} litters, no parental effects were observed with various pup effects in all but the lowest treatment level. The NOEL for this study was established as 0.4 mg moxidectin kg^{-1} day^{-1}. In a three-generation study, treatment levels of 0, 0.07, 0.15, 0.41 or 0.83 mg moxidectin kg^{-1} day^{-1} were tested for a period of 70 days prior to mating. Randomly selected offspring (F_{1b} and F_{2b}) were chosen to produce subsequent generations. Toxic signs were noted in parents and offspring at the highest dose rate only. The NOEL for this study was 0.4 mg moxidectin kg^{-1} day^{-1} for the F_{1b} animals.

Chronic toxicity

Two chronic toxicity/carcinogenicity studies have been conducted with moxidectin. Diets providing 0, 2.5, 5.1 or 12 mg moxidectin kg^{-1} day^{-1} were fed in a 2-year study in mice. After 9 weeks, the highest dose level was reduced to 7.9 mg moxidectin kg^{-1} day^{-1} due to mortality in the original high dose group. Clinical signs of toxicity were limited to the highest dose group in this study with no increased incidence of tumours. The NOEL was 5.1 mg moxidectin kg^{-1} in this study. In rats, dose levels of 0, 0.8, 3.2 or 9.8 mg moxidectin kg^{-1} day^{-1} were tested. After 8 weeks, the highest dose was reduced to 5.1 mg moxidectin kg^{-1} day^{-1} because of increased mortality at the original high dose level. After 2 years, there were no increases in the incidence or type of tumour. The NOEL was established at 5.1 mg moxidectin kg^{-1} day^{-1}. Based on these studies, moxidectin is considered to be non-carcinogenic.

Safety in Target Animals

Moxidectin has been formulated into a variety of presentations. During the global development and registration programmes, the safety of moxidectin in target animals has been extensively evaluated and reported by independent researchers and by company personnel.

Cattle

Moxidectin is formulated as a pour-on and an injectable for use in cattle. The dose rate for the injectable product is 0.2 mg moxidectin kg^{-1} with 0.5 mg moxidectin kg^{-1} registered for the pour-on product. Safety studies have been conducted with each formulation.

In a repeated dose study, animals were given 0, 0.2, 0.4 or 0.8 mg moxidectin kg^{-1} as an injectable on three consecutive days (Wang *et al.*, 1992). Except for two high dose animals, all measurements including clinical, blood, urine, faecal and tissue observations were normal. The clinical signs associated with toxicity in the high dose group included drooping ears, stiff legs and salivation. A study with the pour-on product was conducted at doses of 0.5, 1.5 and 2.5 mg moxidectin kg^{-1} on three consecutive days with no adverse reactions observed, other than minor salivation which was not treatment specific (W.B. Epperson, Princeton, New Jersey, 1993, Company Report). At higher doses with the pour-on formulation (2.5 mg moxidectin kg^{-1} on 5 consecutive days, 5.0 mg moxidectin kg^{-1} on two consecutive days and 12.5 mg moxidectin kg^{-1} in a single application), a similar salivation, which was also not treatment related, was observed, with no other adverse reactions (W.B. Epperson, Princeton, New Jersey, 1993, Company Report). The effect of the injectable product on reproductive performance was studied with treatment of oestral cows (G.T. Wang, Princeton, New Jersey, 1991, Company Report, FD 39-51.00), pregnant heifers (G.T. Wang, Princeton, New Jersey, 1991, Company Report, FD 39-50.00), pregnant cows (G.T. Wang, Princeton, New Jersey, Company Report, FD 39-43.00; Rae *et al.*, 1994) and bulls (G.T. Wang, Princeton, New Jersey, 1992, Company Report, FD 40-04.00) with the pour-on formulation, again with no adverse effect on reproductive performance. The moxidectin injectable formulation has been tested in *Hypoderma*-infected animals (F. Guerino, Princeton, New Jersey, 1995, Company Report) and Murray Grey cattle, a breed known to be sensitive to other macrocyclic lactones (P. Kiernan and R. Cobb, Princeton, New Jersey, 1991, Company Report), with no negative effect.

Sheep

Both an oral formulation and an injectable formulation are registered for use in sheep globally. The dose rate for both products is 0.2 mg moxidectin kg^{-1} for the treatment of internal and external parasites.

With the oral drench product, sheep have been treated with 0.4 and 1.0 mg moxidectin kg^{-1} in a single treatment (P. Kiernan and R. Cobb, Princeton, New Jersey, 1991, Company Report) or 0.5 mg moxidectin kg^{-1} (I. Wood, Princeton, New Jersey, 1991, Company Report) with no adverse effect. An injectable formulation has been administered to sheep at 1.0, 2.0, 2.5 or 3.0 mg moxidectin kg^{-1} in a single dose (F. Guerino, Princeton, New Jersey, 1991, Company Report). No adverse reactions were observed at the two lowest levels. At a dose of 2.5 mg moxidectin kg^{-1}, salivation was observed for an 8-h period which then resolved. At 3.0 mg moxidectin kg^{-1}, salivation, diuresis, muscular tremors, prostration and ataxia were reported for 48–72 h. Young lambs (4–29 days of age) were injected with 0.8 mg moxidectin kg^{-1} or given 2 mg moxidectin kg^{-1} as an oral drench (Tzora and Fthenakis, 1999a). A single injectable group lamb became lethargic but the signs resolved within 12 h. A set amount (1 ml) of injectable was administered to ewes (Tzora and Fthenakis, 1999b) with no adverse injection site reaction.

Dogs

Moxidectin has been shown to be effective in the prevention of heartworm disease in dogs and is formulated as a tablet for monthly use or as an injectable with sustained-release activity.

Doses of 0, 0.17, 0.51 and 0.85 mg moxidectin kg^{-1} as an injectable were administered to dogs with no adverse effects on animal performance, clinical condition, haematology, clinical chemistry and gross of histopathological findings (R.D. Rulli, Princeton, New Jersey, 1999, Company Report). Toy breed puppies were given an oral dose of moxidectin equivalent to 0.6 or 1.0 mg moxidectin with no adverse reactions (van Dassler, Weesp Netherlands, 1999, Company Report). In avermectin-sensitive collies, the tablet formulation of moxidectin was safe when administered at 15 μg moxidectin kg^{-1} three times at 30-day intervals (Paul *et al.*, 1992). At this dose (five times the recommended dose), no abnormal clinical observations, haematology, serum chemistry and urine analysis values were observed. In a subsequent study, Paul *et al.* (2000a) observed no signs of toxicity in avermectin-sensitive collies receiving moxidectin at 30, 60 or 90 μg moxidectin kg^{-1}. This appeared to be a wider margin of safety than for ivermectin or milbemycin in similar tests. When formulated as a sustained-release injectable, moxidectin was also shown to be safe in avermectin-sensitive collies when administered once at dose rates of 0.17, 0.51 and 0.85 μg moxidectin kg^{-1} (Paul *et al.*, 2000b). Both the tablet and injectable formulations have been tested in microfilariae-positive dogs. The tablet was administered at 3 and 15 μg moxidectin kg^{-1} three times at monthly intervals (Hendrix *et al.*, 1992) or a single dose of 0.51 μg moxidectin kg^{-1} of the injectable (Blagburn *et al.*, 2000). There were no negative observations. The safety of the tablet and injectable formulations has been tested in both females and males with no negative reproductive effects. The tablet has been tested at 9 μg moxidectin kg^{-1} day^{-1} when administered prior to breeding, during gestation and during lactation (K.A. Rooney, Princeton, New Jersey, 1993, Company Report). No differences were observed in reproductive performance, blood parameters and growth of pups. Male reproductive parameters were also unaffected at this dose (K.A. Rooney, Princeton, New Jersey, 1993, Company Report). A single injectable dose of 0.51 mg moxidectin kg^{-1} was administered at critical times either prior to mating, or throughout gestation and lactation, with no abnormal effects on female reproductive performance or the health and performance of pups (R.D. Rulli, Princeton, New Jersey, 1999, Company Report). The same dose was administered to males with no adverse effect on reproduction capability (D.A. Peterson, Princeton, New Jersey, 1999, Company Report).

Adverse events have been reported in dogs that have been administered high levels of moxidectin off-label with the horse gel formulation. Though difficult to quantify, these levels may be as high as 50–100 times the recommended tablet dose of 3 μg moxidectin kg^{-1}. Depending on the dose, moxidectin intoxication can cause various abnormal neurological effects including ataxia, weakness, lethargy, stupor, coma, seizures and respiratory arrest (Beal *et al.*, 1999). As with avermectin and milbemycin intoxication, animals usually recover with supportive care.

Horses

Moxidectin is used as a gel in horses 4 months of age and older for the treatment of endoparasite infections. The safety of the product has been demonstrated in foals, breeding mares and breeding stallions. A series of three studies was conducted in foals to evaluate safety. One- to two-week-old foals given 0.4 or 1.2 mg moxidectin kg^{-1} for 3 consecutive days or 1.2 mg moxidectin kg^{-1} in a single dose showed no clinical signs of toxicity. Multiple consecutive doses of 1.2 mg moxidectin kg^{-1} or a single dose of 2.0 mg moxidectin kg^{-1} produced signs of toxicity indicative of a neurological effect including depression, incoordination, tremors, recumbancy and inability to nurse (R.D. Rulli, Princeton, New Jersey, 1996, Company Report). Effects were reversible with supportive care. Similar signs were seen in older foals given single doses of 1.2 and 2.0 mg moxidectin kg^{-1} (R.D. Rulli, Princeton, New Jersey, 1996, Company Report). Single doses up to 2.0 mg moxidectin kg^{-1} have been tolerated in foals 4–21 weeks of age (T. Wilson, Princeton, New Jersey, 1995, Company Report). Clinical signs of toxicosis included ataxia, dragging of limbs, depression and lip reaction (flaccid lips). Animals recovered after supportive care. Single doses of 0.8 and 1.2 mg moxidectin were well tolerated in another study in 4- to 6-month-old foals (J.V. Downer, Princeton, New Jersey, 1996, Company Report). Overdose symptoms in young foals that receive an overdose of moxidectin are neurological in nature (Johnson *et al.*, 1999) and can be treated successfully using supportive care.

The reproductive safety of moxidectin in breeding females was evaluated in a 2-year study at a dose of 1.2 mg moxidectin kg^{-1} given at 2-week intervals (Kivipelto, 1996). No differences were observed in any of the breeding indices calculated during the study. Moxidectin was safe when administered to stallions at a dose of 1.2 mg moxidectin kg^{-1} three times at 7 day intervals (J.V. Downer, Princeton, New Jersey, 1996, Company Report). No clinical signs, differences in semen quality, haematology, clinical chemistry and testicular tissue evaluation indicative of adverse reactions were observed in this study.

Other species

Various moxidectin products have been administered to other species of animals either as part of a registration programme or in discovery/development activities. Dose levels of 0.75, 1.00, 1.25 and 1.50 mg moxidectin kg^{-1} were applied to swine as a pour-on with no adverse effects (Stewart *et al.*, 1998). A 0.2 mg moxidectin kg^{-1} dose of an injectable formula was administered to reindeer (Oksanen and Nieminen, 1998) and alpacas (N. Scherling, Princeton, New Jersey, 2001, Company Report). Levels of 0.2–1.6 mg moxidectin kg^{-1} were administered to chickens as an injection, *per os* or as a pour-on with no adverse effect (Vajna and Varga, 1996). Toxic symptoms were observed with 6 mg moxidectin kg^{-1} by subcutaneous injection. Moxidectin had no adverse effects when applied as a pour-on to red deer fawn at a dose of 2.5 mg moxidectin kg^{-1} (P. Kiernan and R. Cobb, Princeton, New Jersey, 1992, Company Report). Moxidectin administered topically or orally (0.2 mg moxidectin kg^{-1}) to native ostriches was shown to be safe (A. duPlessis, Princeton, New Jersey, 1996, Company Report).

References

Afzal, J., Stout, S.J., daCunha, A.R. and Miller, P. (1994) Moxidectin: absorption, tissue distribution, excretion, and biotransformation of ^{14}C-labeled moxidectin in sheep. *Journal of Agriculture and Food Chemistry* 42, 1767–1773.

Afzal, J., Burke, A.B., Batten, P.L., DeLay, R.L. and Miller, P. (1997) Moxidectin: metabolic fate and blood pharmacokinetics of ^{14}C-labeled moxidectin in horses. *Journal of Agriculture and Food Chemistry* 45, 3627–3633.

Ahmed, Z.H., Fiala, R.R. and Bullock, M.W. (1993) Incorporation of carbon-14 in the biosynthesis of the macrolide antibiotic, LL-F28249-α. *Journal of Antibiotics* 614–622.

Ali, M.S., Sun, T., McLeroy, G.E. and Phillippo, E.T. (2000) Confirmation of eprinomectin, moxidectin, abamectin, doramectin and ivermectin in beef liver by liquid chromatography/positive in atmospheric pressure chemical ionization mass spectrometry. *Journal of the Association of Official Analytical Chemists International*, 83, 31–38 and 39–52.

Alvinerie, M., Sutra, J.F., Badri, M. and Galtier, P. (1995) Determination of moxidectin in plasma by high performance liquid chromatography with automated solid phase extraction and fluorescent detection. *Journal of Chromatography B* 674, 119–124.

Alvinerie, M., Sutra, J.F., Capela, D., Galtier, P., Fernandez-Suarez, A., Mome, E. and O'Keeffe, M. (1996) Matrix solid-phase dispersion technique for the determination of moxidectin in bovine tissues. *Analyst* 121, 1469–1472.

Asato, G. and France, D.J. (1988) 23-oxo (keto) and 23-imino derivatives of LL-F28249 compounds with pesticide activity. European Patent Application.

Bain, B.M., Porter, N., Lambeth, P.F., Nobel, M.M., Rosemeyer, A.C., Fetton, R.A., Ward, J.B., Nobel, D. and Sutherland, D.R. (1987) Preparation of milbemycin derivatives as pesticides and bactericides. European Patent Application.

Beal, M.W., Poppenga, R.H., Birdsall, W.J. and Hughes, D. (1999) Respiratory failure attributable to moxidectin intoxication in a dog. *Journal of the American Veterinary Medical Association* 215, 1813–1817.

Blackhall, W., Liu, H.Y., Xu, M., Prichard, R.K. and Beech, R.N. (1998) Selection at a P-glycoprotein gene in ivermectin- and moxidectin-selected strains of *Haemonchus contortus*. *Molecular and Biochemical Parasitology* 90, 42–28.

Blagburn, B.L., Butler, J.M., Vaughn, J.L. and Rulli, R.D. (2000) Clinical observations following the administration of moxidectin canine sustained release (SR) injectable in heartworm positive dogs. In: *Proceedings of American Association of Veterinary Parasitologists*, Salt Lake City, Utah, p. 57.

Cole, L.M. and Casida, J.E. (1992) GABA-gated chloride channel: binding site for 4′-ethynyl-4-n-[2,3–3H2] propylbicycloortho benzoate ([^{3}H]EBOB) in vertebrate brain and insect head. *Pesticide Biochemistry and Physiology* 44, 1–8.

Fisher, M.H. (1997) Structure–activity relationships of the avermectins and milbemycins. *American Chemical Society Symposium Series, Phytochemicals for Pest Control* 17, 220–238.

Hendrix, C.M., Blagburn, B.L., Bowles, J.U., Spano, J.S. and Aguilar, R. (1992) The safety of moxidectin in dogs infected with microfilariae and adults of *Dirofilaria* immitis. In: Soll, M.D. (ed.) *Proceedings of the Heartworm Symposium '92*. American Heartworm Society, Batavia, Illinois, pp. 183–187.

Ishii, R.M.M., Mosheno, Y. and Nakazawa, M. (1998) Simultaneous determination of residual anthelmintic agents in liver and fat tissues by HPLC with fluorescence detection. *Shohuhin Fiseigakie Zasshi* 39, 42–45.

Johnson, P.J., Mrad, D.R., Schwartz, A.J. and Kellam, L. (1999) Presumed moxidectin toxicosis in three foals. *Journal of the American Veterinary Medical Association* 214, 678–680.

Khunachak, A., daCunha, A.R. and Stout, S.J. (1993) Liquid chromatographic determination of moxidectin residues in cattle tissues and confirmation in cattle fat by liquid chromatography/mass spectometry. *Journal of the Association of Official Analytical Chemists International* 76, 1230–1235.

Kivipelto, J., Asquith, R.L., Harvey, J.W. and Wang, G.T. (1996) Safety of oral moxidectin in breeding/pregnant mares and their unborn/newborn foal. In: *Proceedings of the American Association of Veterinary Parasitologists*, Louisville, Kentucky, p. 32.

Miwa, G.T., Walsh, J.S., VandenHeuvel, J.A., Arison, B., Sestokas, E., Buhs, R., Rosegay, A., Avermitilis, S., Lu, A.Y.H., Walsh, M.A.R., Walker, R.W., Taub, R. and Jacob, T.A. (1982) The metabolism of avermectins B_{1a}, H_2B_{1a}, and H_2B_{1b} by liver microsomes. *Drug Metabolism and Disposition* 10, 268–274.

Oksanen, A. and Nieminen, M. (1998) Moxidectin as an endectocide in reindeer. *Acta Veterinaria Scandinavica* 39 483–489.

Paiement, J.P., Leger, C., Ribeiro, P. and Prichard, P.K. (1999) *Haemonchus contortus*; effects of glutamate, ivermectin, and moxidectin on insulin uptake activity in unselected and ivermectin-selected adults. *Experimental Parasitology* 92, 193–198.

Paul, A.J., Tranquilli, W.J., Todd, K.S. and Aguliar, R. (1992) Evaluation of the safety of moxidectin in collies. In: Soll, M.D. (ed.) *Proceedings of the Heartworm Symposium '92*. American Heartworm Society, Batavia, Illinois, pp. 189–191.

Paul, A.J., Tranquilli, W.J. and Hutchens, D.E. (2000a) Safety of moxidectin in avermectin-sensitive collies. *American Journal of Veterinary Research* 61, 482–483.

Paul, A.J., Hutchens, D.E., Cleale, R.M. and Tranquilli, W. (2000b) Clinical observations from the administration of moxidectin canine sustained release injectable in ivermectin sensitive dogs. In: *Proceedings of the American Association of Veterinary Parasitologists*, Salt Lake City, Utah, p. 56.

Rae, D.O., Larsen, R.E. and Wang, G.T. (1994) Safety assessment of moxidectin 1% injectable on reproductive performance in beef cows. *American Journal of Veterinary Research* 55, 251–253.

Roudaut, B. (1998) Multiresidue method for the determination of avermectin and moxidectin residues in the liver using HPLC with fluorescence detection. *Analyst* 123, 2541–2544.

Rudd, B.A.M. and Ramsey, M.V.J. (1989) Microbial manufacture of macrolide compound S541. European Patent Application.

Sangster, N.C., Bannan, S.C., Weiss, A.S. and Nulf, S.C. (1999) *Haemonchus contortus*: sequence heterogeneity of internucleotide binding domains from P-glycoproteins and an association with avermectin/milbemycin resistance. *Experimental Parasitology* 91, 250–257.

Schenck, F.J. and Lagman, L.M. (1999) Multiresidue determination of abamectin, doramectin, ivermectin and moxidectin in milk using liquid chromatography and fluorescence detection. *Journal of the Association of Official Analytical Chemists International* 82, 1340–1344.

Stewart, T.B., Wiles, S.E., Miller, J.E. and Rulli, R.D. (1998) Efficacy of moxidectin 0.5% pour-on against swine nematodes. In: *Proceedings of the American Association of Veterinary Parasitologists*, Baltimore, Maryland, p. 26.

Stout, S.J., daCunha, A.R., Wu, S.S., Zulalian, J. and Afzal, J. (1994) Moxidectin: characterization of cattle, sheep and rat *in vitro* and *in vivo* metabolites by liquid chromatography/tandem mass spectrometry. *Journal of Agriculture and Food Chemistry* 42, 288–392.

Sutherland, D.R., Pereira, O.Z., Noble, M.M., Ramsay, M.V., Ward, J.B., Fletton, R.A., Tiley, E.P., Porter, N. and Nobel, D. (1987) Macrolide compounds, procedure for their preparation and pharmaceuticals and agrochemicals containing them. German patent application.

Turnipseed, S.B., Royal, J.E., Rupp, M.S., Gonzales, S.A., Pfenning, A.P. and Mulbut, J.A. (1999) Confirmation of avermectin residues in food matrixes with negative-ion atmospheric pressure chemical ionization liquid chromatography/mass spectrometry. *Rapid Communications in Mass Spectrometry* 13, 493–499.

Tzora, A. and Fthenakis, G.C. (1999a) Safety of moxidectin 1% injectable solution and 0.1% oral drench for lambs younger that one month. *Small Ruminant Research* 32, 295–287.

Tzora, A. and Fthenakis, G.C. (1999b) Evaluation of the reaction of sheep during or after injection with moxidectin 1% injectable solution. *Small Ruminant Research* 31, 169–171.

Vajna, Z. and Varga, I. (1996) A moxidektin Syngams grachea elleni hatékonysága kísérzetesens fertözött csirkékben. *Magyar Állatorvosok Lapja* 51, 726–728.

Wang, G.T., Rooney, K. and Rock, D. (1992) Efficacy and safety of moxidectin, a new endectocide against ruminant parasites. In: *Proceedings of the 8th Congress of the Federation of Asian Veterinary Associations*, Quezon City, Philippines, pp. 383–389.

Woodward, K. (1996) Moxidectin. *Toxicological Evaluation of Certain Veterinary Drug Residues in Food*. World Health Organization, Geneva, pp. 27–50.

Wu, S.-S., Stout, S.J., daCunha, A.R., Biroc, S., Singh, S.-S., Washington, J., Ranjan, S. and Miller, P. (1993) Absorption, distribution, excretion, and metabolism of moxidectin in the rat. Presented at the *Fifth North American ISSX Meeting*, Tucson, Arizona, p. 190.

Zulalian, J., Stout, S.J., daCunha, A.R., Garces, T. and Miller, P. (1994) Absorption, tissue distribution, metabolism and excretion of moxidectin in cattle. *Journal of Agriculture and Food Chemistry* 42, 381–7.

Chapter 2

Pharmacokinetics of the Macrocyclic Lactones: Conventional Wisdom and New Paradigms

D.R. Hennessy and M.R. Alvinerie

Molecular structure and formulation

The naturally occurring avermectin and milbemycin compounds are closely related 16-membered macrocyclic lactones (MLs) (Burg *et al.*, 1979; Takiguchi *et al.*, 1980). Both chemical classes of endectocides are produced through fermentation by soil actinomycetes from the genus *Streptomyces* and have similar spectra of activity. The major difference between avermectins and milbemycins is a disaccharide group attached at the C-13 of avermectin whereas that position is unsubstituted in milbemycins.

The avermectins are produced as a mixture of eight different components from fermentation of *Streptomyces avermitilis* (Campbell *et al.*, 1983); these compounds are denoted as A_{1a}, A_{1b}, A_{2a}, A_{2b}, B_{1a}, B_{1b}, B_{2a} and B_{2b}. The 'A' components have a methoxyl group at the C-5 position, whereas the 'B' components have a hydroxyl group; the '1' components have a double bond between the C-22 and C-23 position, whereas the '2' components have a single bond with a hydroxyl group at the C23 position. The 'a' components have a secondary butyl substituent at the C-25 position, while the 'b' components have an isopropyl substituent at the C-25 position (Shoop *et al.*, 1995)

A catalytic regioselective hydrogenation of the naturally occurring avermectin B_1, on the C-22–C-23 double bond yields a C-22–C-23 dihydro derivative, ivermectin. Ivermectin is marketed as a mixture of 22,23-dihydro B_{1a} (>80%) and 22,23-dihydro B_{1b} (<20%) (Fisher and Mrozik, 1989). The presence of a cyclohexyl substituent at C-25 characterizes doramectin (Goudie *et al.*, 1993). Eprinomectin (4″-(epiacetylamino)-4″-deoxyavermectin B_1) is a member of the avermectin class of compounds, derived from the natural product avermectin B_1 (Shoop *et al.*, 1996). Selamectin (25-cyclohexyl-25-de(1-methylpropyl)-5-deoxy-22,23-dihydro-5(hydroxy-imino) avermectin B_1 monosaccharide) is a semisynthetic derivative of doramectin (McTier *et al.*, 1998).

The milbemycins are produced as a mixture of different components from fermentation of *Streptomyces hygroscopicus* and *Streptomyces cyanogriseus*; they could be subdivided into A and B components based on hydroxyl and methoxyl grouping at the C-5 position. Moxidectin is derived from the fermentation product, nemadectin (Carter *et al.*, 1988) and is characterized by the presence of a methoxime substituent on C-23.

The exceptional potency of ML compounds facilitates very low dosage rates contained in a small delivery vehicle volume. Further, their solubility in oils enhances the ability of the drug to be absorbed from an injection site, or from percutaneous delivery. Hence, ML compounds are formulated for oral, injectable and dermal administration. Inclusion of ivermectin in sustained-release intraruminal devices has widened their application potential further. Delivery route does, however, influence systemic availability, and there has been extensive research into the use of mixed oil/solvent/aqueous vehicles to optimize the performance of these compounds. Specific aspects of formulation will be described elsewhere.

Pharmacokinetics of MLs

The relationships between dose route, host species and physiology define the kinetic and dynamic processes of intercompartmental exchange, volumes of distribution, rates and extent of metabolism, residence times and clearance rates which characterize the unique pharmacokinetic behaviour of ML antiparasitic drugs. As shown in Table 2.1, the disposition of ML compounds has been described, in varying detail, in the peripheral plasma ‘central compartment’. This is due largely to the convenient access to sequential blood samples and the homogeneity of the central compartment. Parent drug and metabolites distribute to other body compartments, and it is assumed that the pharmacokinetic disposition in peripheral plasma describes drug availability at extravascular sites, including those of parasite habitat (Lanusse and Prichard, 1993; Baggot and McKellar, 1994; Lifschitz *et al.*, 1999a; Oukessou *et al.*, 1999). There is a complex interaction between pharmacokinetic compartments, and the quantitative and qualitative availability of drug/metabolite in one compartment might, in fact, not be closely related in another. To exacerbate the difficulty in interpretation of ML pharmacokinetics, drug disposition is significantly influenced by the experimental conditions employed. For example, absorption, systemic availability and elimination of orally

Table 2.1. Selected pharmacokinetic parameters describing the disposition of macrocyclic lactone compounds in the peripheral plasma of target animals.

Dose route	Species	Dose (μg kg^{-1})	C_{max} (ng ml^{-1})	T_{max} (h)	AUC ((ng ml^{-1}) × h)	VD_{ss} (l kg^{-1})	$t_{1/2el}$ (h)	MRT (h)	Reference
Ivermectin									
i.v.	Cattle	200	442.0	0	6096	2.2	64.5	67.2	Wilkinson *et al.* (1985)
	Cattle	200			2609	1.2	82.3		Echeverria *et al.* (1997)
	Pig	300	—	—	2040	7.9	55.2	48.0	Friis and Bjorn (1995)
	Pig	300	—	—	1808	5.3	31.8	19.3	Craven *et al.* (2001)
	Sheep	200				4.6	64.8		Fink and Porras (1989)
	Sheep	200	462.0	0	8990	5.3	177.7		Prichard *et al.* (1985)
s.c.	Sheep	200	30.8	60	5718		88.4		Marriner *et al.* (1987)
	Sheep	200	16.3	62.4	6744		168.5	141.1	Atta and Abo-Shehada (2000)
	Cattle	200	42.8	96	11,016	3.4	412.8	176.4	Lanusse *et al.* (1997)
	Cattle	200	54.6	34.8	10,790			156.9	Toutain *et al.* (1988)
	Cattle	200	31.7	95.5	8664		103.7	216.0	Toutain *et al.* (1997)
	Cattle	200	46.1	72					Herd *et al.* (1996)
	Cattle	200	44.0	48			199.2		Fink and Porras (1989)
	Cattle	200	46.4	50.9	6384		132	129.4	Lifschitz *et al.* (1999b)
	Cattle	200	33.1	55.9	7892		137.3		Escheverria *et al.* (1997)
	Pig	300	28.4	27	1714		35.2		Scott and McKellar (1992)
	Pig	300	39.6	22.6	3168		91.2	138.7	Lifschitz *et al.* (1999b)
	Pig	300	13.5	74	2712		100.8	187.2	Friis and Bjorn (1995)
	Horse	200	60.7	80	13,209		88.2	114.7	Marriner *et al.* (1987)
	Goat	200	6.12	68.4	1344			188.4	Alvinerie *et al.* (1993)
	Goat	200	10.5	28.9	831			62.1	Escudero *et al.* (1997)

continued

Table 2.1. *Continued.*

Dose route	Species	Dose (μg kg^{-1})	C_{max} (ng ml^{-1})	T_{max} (h)	AUC ((ng ml^{-1}) × h)	VD_{ss} (l kg^{-1})	$t_{1/2el}$ (h)	MRT (h)	Reference
	Camel	200	1.79	295.9	723			456.7	Oukessou *et al.* (1999)
	Camel	200	3.24	144	1591		365	516	Oukessou *et al.* (1996)
	Deer	200	15.8	19.8					Mackintosh *et al.* (1985)
Oral	Sheep	200	22.0	16.4	2039		61.1		Marriner *et al.* (1987)
	Sheep	200	17.6	23.5	2260		101.9		Prichard *et al.* (1985)
	Goat	200	10.5	28.9	831		28.3	62.2	Escudero *et al.* (1997)
	Goat	200	16.0	8	516				Scott *et al.* (1990)
	Horse	200	82.3	3.3	4822		66.3		Marriner *et al.* (1987)
	Horse	200	44.0	9.2	3184		102	114.7	Perez *et al.* (1999)
Pour-on	Cattle	500	28.2	48					Herd *et al.* (1996)
	Cattle	500	12.2	81.6	2916		127.2	201.6	Gayrard *et al.* (1999)
	Goat	500	4.0	48	317				Scott *et al.* (1990)
Moxidectin									
i.v.	Pig	300			6528	17.9	336.0	317.1	Craven *et al.* (2001)
s.c.	Cattle	200	35.6		3816		213.6	165.6	Lifschitz *et al.* (1999a)
	Cattle	200	39.4	7.7	5208	13.6	348.0	350.4	Lanusse *et al.* (1997)
	Cattle	200	18.3	20.9	6717		69.6	393.6	Oukessou *et al.* (1999)
	Cattle	200	18.3	37.4	6719		347.3	615.8	Alvinerie *et al.* (1997)
	Sheep	200	113.4	3.6	3867		718.6	302.3	Hennessy *et al.* (2000)
	Sheep	200	8.3	21.1	2696		237.6	403.2	Alvinerie *et al.* (1998b)
	Goat	200	24.3	8.6	3280			297.6	Escudero *et al.* (1999)
	Camel	200	8.5	31.4	1760			400.1	Oukessou *et al.* (1997)
	Camel	200	8.7	24	1695			330.2	Oukessou *et al.* (1999)

Oral	Sheep	200	8.1	18.1	1121		385.5	385.2	Hennessy *et al.* (2000)
	Sheep	200	27.1	5.3	2373		504.9	301.2	Alvinerie *et al.* (1998b)
	Goat	200	17.5	8.9	881		288.0	247.2	Escudero *et al.* (1999)
	Horse	400	20.3	6.7	4272		415.2	297.6	Alvinerie *et al.* (1999c)
	Horse	200	70.4	8.9	8726		554.6	442.1	Perez *et al.* (1999)
Pour-on	Cattle	500	2.33	120	549.6			307.2	Sallovitz *et al.* (2000)
Doramectin									
i.v.	Pig	300			5280	6.3	86.4	103.2	Friis and Bjorn (1995)
	Sheep	300	737	0	5546	7.6	92.1		Gottschall (1997)
	Sheep	150	147.5	0	2036	5.7	78.6	101.2	Hennessy *et al.* (2000)
	Cattle	200			11,640	1.7	89.0		Goudie *et al.* (1993)
s.c.	Cattle	200	73.5	60	12,182				Wicks *et al.* (1993)
	Cattle	200	27.8	141.6	10,968		180.0		Nowakowski *et al.* (1995)
	Cattle	200	37.5	144.0	15,048	2.9	150.0	218.2	Lanusse *et al.* (1997)
	Cattle	200	32.6	127.4	12,264		129.4	283.2	Toutain *et al.* (1997)
	Sheep	300	26.3	43.2	6342		112.2		Gottschall (1997)
	Sheep	200	22.7	129.6	9696		273.6	155.8	Atta and Abo-Shehada (2000)
	Goat	200	16.8	41.1	2455		62.4	17.6	Escudero *et al.* (1999)
	Pig	300	22.9	62.4	5472		115.2	206.4	Friis and Bjorn (1995)
Oral	Sheep	150	5.4	29.4	753		155.9		Hennessy *et al.* (2000)
i.m.	Cattle	200	33.1	112.8	11,400		156.0		Nowakowski *et al.* (1995)
	Sheep	300	25.4	55.8	5722		106.6		Gottschall (1997)
Pour-on	Cattle	500	12.2	103.2	4334		235.7	307.2	Gayrard *et al.* (1999)

continued

Table 2.1. *Continued.*

Dose route	Species	Dose (μg kg^{-1})	C_{max} (ng ml^{-1})	T_{max} (h)	AUC ((ng ml^{-1}) × h)	VD_{ss} (l kg^{-1})	$t_{1/2el}$ (h)	MRT (h)	Reference
Selamectin									
i.v.	Dog						14		
	Cat						69		
Pour-on	Dog	24,000	86.5	72					Rowan *et al.* (1999)
	Cat	24,000	5513	15					Rowan *et al.* (1999)
Oral	Dog	24,000	7600	8					Rowan *et al.* (1999)
	Cat	24,000	11900	7					Rowan *et al.* (1999)
Eprinomectin									
Pour-on	Cattle	500	43.8	48.5	5789		48.7	99.8	Alvinerie *et al.* (1999a)
	Goats	500	5.6	61.2	1735		179	226	Alvinerie *et al.* (1999b)

C_{max}, maximum concentration; T_{max}, time of C_{max}; AUC, area under the concentration-with-time curve; VD_{ss}, steady-state volume of distribution; $t_{1/2el}$, elimination half-life; MRT, mean residence time.

administered ivermectin differed significantly with feed quantity (Ali and Hennessy, 1996) or composition (Taylor *et al.*, 1992) in otherwise similarly maintained sheep. Such differences in ivermectin behaviour were compounded in Awassi sheep (Atta and Abo-Shehada, 2000) compared with Finn-Dorset sheep (Marriner *et al.*, 1987). Without standardized methodology, the comparison of kinetic data and interpretation of drug behaviour within a species or dose route is limited. Accepting the variety of experimental or husbandry conditions, this section will compare the disposition of MLs in the peripheral plasma pool of target animals and relate this knowledge to drug behaviour in other compartments of physiological and parasitological importance.

Disposition of MLs in the central compartment

Most reports do not include complete kinetic data; however, selected key pharmacokinetic parameters in plasma of target species at manufacturers' recommended dose rates and routes of administration are presented in Table 2.1. A most significant characteristic of MLs is their high lipophilicity. Regardless of their route of administration, ML compounds are distributed extensively throughout the body and concentrate particularly in adipose tissue. While the magnitude of lipid association and exchange differs among chemical types, the limited vascularization and slow turnover rate of body fat, or the slow rate of release or exchange of drug from these lipid reserves prolongs the residence of drug in the peripheral plasma pool.

Ivermectin

Ivermectin is arguably the least lipophilic ML with the possible exception of eprinomectin. It was the first commercially available ML and has been the most extensively studied. Administered intravenously, ivermectin has a general elimination half-life of 32–65 h. Despite the higher dose rate in pigs (300 μg kg^{-1}) than cattle (200 μg kg^{-1}), a shorter half-life results in maximum concentration (C_{max}) and systemic availability (measured as the area under the concentration time curve in peripheral plasma; AUC) in pigs that are about one-third those in cattle (see Table 2.1). This may be due to a more extensive time for recycling of drug from the slower flowing digesta of a ruminant compared with a monogastrid animal. Subcutaneous injection distributes a much greater proportion of the ivermectin into lipid reservoirs, which increases residence time to 114–216 h, almost three times that following intravenous administration in sheep (Marriner *et al.*, 1987), cattle (Toutain *et al.*, 1988; Lanusse *et al.*, 1997), pigs (Scott and McKellar, 1992; Friis and Bjorn, 1995), horses (Marriner *et al.*, 1987) and goats (Alvinerie *et al.*, 1993). While the elimination half-life after subcutaneous and intravenous administration is of similar duration, the extended release of drug from lipid broadens the concentration–time profile with C_{max} in peripheral plasma of cattle occurring as late as 96 h (Lanusse *et al.*, 1997; Toutain *et al.*, 1997). The C_{max} and AUC in pigs and goats is considerably lower than in cattle, horses and sheep. This could be attributed to a greater apparent volume of distribution in pigs (~6.6 l kg^{-1}) compared with cattle (2.2 l kg^{-1}), given that pigs have more than twice the proportion of body fat to total body weight than cattle (Wood *et al.*, 1983). In addition, there are indications of a more rapid metabolism of ivermectin by pigs (Chiu *et al.*, 1987) and possibly in goats (Sangster *et al.*, 1991). Compared with other ruminants, the appearance of subcutaneously injected ivermectin in peripheral plasma of camels is slow, with a lower C_{max} and protracted time of C_{max} (T_{max}) which might be attributed to a slow turnover of adipose tissue.

When it is administered orally, ivermectin strongly associates with particulate digesta and the low partition into digesta fluid reduces the potential for absorption. Ivermectin C_{max} is many times lower and, due to extended outflow from the rumen, T_{max} is later in the ruminant than in the monogastrid. Once absorbed, the elimination half-life is similar for horses and sheep, but shorter in goats probably due to their higher metabolic activity. Maintenance of low concentrations of ivermectin in the ruminant gastrointestinal tract, and therefore peripheral plasma, provides significant advantages in parasite control. In cattle, the intraruminal sustained-release bolus discharges ivermectin at 12 mg day^{-1} providing plasma concentration of 20 ng ml^{-1} for over 120 days. Alvinerie *et al.* (1998a) report that the AUC of ivermectin over the release period was 59,424 (ng ml^{-1}) × h compared with the AUC of 6384–11,016 (ng ml^{-1}) × h following a single subcutaneous 200 µg kg^{-1} dose (Table 2.1). This indicates that the availability of ivermectin following slow release is less than 13%. Extensive association of ivermectin with digesta material in the rumen and low partition into digesta fluid, together with possible degradation in the gut, might explain this low availability.

Eprinomectin

Few details of the pharmacokinetics of eprinomectin are available, but recent studies indicate that after topical application to cattle, eprinomectin was distributed rapidly, with a higher C_{max}, and AUC than similarly applied ivermectin or doramectin doses. The apparent lack of exchange with lipids reduces residence time of eprinomectin with the elimination half-life being 2–6 times faster than other topically applied ML compounds.

Moxidectin

Moxidectin is about 100 times more lipophilic than ivermectin (Hayes, 1994) and, while non-aqueous oil-based formulations are used to promote absorption from the injection site, a more convenient aqueous-based formulation speeds absorption of moxidectin into the bloodstream. This aqueous base formulation results in moxidectin injectable at 200 μg kg^{-1} T_{max} within 21 h compared with the T_{max} of 43–144 h after subcutaneous administration of ivermectin and doramectin in cattle and sheep (Table 2.1). The fat/liver residue ratio 7 days after subcutaneous administration of moxidectin is 16:7 for moxidectin compared with only 1:7 for ivermectin (Hayes, 1994); the extensive deposition in fat results in a very large apparent volume of distribution. Studies by Lanusse *et al.* (1997) and Craven *et al.* (2001) report moxidectin to be distributed through an apparent volume of 13.6 l kg^{-1} in cattle and 17.6 l kg^{-1} in pigs. Zulalian *et al.* (1994) report subcutaneously administered moxidectin to be distributed extensively in, then cleared from omental and back fat of cattle at half-lives of 288 and 360 h, respectively, compared with 216 h from muscle. Similarly, Afzal *et al.* (1994) reported half-lives of moxidectin in the two fat sources from sheep of 324 and 360 h, respectively. This results in residual moxidectin exchanging from fat into liver for up to 28 days after oral administration to sheep (Afzal *et al.*, 1994) or subcutaneous administration to cattle (Zulalian *et al.*, 1994), reflecting the protracted concentrations of moxidectin in blood. The increased moxidectin concentration with time profile in sheep is possibly due to more extensive lipid reserves, or a slower lipid turnover, in sheep than in cattle. Although of different magnitude, the low C_{max} and later T_{max} of moxidectin in camels, as previously observed for ivermectin, is probably related to slow lipid turnover. Sallovitz *et al.* (2000) reported the pharmacokinetic profile of moxidectin topical application in cattle. Application of 500 μg kg^{-1} along the backline resulted in a C_{max} of 2.3 ng ml^{-1} at a T_{max} of 120 h and an AUC of 549.6 (ng ml^{-1}) × h with a mean resistance time (MRT) of 307.2 h. These data would indicate that moxidectin pour-on had a C_{max} and AUC lower and a T_{max} later than the avermectin pour-ons, with an MRT equivalent to doramectin pour-on in cattle.

Doramectin

Being less lipophilic than moxidectin, but more than ivermectin or eprinomectin, doramectin requires an oil-based formulation, the content of which can significantly affect drug availability (Wicks *et al.*, 1993). The oil formulation assisted the subcutaneously administered doramectin to distribute into lipid reserves and, although the apparent volume of distribution of doramectin is less than half that of moxidectin, it is still greater than for ivermectin. The slower half-life of absorption of subcutaneously administered doramectin (56.4 h) compared with ivermectin (39.4 h) and particularly moxidectin (1.3 h) (Lanusse *et al.*, 1997) results in a later T_{max} of doramectin (29–144 h) than the other two MLs, even though the C_{max} is generally similar in a given species for the three compounds. The larger systemic availability of doramectin is due to a low distribution volume (1.7 l kg^{-1}) and slow rate of clearance (0.013 l kg^{-1} h^{-1}) compared with the same parameters for ivermectin (2.4 l kg^{-1} and 0.035 l kg^{-1} h^{-1}, respectively). The elimination half-life of doramectin is longer than that of ivermectin but of shorter duration than that of moxidectin which produces a concentration–time profile which has a more 'square' shape, providing clinically effective plasma concentration of doramectin in cattle for up to 12 days but without an extended 'tail' of potentially subtherapeutic concentrations. Equivalent systemic availability of subcutaneous and intramuscular administration has been demonstrated for cattle (Nowakowski *et al.*, 1995)

Orally administered to sheep, doramectin was absorbed with a half-life of 17 h, reaching T_{max} after 27 h. Gottschall (1997) and Hennessy *et al.* (2000) reported the doramectin volume of distribution in sheep to be 5.7–7.6 l kg^{-1}, somewhat greater than in cattle, probably reflecting the large body fat:body weight ratio in sheep. The elimination half-life of 156 h of orally administered doramectin is considerably longer than the 79 h following intravenous administration and is due to the extended absorption of doramectin from the digesta–drug complex as it passes through the gastrointestinal tract. While the elimination half-life of doramectin appears to be shorter in sheep than in cattle, the MRT is not dissimilar and, as a general observation, for a given ML compound and route of administration, the respective MRT does not appear to be greatly species dependent.

Selamectin

The lack of efficacy of MLs against key parasites (ticks, fleas) of companion animals and isolated toxicity against ML-susceptible breeds of dogs precipitated development of selamectin, a semisynthetic monosaccharide oxime derivative of doramectin. Applied topically to dogs, selamectin has a poor systemic availability of only 4.4% explaining the high dose rate of 6 mg kg^{-1}. Interestingly, the systemic availability of topical application is significantly greater in cats (74%), probably due to ingestion during grooming (Rowan *et al.*, 1999). Notwithstanding this, topically applied selamectin is believed to distribute systemically to concentrate in sebaceous glands, with its extended release to the skin producing 28-day protection against skin-dwelling parasites of dogs and cats.

Topical application

Only the ML class of compounds, with their high activity and lipophilicity, has facilitated topical application. Wool and wool grease form an almost impenetrable barrier precluding this administration to sheep, but it is widely used in hair coat animals. While pour-on formulations are regarded as a more convenient formulation, they also have greater between-animal variability compared with oral or subcutaneous administration. This is of therapeutic concern and, for less-sensitive parasites species, the use of a formulation with a highly variable systemic availability can lead to a higher percentage of treatment failures and/or lack of persistence in preventing reinfection than a formulation with the same availability but lower inter-animal variability. Topical application utilizes a dose rate of 500 $\mu g\ kg^{-1}$, the drug rapidly partitioning into the stratum corneum and diffusing through the skin to accumulate in the subcutaneous depots. Systemic distribution to target tissues is time consuming and is influenced by physico-chemical properties of the active and administration vehicle, skin surface geography and area of application. Judicious use of vehicle combinations influence ML kinetic disposition. By modifying the lipophilicity of the ivermectin formulation using combinations of Miglyol (caprylic and capric acid triglyceride) with Transcutol (diethylene glycol monoethyl ether; DGME), the latter vehicle known to form subcutaneous depots, Yazdanian and Chen (1995) enhanced ivermectin flux across the skin with greater quantitative retention of ivermectin in subcutaneous depots.

After association with subcutaneous fat, the rate of partition of drug from the fat depends on the affinity of the active form for the fat or tissue into which the drug is dispersed, the rate of diffusion of drug through the fat depot and the rate of fat turnover. This is a rate-determining process, with maximum concentrations of ivermectin in peripheral plasma of goats (Scott *et al.*, 1990) and cattle (Herd *et al.*, 1996; Gayrard *et al.*, 1999) lower, occurring later with the systemic availability as low as 15%, considerably less than after oral or subcutaneous administration, even though the dose has been increased 2.5-fold to 500 $\mu g\ kg^{-1}$. Gayrard *et al.* (1999) attributed the low availability to local degradation or entrapment of drug within the skin lipid layers and speculated that the extended terminal half-lives of topically applied ivermectin (127 h) and doramectin (236 h) reflected an absorption process which continues longer than evidenced from plasma disposition kinetics. Between-animal licking and grooming can also contribute to extended drug presence. Céline *et al.* (2001) demonstrated major differences between lickers and non-lickers following pour-on administration of ivermectin. Prevention of licking resulted in a lower systemic availability (19% vs. 33%) and a higher faecal availability (70% vs. 7%).

On the other hand, the lipophilicity of moxidectin contributes significantly to efficient transdermal passage, supporting the observation of

Liftschitz *et al.* (1999a) of prolonged high concentrations of moxidectin in subcutaneous fat of skin sampled remotely from a subcutaneous injection site. There is an absence of detailed kinetic data on disposition of topically applied moxidectin, but the large apparent volume of distribution observed after subcutaneous administration, attributed to deposition in subcutaneous fat, probably explains its extended protective period for nematode infection compared with a shorter period for an equivalent ivermectin dose for sensitive species.

The structural changes of the ivermectin analogue, eprinomectin, not only affect potency, but also solubility, membrane exchange and receptor binding. This facilitates rapid absorption producing high C_{max} with earlier T_{max} and shorter residence time than other ML compounds (Alvinerie *et al.*, 1999a,b). After extensive studies with eprinomectin, Gogolewski *et al.* (1997) report that the performance of this compound is unaffected by hair coat length, timing of rain before and after application and such environmental conditions as UV exposure, heat and freezing. In comparison, environmental conditions reduced plasma concentrations of ivermectin when topically applied to reindeer when wet (Oksanen *et al.*, 1995), and the efficacy of moxidectin was reduced when cattle were exposed to rain soon after treatment (Guerino *et al.*, 1994).

Disposition in the gastrointestinal tract

The disposition of anthelmintic in the central compartment is considered to reflect its behaviour in extravascular compartments, particularly sites of parasite habitat (Lanusse and Prichard, 1993; Baggot and McKellar, 1994). Lanusse and Prichard (1993) proposed that plasma kinetics could predict efficacy. While there may be some anecdotal evidence for this, there can be large differences in the concentration and duration of drug availability between the vascular and extravascular compartments. To predict accurately the availability of MLs requires description of the drug moiety in the various matrices, and compartments, relevant to parasite habitat. Such analyses, particularly within the gastrointestinal tract, present considerable technical difficulties due to the number, complexity, physico-chemical behaviour and interchange of compounds between these compartments and the circulatory system. Flow rates and composition of digesta differ between neighbouring gastrointestinal compartments; biliary secretion, enterohepatic cycling and the dynamism of the ML metabolism further complicate understanding of drug behaviour. Accurate description of this kinetic behaviour requires surgically cannulating the appropriate compartment and stabilizing or monitoring the kinetic behaviour of the matrix in which the drug is distributed. Meaningful concentration data can be obtained from appropriately spaced post-mortem samples, but sampling from disparate pools provides only limited kinetic information.

For these reasons there are few reports where direct, continuous kinetic comparisons have been made in the blood and extravascular compartments throughout the residence time of the drug under examination.

Administered orally, ivermectin and doramectin are almost completely (97–99%) associated with rumen particulate matter (Ali and Hennessy, 1996; Hennessy *et al.*, 2000). Prichard *et al.* (1985) suggested that orally administered ivermectin degraded in the rumen, but this was discounted by Andrew and Halley (1996). Ruminal clearance of ivermectin and doramectin, at half-lives of 11–15 h (particulates) and 11–12 h (fluid), respectively, is identical to the half-lives of these respective digesta phases (Ali and Hennessy, 1996; Hennessy *et al.*, 2000), which prolongs absorption from the gastrointestinal tract. After intravenous administration of radiolabelled doramectin, very low concentrations of doramectin/metabolites over a similar time profile as those in peripheral plasma are present in rumen particulate digesta and are attributed to secretion in saliva (Hennessy *et al.*, 2000). This is consistent with ivermectin being below detectable levels in rumen fluid sampled 24 h after subcutaneous injection of 200 μg ivermectin kg^{-1} (Bogan and McKellar, 1988). In spite of the likely association of orally administered moxidectin with rumen particulate digesta, Alvinerie *et al.* (1998b) reported a plasma moxidectin T_{max} of only 5.3 h in young sheep, which might suggest that absorption from the rumen occurred. It is not known if the high lipophilicity of moxidectin would facilitate absorption from an organ not regarded as absorptive, but consideration of rumen bypass may contribute to this early T_{max}.

Ivermectin (Ali and Hennessy, 1996) and doramectin (Hennessy *et al.*, 2000) remain almost as strongly (80–94%) associated with digesta particulate material in the abomasum as in the rumen. This drug–particulate digesta complex is not directly available for uptake by the host or parasites but rather functions as a gastrointestinal reservoir for exchange of drug into fluid. Abomasal acidity does not appear to facilitate significant exchange of MLs from particulate into fluid digesta since only minimal concentrations of doramectin and no ivermectin are detected in this fluid. After subcutaneous or intravenous administration, low to negligible quantities of moxidectin or doramectin occur in abomasal particulate or fluid, with doramectin accounting for less than 10% of total radiolabelled metabolites in abomasal fluid within 36 h of oral administration of radiolabelled drug. This proportion is less than the doramectin/total metabolite ratio in blood and indicates that degradation of ML compounds occurs in the abomasum or rumen, as originally reported by Prichard *et al.* (1985) for ivermectin. Passage across the abomasal wall is indicated because high concentrations of ivermectin (Bogan and McKellar, 1988) and moxidectin (Liftschitz *et al.*, 1999a) were detected in abomasal mucosa after subcutaneous administration to sheep and cattle. It is not clear if these residues derive from absorption from the abomasal lumen, but their presence in abomasal secretions is the more likely pathway.

High concentrations of ivermectin are present in digesta sampled from the distal intestine, implicating significant biliary secretion of drug. Bogan and McKellar (1988) and Chiu *et al.* (1990) reported high concentrations of ivermectin in the bile of necropsied sheep and cattle. Recently Lifschitz *et al.* (1999b) reported high concentrations of moxidectin in post-mortem bile of subcutaneously treated cattle. Toutain *et al.* (1997) and Lanusse *et al.* (1997) indicated that biliary secretion was an important pathway for clearance of ML compounds, with this pathway having been demonstrated conclusively for clearance of benzimidazole compounds (Hennessy *et al.*, 1993). Details of quantitative drug secretion in bile, enterohepatic recycling, residence time or contribution of biliary-derived compounds in the gastrointestinal tract cannot be obtained from post-mortem samples. The physiological effect of *in vivo* cystic duct cannulation of an anaesthetized animal with complete removal of bile make estimations of bile flow rate or quantitative metabolite secretion/enterohepatic cycling questionable. Importantly, this preparation cannot be maintained in target species for the extended residence time of ML compounds.

Using a re-entrant cannulation/pump technique, Hennessy *et al.* (2000) observed that after oral administration, biliary secretion of radiolabelled doramectin in bile reached a maximum by 24–30 h and followed a similar time course, but at a much greater concentration as that in peripheral plasma. For example, by 240 h after oral administration, when peripheral plasma concentrations of radiolabelled doramectin metabolites had reduced to less than 2 ng ml^{-1}, such metabolites were still being secreted in bile at a rate of 50 ng min^{-1}. Considering that the systemic availability of orally administered radiolabelled doramectin is 30–35%, biliary secretion of 24% of the dose represents the major clearance pathway for absorbed doramectin.

Extremely high biliary concentrations of radiolabelled doramectin occurred within 4 h of intravenous administration, amounting to 132% of the dose over the 28 day study period. Extensive enterohepatic recycling of moxidectin also occurs, with Hennessy (unpublished data) accounting for about 150% of the subcutaneous, and 50% of the oral dose, in bile. Discounting clearance of metabolites in urine, abomasal (gastric), saliva and other non-specific secretions, as much as one-third of biliary secreted doramectin metabolites are recycled enterohepatically. This extended high concentration in bile is influenced by prolonged exchange of drug from lipid reserves and the enterohepatic recycling of biliary compounds through the portal and biliary pools. There appears to be insignificant exchange of reabsorbed biliary secreted drug with the peripheral circulation, which might explain the differential metabolite profile between the two compartments.

The large amounts of ML compounds in intestinal mucus (Bogan and McKellar, 1988; Lifschitz *et al.*, 1999b) are likely to derive from absorption of biliary metabolites and other active intestinal secretions mediated by P-glycoprotein processes (Alvinerie, unpublished data).

The association of biliary secreted parent drug and metabolites with particulate digesta does not seem to be as complete in the distal compared with the proximal gastrointestinal tract. Bogan and McKellar (1988) and Liftschitz *et al.* (1999b) observed greater ivermectin and moxidectin concentrations in ileal compared with abomasal fluid, while Scott and McKellar (1992) reported a three- to fourfold increase in ivermectin concentration in digesta fluid of the lower intestine compared with the stomach and duodenum of pigs. About half of the doramectin metabolites considered to derive from biliary secretion remained in ileal digesta fluid (Hennessy *et al.*, 2000), but comminution of particulate material in the distal intestine results in the presence of very fine particulate material, and their high surface area will strongly associate with ML compounds and might contribute to the high proportion of compounds attributed to 'fluid' digesta.

Metabolism and distribution of MLs in tissues

Details of the metabolism and disposition of parent compound and metabolites generally requires the administration of radiolabelled parent and identification of respective radiolabelled components in post-mortem tissue. The distribution and metabolism of ivermectin has been well documented by Chiu *et al.* (1987, 1988, 1990) with residues highest in liver and fat and lowest in muscle; all higher in cattle than in sheep. In liver, ivermectin accounted for most (48–60%) of the radiolabelled drug, with the H_2B_{1b} isomer metabolized at a faster rate ($t_{1/2} = 11$–18 h) than the H_2B_{1a} ($t_{1/2} = 25$–34 h). Large quantities of 24-OH-H_2B_{1a} and the monosaccharide were present in liver, with lesser amounts of 24-OH-H_2B_{1b} due to the more rapidly cleared parent isomer. Significant ivermectin and non-polar products were present in fat, and Chiu *et al.* (1988) proposed that the latter derived from fatty acid esterification of polar hydroxy metabolites. Low vascularization and slow turnover of adipose tissue resulted in the half-life of ivermectin residues in fat being longer in fat (182 h) than in liver (118 h). Compared with ruminants, ivermectin was less dominant in pig liver and fat and the absence of hepatic ester conjugation resulted in the presence of the 3″-*O*-desmethyl H_2B_{1a} and 3″-*O*-desmethyl H_2B_{1b} metabolites rather than the 24-hydroxymethyl metabolites. The lack of a functional hydroxyl group makes the former metabolite less available for esterification, which reduces its propensity for deposition in fat. Accordingly, elimination half-lives of ivermectin residues in pig fat and liver were of similar duration (5.1–5.2 days)

In the study conducted by Hennessy *et al.* (2000), doramectin accounted for almost all of the administered radioactivity in sheep plasma during the first 24 h, but this proportion decreased to 63 and 54% of intravenously and orally administered radiolabel, respectively, over the drug residence time. Doramectin accounted for 24% of orally administered radiolabel in abomasal digesta and, since there is minimal secretion of doramectin into the abomasum, the low proportion in digesta is indicative of degradation or metabolism. The similar doramectin:total radiolabelled metabolite concentration ratio in plasma and bile suggests that there was no further significant metabolism of doramectin after secretion in bile, supporting the notion of significant doramectin metabolism within the gastrointestinal tract. Gottschall (1997) accounted for 67–92% of radioactivity as unchanged parent doramectin in marker tissues, with the predominant metabolite being 3″-*O*-desmethyl doramectin. This metabolite was also present in faeces, but its derivation from degradation in the gut biliary secretion is unclear. After intravenous or oral administration to sheep, doramectin only accounted for 55–66% of total radioactivity secreted in bile, the 3″-*O*-desmethyl metabolite probably contributing much of the balance of the radioactivity.

Recently, Liftschitz *et al.* (1999a) compared concentrations of subcutaneously administered moxidectin in plasma and post-mortem tissues, observing high concentrations in abomasal and intestinal mucosa, bile and faeces, demonstrating significant biliary involvement in moxidectin metabolism. More detail was provided by Afzal *et al.* (1994) and Zulalian *et al.* (1994) where very high concentrations of radiolabel, attributed almost entirely to moxidectin, were present in back and omental fat for up to 28 days after subcutaneous administration of radiolabelled drug. A high concentration of radiolabel was present in bile and, while moxidectin predominated, detection of the monohydroxy and dihydroxy metabolites demonstrated that after absorption, moxidectin is stored in fat, and, after limited hepatic metabolism, is secreted in bile.

Secretion in milk

Partition into milk is a complex process relating to physico-chemical characteristics and membrane interactions establishing a reversible equilibrium of drug between milk and plasma, reflecting similar elimination profiles from these two compartments. The high lipophilicity of the MLs is highly conducive for partitioning into milk, and Alvinerie *et al.* (1987, 1993) and Bogan and McKellar (1988) reported unity milk/plasma concentration ratio (K(M/P)) of ivermectin in cattle and sheep, with some 4% of the ivermectin dose secreted in milk. In a more detailed cattle study, Toutain *et al.* (1988) observed ivermectin C_{max} in plasma of 54.6 ng ml^{-1} and 40.5 ng ml^{-1} in milk within 48 h of subcutaneous administration, the drug being detectable in milk for over 17 days, amounting to 5% of dose. The K(M/P) of 0.77 was consistent with the findings of Bogan and McKellar (1988) and demonstrated considerable partitioning of ivermectin into cattle milk. The higher lipophilicity of moxidectin will facilitate even greater partition into milk, with Oukessou *et al.* (1996) reporting that the C_{max} and AUC of moxidectin in milk and plasma of lactating camels were 35.2 ng ml^{-1} and 7761 (ng ml^{-1}) × h and 8.5 ng ml^{-1} and 1760 (ng ml^{-1}) × h, respectively. The elimination half-life of moxidectin was similar in both fluids, but the K(M/P) of 4.13 demonstrated extensive partition into milk, and calves suckling from the moxidectin-treated cow contained significant concentrations of moxidectin in their plasma (Alvinerie *et al.*, 1997). Moxidectin has a very low mammalian toxicity, and residual concentrations in milk are below toxic limits, resulting in a nil withholding period in many countries.

Looking to increase ML use in dairy animals, Shoop *et al.* (1996) reported that saturation of the C-22,23 position of the macrocyclic lactone molecule facilitated partition into milk with residual concentrations higher than in plasma, that is a K(M/P) greater than 1.0. Compounds that were not saturated at the C-22,23 position, such as the 4″-epi-acetylamino or 5-oxime derivative of ivermectin had K(M/P) ratios of less than 1.0, an observation that led to the development of eprinomectin which is used in lactating dairy cattle with zero milk withholding time. Alvinerie *et al.* (1999a) confirmed the eprinomectin K(M/P) of 0.1 and accounted for only 0.1% of the dose in milk of lactating cattle.

Clearance

Secretion of large molecular weight ML compounds in bile results in low urinary excretion and, while there appears to be quantitatively greater urinary excretion following subcutaneous or intravenous route compared with oral, total urinary excretion did not exceed 3% of dose of doramectin in sheep (Gottschall, 1997; Hennessy *et al.*, 2000), moxidectin in cattle (Zulalian *et al.*, 1994) and sheep (Afzal *et al.*, 1994), and ivermectin in sheep (Halley *et al.*, 1989; Ali and Hennessy, 1996), cattle and swine (Halley *et al.*, 1989). Quantitative collection of radiolabelled ivermectin in faeces amounted to 96% of the dose given orally, intraruminally or subcutaneously to cattle, sheep and pigs (Halley *et al.*, 1989; Ali and Hennessy, 1996), predominantly as ivermectin but including monosaccharide, aglycone and 24-hydroxymethyl ivermectin. Up to 28 days after either subcutaneous administration to cattle (Zulalian *et al.*, 1994) or orally to sheep (Afzal *et al.*, 1994), up to 58% of radiolabelled moxidectin was excreted in faeces, the parent predominating, but 25–34% of radioactivity was attributed to the monohydroxy derivative. Over 14 days after a subcutaneous dose, Gottschall (1997) collected 51% of the administered radioactivity in sheep faeces, of which the majority was doramectin with small quantities of the 3′-*O*-desmethyl derivative of doramectin and hydroxymethyl doramectin.

Hennessy *et al.* (2000) accounted for 73 and 75%, respectively, of orally and intravenously administered doramectin in sheep faeces.

There is significant ivermectin presence in faeces within 1 day of application (Sommer and Stephansen, 1993), probably due to secretion in bile, but peak excretion in faeces occurred 3–5 days after subcutaneous administration of ivermectin in cattle (Sommer *et al.*, 1992; Bernal *et al.*, 1994) and 3–7 days for doramectin in sheep (Gottschall, 1997). Because the rumen empties with a half-life of 12–14 h, virtually complete passage of orally administered drug from the rumen will be attained after about 3 days. About 24 h is needed for the remnants of the oral dose to traverse the gastrointestinal tract, which accounts for the majority of excretion within 4–5 days. Thereafter, excretion is slower, probably due to extended secretion of lipid-associated drug in bile. After intravenous administration of doramectin, 6–12 h elapses for digesta to pass from the site of entry of biliary metabolites in the upper small intestine before drug presence in faeces. The prolonged secretion of biliary metabolites following parenteral administration produces a slower, but sustained, excretion in faeces over some 14 days for the cumulative faecal radiolabelled drug profile to plateau at the same quantity as that following oral administration. Even though the kinetic and dynamic relationship of MLs with digesta material will differ throughout the gastrointestinal tract, the quantitative excretion in faeces appears largely independent of the route, or the species, in which the compound was administered.

How can Pharmacokinetics be Used to Improve Product Performance?

Improved analytical and pharmacokinetic techniques have greatly assisted in our ability to characterize therapeutic performance. Notwithstanding this, disposition of drug in the 'whole body' situation is considerably more complex than a set of pharmacokinetic parameters described in the peripheral circulation. Improved drug performance requires knowledge of drug behaviour in the multicompartmental system, including the complex interaction between formulation and route of administration, physico-chemical properties of the compound and physiological conditions of the compartment into which the drug is distributed. Exemplifying this, Wicks *et al.* (1993) noted significantly different parasitological effects after subcutaneous administration of two oil-based formulations of doramectin, even though the plasma pharmacokinetic profiles of the two formulations were almost identical.

The quantitative disposition of equivalent dose rate and route of administration can differ markedly between species, individuals and physical conditions of the recipient. The variation in ML response is evident in Table 2.1 and, while some differences are minimal, others, including that in goats, are very noticeable. MLs have been administered to goats at the sheep dose rate, but a more robust hepatic metabolism and/or lower storage in fat contributed to potentially subtherapeutic parasite exposure in the former animals. Rather than select a specific dose rate for goats, repeated administration separated by the approximate product half-life can significantly increase drug performance (Rolfe *et al.*, 1994).

Husbandry Factors that Influence Pharmacokinetic Behaviour

Pharmacokinetic studies invariably are conducted in an 'ideal' situation, where the animal is usually maintained indoors where ambient temperature, day length, accommodation and social contact, in addition to physiological effects of exercise, body condition and the quantity, type and regimen of feed intake are controlled. Between-animal variations in drug response are minimized but these differences can be significant in the less controlled field situation. The specific mode of action of MLs to reduce parasite nutrient uptake is a time-dependent process, and husbandry practices can be used to prolong the availability of therapeutic concentrations of anthelmintic.

Rumen effect

The association of ML products with digesta particulate material forms an important process in drug availability. While the majority of dose is excreted in faeces, the presence of drug in particulate digesta provides the reservoir for exchange into the fluid phase for absorption. Prichard and Hennessy (1981) conclusively demonstrated that orally administered anthelmintics can by-pass the rumen and lodge in the abomasum. Presentation of the fluid drench in the buccal cavity can stimulate closure of the oesophageal groove or reflex, even in older weaned sheep, and direct the dose past the rumen. Without the 'reservoir' effect of the rumen the duration for drug absorption is shortened. Prichard *et al.* (1985) reported a three- to fourfold increase in C_{max} and AUC of ivermectin after intra-abomasal compared with intraruminal administration but, significantly, ivermectin T_{max} was reduced from 23 to 4 h with the former delivery. More recently, Hennessy (unpublished data) confirmed a fourfold increase in plasma C_{max} and AUC of ivermectin and moxidectin, with the T_{max} of both compounds reduced from 24–30 h after intraruminal to 4 h after intra-abomasal administration. Ensuring that an oral drench is directed over the tongue and not deposited in the buccal cavity greatly assists in reducing the likelihood of rumen by-pass to provide a more reproducible, prolonged ML pharmacokinetic profile.

Feed intake effect

Extending the duration of drug–digesta passage prolongs drug availability in the gastrointestinal tract. Reducing feed intake of sheep increases the half-life of rumen fluid and particulate digesta from 6.5 to 11 h and from 12 to 18 h, respectively (Ali and Hennessy, 1996), to prolong drug absorption, whether originating from passage from the rumen or after secretion in gastric, biliary and enteric secretions. Compared with sheep consuming full feed intake, orally administering ivermectin 24 h after feed withdrawal extended the T_{max} from 22 to 38 h, increased the AUC from 1399 to 2085 (ng ml^{-1}) × h and extended the elimination half-life from 12 to 34 h. Similarly, Alvinerie *et al.* (2000) significantly increased the availability of moxidectin when horses were fasted before oral administration. The practice of feed withdrawal before oral treatment broadens the pharmacokinetic profile, significantly increasing anthelmintic efficacy.

Body condition effect

The residence time of subcutaneously administered ML compounds is influenced by the body condition of the animal. The volume of distribution and elimination half-lives of ML compounds, particularly moxidectin, are generally larger in animals with high body fat content such as pigs and sheep, compared with cattle. Examining the disposition of ivermectin and moxidectin, Craven *et al.* (2001) observed that despite having less body fat than fat pigs, lean animals exhibited similar apparent volumes of distribution (5.1 versus 5.3 l kg^{-1}, respectively) and while mean residence time in tissue was shorter in lean pigs, body condition had no other effect on ivermectin pharmacokinetics. Similarly, the quantity of fat binding sites had no effect on the volume of distribution of the more lipophilic moxidectin (17.9–18.7 l kg^{-1}) even though it was much greater than for ivermectin. However, the mean transit time of moxidectin through the central and tissue compartments was shorter in lean than fat pigs, most probably due to a faster turnover of fat in the former animals. Alvinerie (unpublished data) similarly observed a reduced moxidectin AUC in goats of low, compared with high body fat content, whereas the AUC of eprinomectin was similar in these animals. Thus, the duration for action of moxidectin may be reduced in times when the nutritional stress of parasitic disease impacts on body condition, and the less lipophilic ivermectin or eprinomectin might be a better choice of chemotherapeutic.

New Opportunities Arising from Pharmacokinetic Studies

Identifying aspects of pharmacokinetic disposition that might be available, or conducive, for manipulation can lead to improvement in drug performance. Eprinomectin is an example where saturation of the C-22,23 position influences partition into milk, and 4′-epi-acetylamino substitution of dihydroavermectin B_1 affected membrane partition and reversed the plasma/milk concentration ratio from that of the closely related ivermectin. This modification, derived from pharmacokinetic studies, reduced exchange into milk while retaining high anthelmintic potency, and widened the spectrum of use.

Important opportunities exist to access the significant proportion of dose that is associated with particulate digesta, which functions as a reservoir from which these compounds exchange into digesta fluid. The virtually negligible concentrations of ML compounds in abomasal and intestinal fluid suggest that minimal partition occurs or there is the possibility of degradation within the gut. Rather than relying on the physico-chemical/time process for exchange of particulate-bound, ruminally administered drug into digesta fluid for absorption, more direct use of the particulate-bound drug can be made. Hennessy *et al.* (1994) formulated avermectin into a lipid-in-protein microsphere with the protein surface chemically denatured to provide a robust particle that withstood degradation or particulate binding in the rumen but retained the same digesta flow dynamics as rumen digesta. Its presence in the acidic abomasum partially degraded the protective protein surface to produce a staged release of about half of the avermectin payload. The T_{max} in abomasal fluid was unchanged, but C_{max} increased from 1.33 to 3.58 ng ml^{-1} and AUC from 48.4 to 129.1 (ng ml^{-1}) × h for conventional oral formulations and the staged-release formulation, respectively. Enzymatic degradation in the upper intestine released the remainder of the dose for intestinal absorption.

Potentiating effects

Verapamil, a multidrug resistance-reversing agent, has been reported to impact on the pharmacokinetic behaviour of ivermectin in rats (Alvinerie *et al.*, 1999c) and sheep (M.B. Molento, 2000, personal communication). In both species, the presence of verapamil may have interfered with the P-glycoprotein-mediated elimination mechanism, significantly increasing (40–54%) the systemic availability of co-administered ivermectin. Administration of loperamide, an opioid derivative that inhibits gastrointestinal motility, has been shown by Lifschitz *et al.* (2000) to increase the systemic availability of co-administered moxidectin. Slowing gastric transit time increases absorption and recycling of ML compounds (Ali and Hennessy,

1996) and similarly enhanced moxidectin availability to that reported with loperamide co-administration can be achieved with pre-treatment feed withdrawal (Alvinerie *et al.*, 2000).

Co-administration of metabolic inhibitors provides an interesting means of improving ML availability, but the potential use of such combinations, in food-producing animals, must be viewed with trepidation. Registration will require increased toxicological/residue and safety studies of the enhanced availability of the active drug and the potentiating compound. Considering the high costs associated with registration of a single antiparasitic, let alone that of a compound that specifically inhibits xenobiotic metabolism by the host animal, commercial use of such combinations is most unlikely.

Conclusion

This chapter has integrated the known behaviour of ML compounds in the central and associated pharmacokinetic compartments of the target animal. In essence, the drug's kinetic disposition, from recommended routes of administration, has been described in what could be regarded as a 'whole body' environment. In confronting the challenge of maintaining effective parasite treatment, developing such a comprehensive understanding of ML behaviour allows opportunities to be identified where the performance of this class of antiparasitic compounds might be maintained or enhanced further.

References

Afzal, J., Stout, S.J., da Cunha, A. and Miller, P. (1994) Moxidectin; absorption, tissue distribition, excretion and biotransformation of ^{14}C-labeled moxidectin in sheep. *Journal of Agricultural and Food Chemistry* 42, 1767–1773.

Ali, D.N. and Hennessy, D.R. (1996) The effect of level of feed intake on the pharmacokinetic disposition and efficacy of ivermectin in sheep. *Journal of Veterinary Pharmacology and Therapeutics* 19, 89–94.

Alvinerie, M., Sutra, J.F., Galtier, P. and Toutain, P.-L. (1987) Determination of ivermectin in milk by high performance liquid chromatography. *Annales Recherchers Veterinaires* 18, 269–274.

Alvinerie, M., Sutra, J.F. and Galtier, P. (1993) Ivermectin in goat plasma and milk after subcutaneous injection. *Annales Recherchers Veterinaires* 24, 417–421.

Alvinerie, M., Sutra, J.F., Lanusse, C. and Galtier, P. (1997) Plasma profile study of moxidectin in a cow and its suckling calf. *Veterinary Research* 27, 545–549.

Alvinerie, M., Sutra, J.F., Galtier, P., Liftschitz, A., Virkel, G., Sallovitz, J. and Lanusse, C. (1998a) Persistence of ivermectin in plasma and faeces following administration of a sustained-release bolus to cattle. *Research in Veterinary Science* 66, 57–61.

Alvinerie, M., Escudero, E., Sutra, J.-F., Eeckhoutte, C. and Galtier, P. (1998b) The pharmacokinetics of moxidectin after oral and subcutaneous administration to sheep. *Veterinary Research* 29, 113–118.

Alvinerie, M., Sutra, J.F., Galtier, P. and Mage, C. (1999a) Pharmacokinetics of eprinomectin in plasma and milk following topical administration to lactating cattle. *Research in Veterinary Science* 67, 229–232.

Alvinerie, M., Lacoste, E., Sutra, J.F. and Chartier, C. (1999b) Some pharmacokinetic parameters of eprinomectin in goats following topical application. *Veterinary Research Communications* 23, 449–455.

Alvinerie, M., Dupuy, J., Eeckhoutte, C. and Sutra, J.F. (1999c) Enhanced absorption of ivermectin pour-on formulation in rats by co-administration of the multidrug reversing agent verapamil. *Parasitology Research* 85, 920–922.

Alvinerie, M., Sutra, J.F., Cabezas, I., Rubilar, L. and Perez, R. (2000) Enhanced plasma availability of moxidectin in fasted horses. *Journal of Equine Veterinary Science* 20, 575–578.

Andrew, N.W. and Halley, B.A. (1996) Stability of ivermectin in rumen fluids. *Journal of Veterinary Pharmacology and Therapeutics* 19, 295–299.

Atta, A.H and Abo-Shehada, M.N. (2000) Comparative pharmacokinetics of doramectin and ivermectin in sheep. *Journal of Veterinary Pharmacology and Therapeutics* 23, 49–53.

Baggot, J.D. and McKellar, Q.A. (1994) The absorption, distribution and elimination of anthelmintic drugs: the role of pharmacokinetics. *Journal of Veterinary Pharmacology and Therapeutics* 17, 409–419.

Bernal, J.L., Del Nozal, M.J., Salas, M., Galante, E. and Liumaret, J.P. (1994) HPLC determination of residual ivermectin in cattle dung following subcutaneous injection. *Journal of Liquid Chromatography* 17, 2429–2444.

Bogan, J.A. and McKellar, Q.A. (1988) The pharmacodynamics of ivermectin in sheep and cattle. *Journal of Veterinary Pharmacology and Therapeutics* 11, 260–268.

Burg, R.W., Miller, B.M, Baker, E., Birnbaum, J. and Currie, S.A. (1979) Avermectins, a new family of potent anthelminthics agents: producing organism and fermentation. *Antimicrobial Agents and Chemotherapy* 15, 361–367.

Campbell, W.C., Fisher, M.H., Stapley, A., Albers-Schonberg, G. and Jacobs, T.A. (1983) Ivermectin: a potent new antiparasitic agent. *Science* 221, 823–828.

Carter, G.T., Nietsche, J.A. and Hertz, M.R. (1988) LL-F28249, antibiotic complex: a new family of antiparasitic macrocyclic lactones, isolation, characterisation and structures. *Journal of Antibiotics* 41, 519–529.

Céline, M.L., Alvinerie, M., Bousquet-Mélou, A. and Toutain, P.L. (2001) Licking behaviour and environmental contamination arising from pour-on intermectin for cattle. *International Journal for Parasitology* 31, 1687–1692.

Chiu, S.-H.L., Taub, R., Sestokas, E., Lu, A.Y.H. and Jacob, T.A. (1987) Comparative *in vivo* and *in vitro* metabolism of ivermectin in steers, sheep, swine and rat. *Drug Metabolism Reviews* 18, 289–302.

Chiu, S.-H.L., Carlin, J.R., Taub, R., Sestokas, E., Lu, A.Y.H., Zweig, J., Vandenheuvel, W.J.A. and Jacob, T.A. (1988) Comparative metabolic disposition of ivermectin in fat tissues of cattle, sheep and rats. *Drug Metabolism and Disposition* 16, 728–736.

Chiu, S.-H.L., Green, M.L., Baylis, F.P., Eline, D., Rosegay, A., Meriwether, H. and Jacob, T.A. (1990) Absorption, tissue distribution and excretion of tritium-labelled ivermectin in cattle, sheep, swine and rat. *Journal of Agriculture and Food Chemistry* 38, 2072–2078.

Craven, J., Bjorn, H., Hennessy, D.R., Friis, C. and Nansen, P. (2001) Pharmacokinetics of ivermectin and moxidectin following intravenous injection in pigs with different body compositions. *Journal of Veterinary Pharmacology and Therapeutics* 24, 99–104.

Echeverria, J., Mestorino, N., Giorgieri, S., Turic, E., Alt, M. and Errecalde, J. (1997) Pharmacokinetics of ivermectin after intravenous and subcutaneous administration to cattle. *Journal of Veterinary Pharmacology and Therapeutics* 20, 77–78.

Escudero, E., Carceles, C.M., Galtier, P. and Alvinerie, M. (1997) Influence of fasting on the pharmacokinetics of ivermectin in goats. *Journal of Veterinary Pharmacology and Therapeutics* 20, 71–72.

Escudero, E., Carceles, C.M., Diaz, M.S., Sutra, J.F., Galtier, P. and Alvinerie, M. (1999) Pharmacokinetics of moxidectin and doramectin in goats. *Research in Veterinary Science* 67, 177–181.

Fink, D.W. and Porras, A.G. (1989) Pharmacokinetics of ivermectin in animals and humans. In: Campbell, W.C. (ed.) *Ivermectin and Abamectin*. Springer Verlag, New York, pp. 113–130.

Fisher, M.H. and Mrozik, H. (1989) Chemistry. In: Campbell, W.C. (ed.) *Ivermectin and Abamectin*. Springer Verlag, New York, pp. 1–23.

Friis, C. and Bjorn, H. (1995) Pharmacokinetics of doramectin and ivermectin in pigs. *Proceedings of the Pfizer Symposium 'Doramectin: a novel long acting endectocide for use in swine' 15th International Conference of the World Association for the Advancement of Veterinary Parasitology*, Yokahama, Japan, 31 August 1995.

Gayrard, V., Alvinerie, M. and Toutain, P.L. (1999) Comparison of pharmacokinetic profiles of doramectin and ivermectin pour-on formulations in cattle. *Veterinary Parasitology* 81, 47–55.

Gogolewski, R.P., Allerton, G.R., Pitt, S.R., Thompson, D.R., Langholff, W.K., Hair, J.A., Fulton, R.K. and Eagleson, J.S. (1997) Effect of simulated rain, coat length and exposure to natural climatic conditions on the efficacy of a topical formulation of eprinomectin against endoparasites of cattle. *Veterinary Parasitology* 69, 95–102.

Gottschall, D.W. (1997) A comparison of the pharmacokinetics and tissue residues of doramectin after intravenous, subcutaneous and intramuscular administration to sheep. In: *Innovation in Ovine Ectoparasite Control*. Held at 16th International Conference of the World Association for the Advancement of Veterinary Parasitology, South Africa, August 1997.

Goudie, A.C., Evans, N.A., Gration, K.A.F., Bishop, B.F., Gibson, S.P., Holdom, K.S., Kaye, B., Wicks, S.R., Lewis, D., Weatherly, A.J., Bruce, C.I., Herbert, A. and Seymour, D.J. (1993) Doramectin – a potent novel endectocide. *Veterinary Parasitology* 49, 5–15

Guerino, F., Clymer, B.C. and Janes, T. (1994) Evaluation of the effect of simulated rainfall on the efficacy of moxidectin pour-on against faecal output of nematode eggs in cattle. In: *Proceedings of the 30th Annual scientific Meeting of the Australian Society for Parasitology*, Nelson Bay, Australia, August 1994, p. 31.

Halley, B.A., Nessel, R.J. and Lu, A.Y.H. (1989) Environmental aspects of ivermectin usage in livestock: general considerations. In: Campbell, W.C. (ed.) *Ivermectin and Abamectin*. Springer, New York, pp. 162–172.

Hayes, P.W. (1994) Moxidectin: understanding the unique persistent anthelmintic activity of this second generation macrocyclic lactone. In: *Australian Veterinarians in Industry, Australian Veterinary Association, 1994 Annual Conference*, Canberra, Australia.

Hennessy, D.R., Steel, J.W. and Prichard, R.K. (1993) Biliary secretion and enterohepatic recycling of fenbendazole metabolites in sheep. *Journal of Veterinary Pharmacology and Therapeutics* 16, 132 –140.

Hennessy, D.R., Ashes, J.R., Scott, T.R., Gulati, S.K. and Steel, J.W. (1994) Antiparasitic composition. Patent No PCT /AU94/00272.

Hennessy, D.R., Page, S.W. and Gottschall, D. (2000) The behaviour of doramectin in the gastrointestinal tract, its secretion in bile and pharmacokinetic disposition in the peripheral circulation after oral and intravenous administration to sheep. *Journal of Veterinary Pharmacology and Therapeutics* 23, 203–214.

Herd, R.P., Sams, R.A. and Ashcroft, S.M. (1996) Persistence of ivermectin in plasma and faeces following treatment of cows with ivermectin sustained-release, pour-on or injectable formulations. *International Journal for Parasitology* 26, 1087–1093.

Lanusse, C. and Prichard, R.K. (1993) Relationship between pharmacological properties and clinical efficacy of ruminant anthelmintics. *Veterinary Parasitology* 49, 123–158.

Lanusse, C., Lifschitz, A., Virkel, G., Alverez, L., Sanchez, S., Sutra, J.F., Galtier, P. and Alvinerie, M. (1997) Comparative plasma disposition kinetics of ivermectin, moxidectin and doramectin in cattle. *Journal of Veterinary Pharmacology and Therapeutics* 20, 91–99.

Lifschitz, A., Virkel, G., Imperiale, F., Sutra, J.F., Galtier, P., Lanusse, C. and Alvinerie, M. (1999a) Moxidectin in cattle: correlation between plasma and target tissue disposition. *Journal of Veterinary Pharmacology and Therapeutics* 22, 266–273.

Lifschitz, A., Pis, P., Alvarez, L., Virkel, G., Sanchez, S., Sallovitz, J., Kujanek, R. and Lanusse, C. (1999b) Bioequivalence of ivermectin formulations in pigs and cattle. *Journal of Veterinary Pharmacology and Therapeutics* 22, 27–34.

Lifschitz, A., Sallovitz, J., Virkel, G., Imperiale, F., Pis, A. and Lanusse, C. (2000) Loperamide-induced enhancement of moxidectin availability in cattle. *Proceedings of 8th European Association of Veterinary Pharmacology and Therapeutics Congress*, Jerusalem, August 2000.

Mackintosh, C.G., Mason, P.C., Manley, T., Baker, K. and Littlejohn, R. (1985) Efficacy and pharmacokinetics of febantel and ivermectin in red deer (*Cervus elaphus*). *New Zealand Veterinary Journal* 33, 127–131.

Marriner, S.E., McKinnon, I. and Bogan, J.A. (1987) The pharmacokinetics of ivermectin after oral and subcutaneous administration to sheep and horses. *Journal of Veterinary Pharmacology and Therapeutics* 10, 175–179

McTier, T.L., McCall, J.W., Jernigan, A.D., Rowan, T.G., Giles, C.J., Bishop, B.F. and Bruce, C.L. (1998) Efficacy of UK-124,114, a novel avermectin for the prevention of heartworms in dogs and cats. *Proceedings of the American Veterinary Association*, Baltimore, Maryland, 25–28 July.

Nowakowski, M.A., Lynch, M.J., Smith, D.G., Logan, N.B., Mouzin, D.E., Liukaszewitz, J., Ryan, N.I., Hunter, R.P. and Jones, R.M. (1995) Pharmacokinetics and bioequivalence of parenterally administered doramectin in cattle. *Journal of Veterinary Pharmacology and Therapeutics* 18, 290–298.

Oksanen, A., Norberg, H., Niemenen, M. and Bernstad, S. (1995) Influence of route of administration on the plasma concentrations of ivermectin in reindeer. *Research in Veterinary Science* 58, 286–287.

Oukessou, M., Badri, M., Sutra, J.F., Galtier, P. and Alvinerie, M. (1996) Pharmacokinretics of ivermectin in the camel (*Camelus dromedarius*). *Veterinary Record* 139, 424–425.

Oukessou, M., Sutra, J.F., Galtier, P. and Alvinerie, M. (1997) Plasma and milk pharmacokinretics of moxidectin in the camel (*Camelus dromedarius*). *Journal of Veterinary Pharmacology and Therapeutics* 20, 81.

Oukessou, M., Berrag, B. and Alvinerie, M. (1999) A comparative kinetic study of ivermectin and moxidectin in lactating camels (*Camelus dromedarius*). *Veterinary Parasitology* 83, 151–159.

Perez, R., Cabezas, I., Garcia, M., Rubilar, L., Sutra, J.F., Galtier, P. and Alvinerie, M. (1999) Comparison of the pharmacokinetics of moxidectin (Equest®) and ivermectin (Eqvalan®) in horses. *Journal of Veterinary Pharmacology and Therapeutics* 22, 174–180.

Prichard, R.K. and Hennessy, D.R. (1981) Effect of oesophageal groove closure on the pharmacokinetic behaviour and efficacy of oxfendazole in sheep. *Research in Veterinary Science* 30, 22–27.

Prichard, R.K., Steel, J.W., Lacey, E. and Hennessy, D.R. (1985) Pharmacokinetics of ivermectin in sheep following intravenous, intra-abomasal or intraruminal administration. *Journal of Veterinary Pharmacology and Therapeutics* 8, 88–94.

Rolfe, P.F., Evers, J. and Searson, J. (1994) Moxidectin resistance in *Ostertagia* spp. in goats imported from New Zealand. *Australian Society for Parasitology Annual Scientific Conference*, Nelson Bay NSW Australia, 26–30 September 1994, p. 22.

Rowan, T.G., Novotny, M.J., Krautmann, J., Sarasola, P., Inskeep, P.B., Pillai, U.A., Reese, C.P., Shanks, D.J., Smith, D.G., Smothers, C.D. and Jernigan, A.D. (1999) Selamectin in dogs and cats: an overview of safety. In *Proceedings of the Pfizer Symposium* 'Ectoparasitic activity of selamectin', *17th International Conference of the World Association for the Advancement of Veterinary Parasitology*, Copenhagen, Denmark, August 1999.

Sallovitz, J., Imperiale, F., Lifschitz, A. and Lanusse, C. (2000) Pour-on administration of moxidectin to cattle: plasma kinetics and distribution to target tissues. *Journal of Veterinary Pharmacology and Experimental Therapeutics* 23, Supplement 1, Abstract H17.

Sangster, N.C., Rickard, J.M., Hennessy, D.R., Steel, J.W. and Collins, G.H. (1991) Disposition of oxfendazole in goats and efficacy compared to sheep. *Research in Veterinary Science* 51, 258–263.

Scott, E.W. and McKellar, Q.A. (1992) The distribution and some pharmacokinetic parameters of ivermectin in pigs. *Veterinary Research Communications* 16, 139–146.

Scott, E.W., Kinabo, L.D. and McKellar, Q.A. (1990) Pharmacokinetics of ivermectin after oral or percutaneous administration to adult milking goats. *Journal of Veterinary Pharmacology and Therapeutics* 13, 432–435

Shoop, W.L., Mrozik, H. and Fisher, M.H. (1995) Structure and activity of avermectins and milbemycins in animal health. *Veterinary Parasitology* 59, 139–156.

Shoop, W.L., Demontigny, P., Fink, D.W., Williams, J.B., Egerton, J.R., Mrozik, H., Fisher, M.H., Skelly, B.J. and Turner, M.J. (1996) Efficacy in sheep and pharmacokinetics in cattle that led to the selection of eprinomectin as a topical endectocide for cattle. *International Journal for Parasitology* 26, 1227–1235.

Sommer, C. and Stephanson, B. (1993) Changes with time after treatment in the concentrations of ivermectin in fresh cow dung and in cow pats aged in the field. *Veterinary Parasitology* 48, 67–73

Sommer, C., Steffansen, B., Overgaard Nielsen, B., Gronvold, J., Vagn Jensen, K.M., Brocher Jespersen, J., Springborg, J. and Nansen, P. (1992) Ivermectin excreted in cattle dung after subcutaneous injection or pour-on treatment. *Bulletin of Entomological Research* 82, 257–264.

Takiguchi, Y., Mishima, H., Okuda, M., Terrao, M. and Fukuda, R. (1980) Milbemycins a new family of macrolid antibiotics: fermentation, isolation, and physico-chemical properties. *Journal of Antibiotics* 33, 1120–1127.

Taylor, S.M., Mallon, T.R., Blanchflower, W.J., Kennedy, D.G. and Green, W.P. (1992) Effects of diet on plasma concentrations of oral anthelmintics for cattle and sheep. *Veterinary Record* 130, 264–268.

Toutain, P.L., Campan, M., Galtier, P. and Alvinerie, M. (1988) Kinetic and insecticidal properties of ivermectin residues in the milk of dairy cows. *Journal of Veterinary Pharmacology and Therapeutics* 11, 288–291

Toutain, P.L., Upson, D.W., Terhune, T.N. and McKenzie, M.E. (1997) Comparative pharmacokinetics of doramectin and ivermectin in cattle. *Veterinary Parasitology* 72, 3–8.

Wicks, S.R., Kaye, B., Weatherly, A.J., Lewis, D., Davidson, E., Gibson, S.P. and Smith, D.G. (1993) Effect of formulation on the pharmacokinetics and efficacy of doramectin. *Veterinary Parasitology* 49, 17–26.

Wilkinson, P.K., Pope, D.G. and Baylis, F.P. (1985) Pharmacokinetics of ivermectin administered intravenously to cattle. *Journal of Pharmaceutical Sciences* 10, 1105–1107.

Wood, J.D., Whelehan, O.P., Ellis, M., Smith, W.C. and Laird, R. (1983) Effects of selection for low backfat thickness in pigs on the sites of tissue deposition in the body. *Animal Production* 36, 389–397.

Yazdanian, M. and Chen, E. (1995) The effect of diethylene glycol monoethyl ether as a vehicle for topical delivery of ivermectin. *Veterinary Research Communications* 19, 309–319.

Zulalian, J., Stout, S.J., da Cunha, A.R., Garces, T. and Miller, P. (1994) Absorption, tissue distribution, metabolism and excretion of moxidectin in cattle. *Journal of Agricultural and Food Chemistry* 42, 381–387.

Chapter 3

Mode of Action of the Macrocyclic Lactones

R.J. Martin, A.P. Robertson and A.J. Wolstenholme

Introduction

The avermectins and related milbemycins (here referred to as the macrocyclic lactones (MLs)) are derived from *Streptomyces* micro-organisms. They are a group of very hydrophobic compounds, which have broad-spectrum antinematodal (Campbell and Benz, 1984) and antiarthropodal properties. The group includes: abamectin, doramectin, eprinomectin, ivermectin, and selamectin, derived from *Streptomyces avermitilis*; milbemycin D, derived from *S. hygroscopicus*; and moxidectin, derived from *S. cyanogriseus*. Application of these compounds to a number of insect and nematode preparations produces a reduction in motor activity and paralysis. A clinical cure is effected as the paralysed parasites fall from their host or are swept away out of the intestinal tract. Although the MLs have selective toxic effects on insects, acarines and nematodes, they have little effect on mammals and, interestingly, little effect on tapeworms or flukes.

When ivermectin first appeared on the market, in around 1980, electrophysiological observations revealed that the relaxation of invertebrate muscles was associated with an increase in the chloride (Cl) conductance of the membrane (for a review, see Cleland, 1996). It was also noticed that when application of ivermectin increased the Cl conductance of the membrane, high concentrations of picrotoxin blocked this increase. It was presumed from this effect of picrotoxin that ivermectin was acting as a γ-aminobutyric acid (GABA) agonist. The hypothesis that ivermectin acts as a GABA agonist was then suggested to be the mode of action of the macrocyclic lactones. This GABA agonist hypothesis could explain: (i) why ivermectin lacked an effect on flatworms (fluke) and tapeworms – they do not possess GABA receptors; and (ii) why ivermectin does not affect mammalian hosts – MLs do not cross the blood–brain barrier to get to GABA receptors in the central nervous system (CNS). This hypothesis of the mode of action of ivermectin still persists in some pharmacology and parasitology texts, but which now should be updated and replaced.

There are a number of observations inconsistent with the GABA agonist hypothesis for the mode of action of MLs. Scott and Duce (1987) observed that specific locust muscles, which were not sensitive to GABA, responded to dihydroavermectin (ivermectin) with an increase in membrane Cl conductance. Martin and Pennington (1988) observed that ivermectin did not open GABA channels in *Ascaris* muscle but reduced their opening. It was also known that there are peripheral GABA receptors in mammals, in gastrointestinal neurones and on autonomic ganglia, so that the hypothesis of the selective effect of ivermectin on mammals would then have to explain why there was no effect on the peripheral GABA receptors of mammalian hosts. The GABA agonist hypothesis then, was not supportable. What emerged was a recognition that MLs have a potent effect on a group of ligand-gated Cl channels found only in invertebrates, and these Cl channels are not gated by GABA but by glutamate – they are known as glutamate-gated chloride channels or GluCl channels.

The Discovery of Effects of Macrocyclic Lactones on GluCl Channels

The quest for the ivermectin receptor

The successful commercial development of ivermectin by Merck and knowledge of the very potent effects against nematodes led to a search for the mode of action of these compounds. One of the initial helpful observations in the discovery was that ivermectin was found to bind specifically to proteins in membrane preparations of the model nematode *Caenorhabditis elegans* (Schaeffer and Haines, 1989). Interestingly, at this time, it was not possible to demonstrate specific binding in some parasitic nematodes, such as *Ascaris suum*. The lipophilic nature of the avermectins makes the detection of specific binding to receptors present in low amounts very difficult and, in some species, such as *Ascaris*, it was not possible. A number of binding studies followed in *C. elegans* (Schaeffer *et al.*, 1989, 1990; Arena *et al.*, 1995), leading to a degree of confidence that there was in fact an 'ivermectin receptor' in nematodes.

The advantage of using *C. elegans* for mode of action studies was that large quantities could be grown in fermentation tanks to produce significant amounts of RNA for extraction and expression. The technique used by the Merck scientists at Rahway was to collect and fractionate *C. elegans* mRNA for expression screening. The GluCl ion channels in nematodes were first recognized by expression of a glutamate-activated chloride current, sensitive to avermectins, in *Xenopus* oocytes injected with *C. elegans* RNA (Arena *et al.*, 1991, 1992). Expression cloning then led to the recognition of two *C. elegans* channel subunits, GluClα1 on chromosome V and GluClβ on chromosome I. When these were expressed singly and separately in *Xenopus* oocytes, it was found that the GluClα subunit was responsible for the sensitivity to ivermectin and the GluClβ subunit was sensitive to glutamate (Cully *et al.*, 1994). Subsequently, molecular genetic and cloning approaches have led to the recognition of additional GluCl subunits from *C. elegans* (summarized in Table 3.1). It appears that the sensitivity to ivermectin is associated with the α-subunits of the GluCl receptor.

GluCl subunit structure and ion channels

The GluCl subunits are each approximately 500 amino acids in length (Fig. 3.1). The presence of N-terminal cysteines in the GluCl subunits and hydrophobicity analysis suggest that GluCl subunits have similar motifs common to all cysteine loop ligand-gated channels. The GluCl subunits possess a second pair of cysteine residues in the N-terminal extracellular domain that seem to be diagnostic for this family and for the vertebrate glycine receptors (GlyRs). This information and sequence similarity have led to the suggestion that the invertebrate GluCls are in fact orthologous (connected) to the vertebrate GlyR. There is a large extracellular N-terminal domain carrying the ligand-binding site attached to four membrane-spanning α-helices with a long cytoplasmic loop between M3 and M4. Interestingly, the cytoplasmic loop in several subunits possesses consensus (predicted) sites for phosphorylation by protein kinase C (Cully *et al.*, 1994) that may be involved in receptor desensitization. It is assumed that five GluCl subunits come together (Fig. 3.1) as do the subunits of the nicotinic acetylcholine receptors (nAChRs), to produce the GluCl ion channel, but the stoichiometric arrangement has not been determined for any native receptor. It is known that GluClα1 and GluClβ subunits may form homo-oligomeric (identical subunits grouping together to form channels) channels as well as heteromeric (dissimilar subunits grouping together to form channels) ion channels when expressed in *Xenopus* oocytes (Cully *et al.*, 1994). The oocytes injected with RNA were examined using a two-microelectrode voltage clamp to examine the inhibitory Cl currents produced following the application of different analogues of ivermectin (Arena *et al.*, 1991, 1992, 1994, 1995; Arena, 1994; Cully *et al.*, 1994).

Table 3.1. Cloned glutamate-gated chloride channel subunits from *Caenorhabditis elegans*.

Gene	WormPep	EMBL/Genbank Accession No.	Other name for subunit	*In vitro* expression data[a]	*In vivo* expression data[b]	Reference
glc-1	F11A5.10	U14524	GluClα	Forms ivermectin-gated channels		Cully *et al.* (1994)
glc-2	F25F8.2	U14525	GluClβ	Forms glutamate-gated channels	Pharyngeal muscle pm4 cells	Cully *et al.* (1994) Laughton *et al.* (1997)
glc-3	ZC317.3	AJ243914		Forms glutamate- and ivermectin-gated channels		Horoszok *et al.* (1999)
avr-14 (*gbr-2*)	B0207.12	U40573 U41113	GluClα3A GluClα3B	AVR-14B forms glutamate- and ivermectin-gated channels AVR-14A does not form functional channels	Multiple neurones in the nerve ring. Ventral cord motor neurones and mechanosensory neurones	Dent *et al.* (2000) Laughton *et al.* (1997)
avr-15	T10G3.7	AJ000537	GluClα2A GluClα2B	AVR-15L forms glutamate- and ivermectin-gated channels	Pharynx – pm4 and pm5 cells. RME and RMG ring motorneurones, DA9 and VA12 and other nerve cord neurones	Dent *et al.* (1997) Vassilatis *et al.* (1997)
	C27H5.5	U14635		Does not form functional channels		Cully *et al.* 1996

[a]Expression in *Xenopus* oocytes.
[b]Reporter gene data.

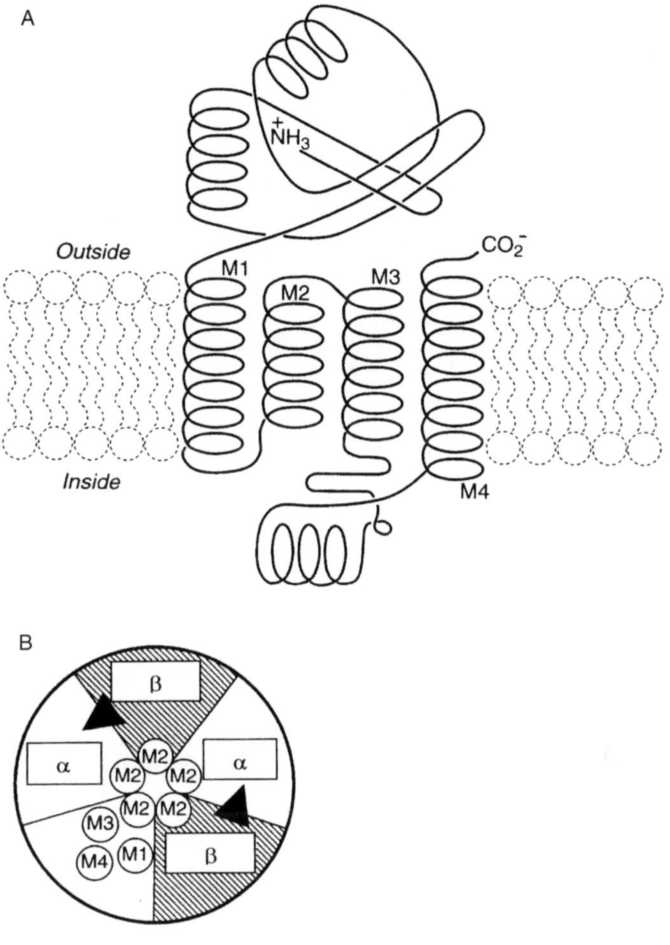

Fig. 3.1. (A) Diagram of an α-subunit and a β-subunit of a glutamate-gated chloride channel receptor (GluClR) showing the long extracellular N-terminal region, the four transmembrane regions, M1, M2, M3, M4, and the cytoplasmic loop between M3 and M4. (B) Diagram of the putative arrangement of five subunits comprising α-subunits and β-subunits forming the GluClR. The arrangement of the four transmembrane regions is such that the M2 region forms the lining of the pore of the channel. The binding sites of the agonist (◄) cross the interface between the β-subunit and the adjacent subunit.

Genes encoding GluClα subunits

In *C. elegans*, a family of genes encodes the α-type subunits: *glc-1* encodes the subunit GluClα1; *avr-15* encodes GluClα2 subunits; *avr-14* encodes GluClα3A and GluClα3B (Dent *et al.*, 1997, 2000); and *glc-3* encodes a fourth GluClα. The *avr-15* gene encodes two alternatively spliced channel subunits that are present on pharyngeal muscle and in motoneurones of the nerve ring and nerve cords. The *avr-15*-encoded subunit (the long form) can form a homeric channel that is ivermectin sensitive and glutamate gated when expressed in *Xenopus* oocytes. The *avr-14* subunits are expressed in a subset of some 40 extrapharyngeal neurones in the ring ganglia of the head and some motor neurones of the ventral cord and mechanosensory neurones (though which of the two subunits is expressed where we still do not know). *glc-1* appears to be represented in the extrapharyngeal neurones (Dent *et al.*, 2000). The expression of *glc-3* is still unknown.

Other GluCl subunits

The location of the GluClβ subunit was studied by expression of a *lacZ* reporter gene and was found on the pm4 pharyngeal muscle cells of *C. elegans* (Laughton *et al.*, 1997). The genome sequence predicts a sixth GluCl gene, C27H5.8. This has been cloned and injected into *Xenopus* oocytes; however, no functional channels were formed. There is no information on the expression pattern of this gene.

Electrophysiology of ivermectin in nematode parasites

Avermectin-sensitive sites in *A. suum* have been identified on pharyngeal muscle using a two-microelectrode current clamp technique (Fig. 3.2). Avermectins produce hyperpolarization and an increase in Cl^- conductance when either bath-applied or pressure-ejected on to the pharyngeal preparation, and usually the response is irreversible. Figure 3.3 illustrates electrophysiological traces from an experiment demonstrating the effects of glutamate and ivermectin. Glutamate and MLs increase the total open-time of expressed GluCl receptor channels (Cully *et al.*, 1994) and in *A. suum* pharyngeal muscle (Martin, 1995). Figure 3.3 shows that tracings become narrower as the resistance of the cell membrane reduces as the ion channel receptors open in the membrane. Notice that glutamate and ivermectin make the traces narrower, they open the same ion channels.

The split chamber technique described by Kass *et al.* (1980, 1982, 1984) has permitted the selective application of avermectin to the dorsal and ventral halves of *Ascaris*. It was found that avermectin blocked the motor neurone dorsal excitatory 1 (DE1) response to indirect stimulation but not direct stimulation. The observations were interpreted as suggesting that avermectins block transmission between interneurones in the nerve and the excitatory neurones. It was also found by these authors that the hyperpolarizing response of muscle following direct stimulation of the motorneurone VI was blocked by avermectin, an observation consistent with the antagonism of muscle GABA receptors described above. These

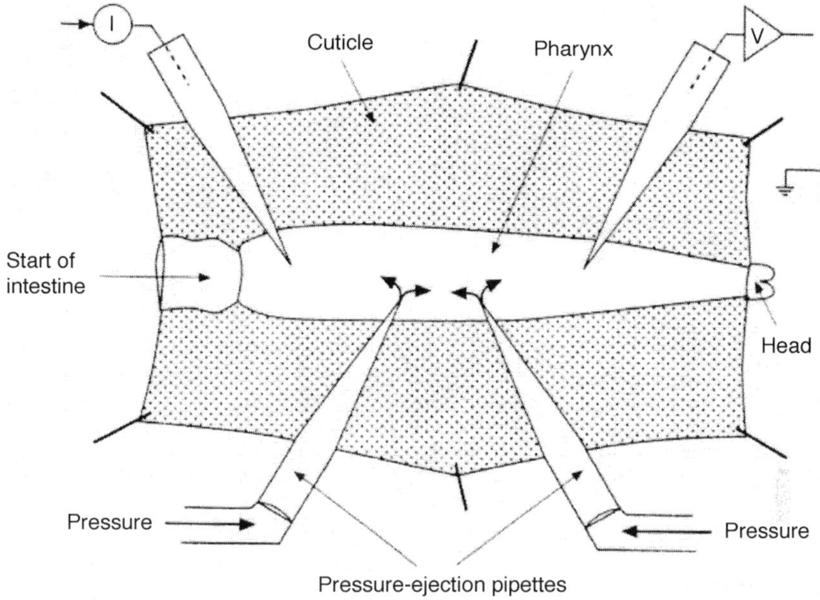

Fig. 3.2. Diagram of the two-microelectrode current clamp recording technique that we have used to examine electrophysiological effects of focal application of glutamate and ivermectin to the pharyngeal muscle of *Ascaris suum*. One micropipette (labelled V) is placed in the pharyngeal muscle to record the membrane potential, and a second micropipette (I) is also placed in the muscle to inject hyperpolarizing current pulses. Two additional micropipettes are shown, which are used for local application of glutamate and ivermectin. A short burst or 'puff' of pressure is used to apply the drugs.

observations, together with the observations referred to above, show that avermectins have more than one site of action and that these include ion channels in muscle membrane and in neuronal membranes. These results are consistent with the immunological identification of GluCl subunits on the nerve cords and motoneurone commisures of *Haemonchus contortus* and *A. suum* (Table 3.2). There is also evidence that there are inhibitory glutamate receptors on the muscle cells of the female reproductive tract of *Ascaris* (Fellowes *et al.*, 2000). Inhibitory effects of MLs on egg laying and fertility by nematodes are known, and may account for much of the efficacy of ivermectin treatment in controlling infections by *Onchocerca volvulus* and other filarial parasites (see Fig. 3.4, p. 134).

Distribution of a fluorescent ivermectin analogue in Ascaris

A fluorescent derivative of ivermectin (4′-5,7 dimethyl bodipy propriony livermectin) has been prepared and injected into adult *Ascaris* where

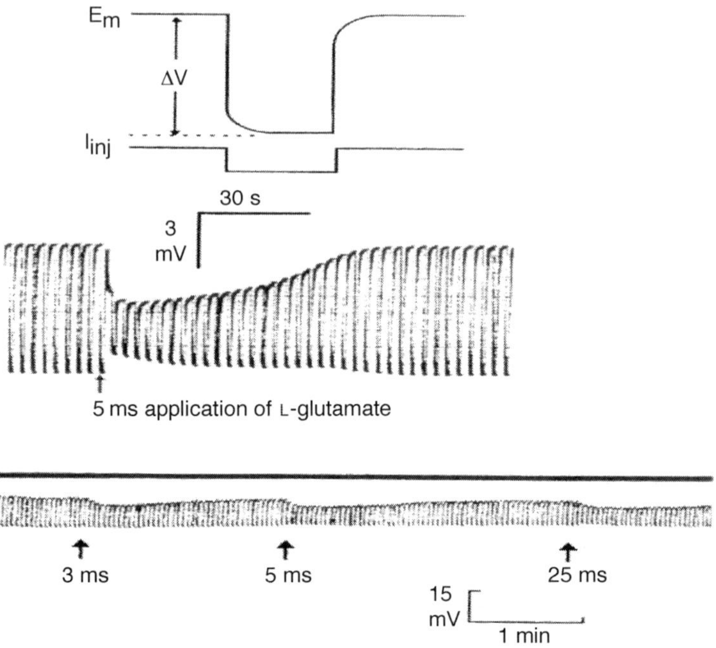

Fig. 3.3. Top trace: diagram of the effect of injecting a rectangular current pulse (I_{inj}) on the membrane potential (E_m) using a two-microelectrode clamp. The current produces an exponentially increasing hyperpolarization that settles after sufficient time to a change of ΔV mV. The input conductance of the pharynx can be determined from Ohm's law because the current–voltage relationship is linear. Middle trace: the trace shows that brief application of glutamate using a 5-ms 'puff' or pressure application from a micropipette filled with 0.5 M L-glutamate leads to a hyperpolarization of the membrane potential and an increase in the input conductance to 200 μS. As the ion channels open up and carry Cl ions, the resistance of the membrane decreases and the membrane potential hyperpolarizes. Lower trace: effects of glutamate and ivermectin on pharyngeal membrane potential and conductance. The trace shows the membrane potential and input conductance response to a continuous application of ivermectin (horizontal bar), and 3, 5 and 25 ms 'puffs' of 0.5 M glutamate from a micropipette.

it produces dose-dependent immobilization (Martin and Kusel, 1992; Martin *et al.*, 1992). Fluorescent microscopy of frozen sections has revealed the distribution of the probe in the whole nematode. The probe accumulates in muscle membranes and within the nerve cord; these two sites are consistent with an action of avermectins on muscle and nerve membrane. It also accumulates under the hypodermis (Martin and Kusel, 1992), a site consistent with excretion from the parasite, perhaps by a P-glycoprotein transporter (Xu *et al.*, 1998).

Table 3.2. Cloned GluCl subunits from parasitic nematodes.

Species	Clone	Accession number	Other name	Closest *C. elegans* gene	Expression data[a]	References
Haemonchus contortus	Hc-gbr2A Hc-gbr2B	Y14233 Y14234	HG2 HG3	*avr-14* (*gbr-2*)	Hc-gbr2B binds ivermectin. Present in the nerve ring, on ventral and dorsal nerve cords, the anterior dorsal sublateral cord and motor neurone commissures[c]	Jagannathan *et al.* (1999)
Haemonchus contortus	HcGluClβ	Y09796	HG4	*glc-2*	Does not bind ivermectin. Expressed on motor neurone commissures	Delany *et al.* (1998)
Haemonchus contortus	HcGluClα	AF076682 AJ131347	HG5	*avr-14* (*gbr-2*)[b]	Binds ivermectin. Expressed at higher levels in adults than larvae. Present on motor neurone commissures	Forrester *et al.* (1999)
Ascaris suum	As-gbr2	Y18347		*avr-14* (*gbr-2*)	Present on nerve cords	Jagannathan *et al.* (1999)
Onchocerca volvulus	GluClX	U59745		*avr-14* (*gbr-2*)		Cully *et al.* (1996)
Onchocerca ochengi	Putative α-subunit	AF054632		*glc-1*		
Onchocerca ochengi	Putative β-subunit	AF054631		*glc-2*		
Dirofilaria immitis	GluClX	U59744		*avr-14* (*gbr-2*)		Cully *et al.* (1996)

[a]Localization data obtained using immunofluorescence.
[b]The level of identity between this subunit and AVR-14 is lower than for the Hc-GBR2, As-GBR2 and GluClX subunits.
[c]The antibody used did not distinguish between Hc-gbr2A and 2B.

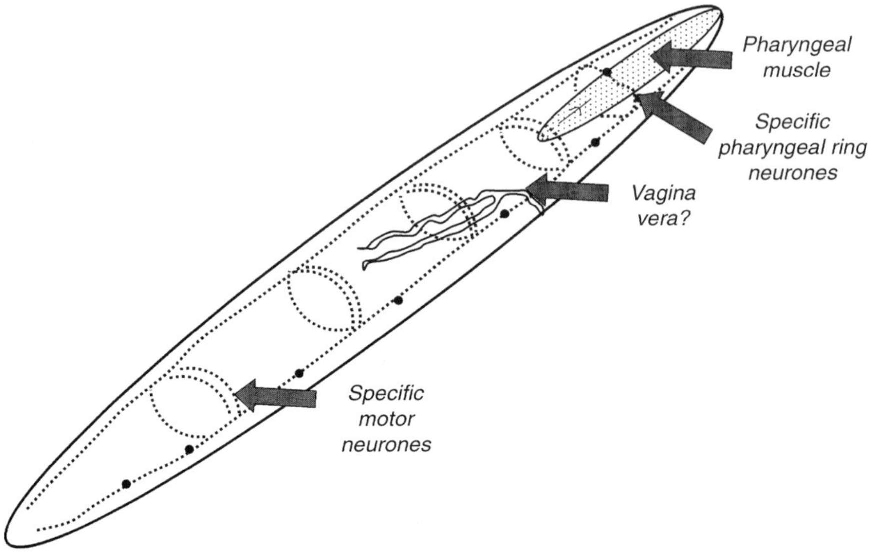

Fig. 3.4. Diagram of the putative 'ivermectin receptor sites'. The illustration shows the two nerve cords and connecting commissures, the pharynx and the opening of the oviduct via the vagina vera. Each receptor site is marked with an arrow.

Molecular cloning of GluCl subunits from parasitic nematodes

Compared with *C. elegans*, we know rather less about the molecular biology of the GluCl from parasitic species. Our current knowledge is summarized in Table 3.2. Molecular cloning experiments have revealed the existence of three GluCl genes in *H. contortus*, a species very closely related to *C. elegans*. Two of these are clearly orthologous to the *C. elegans glc-2* and *avr-14* genes, and the pattern of alternative splicing of *avr-14* is conserved between the two species. Not only that, but the expression of *avr-14* on ring and cord motoneurones seems similar in both *C. elegans* and *H. contortus*, and the GluClα3B subunit from both species, but not the GluClα3A subunit, is capable of binding ivermectin when expressed alone. *avr-14* is the gene identified most widely in parasitic nematodes, as it is also present in *A. suum* and the filaria, *O. volvulus* and *Dirofilaria immitis*, though the alternative splicing does not appear to be conserved in these species.

The other *H. contortus* GluClα gene is interesting in that it does not appear to be obviously orthologous with any of the four *C. elegans* GluClα genes, as it shares about 55–60% sequence identity with all of them. It is also expressed at higher levels in adult worms than in larvae, raising the interesting possibility that this subunit may be most important in the parasitic stages of the life cycle. This subunit can also bind ivermectin and is found on motoneurones. It may represent an important target for the drug *in vivo*.

The *H. contortus* GluClβ subunit, like its *C. elegans* counterpart, does not bind ivermectin. Though the pharmacology of the two subunits is similar, their distribution may be very different. Antibody studies have shown that the *H. contortus* GluClβ subunit is found on motoneurone commissures in the anterior region of the adult parasite, whereas the *C. elegans* subunit was only found on pharyngeal muscle cells. This may represent an important difference between the model species and the parasite, though the results may also be due, in part, to the antibody staining method used being unable to reveal pharyngeal expression in the parasite and/or the reporter gene construct lacking important signals driving motoneurone expression. A GluClβ subunit has also been partially cloned from *Onchocerca ochengi*, though no other data have been reported on this.

What Do Mode of Action Studies Predict for ML Resistance?

The genetics of ivermectin resistance has been studied in *C. elegans* (Dent *et al.*, 2000). A number of genes are known to be involved in ivermectin resistance. They include: *avr-15, avr-14, glc-1, unc-7, unc-9* and *dyf* (dye filling defective) genes. Dent *et al.* (2000) have examined interactions of these genes that are required for the establishment of high levels of ivermectin resistance. They observed that simultaneous mutations of the GluClα genes *avr-15, avr-14* and *glc-1* were required for the establishment of high levels of immunityof more than 4000-fold. Little or no resistance is seen if there are mutations in single genes. A combination of two of the genes (*avr-14* and *avr-15*) produced a 13-fold increase in resistance. It was also noticed that the genes *unc-7* and *unc-9*, that encode innexins (gap junction proteins), and the *dyf* gene, *osm-1*, were connected and involved in resistance.

Figure 3.5 summarizes the information relating to the genetics of resistance to the MLs in nematodes and is derived from information mostly from *C. elegans*. The entry of ivermectin into the nematode is facilitated by sensory (amphidial) neurones on the head. Once the drug has gained entry across the cuticle, it is then able to interact with the GluClα receptors. There are at least three different genes encoding at least three different GluClα subunits that form inhibitory ion channels on muscle of the pharynx, motor neurones and other neurones and, in addition, perhaps on the female reproductive tract. The neurones that possess the GluCl channels connect via gap junctions that are made up of innexins, coded for by *unc-7* and *unc-8* genes. Thus the inhibitory effect of ivermectin on the pharynx may be direct via the GluClα2 subunit, or indirect, via the GluClα3 and GluClα1 subunits on extrapharyngeal

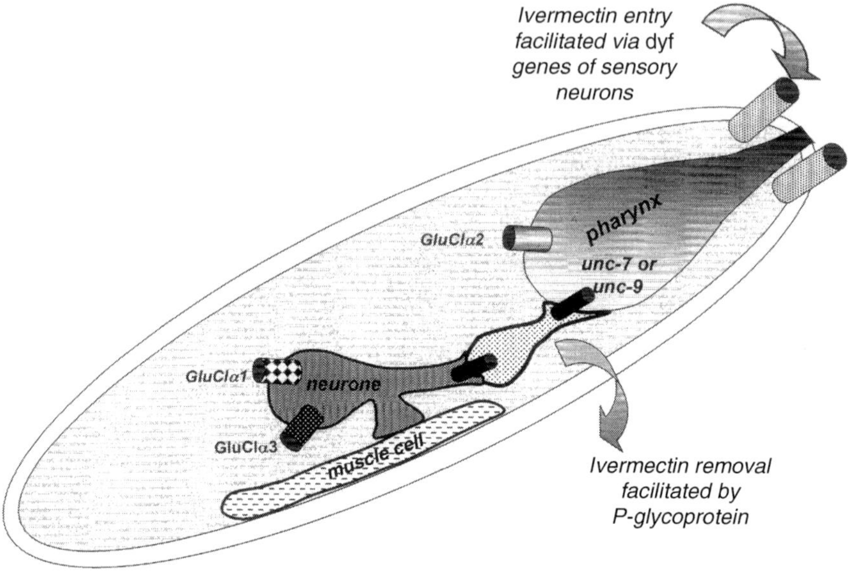

Fig. 3.5. Diagram showing the interaction of 'ivermectin resistance' genes. Note that the physiology and location of the target site determine how resistance genes may interact. For example, the genes *unc-7* or *unc-9* form gap junctions that are able to pass on potential changes (hyperpolarization) associated with stimulation of GluClα1 and GluClα3 subunits in inhibitory channels.

neurones, and require that effect to be mediated across gap junctions (*unc-7* and *unc-9*) to the pharynx. Removal of ivermectin and other MLs from the body of nematodes appears to be mediated by P-glycoprotein excretion (Sangster, 1994). The mode of action and genetics of resistance illustrate that the development of resistance requires the simultaneous mutation of several genes to develop a high level of resistance. Factors that increase the concentration of MLs in the nematode will increase susceptibility; genes (*unc-7* and *unc-9*) which increase the electrical effects of stimulation of the GluCl ion channel will increase susceptibility; and the presence of genes coding for GluClα subunits will increase susceptibility.

To Be the MISER is Best

'. . . like a miser, sir, in a poor house, as your pearl . . .' As You Like It, William Shakespeare

The molecular and physiological evidence shows us that there are multiple sites of action for the MLs (the different GluClα gene products). The presence of multiple target genes will require that resistance is polygenic.

If genes for different target sites are inherited independently, then more genes are required for the nematode parasite to show resistance, thus the probability of resistance developing will be slower.

We can illustrate this point with *C. elegans*, where there are three (*avr-15, avr-14* and *glc-1*) independent, presumably recessive resistance genes. If only fractions, $f_{avr\text{-}15} = 10^{-4}$, $f_{avr\text{-}14} = 10^{-4}$, $f_{glc\text{-}1} = 10^{-4}$, of a particular nematode parasite population carry single copies of the respective recessive genes, the probability of observing the resistant phenotype (requiring the simultaneous presence of two copies of all of the recessive genes) will then be $f_{avr\text{-}15}{}^2 \times f_{avr\text{-}14}{}^2 \times f_{glc\text{-}1}{}^2 = 10^{-24}$. We can see that the more genetically independent target sites there are for an anthelmintic, then the lower will be the probability for development of resistance to that anthelmintic. One consequence is that resistance would be expected to develop more slowly in nematode parasites where an anthelmintic has multiple independent sites of action. This concept may be emphasized by referring to it as the MISER (multiple independent sites of action evading resistance) principle. MISER anthelmintics such as the MLs are expected to produce resistance more slowly than those anthelmintics that have only one gene coding for its target site. Further discussion of delaying the development of anthelmintic resistance in the field is given by Jackson and Coop (2000).

Conclusions

- We have covered evidence that shows that MLs activate glutamate-gated Cl channels that are found on membranes of the pharynx and particular neurones. There is a good correlation between the presence of GluCl on neurones and muscle cells and the observed effects of the MLs.
- There is evidence from *C. elegans* and parasitic species that there are several genes that encode the α-subunits of these ion channels, and it is the α-subunits that are sensitive to the MLs.
- This implies that nematode parasite resistance to these compounds will be polygenic and will require the simultaneous presence of more than one gene for the whole nematode to show resistance. In *C. elegans*, there are three main genes, *avr-15, avr-14* and *glc-1*, that have to be present simultaneously for the development of a high level of resistance. It will be of interest to determine if similar genes are involved in the development of resistance in parasitic nematodes.
- The GluCl channels are found in invertebrates (insects, nematodes and crustacea) but not in vertebrates. The selective toxicity of MLs as an anthelmintic may be explained by the action on GluCl channels that are not present in host animals.

- Inhibitory effects of MLs on the reproductive system of females may explain the observed reductions in egg production. If receptors in the female reproductive tract are more sensitive to MLs than other receptors on nerve or muscle, then temporary reductions in egg production will be seen if the parasite is not killed or removed by drug therapy.

References

Arena, J.P. (1994) Expression of *Caenorhabditis-elegans* messenger-RNA in *Xenopus*-oocytes – a model system to study the mechanism of action of avermectins. *Parasitology Today* 10, 35–37.

Arena, J.P., Liu, K.K., Paress, P.S. and Cully, D.F. (1991) Avermectin-sensitive chloride currents induced by *Caenorhabditis elegans* RNA in *Xenopus* oocytes. *Molecular Pharmacology* 40, 368–374.

Arena, J.P., Liu, K.K., Paress, P.S., Schaeffer, J.M. and Cully, D.F. (1992) Expression of a glutamate-activated chloride current in *Xenopus* oocytes injected with *Caenorhabditis elegans* RNA: evidence for modulation by avermectin. *Molecular Brain Research* 15, 339–348.

Arena, J.P., Liu, K.K., Vassilatis, D.K., Paress, P.S. and Cully, D.F. (1994) Properties of the *Caenorhabditis elegans* glutamate avermectin-sensitive chloride channel expressed in *Xenopus*-oocytes. *Abstracts of Papers of The American Chemical Society* 207, 196-AGRO.

Arena, J.P., Liu, K.K., Paress, P.S., Frazier, E.G., Cully, D.F., Mrozik, H. and Schaeffer, J.M. (1995) The mechanism of action of avermectins in *Caenorhabditis elegans*: correlation between activation of glutamate-sensitive chloride current, membrane-binding, and biological-activity. *Journal of Parasitology* 81, 286–294.

Campbell, W.C. and Benz, G.W. (1984) Ivermectin: a review of efficacy and safety. *Journal of Veterinary Pharmacology and Therapeutics* 7, 1–16.

Cleland, T.A. (1996) Inhibitory glutamate-receptor channels. *Molecular Neurobiology* 13, 97–136.

Cully, D.F., Vassilatis, D.K., Liu, K.K., Paress, P.S., Vanderploeg, L.H.T. and Schaeffer, J.M. (1994) Cloning of an avermectin-sensitive glutamate-gated chloride channel from *Caenorhabditis-elegans*. *Nature* 371, 707–711.

Cully, D.F., Wilkinson, H., Vassilitis, D.K., Etter, A. and Arena, J.P. (1996) Molecular biology and electrophysiology of glutamate-gated chloride channels of invertebrates. *Parasitology* 114, S191-S200.

Delany, N.S., Laughton, D.L. and Wolstenholme, A.J. (1998) Cloning and localisation of an avermectin receptor-related subunit from *Haemonchus contortus*. *Molecular and Biochemical Parasitology* 97, 177–187.

Dent, J.A., Davis, M.W. and Avery, L. (1997) Avr-15 encodes a chloride channel subunit that mediates inhibitory glutamatergic neurotransmission and ivermectin sensitivity in *Caenorhabditis elegans*. *EMBO Journal* 16, 5867–5879.

Dent, J.A., Smith, M.M., Vassilatis, D.K. and Avery, L. (2000) The genetics of ivermectin resistance in *Caenorhabditis elegans*. *Proceedings of the National Academy of Sciences of the USA* 97, 2674–2679.

Fellowes, R.A., Maule, A.G., Martin, R.J., Kimber, M.J., Marks, N.J. and Halton, D.W. (2000) Classical neurotransmitters in the ovijector of *Ascaris suum*: localisation and modulation of muscle activity. *Parasitology* 121, 325-336.

Forrester, S.G., Hamdan, F.F., Prichard, R.K. and Beech, R.N. (1999) Cloning, sequencing, and developmental expression levels of a novel glutamate-gated chloride channel homologue in the parasitic nematode *Haemonchus contortus*. *Biochemical and Biophysical Research Communications* 254, 529–534.

Horoszok, L., Sattelle, D.B. and Wolstenholme, A.J. (1999) A new glutamate-gated chloride channel subunit from *Caenorhabditis elegans*. *Society for Neuroscience Abstracts* 24, 1704.

Jackson, F. and Coop, R.L. (2000) Electrophysiological investigation of anthelmintic resistance. *Parasitology* 120, S95-S108.

Jagannathan, S., Laughton, D.L., Critten, C.L., Skinner, T.M., Horoszok, L. and Wolstenholme, A.J. (1999) Ligand-gated chloride channel subunits encoded by the *Haemonchus contortus* and *Ascaris suum* orthologues of the *Caenorhabditis elegans gbr-2* (*avr-14*) gene. *Molecular and Biochemical Parasitology* 103, 129–140.

Kass, I.S., Wang, C.C., Waldrond, J.P. and Stretton, A.O.W. (1980) Avermectin B1a, a paralysing anthelmintic that affects interneurones and inhibitory motoneurones in *Ascaris*. *Proceedings of the National Academy of Sciences of the USA* 77, 6211–6215.

Kass, I.S., Larsen, D.A., Wang, C.C. and Stretton, A.O.W. (1982) *Ascaris suum*: differential effects of avermectin B1a on the intact animal and neuromuscluar strip preparations. *Experimental Parasitology* 54, 166–174.

Kass, I.S., Stretton, A.O.W. and Wang, C.C. (1984) The effects of avermectin and drugs related to acetylcholine and 4-aminobutyric acid on neurotransmission in *Ascaris suum*. *Molecular and Biochemical Parasitology* 13, 213–225.

Laughton, D.L., Lunt, G.G. and Wolstenholme, A.J. (1997) Reporter gene constructs suggest that the *Caenorhabditis elegans* avermectin receptor beta-subunit be expressed solely in the pharynx. *Journal of Experimental Biology* 200, 1509–1514.

Martin, R.J. (1995) An electrophysiological preparation of *Ascaris suum* pharyngeal muscle reveals a glutamate-gated chloride channel sensitive to the avermectin analogue milbemycin D. *Parasitology* 212, 247–252.

Martin, R.J. and Kusel, J.R. (1992) On the distribution of a fluorescent ivermectin probe (4″ 5,7 dimethyl-bodipy proprionylivermectin) in *Ascaris* membranes. *Parasitology* 104, 549–555.

Martin, R.J. and Pennington, A.J. (1988) Effect of dihydroavermectin-b1a on CI single-channel currents in *Ascaris*. *Pesticide Science* 24, 90–91.

Martin, R.J., Kusel, J.R., Robertson, S.J., Minta, A. and Haugland, R.P. (1992) Distribution of a fluorescent ivermectin probe, bodipy ivermectin, in tissues of the nematode parasite *Ascaris suum*. *Parasitology Research* 78, 341–348.

Sangster, N.C. (1994) P-glycoproteins in nematodes. *Parasitology Today* 10, 319–322.

Schaeffer, J.M. and Haines, H.W. (1989) Avermectin binding in *Caenorhabditis elegans*: a 2-state model for the avermectin binding-site. *Biochemical Pharmacology* 38, 2329–2338.

Schaeffer, J.M., Stiffey, J.H. and Mrozik, H. (1989) A chemi-luminescent assay for measuring avermectin binding-sites. *Analytical Biochemistry* 177, 291–295.

Schaeffer, J.M., Frazier, E.G., Bergstrom, A.R., Williamson, J.M., Liesch, J.M. and Goetz, M. (1990) Cochlioquinone-a, a nematocidal agent which competes for specific [h-3] ivermectin binding-sites. *Journal of Antibiotics* 43, 1179–1182.

Scott, R.H. and Duce, I.R. (1987) Pharmacology of GABA receptors on skeletal muscle fibres of the locust (*Schistocerca gregaria*). *Comparative Biochemistry and Physiology* C 86, 305–311.

Vassilatis, D.K., Arena, J.P., Plasterk, R.H.A., Wilkinson, H.A., Schaeffer, J.M., Cully, D.F. and Vanderploeg, L.H.T. (1997) Genetic and biochemical evidence for a novel avermectin-sensitive chloride channel in *Caenorhabditis elegans* – isolation and characterization. *Journal of Biological Chemistry* 272, 33167–33174.

Xu, M., Molento, M., Blackhall, W., Ribeiro, P., Beech, R. and Prichard, R. (1998) Ivermectin resistance in nematodes may be caused by alteration of p-glycoprotein homolog. *Molecular and Biochemical Parasitology* 91, 327–335.

Chapter 4

Ecological Impact of Macrocyclic Lactones on Dung Fauna

J.W Steel and K.G Wardhaugh

Introduction

Concern over the use of chemical additives in livestock feedstuffs and the potential consequences of insect-free manure was first expressed by Anderson (1966). Since then, the chemotherapeutic revolution in livestock management has continued, particularly in the control of internal and external parasites with the deployment of new, highly active compounds. These include synthetic pyrethroids, insect growth regulators and macrocyclic lactones (MLs), which are used at extremely low dosage rates often in formulations and delivery systems designed to extend drug residence time in the animal and hence the period of protection. The potential ecological significance of these developments passed largely unnoticed until Wall and Strong (1987) reported effects on non-target dung insects and dung dispersal after cattle were treated with an experimental sustained-release device containing ivermectin. However, this latter development, together with a wide-reaching review of avermectin usage in agriculture (Strong and Brown, 1987), had two unintentional effects on research to assess the ecological consequences of widespread adoption of MLs by the livestock industry. First, it generated many unproductive and, at times, vitriolic claims and counter-claims between those committed to the therapeutic value of these anthelmintics in livestock production and those intent on evaluating the ecological and environmental costs. Diverse opinions have been presented on the potential impact of faecal ML residues on the pasture ecosystem (Herd, 1995; Bulman *et al.*, 1996; Wratten and Forbes, 1996; McKellar, 1997; Spratt, 1997). Alluding to these sometimes uncompromising exchanges, Spratt (1997) pleaded for avoidance of 'the dialogue of the deaf' to improve interdisciplinary research between ecology and animal production. Secondly, concern became so

(eds J. Vercruysse and R.S. Rew)

focused on ivermectin that other MLs, such as moxidectin, doramectin and eprinomectin, have received scant attention.

The effects of the various ML chemicals on dung-feeding arthropods were reviewed comprehensively by Strong (1993) and Steel (1998); the latter included unpublished information on the newer ML products submitted by pharmaceutical companies to a Special Review of Macrocyclic Lactones, commissioned by the Australian National Registration Authority for Agricultural and Veterinary Chemicals. The present paper draws extensively on this material and on more recent unpublished information relating to eprinomectin and the milbemycins (K.G. Wardhaugh, Canberra, 1995, 1997, personal communications).

In considering the ecotoxicology of anthelmintic residues, McKellar (1997) summarized the following as contributory factors to environmental impact.

- Activity, both direct and indirect, of an excreted agent or its metabolites on non-target fauna.
- Amount and temporal nature of excretion of the active agent into the environment.
- Stability and persistence of the excreted residues after they enter the environment.
- Environmental influences, such as sunlight, temperature, rainfall and mechanical disruption, on the processes of physical degradation of excreta.

These factors will be considered in this chapter with respect to the toxicity of the various ML compounds and formulations against individual species of dung-dwelling and dung-dispersing insects, together with the overall impact on population dynamics in pasture ecosystems.

Excretion of ML residues into the environment

In Chapter 2 of this book, Hennessy and Alvinerie have detailed the differences in pharmacokinetic and metabolic behaviour of the MLs, which, in turn, are reflected in the differing faecal excretion profiles for the various formulations and delivery routes of these compounds in cattle and sheep. The pattern of excretion in cattle faeces of ivermectin following subcutaneous dosing at 200 $\mu g\ kg^{-1}$ indicates that peak concentrations may occur from day 1 (Lifschitz *et al.*, 2000) to days 6 or 8 (Cook *et al.*, 1996). These variations may be attributable to dietary differences between studies; Cook *et al.* (1996) showed that ivermectin concentrations in the faeces of pasture-fed cattle were lower throughout the 14 day post-injection period and peaked later than in grain-fed animals. Lifschitz *et al.* (2000) estimated a mean residence time of 6.3 days for excretion of ivermectin in faeces of 10-month-old calves after subcutaneous administration, compared with 7.7 days for doramectin under the same pasture conditions. In this same study, maximum concentrations of doramectin in cattle faeces occurred 4 days after injection and were almost 2.5 times the peak value for ivermectin. In a separate study done under the same pasture conditions with similar calves, Lifschitz *et al.* (1999) found that maximum faecal concentrations of moxidectin occurred during the first day after drug administration at values about 30% higher than for ivermectin; excretion of moxidectin in faeces had a mean residence time of 10.7 days. The parent compound of ivermectin, doramectin and moxidectin was still detectable in faeces 58 days post-treatment at the conclusion of sampling (Lifschitz *et al.*, 1999, 2000).

Percutaneous, or pour-on, administration of ivermectin at 500 $\mu g\ kg^{-1}$ resulted in higher initial concentrations in faeces of cattle, but by 5–7 days these were similar to those following subcutaneous treatment (Sommer and Steffanson, 1993; Herd *et al.*, 1996). In contrast, pour-on treatment of cattle with moxidectin at 500 $\mu g\ kg^{-1}$ resulted in substantially lower faecal residue concentrations than after subcutaneous dosing, and peak levels were not attained until 11 days (see Steel, 1998).

The ivermectin sustained-release bolus for cattle (Ivomec SR bolus, Merial) is designed to deliver ivermectin at 12.7 mg day^{-1} for 135 days in animals weighing between 100 and 300 kg. Following administration of this device to calves under pasture conditions similar to those used by Lifschitz *et al.* (2000), Alvinerie *et al.* (1998) found that faecal ivermectin concentration increased to a peak at 4 days and then declined to a steady-state level at 7 days, which was maintained until 120 days.

Faecal excretion of orally administered ivermectin by sheep is more rapid; by 7 days, cumulative faecal residues range from 69 to >95% of the dose, two-thirds of this being recovered during the first 2 days (Ali and Hennessy, 1996), and 61–69% of these residues are present as the parent drug (Halley *et al.*, 1989b). Doramectin given orally to sheep is also excreted rapidly in the faeces, being about 90% complete by the fifth day (Hennessy *et al.*, 2000). Oral administration of moxidectin to sheep results in initial faecal concentrations ten times higher than those observed in cattle after subcutaneous injection (see Steel, 1998), but by 7 days, drug concentrations in the dung of the two species are similar. At this stage, cumulative excretion accounts for 43% of the dose and moxidectin comprises 25% of the total residues in sheep faeces (Afzal *et al.*, 1994).

Toxicity of ML Residues to Flies and Dung Beetles

Abamectin and ivermectin

Reviews of the toxic effects of MLs on dung-feeding arthropods compiled in the early 1990s dealt almost exclusively with ivermectin and abamectin

(Strong, 1992, 1993). Most of the available information originated from studies in Europe and Australia, and demonstrated that avermectin residues in animal faeces had the potential to affect arthropod development in a variety of ways. Aside from causing mortality of adults and larvae of several species of Diptera and Coleoptera, avermectin residues were implicated in the occurrence of a wide variety of sublethal effects, particularly among the higher Diptera. These included delayed reproductive development, reduced fecundity, disruption of water balance, interference with moulting and emergence, and developmental abnormalities such as fluctuating asymmetry with regard to patterns of wing venation. Among Coleoptera, evidence of sublethal effects was less common and confined mainly to delayed reproductive development and reduced fecundity.

Tables 4.1 and 4.2 summarize results from more recent studies in which dung from ivermectin-treated sheep or cattle was fed to a range of Coleoptera and Diptera species. These studies confirm that the larvae of cyclorrhaphous Diptera generally are more sensitive to ivermectin than are Coleopteran larvae. Mature adult Coleoptera are usually unaffected by ivermectin residues found in dung, and this may be related to differences in the feeding behaviour of adults and larvae. Adult dung beetles are mostly filter feeders and only imbibe very small particles of organic matter (Holter, 2000), whereas their larval stages are bulk feeders and thereby ingest large quantities of organic material. Ivermectin is largely insoluble in water, and attaches strongly to particulate material (Halley *et al.*, 1989a). It is possible, therefore, that adult beetles that feed mainly on bacteria in the fluid portion of dung are exposed to less ivermectin residue than their bulk-feeding larvae.

Data presented in Tables 4.1 and 4.2 indicate considerable variation in sensitivity between species (cf. *Euoniticellus intermedius* and *Onthophagus gazella*) as well as within species (e.g. *E. intermedius*). Within-species differences most probably reflect differences in assay technique, including insect strain and dung quality. However, such variability introduces uncertainty that can be problematic for regulatory authorities; it is also important for those concerned with modelling population effects (see below). In Australia, standardization of assay methodology for the drug registration process in order to minimize the influence of experimental artefacts is currently under consideration (J. Holland, Canberra, 2000, personal communication).

Formulation and delivery route of ivermectin exert a major influence on the period over which the dung of treated livestock may remain toxic. Oral formulations, generally used for sheep, are the least persistent and rarely affect the development of fly or beetle larvae for more than a week post-treatment (Tables 4.1 and 4.2). Subcutaneous treatment of cattle with ivermectin or abamectin increases larval mortality of most species of dung beetle for up to 14 or 21 days; increased fly larval mortality may persist in excess of 28 days depending on the species. Sommer *et al.* (1992) concluded

Table 4.1. Summary of laboratory assays of the effects of residues of abamectin (ABM), ivermectin (IVM), eprinomectin (EPR) and doramectin (DOR) in faeces of cattle and sheep on the development and survival of scarabaeine dung beetles.

Drug	Formulation[a]	Species	Adult mortality	Fecundity	Duration of larval mortality (days)	Source
ABM	SC (cattle)	*Onthophagus binodis*	Yes	Negative	>42	Dadour *et al.* (2000)
IVM	Oral (sheep)	*Euoniticellus fulvus*	Yes/no[b]	Negative (1 day)	2–5	Wardhaugh *et al.* (1993)
		Aphodius constans	—	—	5–7	Kadiri *et al.* (1999)
	SC (cattle)	*Euoniticellus intermedius*	None	No effect	7–14	Fincher (1992)
		Onthophagus gazella	None	No effect	14–21	Fincher (1992)
		Euoniticellus fulvus	None	No effect	<10	Lumaret *et al.* (1993)
		Onthophagus gazella	—	No effect	8–16	Sommer *et al.* (1993)
		Diastellopalpus quinquedens	—	No effect	2–8	Sommer *et al.* (1993)
		Euoniticellus fulvus	—	No effect	7–14	Wardhaugh (1995, personal communication)
		Euoniticellus intermedius	—	No effect	14–21	Kruger and Scholtz (1997)
		Onitis alexis	—	No effect	7–14	Kruger and Scholtz (1997)
		Aphodius haemorrhoidalis	—	—	None	Kadiri *et al.* (1999)
	PO (cattle)	*Euoniticellus intermedius*	None	No effect	7–14	Fincher (1996)
		Onthophagus gazella	None	No effect	14–21	Fincher (1996)
	PO	*Euoniticellus fulvus*	No	No effect	7–14	Wardhaugh (1997, personal communication)
	SR bolus (sheep)	*Onthophagus taurus*	Yes/no[b]	Negative	100	Wardhaugh *et al.* (2001a)
		Euoniticellus fulvus	No	No effect	100	Wardhaugh *et al.* (2001a)
	SR bolus (cattle)	*Onthophagus sagittarius*	Yes	Negative	135	Wardhaugh *et al.* (2001b)
		Aphodius constans	–	–	143	Errouissi *et al.* (2001)
EPR	PO	*Onthophagus taurus*	Yes/no[b]	Negative	7–14	Wardhaugh (1999, personal communication)
DOR	SC	*Dichotomius anaglypticus*	Yes	No effect	—	Galbiati *et al.* (1995)
	Spiked	*Onthophagus gazella*	—	No effect	Yes	Reported in Steel (1998)
	SC	*Onthopagus binodis*	No	Negative	9–18	Dadour *et al.* (2000)

[a]SC, subcutaneous; PO, pour-on; SR, sustained release.
[b]Indicates mortality of newly emerged adults, but not among mature insects.

Table 4.2. Summary of laboratory assays of the effects of residues of abamectin (ABM), ivermectin (IVM), eprinomectin (EPR) and doramectin (DOR) in faeces of cattle and sheep on the survival of species of dung feeding Diptera. See Table 4.1 footnote.

Drug	Formulation	Species	Duration of larval mortality (days)	Source
ABM	SC (cattle)	*M. vetustissima*	16–32	Wardhaugh and Mahon, 1998
IVM	Oral (sheep)	*M. vetustissima*	6–8	Wardhaugh *et al.*, 1993
		N. cornicina	7–9	Kadiri *et al.*, 1999
	Oral (cattle)	*M. vetustissima*	8–16	Wardhaugh and Mahon, 1998
	PO (cattle)	*M. autumnalis*	14	Sommer *et al.*, 1992; Marley *et al.*, 1993;
		M. domestica	7–11	Floate *et al.*, 2001
		H. irritans	14–42	
		N. cornicina	14	
		S. calcitrans	>7	
	SC (cattle)	*M. vetustissima*	16–32	Sommer *et al.*, 1992; Lumaret *et al.*, 1993;
		M. autumnalis	14	Kruger and Scholtz, 1995; Wardhaugh,
		M. domestica	7	1995 (personal communication);
		M. nevilli	49–56	Wardhaugh *et al.*, 1996; Wardhaugh and
		H. irritans	14	Mahon, 1998; Kadiri *et al.*, 1999
		N. cornicina	10–34	
	SR bolus (sheep)	*M. vetustissima*	100	Wardhaugh *et al.*, 2001a
	SR bolus (cattle)	*M. inferior, O. timorensis*	115+	Wardhaugh *et al.*, 2001b
EPR	SC (cattle)	*M. inferior, O. timorensis*	9–15	Wardhaugh *et al.*, 2001b
	PO (cattle)	*H. irritans*	>28	Floate *et al.*, 2001
		M. domestica	>7	
		S. calcitrans	>28	
DOR	SC (cattle)	*M. inferior, O. timorensis*	9–15	Wardhaugh *et al.*, 2001b
		S. pruna	>36	Silva Junior *et al.*, 1997
	PO (cattle)	*H. irritans, M. domestica*	>28	Floate *et al.*, 2001
		S. calcitrans		

that, despite the substantially higher dose rate, the dung of cattle treated with a pour-on ivermectin formulation was less harmful to fly larvae than that of cattle treated with a subcutaneous preparation. There appears to be no such effect on beetles, with similar durations of larval mortality with both formulations (Fincher, 1992, 1996; Sommer *et al.*, 1992).

Maximum effects on dung fauna occur when livestock are treated with sustained-release devices that deliver ivermectin continuously for periods of 100 or more days. These are now available for both sheep and cattle. Wardhaugh *et al.* (2001a) found that the dung of sheep treated with sustained-release capsules was toxic to larvae of the bush fly (*Musca vetustissima*) and two species of dung beetle (*Onthophagus taurus* and *Euoniticellus fulvus*). They also found evidence of mortality among newly emerged beetles, and reduced fecundity in those that survived. Wardhaugh *et al.* (2001b) found that ivermectin residues in the faeces of cattle treated with a sustained-release bolus inhibited fly breeding and caused significant mortality of adult and larval *Onthophagus sagittarius*. More recently, Errouissi *et al.* (2001) have shown that the treatment of cattle with SRIs inhibited the survival of larval of the dung beetle *Aphodius constans* for more than 20 weeks post-treatment.

Several authors have presented evidence of a wide range of important sublethal effects associated with ivermectin usage. Kruger and Scholtz (1997) recorded two- to threefold increases in larval development time among juveniles of *E. intermedius* when reared on dung collected up to 4 weeks after subcutaneous treatment with ivermectin. Delayed larval growth has also been observed in *Onitis alexis* (Kruger and Scholtz, 1997) and *E. fulvus* (Lumaret *et al.*, 1993; K.G. Wardhaugh, Canberra, 1995, 1997, personal communications), although with these species the effect was less marked than observed by Kruger and Scholtz (1997). Sommer *et al.* (1993) have described morphometric abnormalities in the mouthparts of *Diastellopalpus quinquedens* larvae reared in the dung of ivermectin-treated cattle. Floate (1998) and Floate and Fox (1999) have found evidence of cross-trophic effects in which abundance of parasitic wasps was reduced as a consequence of ivermectin-induced reductions of their Dipteran hosts. In a study of the dung fly, *Musca nevilli*, Kruger and Scholtz (1995) showed that whereas dung collected from ivermectin-treated cattle inhibited larval survival for 4–5 weeks after treatment, flies emerging from dung voided in weeks 5–7 sustained up to 60% reduction in fertility.

Doramectin

There is little published information about the toxicity of doramectin against coprophagous beetles and flies (Tables 4.1 and 4.2). Galbiati *et al.* (1995), using dung voided by cattle treated with an injectable formulation of doramectin, observed a significant reduction in the survival of adults of the dung beetle, *Dichotomius anaglypticus*, when fed on faeces collected during the first 10 days after treatment. Steel (1998), citing unpublished work of Clymer and Chappel, noted that in cattle dung spiked with pure doramectin, concentrations in excess of 64 ppb completely inhibited the development of larvae of *O. gazella*. A concentration of 55 ppb was estimated to reduce larval survival by 90%, but concentrations as high as 250 ppb had no detectable effect on the development or survival of adult beetles. Extrapolating from doramectin excretion data, Steel (1998) inferred that faeces of cattle given a subcutaneous formulation of doramectin may remain toxic to larvae of *O. gazella* for at least 2 weeks after treatment. Working with the dung of cattle treated with an injectable preparation of doramectin (200 $\mu g\ kg^{-1}$ live weight), Dadour *et al.* (2000) observed 100% mortality among larvae of *Orthophagus binodis* reared in faeces voided on days 3 and 6 after treatment. There was a significant reduction in larval survival at day 9 post-treatment, but by day 18 differences between treated and control dung were non-significant. Dung collected from cattle dosed subcutaneously with doramectin completely inhibited larval development of the muscid fly, *Sarcopromusca pruna*, for the first 29 days after treatment, and fly emergence at 36 days was still reduced by more than half (Silva Junior *et al.*, 1997). Wardhaugh *et al.* (2001b) measured numbers of fly larvae (mostly *Musca inferior* and *Orthelia timorensis*) in natural pats voided by cattle treated with a pour-on formulation of doramectin and observed reduced larval survival for 9–15 days after treatment. In laboratory bioassays of cattle dung from topically treated cattle, Floate *et al.* (2001) recorded reduced survival of larvae of three species of pest Diptera (*Haematobia irritans*, *Musca domestica* and *Stomoxys calcitrans*) for at least 4 weeks post-treatment.

Eprinomectin

Eprinomectin residues have also been shown to have adverse effects on the survival species of dung-feeding Diptera for periods of 1–4 weeks after treatment (Floate *et al.*, 2001; Wardhaugh *et al.*, 2001b). Wardhaugh *et al.* (2001c) showed that faeces of cattle treated with a pour-on formulation of eprinomectin were toxic to larvae of the dung beetle, *Onthophagous taurus*, for 7–14 days after treatment. Adult beetles were also found to be sensitive to eprinomectin residues, which resulted in increased mortality among newly-emerged insects and reduced fecundity in those that survived.

Moxidectin and milbemycin

Table 4.3 summarizes the published information available for laboratory assays of toxicity of faecal residues of moxidectin. Doherty *et al.* (1994)

Table 4.3. Summary of laboratory assays of the effects of residues of moxidectin and milbemycin in cattle and sheep faeces on the development and survival of dung-breeding Coleoptera and Diptera.

Drug	Formulation[a]	Species	Adult mortality	Fecundity	Duration of larval mortality (days)	Source
		Coleoptera				
Moxidectin	Spiked (cattle)	*Onthophagus gazella*	—	No effect	Yes	Doherty *et al.* (1994)
Moxidectin	Oral (sheep)	*Aphodius constans*	—	—	2–3	Kadiri *et al.* (1999)
Moxidectin	SC (cattle)	*Onthophagus gazella*	None	No effect	None	Fincher and Wang (1992)
		Euoniticellus intermedius	None	No effect	None	Fincher and Wang (1992)
Moxidectin	SC (cattle)	*Euoniticellus fulvus*	None	Reduced	None	Wardhaugh (1997, personal communication)
Moxidectin	SC (cattle)	*Aphodius haemorrhoidalis*	—	—	None	Kadiri *et al.* (1999)
Moxidectin	PO (cattle)	*Onthophagus taurus*	None	No effect	None	Wardhaugh (1999, personal communication)
Milbemycin (CGA 291,046)	SC (cattle)	*Euoniticellus fulvus*	—	Reduced	1–3	Wardhaugh (1995, personal communication)
		Diptera				
Moxidectin	Spiked (cattle)	*Haematobia irritans exigua*	—	—	Yes	Doherty *et al.* (1994)
Moxidectin	Oral (sheep)	*Neomyia cornicina*	—	—	2–3	Kadiri *et al.* (1999)
Moxidectin	SC (cattle)	*Haematobia irritans*	—	—	3	Reported in Steel (1998)
Moxidectin	SC (cattle)	*Musca vetustissima*	—	—	None	Wardhaugh *et al.* (1996)
		Musca domestica	—	—	None	Wardhaugh *et al.* (1996)
Moxidectin	SC (cattle)	*Haematobia irritans*			28	Miller *et al.* (1994)
Moxidectin	SC (cattle)	*Neomyia cornicina*	—	—	10–16	Kadiri *et al.* (1999)
Moxidectin	PO (cattle)	*Haematobia irritans*	—	—	7	Floate *et al.* (2001)
		Musca domestica	—	—	None	Floate *et al.* (2001)
		Stomoxys calcitrans	—	—	None	Floate *et al.* (2001)
Milbemycin (CGA 291,046)	SC (cattle)	*Musca vetustissima*	—	—	3–7	Wardhaugh (1995, personal communication)

[a]SC, subcutaneous; PO, pour-on; SR, sustained release.

added injectable moxidectin formulation to fresh cattle faeces to provide concentrations of moxidectin similar to those expected in naturally voided faeces following subcutaneous injection. Moxidectin had no effect on the fecundity of females of *O. gazella*, and survival of the larval stages was significantly reduced only at the highest concentration of moxidectin (512 ppb). Following oral dosing of moxidectin to sheep, Kadiri *et al.* (1999) observed significant mortality among larvae of *Aphodius constans* when fed on dung collected on days 1 or 2 after treatment; dung collected from day 3 onwards, however, had no effect on survival.

In three other studies using subcutaneous and pour-on formulations in cattle (Table 4.3), moxidectin residues had no adverse effects on survival of dung beetle larvae, although one species, *E. fulvus*, exhibited a slight reduction in fecundity. This species responded similarly to residues of an unregistered milbemycin, CGA 291,046, (K.G. Wardhaugh, Canberra, 1995, personal communication) when given subcutaneously, showing slight reductions in larval survival and development rate, but only in dung collected at day 1 post-treatment. Although Doherty *et al.* (1994) found no effect of moxidectin on pupation by larvae of the buffalo fly, *Haematobia irritans exigua*, in spiked dung until levels exceeded 64 ppb, such concentrations are likely to be exceeded in the faeces of subcutaneously dosed cattle for between 1 and 4 days after treatment (Lifschitz *et al.*, 1999). Strong and Wall (1994) suggested that dung produced on day 2 after treatment with moxidectin might have reduced the survival of cyclorraphous fly larvae in their study, but considered sample sizes too small to be certain. Wardhaugh *et al.* (1996) were unable to detect any effects of moxidectin on the survival of larvae of either *Musca vetustissima* or *Musca domestica* in dung voided from 2 to 35 days after treatment, but in dung collected on day 2 there was a reduction in larval growth rates. Steel (1998) cited the unpublished work of Fincher who recorded no mortality of hornfly larvae, *Haematobia irritans*, beyond day 3 in the dung of treated cattle but, paradoxically, Miller *et al.* (1994) recorded a progressive increase in the mortality of hornfly larvae in dung collected over four successive weeks after moxidectin treatment. Although such prolonged effects appear to be inconsistent with the available faecal excretion data (Lifschitz *et al.*, 1999), Kadiri *et al.* (1999) also found reduced survival of fly larvae (*Neomyia cornicina*) for an extended period (10–16 days) after moxidectin treatment.

The data summarized in Table 4.3 suggest that larval stages of Diptera may be more sensitive to faecally excreted residues of moxidectin than those of the Coleoptera, although there are some species (e.g. *M. vetustissima*) that appear to be unaffected. Overall, the milbemycins appear to be less harmful to fly and beetle larvae tested than the avermectins, but only a limited number of species of dung-inhabiting arthropods have been examined. In a comparative study of the effects of ivermectin, doramectin, eprinomectin and moxidectin on three species of pest Diptera, Floate *et al.*

(2001) concluded that doramectin was more toxic than either ivermectin or eprinomectin (which were roughly equivalent) and all three avermectins were more harmful than moxidectin.

Toxicity of ML Residues to Earthworms

Information on the effects of MLs on earthworms is available only for ivermectin. Halley *et al.* (1989a) used an epigeic earthworm, *Eisenia fetida*, to examine the toxicity of ivermectin in soil at concentrations of 12–200 ppm. They determined the no-effect level to be 12 ppm and concluded that at concentrations likely to occur in the faeces of ivermectin-treated cattle, the drug would have no effect on earthworm survival. This view was challenged by Gunn and Sadd (1994) who observed 100% mortality of *E. fetida* at concentrations of 20 ppm and reduced reproductive success at levels as low as 4 ppm. However, Halley *et al.* (1989a) assayed pure technical grade ivermectin, whereas Gunn and Sadd (1994) used a commercial formulation of ivermectin developed for oral dosing; regrettably, the latter made no allowance for possible effects due to the presence of excipient chemicals. Since *E. fetida* is not normally associated with cattle dung, and neither study considered the possibility of toxicity arising from the actual ingestion of ivermectin, their relevance to the pasture situation and dung-feeding earthworms is most uncertain.

To date, only two systematic studies of the effects of ivermectin on anecic, pasture-living, earthworms have been conducted. Madsen *et al.* (1988), using *Aporrectodea longa* and *Aporrectodea tuberculata*, examined the effects of residues in dung voided 24 h after cattle were treated with an injectable formulation of ivermectin. They found no evidence of an effect due to ivermectin, although their results were compromised, to some degree, by infestations of larvae of the dung-breeding fly, *Scatophaga stercoraria*. Svendsen *et al.* (2000) found no measurable effect on the development and survival of hatchlings of *Lumbricus terrestris* when fed on dung voided by cattle treated with sustained-release boluses of ivermectin.

The results of Svendsen *et al.* (2002) are consistent with available field data on earthworms, all of which indicate a nil effect of ivermectin. Madsen *et al.* (1990) and Barth *et al.* (1994) examined several species of earthworm, including *L. terrestris* and *A. longa*, and could not detect any differences in earthworm abundance in dung voided by cattle dosed with an injectable formulation of ivermectin. Wratten *et al.* (1993) have drawn similar conclusions from field studies involving both injectable and sustained-release formulations of ivermectin. However, it should be noted that the methodology of the latter authors, particularly that used for sampling earthworms, has been the subject of substantive criticism (Holter *et al.*, 1994). Nevertheless, other field studies using an experimental sustained-release bolus showed there was no obvious adverse effect on earthworms due to ivermectin (Wall and Strong, 1987).

Earthworms, especially in temperate climates, are very important in the processes of dung degradation (Holter, 1979, 1983; Hendriksen, 1991a,b). However, their activity is closely dependent on the presence of coprophagous beetles and flies, which are usually the first colonists of fresh dung. Holter (1979) observed that when insects were excluded during the first few days after dung pat deposition, the biomass and abundance of earthworms were reduced, resulting in a decreased rate of dung dispersal. Similar findings have been reported by Gittings *et al.* (1994). Holter (1979, 1983) suggested that the initial feeding by beetles and fly larvae facilitated the process of succession and made dung pats more attractive for earthworm colonization. Thus, although there is no evidence that ivermectin residues exert any direct effect on the development or survival of earthworms, it could be argued that their detrimental effects on other dung-feeding organisms, in particular fly and beetle larvae, may disturb the processes of succession. Madsen *et al.* (1990) attributed observed delays in dung pat degradation to the disruption of early insect activity, particularly that associated with fly larvae. This is an area that merits further study.

Impact of ML Residues on Dung Colonization in the Field

The pioneering field study of Wall and Strong (1987), which drew attention to the potential ecological consequences of long-term ivermectin treatment on the colonization of cattle dung pats on pasture by Coleoptera and Diptera, has generated many investigations on this topic over the past decade. Steel (1998) reviewed the results of 12 field studies in England, Germany, Denmark, Spain, South Africa and Australia of the impact of ivermectin or abamectin treatment given subcutaneously, topically or by an intraruminal sustained-release bolus on the insect fauna of cattle dung. In one of these studies, moxidectin treatment was also examined. Most studies have been undertaken under temperate conditions in the northern hemisphere, but work in South Africa and Spain has provided information of particular relevance to the Scarabaeidae, which are represented in the introduced dung beetle fauna of Australia. Variable responses have been observed in the Coleoptera and Diptera populations colonizing the dung of cattle treated subcutaneously or topically with ivermectin or abamectin, some of which may be attributable to differences in the period of observation after natural or artificial deposition of the dung pat on pasture. However, Steel (1998) drew some general conclusions from the combination of short- and long-term observations extending up to 105 days after dung deposition. Numbers of adult Coleoptera generally were not affected by ivermectin or abamectin residues in dung collected as early as 1 day after subcutaneous or topical treatment; on the other hand, larval abundance generally was reduced in dung deposited up to 17 days after treatment. Reduced numbers of dipteran larvae were also recorded in dung deposited up to 30 days after subcutaneous ivermectin treatment, but differential effects were noted between the suborders of Diptera. Several studies indicated that Nematocera larvae were much less sensitive to ivermectin residues than those of Brachycera and Cyclorrhapha.

Herd *et al.* (1996) suggested that the high dosage rate of the pour-on formulation of ivermectin, which results in high excretion rates during the first few days after treatment, might affect the development or survival of adult beetles exposed during this period. Studies on three species of dung beetle (Fincher, 1996; K.G. Wardhaugh, Canberra, 1997, personal communication) provided no evidence to support this notion. However, in a recent Canadian field study, Floate (1998) recorded significant reductions in the abundance of a diverse array of taxa following the percutaneous treatment of cattle with ivermectin. With several species, decreased emergence was detected in dung collected up to 12–16 weeks after treatment, which is much longer than any previously reported studies.

Studies of dung pat colonization under European conditions during the period of continuous release of ivermectin from an intraruminal bolus (Barth *et al.*, 1993; Strong *et al.*, 1996) showed that effects generally were similar to those observed in dung collected during the first 2–3 weeks following a single subcutaneous or topical dose of ivermectin/abamectin. Larval numbers of Coleoptera were reduced or totally absent, and there was evidence of impaired larval development in Scarabaeidae. Similar effects were observed on the number of Diptera larvae, although again Cyclorrhapha were more adversely affected than Nematocera. In tropical pastures in Malaysia, Wardhaugh *et al.* (2001b) found that naturally voided pats of bolus-treated cattle were devoid of fly larvae (mainly *Musca inferior* and *Orthelia timorensis*). Such findings suggest that the widespread adoption of sustained release technology could have far-reaching effects on the populations of coprophagous arthropods and hence on the processes of dung degradation and nutrient cycling.

In contrast to the effects of ivermectin and abamectin treatments, pats prepared from faeces collected up to 21 days after subcutaneous dosing of cattle with moxidectin did not affect larval numbers of Coleoptera or Diptera species colonizing the dung (Strong and Wall, 1994).

Environmental Stability and Persistence of Excreted ML Residues

The broader impact of excreted residues is likely to vary according to species composition colonizing the dung. In temperate climates, where the dung-degrading fauna is dominated by *Aphodius* spp. and the larvae of dung-feeding Diptera, incorporation of dung into the soil is mainly via earthworms and hence likely to be relatively slow. By comparison, in tropical and subtropical habitats, where dung-burying scarabaeiids tend to be dominant, incorporation of faecal material into the soil is generally rapid and mostly from the base of the pat. Thus, such dung, which may be transported to depths of 20 cm or more, is likely to be little affected by photodegradation or other abiotic factors such as high temperature and heavy rainfall. Because avermectins have low water solubility and bind tightly to particulate material, there is an obvious potential for residues to accumulate in the soil profile. Under conditions of darkness and at temperatures of 22°C or less, ivermectin degrades rather slowly (half-life 93–240 days; Halley *et al.*, 1989a). Even above ground, under tropical conditions in Tanzania (Sommer and Steffansen, 1993) or in European summer conditions (Sommer *et al.*, 1992), no degradation of ivermectin could be detected in pats exposed for 45 days. Hence, effects of residual ivermectin in buried dung, particularly on free-living soil nematodes, could be high. More recently, Sommer and Bibby (2002) have confirmed that residues of ivermectin adversely affect the decomposition rate of dung once it is incorporated in the soil. Whether the observed effects were due to effects on detritivores, microbial decomposers, or both was not determined. Ivermectin residues apparently reduced the numbers of dung-specific nematodes in surface dung (Schaper and Liebisch, 1991; Barth *et al.*, 1993, 1994) but the abundance of these organisms in buried dung was not examined.

Nothing is available in the public domain about the environmental fate of moxidectin or doramectin.

Dung degradation

The effects of avermectin residues on dung degradation and the nutrient economy of pastures have proved to be a very contentious area of research. From an examination of the results of experiments conducted in Denmark, Germany, Scotland, Zimbabwe, Australia, England and France, Steel (1998) concluded that there was no evidence for long-term adverse effects of ivermectin or abamectin residues on the degradation of dung pats or on the accumulation of dung on pasture, despite initial reductions in insect colonization and dung disappearance rate. Much controversy has arisen over the precision with which dung disappearance has been measured, ranging from simple visual assessments to more precise measurements of organic matter loss. Holter (1979) has shown that visual estimates of dung dispersal are highly inaccurate and may fail to detect losses as high as 70%, but, even where the more precise methodology was employed (Madsen *et al.*, 1990; Sommer *et al.*, 1992; Dadour *et al.*, 1999), there was still no consistent evidence that excreted residues of the avermectins have an adverse effect on the processes of dung degradation. The immediate effects of avermectin residues on dung degradation are probably of less ecological consequence than those that may derive from their long-term effects on the populations of dung-degrading arthropods. If these latter effects prove to be substantial, then the processes of dung degradation may change significantly and permanently. This is where further research effort is required.

Long-term Effects on Insect Populations of Pasture Ecosystems

A plethora of information has accumulated on the effects of ML residues on development and survival of dung-feeding arthropods, but little attention has been given to their potential impact on the long-term stability of pastoral ecosystems. Almost 20 years ago, Kunz *et al.* (1983) demonstrated suppression of a population of hornfly, *H. irritans*, via area-wide (960 km^2) treatment of cattle with fenvalerate-impregnated eartags. No such large-scale trial has ever been attempted for the MLs. The most extensive to date is a 2-year study in South Africa by Kruger and Scholtz (1995, 1998a,b) in which effects associated with single, subcutaneous ivermectin treatment of herds of cattle were examined in 80 ha paddocks. In the first year, which coincided with a period of drought, there was evidence of decreased species diversity and evenness for a period of 1–3 months after treatment. Return to the status quo was attributed to the local emergence of insects associated with pre-treatment breeding rather than to immigration from untreated areas. In year 2, which was characterized by above average rainfall, no effect of ivermectin was detected 1 and 3 months after treatment. Kruger and Scholtz (1998b) considered that differences between the years were largely climate-driven, but suggested that the seriousness of the impact of ivermectin also depended on the spatial scale of treatment and numbers of animals treated in a herd.

Because of the high cost of field trials and the almost inevitable uncertainty of their outcome, attention has turned recently to computer modelling, which is being used increasingly in environmental management (see Jorgensen *et al.*, 1995). Models based on a sound ecological knowledge can be powerful tools for understanding the impact of xenobiotics, and often represent the only practical and objective way of assessing likely outcomes at the larger temporal and spatial scales (Sherratt *et al.*, 1998). They are also valuable as tools for setting research priorities and, in the context of the current review, for comparing the effects of different endectocides. Indeed, in their current simplistic form, this might be their greatest asset.

Sherratt *et al.* (1998) and Wardhaugh *et al.* (1998) have designed simple models to assess the impact of endectocidal residues on insect populations. As yet, none of these models makes allowance for sublethal effects associated with particular chemicals, or for the potential of immigration or density-dependent responses to compensate for losses caused by drug treatment. Each model has its own additional limitations, but discussion of these is beyond the scope of the present review.

Sherratt *et al.* (1998) examined, for several typical cattle farming systems in Europe, South Africa and Australia, the quantitative impact of single or multiple treatments of avermectin on a number of dipterous and coleopterous insects. They concluded that cumulative mortality across a breeding season in these systems was unlikely to exceed 25%, which could be a sizeable impost in a season in which factors such as weather are limiting. The model of Wardhaugh *et al.* (1998), which deals only with dung beetle populations associated with a treated herd or flock, has been used to compare the effects of oral and sustained-release formulations of ivermectin, as used in sheep (Wardhaugh *et al.*, 2001a). With oral ivermectin, the model indicates that, even in a worst-case scenario involving four doses at 4-weekly intervals, effects on beetle populations were minimal. In contrast, because the sustained-release capsule can inhibit successful breeding by both local and immigrant beetles over a prolonged period, populations associated with the treated flock may be driven towards local extinction. Simulations indicate that time of treatment in relation to beetle phenology or beetle life-cycle (multivoltine versus univoltine) is a critical determinant of overall effects.

Conclusions

There is now unequivocal evidence to show that the avermectin group of MLs are toxic to larval and young adult stages of several important dung-breeding arthropods. While MLs clearly have the potential to disrupt the ecology of dung fauna if applied at particular stages of the life cycle, and more so if given in persistent formulations or sustained-release devices, our understanding of the impact on local populations is still relatively rudimentary. In a given livestock production system, the magnitude of impact will depend on the frequency and synchrony of livestock treatments, the proportion of animals treated and the kinds of compounds used; climate, species composition, seasonal phenology and compensatory effects associated with density dependence and insect migration will also be important in determining overall effects. In other words, consideration of the principal factors that regulate arthropod populations under natural conditions is needed to quantify any additional risk posed by avermectin usage. Unfortunately, there is still inadequate understanding of many of these complex, interacting processes. However, perhaps improved models can be used to develop simple decision support tools that can enable an appropriate balance at the farm or regional level between optimal management of livestock parasites and minimal impact on pastureland fauna. It is timely for ecologists modelling dung beetle populations and parasitologists modelling worm population dynamics to work more closely together towards this end.

Although the evidence is still limited in comparison with that available for the avermectins, it is apparent that the milbemycin group of MLs, such as moxidectin, are intrinsically less toxic to dung-breeding arthropods tested to date. Hence, their long-term effects on dung beetle populations seem likely to be less than those associated with the avermectins, but studies similar to those of Kunz *et al.* (1983) and Kruger *et al.* (1998a,b) will be needed to confirm this. Moreover, because of differential insecticidal activity of the avermectins and the milbemycins, there is a suggestion that the high potency of the ML moiety against parasites can be retained by structural modifications that simultaneously reduce potency against non-target species. This is another area worthy of more research.

Taking a broader perspective, it is important to recognize that MLs are only one in an armoury of veterinary chemicals that are used on farms to control pests and parasites and should not therefore be considered in isolation when assessing ecological impact on the dung fauna. For example, topical application of synthetic pyrethroids to control external parasites of cattle can have long-term detrimental effects on dung-breeding insect populations (Wardhaugh *et al.*, 1998). Concurrent or sequential use of MLs and synthetic pyrethroids could therefore have profound consequences, both locally and regionally, since such practices diminish the possibility of non-target fauna residing in chemically untreated *refugia* (see Wardhaugh and Ridsdill-Smith, 1998). To enable progress in modelling these possibilities and assessing risks, there is a need to compile information on usage patterns of veterinary chemicals. In Australia, this has commenced for endectocides (Wardhaugh, 2000), and the collection of information on pesticide usage currently is under consideration by government (P. Ryan, Canberra, 2000, personal communication). This approach should establish a register of regional drug usage patterns and help identify animal husbandry systems most likely to promote environmental damage, for example in Australia, dairy and/or beef cattle in areas in which ticks and/or biting flies are endemic; high rainfall sheep/cattle grazing areas where gastrointestinal parasites are endemic.

Finally, there is still a need to understand the effects of MLs on other invertebrate fauna at the dung–soil interface, for example, collembola, mites and free-living nematodes, and to assess 'knock-on' effects in the food chain and on the processes of nutrient cycling. In this context, it is relevant to note the persistence of avermectins in buried dung and therefore the possibility of accumulation within the microenvironment of these fauna. These invertebrate populations play a major role in maintaining the resilience and productivity of our pastoral ecosystems (see Spratt, 1997). When considering the ecotoxic effects likely to flow from the use of MLs in the livestock industry, it needs to be recognized there are great differences within and between countries and continents in agricultural practice and environmental conditions. An interdisciplinary approach, using knowledge of veterinary and agricultural practice, risk management and environmental toxicology, is needed to discriminate between relevant exposure scenarios and far-fetched, worst-case calculations (Montforts *et al.*, 1999).

References

Afzal, J., Stout, S.J., da Cunha, A.R. and Miller, P. (1994) Moxidectin: absorption, tissue distribution, excretion and biotransformation of ^{14}C-labeled moxidectin in sheep. *Journal of Agricultural Food Chemistry* 42, 1767–1773.

Ali, D.N. and Hennessy, D.R. (1996) The effect of level of feed intake on the pharmacokinetic disposition and efficacy of ivermectin in sheep. *Journal of Veterinary Pharmacology and Therapeutics* 19, 89–94.

Alvinerie, M., Sutra, J.F., Galtier, P., Lifschitz, A., Virkel, G., Sallovitz, J. and Lanusse, C. (1998) Persistence of ivermectin in plasma and faeces following administration of a sustained-release bolus to cattle. *Research in Veterinary Science* 66, 57–61.

Anderson, J.R. (1966) Recent developments in the control of some arthropods of public health and veterinary importance. *Bulletin of the Entomological Society of America* 12, 342–348.

Barth, D., Heinze-Mutz, E.M., Roncalli, R.A., Schluter, D. and Gross, S.J. (1993) The degradation of dung produced by cattle treated with an ivermectin slow-release bolus. *Veterinary Parasitology* 48, 215–227.

Barth, D., Heinze-Mutz, E.M., Langholff, W.K., Roncalli, R. and Schluter, D. (1994) Colonization and degradation of dung pats after subcutaneous treatment of cattle with ivermectin or levamisole. *Applied Parasitology* 35, 277–293.

Bulman, G.M., Munoz-Cobenas, M.E. and Ambrustolo, R.R. (1996) The environmental impact of the macrocyclic lactones (endectocides): a comprehensive and comparative update. *Veterinaria Argentina* 13, 490–504.

Cook, D.F., Dadour, I.R. and Ali, D.N. (1996) Effect of diet on the excretion profile of ivermectin in cattle faeces. *International Journal for Parasitology* 26, 291–295.

Dadour, I.R., Cook, D.F. and Neesam, C. (1999) Dispersal of dung containing ivermectin in the field by *Onthophagus taurus* (Coleoptera: Scarabaeidae). *Bulletin of Entomological Research* 89, 119–123.

Dadour, I.R., Cook, D.F. and Hennessy, D. (2000) Reproduction and survival of the dung beetle *Onthophagus binodis* (Coleoptera: Scarabaeidae) exposed to abamectin and doramectin residues in cattle dung. *Environmental Entomology* 29, 1116–1122.

Doherty, W.M., Stewart, N.P., Cobb, R.M. and Kieran, P.J. (1994) *In vitro* comparison of the larvicidal activity of moxidectin and abamectin against *Onthophagus gazella* (F.) (Coleoptera: Scarabaeidae) and *Haematobia irritans exigua* De Meijere (Diptera: Muscidae). *Journal of the Australian Entomological Society* 33, 71–74.

Errouissi, F.R., Alvinerie, M., Galtier, P., Kerboeuf, D. and Lumaret, J.P. (2001) The negative effects of the residues of ivermectin in cattle dung using a sustained-release bolus on *Aphodius constans* (Duft.) (Coleoptera: Aphodiidae). *Veterinary Research* 32, 421–427.

Fincher, G.T. (1992) Injectable ivermectin for cattle: effects on some dung-inhabiting insects. *Environmental Entomology* 21, 871–876.

Fincher, G.T. (1996) Ivermectin pour-on for cattle: effects on some dung-inhabiting insects. *Southwestern Entomologist* 21, 445–450.

Fincher, G.T. and Wang, G.T. (1992) Injectable moxidectin for cattle: effects on two species of dung burying beetles. *Southwestern Entomologist* 17, 303–306.

Floate, K.D. (1998) Off-target effects of ivermectin on insects and on dung degradation in southern Alberta, Canada. *Bulletin of Entomological Research* 88, 25–35.

Floate, K.D. and Fox, A.S. (1999) Indirect effects of ivermectin residues across trophic levels: *Musca domestica* (Diptera: Muscidae) and *Muscidifurax zarapator* (Hymenoptera: Pteromalidae). *Bulletin of Entomological Research* 89, 225–229.

Floate, K.D., Spooner, R.W. and Colwell, D.D. (2001) Larvicidal activity of endectocides against pest flies in dung of treated cattle. *Medical and Veterinary Entomology* 15, 117–120.

Galbiati, C., Conceicao, C.H.C., Florcovski, J.L., Calafiori, M.H. and Tobias, A.C.T. (1995) Effect of anthelmintic injection into cattle on the coprophagous beetle *Dichotomius anaglypticus* (Mann., 1829). *Ecossistema* 20, 100–108.

Gittings, L., Giller, P.S. and Stakelum, G. (1994) Dung decomposition in contrasting temperate pastures in relation to dung beetle and earthworm activity. *Pedobiologia* 38, 455–474.

Gunn, A. and Sadd, J.W. (1994) The effect of ivermectin on the survival, behaviour and cocoon production of the earthworm *Eisenia fetida*. *Pedobiologia* 38, 327–333.

Halley, B.A., Jacob, T.A. and Lu, A.Y.H. (1989a) The environmental impact of the use of ivermectin: environmental effects and fate. *Chemosphere* 18, 1543–1563.

Halley, B.A., Nessel, R.J. and Lu, A.Y.H. (1989b) Environmental aspects of ivermectin usage in livestock: general considerations. In: Campbell, W.C. (ed.) *Ivermectin and Abamectin*. Springer, New York, pp. 162–172.

Hendriksen, N.B. (1991a) Consumption and utilisation of dung by detritivorous and geophagous earthworms in a Danish pasture. *Pedobiologia* 35, 65–70.

Hendriksen, N.B. (1991b) The effects of earthworms on the disappearance of particles from cattle dung pats during decay. *Pedobiologia* 35, 139–146.

Hennessy, D.R., Page, S.W. and Gottschall, D. (2000) The behaviour of doramectin in the gastrointestinal tract, its secretion in bile and pharmacokinetic disposition in the peripheral circulation after oral and intravenous administration to sheep. *Journal of Veterinary Pharmacology and Therapeutics* 23, 203–213.

Herd, R. (1995) Endectocidal drugs: ecological risks and counter-measures. *International Journal for Parasitology* 25, 875–885.

Herd, R.P., Sams, R.A. and Ashcraft, S.M. (1996) Persistence of ivermectin in plasma and faeces following treatment of cows with ivermectin sustained release, pour-on or injectable formulations. *International Journal for Parasitology* 26, 1087–1093.

Holter, P. (1979) Effect of dung beetles (*Aphodius* spp.) and earthworms on the disappearance of cattle dung. *Oikos* 32, 393–402.

Holter, P. (1983) Effect of earthworms on the disappearance rate of cattle droppings. In: Satchell, J.E. (ed.) *Earthworm Ecology*. Chapman and Hall, London, pp. 49–57.

Holter, P. (2000) Particle feeding in *Aphodius* dung beetles (Scarabaeidae): old hypotheses and new experimental evidence. *Functional Ecology* 14, 631–637.

Holter, P., Strong, L., Wall, R., Wardhaugh, K. and Herd, R. (1994) Effects of ivermectin on pastureland ecology. *Veterinary Record* 135, 211–212.

Jorgensen, S.E., Halling-Sorenson, B. and Nielsen, S.N. (1995) *Handbook of Environmental and Ecological Modeling*. Lewis, Boca Raton, Florida.

Kadiri, N., Lumaret, J.P. and Janati-Idrissi, A. (1999) Macrocyclic lactones: impact on non-target fauna in pastures. *Annales de la Societe Entomologique de France* 35, 222–229.

Kruger, K. and Scholtz, C.H. (1995) The effect of ivermectin on the development and reproduction of the dung-breeding fly *Musca nevilli*. Kleynhans (Diptera, Muscidae). *Agriculture, Ecosystems and Environment* 53, 13–18.

Kruger, K. and Scholtz, C.H. (1997) Lethal and sublethal effects of ivermectin on the dung-breeding beetles *Euoniticellus intermedius* (Reiche) and *Onitis alexis* Klug (Coleoptera, Scarabaeidae). *Agriculture, Ecosystems and Environment* 61, 123–131.

Kruger, K. and Scholtz, C.H. (1998a) Changes in the structure of dung insect communities after ivermectin usage in a grassland ecosystem – I – impact of ivermectin under drought conditions. *Acta Oecologica* 19, 425–438.

Kruger, K. and Scholtz, C.H. (1998b) Changes in the structure of dung insect communities after ivermectin usage in a grassland ecosystem – II – impact of ivermectin under high rainfall. *Acta Oecologica* 19, 439–451.

Kunz, S.E., Kinzer, H.G. and Miller, J.A. (1983) Area wide cattle treatments on populations of horn flies (Diptera: Muscidae). *Journal of Economic Entomology* 76, 525–528.

Lifschitz, A., Virkel, G., Imperiale, F., Sutra, J.F., Galtier, P., Lanusse, C. and Alvinerie, M. (1999) Moxidectin in cattle: correlation between plasma and target tissues. *Journal of Veterinary Pharmacology and Therapeutics* 22, 266–273.

Lifschitz, A., Virkel, G., Sallovitz, J., Sutra, J.F., Galtier, P., Alvinerie, M. and Lanusse, C. (2000) Comparative distribution of ivermectin and doramectin to parasite location tissues in cattle. *Veterinary Parasitology* 87, 327–338.

Lumaret, J.P., Galante, E., Lumbreras, C., Mena, J., Bertrand, M., Bernal, J.L., Cooper, J.F., Kadiri, N. and Crowe, D. (1993) Field effects of ivermectin residues on dung beetles. *Journal of Applied Ecology* 30, 428–436.

Madsen, M., Gronvold, J., Nansen, P. and Holter, P. (1988) Effects of treatment of cattle with some anthelmintics on the subsequent degradation of their dung. *Acta Veterinaria Scandinavica* 29, 515–517.

Madsen, M., Overgaard Nielsen, B.O., Holter, P., Pedersen, O.C., Jespersen, J.B., Jensen, K.M.V., Nansen, P. and Gronvold, J. (1990) Treating cattle with ivermectin: effects on the fauna and decomposition of dung pats. *Journal of Applied Ecology* 27, 1–15.

Marley, S.E., Hall, R.D. and Corwin, R.M. (1993) Ivermectin cattle pour-on: duration of a single late spring treatment against horn flies, *Haematobia irritans* (L) (Diptera: Muscidae) in Missouri, USA. *Veterinary Parasitology* 51, 167–172.

McKellar, Q.A. (1997) Ecotoxicology and residues of anthelmintic compounds. *Veterinary Parasitology* 72, 413–435.

Miller, J.A., Oehler, D.D. and Scholl, P.J. (1994) Moxidectin: pharmacokinetics and activity against horn flies (Diptera: Muscidae) and trichostrongyle nematode egg production. *Veterinary Parasitology* 53, 131–143.

Montforts, M.H.M.M., Kalf, D.F., vanVlaardingen, P.L.A. and Linders, J.B.H.J. (1999) The exposure assessment for veterinary medicinal products. *Science of the Total Environment* 225, 119–133.

Schaper, R. and Liebisch, A. (1991) Effect of treating grazing cattle with a systemic parasiticide (Ivermectin) on the fauna and decomposition of their dung. *Tierarztliche Umschau* 46, 12–18.

Sherratt, T.N., Macdougall, A.D., Wratten, S.D. and Forbes, A.B. (1998) Models to assist the evaluation of the impact of avermectins on dung insect. *Ecological Modelling* 110, 165–173.

Silva Junior, V.P. da, Borja, G.E.M. and Sanavria, A. (1997) Residual effect of doramectin and ivermectin on the development of *Sarcopromusca pruna* (Shannon & Del Ponte, 1926) (Diptera: Muscidae) in cattle faeces. *Arquivos da Faculdade de Veterinaria, UFRGS* 25, 85–91.

Sommer, C. and Bibby, B.M. (2002) The influence of veterinary medicines on the decomposition of dung organic matter in soil. *European Journal of Soil Biology* (in press).

Sommer, C. and Steffansen, B. (1993) Changes with time after treatment in the concentrations of ivermectin in fresh cow dung and in cow pats aged in the field. *Veterinary Parasitology* 48, 67–73.

Sommer, C., Steffansen, B., Overgaard Nielsen, B., Gronvold, J., Vagn Jensen, K.-M., Brochner Jespersen, J., Springborg, J. and Nansen, P. (1992) Ivermectin excreted in cattle dung after subcutaneous injection or pour-on treatment: concentrations and impact on dung fauna. *Bulletin of Entomological Research* 82, 257–264.

Sommer, C., Gronvold, J., Holter, P. and Nansen, P. (1993) Effects of ivermectin on two afrotropical dung beetles, *Onthophagus gazella* and *Diastellopalpus quinquedens* (Coleoptera: Scarabaeidae). *Veterinary Parasitology* 48, 171–179.

Spratt, D.M. (1997) Endoparasite control strategies: implications for biodiversity of native fauna. *International Journal of Parasitology* 27, 173–180.

Steel, J.W. (1998) Assessment of the effects of the macrocyclic lactone class of chemicals on dung beetles and dung degradation in Australia. In*: NRA Special Review of Macrocyclic Lactones.* National Registration Authority for Agricultural and Veterinary Chemicals, Canberra, Australia, pp. 15–79.

Strong, L. (1992) Avermectins: a review of their impact on insects of cattle dung. *Bulletin of Entomological Research* 82, 265–274.

Strong, L. (1993) Overview: the impact of avermectins on pastureland ecology. *Veterinary Parasitology* 48, 3–17.

Strong, L. and Brown, T.A. (1987) Avermectins in insect control and biology: a review. *Bulletin of Entomological Research* 77, 357–389.

Strong, L. and Wall, R. (1994) Effects of ivermectin and moxidectin on the insects of cattle dung. *Bulletin of Entomological Research* 84, 403–409.

Strong, L., Wall, R., Woolford, A. and Djeddour, D. (1996) The effect of faecally excreted ivermectin and fenbendazole on the insect colonisation of cattle dung following the oral administration of sustained-release boluses. *Veterinary Parasitology* 62, 253–266.

Svendsen, T.S., Sommer, C., Holter, P. and Grønvold, J. (2002) Survival and growth of *Lumbricus terrestris* fed on dung from cattle given sustained-release boluses of ivermectin or fenbendazole. *European Journal of Soil Biology* (in press).

Wall, R. and Strong, L. (1987) Environmental consequences of treating cattle with the antiparasitic drug ivermectin. *Nature* 327, 418–421.

Wardhaugh, K.G. (2000) Parasiticides registered for use in cattle in Australia – an annotated bibliography and literature guide prepared for the National Dung Beetle Planning Forum. *CSIRO Entomology Contracted Report No 56.* CSIRO Entomology, Canberra, Australia.

Wardhaugh, K.G. and Mahon, R.J. (1998) Comparative effects of abamectin and two formulations of ivermectin on the survival of larvae of a dung-breeding fly. *Australian Veterinary Journal* 76, 270–272.

Wardhaugh, K.G. and Ridsdill-Smith, T.J. (1998) Antiparasitic drugs, the livestock industry and dung beetles – cause for concern? *Australian Veterinary Journal* 76, 259–261.

Wardhaugh, K.G., Mahon, R.J., Axelsen, A., Rowland, M.W. and Wanjura, W. (1993) Effects of ivermectin residues in sheep dung on the development and survival of the bush fly, *Musca vetustissima* Walker and a scarabaeine dung beetle, *Euoniticellus fulvus* Goeze. *Veterinary Parasitology* 48, 139–157.

Wardhaugh, K.G., Holter, P., Whitby, W.A. and Shelley, K. (1996) Effects of drug residues in the faeces of cattle treated with injectable formulations of ivermectin and moxidectin on larvae of the bush fly, *Musca vetustissima* and the house fly *Musca domestica. Australian Veterinary Journal* 74, 370–374.

Wardhaugh, K.G., Longstaff, B.C and Lacey, M.J. (1998) Effects of residues of deltamethrin in cattle faeces on the development and survival of three species of dung-breeding insect. *Australian Veterinary Journal* 76, 273–280.

Wardhaugh, K.G., Holter, P. and Longstaff, B.C. (2001a) The development and survival of three species of coprophagous insect after feeding on the faeces of sheep treated with controlled-release formulations of ivermectin or albendazole. *Australian Veterinary Journal* 79, 125–132.

Wardhaugh, K.G., Mahon, R.J. and Hamdan bin Ahmad (2001b) Efficacy of macrocyclic lactones for the control of larvae of the Old World Screw-worm fly, *Chrysomya bezziana. Australian Veterinary Journal* 79, 120–124.

Wardaugh, K.G., Longstaff, B.C. and Morton, R. (2001c) A comparison of the development and survival of the dung beetle, *Onthophagus taurus* (Schreb.), when fed on faeces of cattle treated with pour-on formulations of eprinomectin or moxidectin. *Veterinary Parasitology* 99, 155–168.

Wratten, S.D. and Forbes, A.B. (1996) Environmental assessment of veterinary avermectins in temperate pastoral ecosystems. *Annals of Applied Biology* 128, 329–348.

Wratten, S.D., Mead-Briggs, M., Gettinby, G., Ericsson, G. and Baggott, D.G. (1993) An evaluation of the potential effects of ivermectin on the decomposition of cattle dung pats. *Veterinary Record* 133, 365–371.

Chapter 5

Resistance Against Macrocyclic Lactones

R.K. Prichard

Introduction

Resistance to macrocyclic lactone (ML) anthelmintics is a serious problem for the control of trichostrongylid nematodes of sheep and goats and is becoming a problem in trichostrongylid nematodes of cattle in some parts of the world, particularly in Latin America. Selection for resistance is a natural evolutionary process, and the speed with which resistance becomes apparent depends on the genetic diversity of the different parasite species under selection, the selection pressure and time.

Parasites in a population do not respond uniformly to treatment. A dose titration assay will almost always reveal a gradient of responses to increasing anthelmintic concentration until high efficacy is achieved. This is due to genetic diversity. Furthermore, treatment changes the genetic composition by eliminating parasites whose genotype renders them susceptible, and the eliminated worms cannot pass on their 'susceptibility' genes to succeeding generations. However, parasites that are resistant to the dose survive, and can pass on their 'resistance' genes. Generally, when we refer to 'susceptibility' or 'resistance' genes, we are referring to alleles of relevant genes that confer susceptibility or resistance to an antiparasiticide. When a new class of anthelmintic is discovered, it may be developed because the majority of the target parasites have 'susceptibility' alleles and the drug is highly effective. The frequency and intensity (dose dependent) of treatment and the extent of dilution of 'resistance' alleles by 'susceptibility' alleles in the reproducing population, by parasites that establish in the host after the treatment, will determine the rate of selection for resistance phenotype.

Genetic Diversity in Nematodes

The genetic composition of an organism includes the nuclear and mitochondrial (mt) DNA. The complete sequence of mtDNA is known for the nematodes *Ascaris suum*, *Caenorhabditis elegans* (Okimoto *et al.*, 1992) and *Onchocerca volvulus* (Keddie *et al.*, 1998), and sequence information on a number of trichostrongylid nematodes have been compared (Blouin *et al.*, 1995). Usually the mtDNA is more variable than nuclear DNA. Nematode mtDNA tends to be very AT rich (up to 80%), and it has been thought that this bias is due to a high rate of mutation (Hugall *et al.*, 1997). This conclusion has been reinforced by analyses of nematode phylogenetic trees, which indicate a high rate of mtDNA evolution relative to other phyla (Anderson *et al.*, 1998), and mtDNA diversity in trichostrongylid nematodes of ruminants is up to ten times higher than that found in vertebrates (Blouin *et al.*, 1995).

Haemonchus contortus is the most studied trichostrongylid nematode. Numerous studies on nuclear DNA suggest that *H. contortus* is extremely diverse (Beech *et al.*, 1994; Blackhall *et al.*, 1998a,b; Hoekstra *et al.*, 1999; Sangster *et al.*, 1999; Blackhall, 2000). Genetic diversity can be within a population or between populations (geographically separated or drug selected). Studies in four species of trichostrongylid nematodes indicated that 96–99% of nucleotide diversity is found within populations (Blouin *et al.*, 1992). *H. contortus* shows great genetic diversity both within populations and between distinct isolates. Polymorphism depends on the mutation rate, the effective size of the population and, in the case of parasites, the rates of host migration.

Trichostrongylid nematodes are extremely successful parasites, found in many domesticated host species from the humid tropics to cool temperate climates. Ruminants have been domesticated throughout the world and represent a huge host population for trichostrongylid nematodes. Each ruminant may harbour hundreds to thousands of trichostrongylid nematodes. These parasites are prolific breeders, with one female worm typically producing 1000–10,000 eggs per day, depending on the particular species. The population size of these nematodes on pasture is often much greater than that within ruminant hosts. These factors mean that usually the effective population size of trichostrongylid nematodes on a farm is huge. This huge population size, high reproduction rate and the enormous range of habitats are conducive to a high level of genetic diversity. As a result, the repeated use of ML anthelmintics will select for those rare individual nematodes that have an appropriate combination of alleles to be able to survive the normal dose rate, reproduce and generate a resistant strain of parasites.

In order to determine the mechanisms of resistance to MLs, it is necessary to have an appreciation of the way ML anthelmintics exert their effects and of the genes likely to be under selection pressure.

Genes Involved in the Mechanisms of Action of Macrocyclic Lactone Anthelmintics

The MLs paralyse the pharynx, the body wall and the uterine muscles of nematodes. In adult *H. contortus*, the pharynx appears to be the most sensitive of these muscles (Geary *et al.*, 1993). Early studies on the mechanism of action of avermectins showed that avermectins are γ-aminobutyric acid (GABA) receptor antagonists, blocking hyperpolarization of nematode somatic muscle membranes (Holden-Dye *et al.*, 1988; Holden-Dye and Walker, 1990). Recently, GABA receptors have been implicated in the action of ivermectin on insects (Ludmerer *et al.*, 2002). A putative GABA receptor subunit (HG1) has been localized along the ventral nerve cord and in several neurones in the head, possibly nerve ring neurones, in *H. contortus* (Skinner *et al.*, 1998). If such a ligand-gated receptor is involved in the action of MLs, it would be consistent with earlier work (Kass, 1984), which indicated that paralysis of nematode body muscle by MLs resulted from a hyperpolarization of neurones, and thus from an inhibition of excitatory signals to the muscles rather than from a direct inhibition of body muscle cells.

Another class of chloride ion channels in invertebrates is the glutamate-gated chloride (GluCl) channels (Cull-Candy and Usherwood, 1973), which were first cloned from the free-living nematode *C. elegans* (Cully *et al.*, 1994). The GluClα-subunit gene *avr-15* is expressed in the pharynx of *C. elegans* (Dent *et al.*, 1997; Laughton *et al.*, 1997) and has ivermectin and glutamate receptors (Dent *et al.*, 1997; Vassilatis *et al.*, 1997). Two alternatively spliced GluCl subunit cDNAs (*HcGluCla* and *HcGluClb*) have been cloned from *H. contortus* (Forrester *et al.*, 1999), and the longer sequence has been expressed and found to bind glutamate and, separately, ivermectin with high affinity (K_d ~10^{-10} M) (Forrester *et al.*, 2000). Three other *H. contortus* GluCl subunit cDNAs have been sequenced, *Hc-GBR2A* and its alternatively spliced *Hc-GBR2B* (Jagannathan *et al.*, 1999), which have high homology to *avr-14* from *C. elegans*, and *HG4* (Delany *et al.*, 1998). However, it is not yet known whether *HcGRB2* or *HG4* have glutamate- or ivermectin-binding sites. *HcGRB2* expression has been localized to the nerve ring, the ventral and dorsal nerve cords, the anterior portion of the dorsal sublateral cord and motor neurone commissures in *H. contortus* (Jagannathan *et al.*, 1999), while antibody to *HG4* localized its expression to motor neurone commissures in the anterior portion of *H. contortus* from the nerve ring to just anterior of the vulva, including possible nerve cord staining, but no expression on pharyngeal muscle was detected (Delany *et al.*, 1998).

Different GluCl subunits may show variable sensitivity to MLs and different sites of expression, which could account for the paralytic effects of MLs on different neuromuscular systems at different MLs concentrations. This may account for differential sensitivity of nematode life cycle stages and perhaps, in part, differential potency between avermectins and milbemycins. The level of expression of *H. contortus* GluClα was found to be adult > L_3 > eggs (Forrester *et al.*, 1999). It has been found that feeding is inhibited by MLs in adult *H. contortus* at about 10^{-9} M (Geary *et al.*, 1993; Paiement *et al.*, 1999a) and in developing larvae (L_1 and L_2) by 10^{-8} M (Kotze, 1998; Sangster and Gill, 1999), whereas migration of *H. contortus* L_3 is inhibited by ML at 3×10^{-7} M (Gill *et al.*, 1998; Molento, 2000). *In vivo*, susceptible *H. contortus* are expelled 8–10 h after ivermectin treatment, suggesting that paralysis of body wall muscle may be critical for this rapid expulsion, even though paralysis of pharyngeal muscle is more sensitive (Sangster and Gill, 1999). However, ML anthelmintics persist, at decreasing concentrations, for several days. As the ML concentration decreases, motility may be regained, but paralysis of the pharynx and resultant inhibition of feeding may endure longer than body muscle paralysis and contribute to worm deaths. These differences in sensitivity to MLs between pharynx and body muscle may be due to different GluCl subunits or GABA receptor subunits being expressed differentially in different tissues and life cycle stages. The accumulating evidence suggests that the products of several genes are involved in the mechanism of action of the MLs.

Macrocyclic Lactone Resistance Studies in *C. elegans*: Implications for Parasitic Nematodes

A great deal of information on the targets for ML action and on the involvement of different genes in the ML-resistance phenotype has been gained by studying gene knockouts and mutations in the free-living nematode *C. elegans*. This approach has been used to identify over 20 loci in this non-parasitic nematode which can confer low level ML resistance, while deletion or mutation of additional loci can confer high level ML resistance (Blaxter and Bird, 1997). Recent findings in *C. elegans* have led to the following hypothesis for ivermectin action on the pharynx of this worm (see Fig. 5.1), resulting in starvation, and for ivermectin resistance in *C. elegans* (Dent *et al.*, 2000). Three GluClα-subunit products of the genes *avr-15* (*GluClα2*), *avr-14* (*GluClα3*) and *GLC-1* (*GluClα1*) may respond to ivermectin. *avr-15* is located in the pharyngeal muscle, allowing ivermectin to act directly on the pharynx. In contrast, *avr-14* and *GLC-1* are located on extrapharyngeal neurones, which are connected to the pharyngeal cells via linking neurones in which the gap junction innexins, produced by the genes *unc-7* and *unc-9*, which do not bind ivermectin themselves, convey the hyperpolarization signal from the extrapharyngeal neurones to the pharynx. In this model, the amphid dye filling Dyf gene, *osm-1*, and other Dyf genes may act additively to regulate ivermectin uptake and access to the various GluCl/ivermectin receptors, especially AVR-14 (Dent *et al.*, 2000).

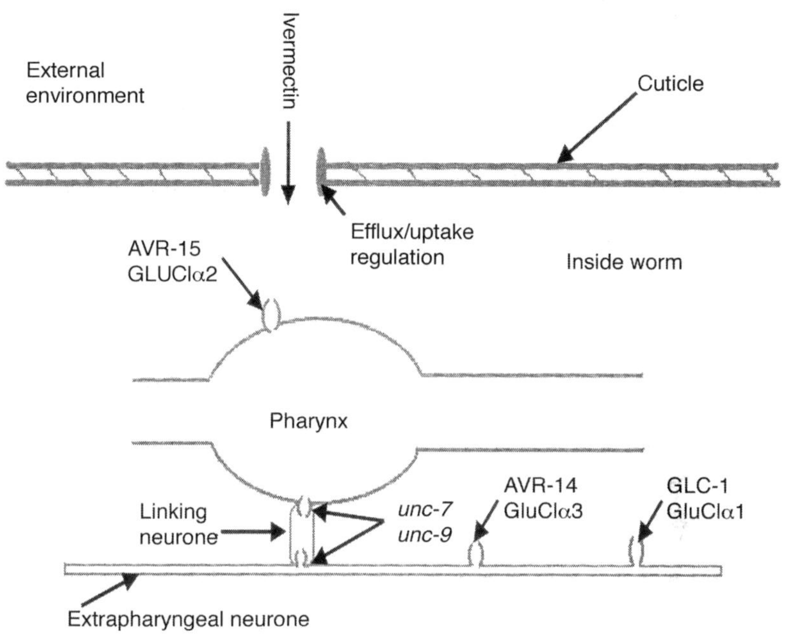

Fig. 5.1. Proposed model of mechanism of ivermectin action on the pharynx of *C. elegans* (modified from Dent *et al.*, 2000). Ivermectin in the worm acts on various glutamate-gated chloride (GluCl) channels. AVR-15 (GluClα2) is essential for the direct action of ivermectin on paralysis of the pharynx. AVR-14 (GluClα3) and GLC-1 (GluClα1) are not essential for the action of ivermectin on the pharynx, but contribute to its effects.

Many of the *C. elegans* mutants that are resistant to MLs have mobility, egg laying, amphid and pharyngeal dysfunctions that would be lethal in a parasite and thus be irrelevant to resistance in parasites. High level ML resistance in *C. elegans* is also not likely to be relevant to resistance in parasites in which an increased dose of ivermectin or the recommended dose of moxidectin frequently are sufficient to eliminate populations of parasitic nematodes resistant to the recommended dose rate of ivermectin. Gene knockouts and induced mutations cannot be used readily to study genes associated with resistance in parasites. However, *C. elegans* can help our understanding of genetic associations with resistance in parasitic nematodes by identifying possible candidate gene homologues for further study in parasites. These studies can provide insight into the possible mechanisms of ML resistance in parasitic nematodes, but will not necessarily replicate the mechanisms of resistance important in parasitic nematodes. Susceptibility in *C. elegans* is represented by death of the nematode living on a nutrient-rich agar plate when the free-living worms are exposed to an ML. The death of the *C. elegans* is most probably due to paralysis of the pharynx by the ML and the starvation of the worm.

Susceptibility in a parasitic nematode, for example in the gastrointestinal tract, may be represented by a similar pharyngeal paralysis and starvation, but is more likely to be due to paralysis of body wall muscle and removal of the parasite by the peristaltic action of the gastrointestinal tract. As noted above, *H. contortus* may be expelled within 8 h after ivermectin treatment (Sangster and Gill, 1999), suggesting that body wall paralysis may be more important for initial removal of parasitic worms in the gastrointestinal tract than inhibition of feeding. In the case of filarial nematodes living in the tissues, in which the females move very little and nutrient absorption occurs through the cuticle, a major effect of MLs on adult worms is probably paralysis of the uterine muscle resulting in disruption of reproduction (Grant, 2000; Prichard, 2000).

Determination of Genes Associated with Macrocyclic Lactone Resistance in Parasitic Nematodes

Some methods for assessing genetic variation in parasitic nematodes have been reviewed previously (Grant, 1994). However, a number of recent approaches are particularly relevant to the study of anthelmintic resistance. A common approach to studying associations between different loci and resistance has been to look for DNA sequence differences in independent drug-sensitive and resistant populations of parasitic nematodes by methods such as restriction fragment length polymorphism (RFLP), single-strand conformational polymorphism (SSCP) and direct sequencing. While this approach can indicate that different populations differ at a particular locus, it is not specific for anthelmintic resistance unless the resistant population is derived from the susceptible population. Differences in a locus between susceptible and resistant populations that are genetically independent, for example from different geographic locations, may simply be a reflection of the genetic isolation of the populations. However, when a resistant population has been derived from a susceptible population solely by drug treatment, the resistant and susceptible populations will share the same genetic background except for the effects of the drug selection. Such comparisons of resistant strains with parental susceptible strains have been used successfully to compare polymorphism between ivermectin- and moxidectin-selected strains, and parental unselected strains (Blackhall *et al.*, 1998a,b) to elucidate genes associated with ML resistance in *H. contortus*.

Another approach to identifying genes associated with anthelmintic resistance in *H. contortus* has been to produce hybrids by crossing anthelmintic-resistant *H. contortus* with susceptible *H. placei* (Le Jambre *et al.*, 1999). Hybrid males are sterile, but females can be backcrossed with *placei* to give a background of *placei*, but with resistance-associated genes coming from the *contortus* ancestor following anthelmintic treatment.

Once the backcrosses have been made, the anthelmintic-selected hybrids can be screened for any candidate gene to see if that gene is *contortus* or *placei* by sequence analysis and if the *contortus* sequence occurs at a higher frequency than expected. This can be a definitive means of showing the association of a gene with resistance. Results must, however, be interpreted with some qualifications. The association does not necessarily mean that the gene, showing a frequency significantly different from that expected from random matings, is itself involved in resistance; it may be located near a gene that is involved in resistance and be linked during segregation. Another qualification is that the sequence analysed always be uniquely different between *contortus* and *placei*. In the analysis conducted by Le Jambre *et al.* (1999), approximately 700 bp were sequenced. No polymorphism at all was described for *H. placei* in this fragment, even though the *H. placei* population was not constricted by drug selection. This is surprising given the high diversity of unselected *H. contortus* and other parasitic nematodes. The anthelmintic-resistant *H. contortus* showed two polymorphs for this fragment. This limited polymorphism for the *H. contortus* could be expected, as the CAVRS strain of *H. contortus* used previously had been subjected to severe bottlenecking during the experimental ivermectin resistance selection protocol. On the basis of the polymorphism observed in the resistant *H. contortus* strain and the *H. placei*, two restriction sites were selected for analysis, a *Cla*I site that was apparently unique to the single *placei* polymorph identified and another *Bam*HI site that mostly occurred in *placei*, but was also present in the less frequent *contortus* allele. Should the *H. placei* be more diverse in the fragment analysed and contain a low frequency of alleles without the *Cla*I site, these alleles would be counted as *contortus* rather than *placei*, leading to an overestimate of *contortus* alleles in the anthelmintic-selected backcross population. This could produce a misleading result. This method, therefore, requires a careful analysis of the polymorphism in the fragment being analysed in each population for results to be interpreted unequivocally. A third qualification is that, as with most of the methods discussed for analysing associations between different genes and anthelmintic resistance, the results are relevant only to the populations being investigated. Nevertheless, the hybridization technique can be a powerful tool for investigating genetic associations with anthelmintic resistance.

Macrocyclic Lactone Resistance-associated Genes in *H. contortus*

Genes involved in the mechanism of resistance may be, but will not necessarily be involved in the mechanism of action of an anthelmintic. Resistance mechanisms can broadly be considered as due to changes in the drug receptor, or modulation of drug concentration. Genes that code for transport or metabolism, for example, can be involved in drug resistance but have no direct role in the mechanism of action. If the mechanism of resistance is due solely to mutation(s) in genes involved in the mode of action, anthelmintics with the same mode of action are likely to share the same mechanism of resistance. However, if the mechanism of resistance is entirely or partly due to modulation of drug concentration, different anthelmintics in the same mode-of-action class may not necessarily show the same level of resistance because of the effect of different chemical substituents on transport or metabolism.

Different avermectin-resistant strains of *H. contortus* show different phenotypes (Gill *et al.*, 1998), and it has been argued that these differences may reflect differences in selection pressures due either to treatment at the recommended dose rate, treatment at the LD_{95} (or other level below the recommended dose rate), low continuous exposure from a long-acting bolus formulation or due to pressure on larvae rather than on adult stages being the target of selection (Gill and Lacey, 1998). These findings suggest that, as in *C. elegans*, more than one gene may contribute to ML resistance in *H. contortus*. Selection pressure may well influence the selection on different genes involved in avermectin resistance. However, other factors such as the genetic diversity of different populations under selection and the ML used in the selection may also influence which genes respond to ML selection in a given parasite population. Thus different ML-resistant populations may not show identical genetic responses.

Crossing the avermectin-resistant CAVRS strain of *H. contortus* with a susceptible strain and then assessing phenotype with the larval development assay has shown that resistance in this strain is completely dominant and mainly under the control of a major gene. However, in the larvae, this gene mapped to an autosomal locus, whereas in adult *H. contortus* expression of resistance was sex linked, being stronger in females than males (Le Jambre *et al.*, 2000). The differences between the genetics of resistance in larvae and adults suggest that resistance may involve more than a single gene, as has been concluded for this strain. The CAVRS resistant strain was selected experimentally in which the progeny of the very few survivors of a treatment with ivermectin at 0.2 mg kg^{-1} were cultured in the laboratory and used to infect worm-free sheep in a closed experimental system. Such an experimentally selected strain should be distinguished on two grounds from isolates in the field that are ML resistant. First, the number of survivors of this closed selection process was very small and their genetic diversity would have been greatly restricted, not only in terms of any genes that may be involved with avermectin resistance but also in the diversity of all genes. This process is known as bottlenecking. Secondly, if one allele by itself can confer sufficient resistance to allow some of the adult worms to survive, only a single gene will appear to be responsible for resistance. However, there may be several genes that can contribute to ML resistance and, in a field selection process in which several generations of ML selection will have taken place before a lack of control is apparent and resistance is confirmed, breeding between different generations of survivors of treatment and nematodes *in refugia* at the time of treatment, will have occurred. More than a single allele on a single gene may have been selected in ML-resistant worms. Where ivermectin resistance has been found in the field, typically there have been 5–30 treatments with a ML anthelmintic (Shoop, 1993).

One can also question whether another experimental selection technique, in which several generations of selection occurred, as typically occurs in field isolates, but with doses of ML used below the recommended dose rate (Egerton *et al.*, 1988; Wang *et al.*, 1999), also exactly reflects the situation of ML resistance found in the field. Such selection with sub-LD_{95} dose rates is likely to select for all of the alleles on all of the genes that can contribute to ML resistance in the population under selection. Analyses of strains selected in this way is likely to reveal all of the potential resistance-associated genes, but fail to distinguish which gene may have the largest effect in field-selected ML resistance.

P-Glycoproteins

The first gene found to be associated with ML resistance in *H. contortus* was a P-glycoprotein (Pgp) gene, PGP-A (Blackhall *et al.*, 1998a; Xu *et al.*, 1998), found in two strains of ivermectin- (IVC and IVF17) and one strain of moxidectin-selected (MOF17) worms compared with their unselected parental strains (BBH in the case of IVC, and PF17 in the case of IVF17 and MOF17, respectively). These three selected strains were produced in the USA (Rohrer *et al.*, 1994; Wang *et al.*, 1999) by repeated treatment with sub-LD_{95} dose rates of ivermectin or moxidectin. Subsequently, it was found independently that there was selection on the same PGP-A gene (referred to as A27) and another Pgp gene (A28) in two experimentally selected ivermectin-resistant strains (Warren and CAVRS) in Australia (Sangster *et al.*, 1999). Another Pgp gene in *H. contortus*, hcpgp-1, has been investigated for association with ivermectin resistance. A limited RFLP analysis, on a multidrug-resistant (benzimidazole/ivermectin/closantel) isolate (RSA) from South Africa, indicated that hcpgp-1 was not associated with resistance (Kwa *et al.*, 1998). However, using the *H. contortus*/*H. placei* backcross method, it was found that hcpgp-1 has been selected by ivermectin in the experimentally produced CAVRS resistant strain (Le Jambre *et al.*, 1999). However, the latter authors concluded that although this gene is associated with ivermectin resistance in CAVRS, it is not the major gene responsible for resistance in this strain. It is interesting that Pgp genes seem to be closely linked with each other in *C. elegans* (Lincke *et al.*, 1992), thus selection on one or more Pgp genes during treatment with ML anthelmintics may select certain alleles of other Pgp genes because they will segregate with the Pgp gene(s) that contribute to ML resistance. Pgp may play a role in effluxing the ML anthelmintics from cells containing receptors involved in the action of this class of anthelmintics. It is therefore of interest that ivermectin is a potent ligand for Pgps (Didier and Loor, 1996; Pouliot *et al.*, 1997) and that multidrug resistance (mdr)-reversing agents, which inhibit the transport functions of some Pgps, can partially reverse ML resistance in *H. contortus* (Xu *et al.*, 1998; Molento and Prichard, 1999). The overwhelming evidence for an association between Pgp genes and ML-resistance suggests that these genes will be useful markers for this type of resistance in parasitic nematodes.

Changes in Genes Associated with Putative Mode of Action Sites in Macrocyclic Lactone-resistant Parasitic Nematodes

Comparison of the polymorphism in ML-selected strains, IVC, IVF17 and MOF17, compared with their parental unselected strains, BBH and PF17, revealed that there was selection on a GluCl channel gene (Blackhall *et al.*, 1998b), which subsequently was characterized as two alternatively spliced cDNAs, *HcGluCla* and *HcGluClb* (Forrester *et al.*, 1999). Membranes from ivermectin-selected strains had a higher density of glutamate receptors (B_{max}) than did unselected strains (Paiement *et al.*, 1999b; Hejmadi *et al.*, 2000), and ivermectin decreased the B_{max} for glutamate binding in a susceptible strain (PF17) but not in the IVF17 selected strain. Furthermore, glutamate attenuates the inhibitory effect on pharyngeal pumping of moxidectin, but not ivermectin in susceptible *H. contortus*, and of both moxidectin and ivermectin in ivermectin-selected *H. contortus* (Paiement *et al.*, 1999a). These studies suggest that an up-regulation of glutamate binding is involved in ML resistance, while in susceptible *H. contortus*, ivermectin decreased glutamate binding. In *C. elegans* and *Drosophila melanogaster* (Arena *et al.*, 1992; Cully *et al.*, 1996), ivermectin potentiates the action of glutamate on channel opening. Nematodes have glutamate re-uptake transporters (Davis, 1998), which could relieve the effects of glutamate on GluCl channels and also constitute part of the glutamate binding in crude membrane preparations. Part of the action of ivermectin appears to be the potentiation of glutamate-gated opening of the GluCl channels, and a component of the avermectin-resistance mechanism could be higher levels of glutamate re-uptake in the vicinity of the GluCl receptors. In this context, it is interesting that glutamate significantly potentiates the binding of ivermectin and moxidectin to the *H. contortus* GluCl channel subunit (*HcGluCla*) (Forrester *et al.*, 2002), so that any increase in glutamate re-uptake in ML-resistant nematodes would reduce the potentiation of ML binding to GluCl channel subunits by glutamate and render these GluCl subunits less susceptible to the effects of MLs.

Such a model is consistent with the glutamate binding data, but further work needs to be done to test this hypothesis.

HG1, a putative member of the amino acid-gated anion channel subunit family (Laughton *et al.*, 1994), has also been found to be selected by ivermectin (strains IVC and IVF17) and moxidectin (strain MOF17) compared with their parental unselected strains (BBH and PF17, respectively) (Blackhall, 2000). Further work is required to characterize the function of this gene. A number of other genes have been investigated for association with ML selection in *H. contortus*, including a GluClβ-subunit, an *N*-acetylcholine receptor, phosphoenolpyruvate carboxykinase and phosphofructokinase (Blackhall *et al.*, 1998b; Blackhall, 2000), and none of these genes was linked with ML selection. Another gene of potential interest for ML resistance is *Hc-gbr2* which has high homology to *C. elegans avr-14* (*gbr-2*) (Jagannathan *et al.*, 1999).

The Problem of Anthelmintic Resistance

In nematodes of small ruminants, and especially in *H. contortus*, anthelmintic resistance has reached catastrophic proportions in many parts of the world (Prichard *et al.*, 1980; Prichard, 1990, 1994; Condor and Campbell, 1995; Waller *et al.*, 1996; Waller, 1997; Sangster, 1999; Jackson and Coop, 2000), and reports of resistance to multiple classes of anthelmintics are increasing. Resistance to ivermectin and other MLs is also found in *Teladorsagia circumcincta* and *Trichostrongylus colubriformis* in sheep and goats (see, for example, Gopal *et al.*, 1999; Sangster and Gill, 1999), and, recently, Besier (2000) reported that ML resistance in *T. circumcincta* was present on 38% of sheep farms tested in the southwest of Western Australia despite the fact that ML anthelmintics had only been used for 5–6 years at 1–3 treatments per year. However, the author attributed the rapid selection of resistance to the 'summer drenching' control programme that has been advocated for the region. In this programme, sheep are treated during the dry season when few larvae remain on pasture.

Resistance to all of the available anthelmintic classes, with the exception of the MLs, is also widespread in cyathostome nematodes of horses (Slocombe, 1991; Ihler, 1995), and cyathostomes are now considered the principal parasitic pathogens of horses (Love *et al.*, 1999). Perhaps surprisingly, ML-resistance has not been demonstrated in cyathostomes, despite many years of intensive treatment of horses with ivermectin. However, it should be recognized that ML resistance in cyathostomes is unlikely to be noted until it is flagrant because it has not been investigated. Recently, ML resistance has been found in *Parascaris equorum* in horses (Boersema *et al.*, 2002). There are also limited reports of ivermectin resistance in *Oesophagostomum dentatum* in pigs (Várady *et al.*, 1997).

In cattle nematodes, anthelmintic resistance occurs, but the problem has not been considered serious. However, recently there have been reports of ivermectin resistance in *Cooperia oncophora* in cattle in New Zealand (Vermunt *et al.*, 1995, 1996), the UK (Coles *et al.*, 1998), Brazil (Echevarria and Pinheiro, 1999) and Argentina (Fiel *et al.*, 2000). Anthelmintic formulation may affect the expression of resistance against dose-limiting parasites such as *C. oncophora*. In this context, it is interesting that Hooke *et al.* (1997) reported an 80–86% efficacy with ivermectin and moxidectin pour-on formulations whereas doramectin injectable was 99–100% efficacious against primarily *Cooperia* in cattle in New Zealand. These workers assumed that the higher C_{max} with injectable formulations was responsible for the higher efficacy and that efficacy differences were not inherent differences among the molecules.

Ivermectin resistance has also been reported in *Cooperia punctata* and *H. placei* (Paiva and Menz, 2000) in cattle in Brazil. Ivermectin and doramectin resistance were seen in *T. colubriformis* and *T. longispicularis* (Fiel *et al.*, 2000) in the Pampa, Argentina. In the survey by Echevarria and Pinheiro (1999) in southern Brazil, 20% of the ranches had resistance to ivermectin in their cattle parasites. In the report of Fiel *et al.* (2000) in Argentina, resistance was seen in *C. oncophora*, *T. colubriformis* and *T. longispicularis* to a long-acting formulation of ivermectin (40 μg kg^{-1} day^{-1}) as well as to injectable ivermectin (200 μg kg^{-1}); however, moxidectin injectable at 200 μg kg^{-1} remained effective.

Ostertagia ostertagi is the major pathogen of cattle in temperate regions and, so far, no ML resistance has been reported in this species. However, it is likely that ML resistance in nematode parasites of cattle is more widespread than we realise and needs more investigation. There is no reason to believe that the genetic diversity, encompassing the existence of alleles that confer resistance, is not as great in some species of nematodes of cattle as occurs in some of the nematodes species that now cause problems for anthelmintic control in small ruminants.

There are factors that may increase the proportion of a nematode population *in refugia* in the case of cattle parasites, which have mitigated against rapid selection for anthelmintic resistance in this host. In addition, treatment frequencies have tended to be greater in small ruminants than in cattle. Adult cattle, unlike small ruminants, develop strong resistance to helminth infection, so that traditionally young, but not adult cattle, have been treated for parasites. However, anthelmintic use is increasing in cattle. The MLs can be administered easily to cattle of all ages as pour-on or injectable formulations, whereas the older benzimidazoles had to be given orally, which is more difficult in cattle than in small ruminants. The increased ease of administration of anthelmintics to cattle and the greater spectrum of the MLs compared with the benzimidazoles or levamisole/ morantel are leading to more frequent treatment of cattle. In addition, the development of treatments such as the long-acting ivermectin bolus and long-acting depot formulations of MLs are increasing selection pressure for anthelmintic resistance in helminths of cattle. Treatment of large human populations to break transmission of filarial nematodes, and the tendency to increase the frequency of ivermectin treatment in onchocerciasis from one per year to up to two or more times per year will increase selection pressure for resistance.

Future Prospects for Maintaining Parasite Control Despite the Development of Anthelmintic Resistance

Once resistance to use-level anthelmintics has been found, it is difficult to do very much about it. Experience suggests that withdrawal of the use of the anthelmintic to which the worms have become resistant does not lead to reversion to susceptibility in the case of ML or benzimidazole resistance. If another class of anthelmintic remains effective, it can be used to maintain parasite control until resistance eventually also develops to this class. In some parts of South Africa, the management of multidrug-resistant strains of parasites has become so difficult (van Wyk *et al.*, 1997b) that strenuous efforts are made to remove the resistant worms, using drug combinations, slaughter of infected animals, spelling of pasture for several seasons or ploughing and re-seeding pastures, followed by the introduction and dissemination of a strain of worms known to be susceptible to anthelmintics. These heroic efforts to allow livestock producers to remain in business may help in the short term. However, unless better methods of monitoring the frequency of resistance alleles are developed, sensible management of our few classes of anthelmintics will be difficult. We need to be able to detect resistance alleles while their frequency in a population is low, so that changes in parasite management can be instigated before a resistance problem occurs.

To reduce the rate of resistance selection, we need to reduce selection pressure and reduce the evolutionary advantage drug-resistant worms have over susceptible worms while we adequately manage and prevent parasitic disease. Advice has often been given to reduce the number of anthelmintic treatments per year. As part of efforts to reduce the number of anthelmintic treatments per year, strategic control programmes have been devised for some environments, such as the one or two summer treatments per year recommended in parts of Australia with a Mediterranean climate. This strategy relies on the fact that in the hot dry summer, very few larvae survive on pasture, so that animals do not become reinfected rapidly, reducing the number of treatments required each year. However, as discussed above, this may put intense selection pressure on the worm population because virtually the whole worm population will be under selection (very few larvae on pasture and *in refugia*), and survivors of treatment (with resistance alleles) will not be significantly diluted by newly ingested larvae. This can result in the rapid selection of resistance, despite a low treatment frequency (Besier, 2000). Serious consideration should be given to only treating animals that are at risk of disease or significant production loss. This can be readily assessed in the case of *H. contortus* infection by noting the colour of the conjunctiva (Van Wyk *et al.*, 1997a). Animals with moderate or heavy infection with *H. contortus* will be anaemic and their conjunctiva will be pale. These animals should be treated with an anthelmintic to remove *H. contortus*. Healthy animals will have a pink conjunctiva and probably can be left untreated if *H. contortus* is the principal helminth pathogen in the environment.

Drug combinations may be used to slow the development of a resistance problem, provided the anthelmintics used in the combination are both effective and select for different resistance mechanisms. However, we really need new anthelmintic classes and other means of parasite control, such as biological control or antiparasite vaccines.

Some means exist to give a short-term extension of parasite control in the face of a resistance problem. For example, when resistance to the recommended dose rate of ivermectin first appears, moxidectin, at its recommended dose rate, is usually still effective. However, there is cross-resistance between the avermectins and the milbemycins, and continued anthelmintic pressure by the MLs will confer use level resistance to moxidectin as well. An experimental approach has been to co-administer an mdr-reversing agent, such as verapamil, with ivermectin or moxidectin in order to block the P-glycoprotein efflux of the anthelmintic (Molento and Prichard, 1999). This has allowed a limited increase in efficacy of MLs against resistant *H. contortus*, but is unlikely to be of general use.

Concluding Remarks

ML resistance has been found to be associated with P-glycopoteins in parasitic helminths and possibly some, but not all, of the diverse amino acid-gated anion channel subunit genes found in nematodes, and possibly with other genes involved with glutamate neuroreceptors/transporters. Genes associated with amphid structure and function have been suggested as being involved in avermectin resistance in *C. elegans*, but no published evidence exists for parasitic nematodes. Considerable progress has been made recently in obtaining a better understanding of anthelmintic resistance in parasitic nematodes, primarily with research on resistant and susceptible strains of *H. contortus*. However, the gaps in our knowledge of the major genes involved in ML resistance require further research. Research on the genes involved in anthelmintic resistance is essential if we are to develop sensitive probes for monitoring the frequency of resistance alleles and provide advice on appropriate parasite control before resistance becomes problematic. The existing biological methods of monitoring resistance, such as larval development assays (Gill *et al.*, 1995), egg reduction assays and worm reduction assays, are too insensitive to monitor the development of resistance before we have a resistance problem. DNA-based assays will provide information on the absolute frequency of resistance alleles even when their frequency is not high. This will allow appropriate advice to be given about control of designated populations of parasites with full knowledge of the implications for selection for resistance. We need to take action to reduce selection for anthelmintic resistance in nematodes of hosts, such as cattle, where ML use is increasing, and in human populations where widespread treatment programmes for gastrointestinal nematodes, lymphatic filaria and *O. volvulus* using ivermectin have commenced. To do this, we need DNA-based tools for anthelmintic resistance in each of the main target parasite species. In addition to providing us with the appropriate detection tools for resistance alleles, molecular studies on anthelmintic resistance are likely to provide insights to new chemotherapeutic and immunotherapeutic means to control parasites, including strains already resistant to conventional therapy.

References

Anderson, T.J., Blouin, M.S. and Beech, R.N. (1998) Population biology of parasitic nematodes: applications of genetic markers. *Advances in Parasitology* 41, 219–283.

Arena, J.P., Liu, K.K., Paress, P.S., Schaeffer, J.M. and Cully, D.F. (1992) Expression of a glutamate-activated chloride current in *Xenopus* oocytes injected with *Caenorhabditis elegans* RNA: evidence for modulation by avermectin. *Molecular Brain Research* 15, 339–348.

Beech, R.N., Prichard, R.K. and Scott, M.E. (1994) Genetic variability of the β-tubulin genes in benzimidazole-susceptible and -resistant strains of *Haemonchus contortus*. *Genetics* 138, 103–110.

Besier, R.B. (2000) Macrocyclic lactone resistance in *Ostertagia* in Western Australia: new evidence explains the high prevalence. *Joint Meeting of the New Zealand Society for Parasitology and Australian Society for Parasitology*, Wellington, New Zealand, 24–28 September, Abstract C14.

Blackhall, W.J. (2000) Genetic variation and multiple mechanisms of anthelmintic resistance in *Haemonchus contortus*. PhD thesis, McGill University, Montreal.

Blackhall, W., Liu, H.Y., Xu, M., Prichard, R.K. and Beech, R.N. (1998a) Selection at a P-glycoprotein gene in ivermectin- and moxidectin-selected strains of *Haemonchus contortus*. *Molecular and Biochemical Parasitology* 95, 193–201.

Blackhall, W.J., Pouliot, J.-F., Prichard, R.K. and Beech, R.N. (1998b) *Haemonchus contortus*: selection at a glutamate-gated chloride channel gene in ivermectin- and moxidectin-selected strains. *Experimental Parasitology* 90, 42–48.

Blaxter, M. and Bird, D. (1997) Parasitic nematodes. In: Riddle, D.L., Blumenthal, T., Meyer, B.J. and Priess, J.R. (eds) *C. elegans II*. Cold Spring Harbor Laboratory Press, Plainview, New York, pp. 851–878.

Blouin, M.S., Dame, J.B., Tarrant, C.A. and Courtney, C.H. (1992) Unusual population genetics of a parasitic nematode: mtDNA variation within and among populations. *Evolution* 46, 470–476.

Blouin, M.S., Yowell, C.A., Courtney, C.H. and Dame, J.B. (1995) Host movement and the genetic structure of populations of parasitic nematodes. *Genetics* 141, 1007–1014.

Boersema, J.H., Eysker, M. and Nas, J.W.H. (2002) Apparent resistance of *Parascaris equorum* to macrocyclic lactones. *Veterinary Record* 150, 279–281.

Coles, G.C., Stafford, K.A. and MacKay, P.H. (1998) Ivermectin-resistant *Cooperia* species from calves on a farm in Somerset. *Veterinary Record* 142, 255–256.

Conder, G.A. and Campbell, W.C. (1995) Chemotherapy of nematode infections of veterinary importance, with special reference to drug resistance. *Advances in Parasitology* 35, 1–84.

Cull-Candy, S.G. and Usherwood, P.N.R. (1973) Two populations of L-glutamate receptors on locust muscle fibres. *Nature* 246, 62–64.

Cully, D.F., Vassilatis, D.K., Liu, K.K., Paress, P.S., Van der Ploeg, L.H., Schaeffer, J.M. and Arena, J.P. (1994) Cloning of an avermectin-sensitive glutamate-gated chloride channel from *Caenorhabditis elegans*. *Nature* 371, 707–711.

Cully, D.F., Paress, P.S., Liu, K.K., Schaeffer, J.M. and Arena, J.P. (1996) Identification of a *Drosophila melanogaster* glutamate-gated chloride channel sensitive to the antiparasitic agent avermectin. *Journal of Biological Chemistry* 271, 20187–20191.

Davis, R.E. (1998) Action of excitatory amino acids on hypodermis and the motor-nervous system of *Ascaris suum*: pharmacological evidence for a glutamate transporter. *Parasitology* 116, 487–500.

Delany, N.S., Laughton, D.L. and Wolstenholme, A.J. (1998) Cloning and localisation of an avermectin receptor-related subunit from *Haemonchus contortus*. *Molecular and Biochemical Parasitology* 97, 177–187.

Dent, J.A., Davis, M.W. and Avery, L. (1997) *Avr-15* encodes a chloride channel subunit that mediates inhibitory glutamatergic neurotransmission and ivermectin sensitivity in *Caenorhabditis elegans*. *EMBO Journal* 16, 5867–5879.

Dent, J.A., Smith, M.M., Vassilatis, D.K. and Avery, L. (2000) The genetics of ivermectin resistance in *Caenorhabditis elegans*. *Proceedings of the National Academy of Sciences of the USA* 97, 2674–2679.

Didier, A. and Loor, F. (1996) The abamectin derivative ivermectin is a potent P-glycoprotein inhibitor. *Anti-Cancer Drugs* 7, 745–751.

Echevarria, F. and Pinheiro, A. (1999) Eficiência de anti-helmínticos em bovinos. XI *Seminário Brasileiro de Parasitologia Veterinária*, 24–28 October, Abstract TL-HB-274.

Egerton, J.R., Suhayda, D. and Eary, C.H. (1988) Laboratory selection of *Haemonchus contortus* for resistance to ivermectin. *Journal of Parasitology* 74, 614–617.

Fiel, C.A., Saumell, C.A., Steffan, P.E., Rodrigues, E.M. and Salaberry, G. (2000). Resistance of *Cooperia* and *Trichostrongylus* species to ivermectin treatments in grazing cattle of the Humid Pampa-Argentina. *Revista de Medicina Veterinaria (Buenos Aires)* 81, 310–315.

Forrester, S.G., Hamdan, F.F., Prichard, R.K. and Beech, R.N. (1999) Cloning, sequencing and developmental expression levels of a novel glutamate-gated chloride channel homologue in the parasitic nematode *Haemonchus contortus*. *Biochemical and Biophysical Research Communications* 254, 529–534.

Forrester, S.G., Prichard, R.K. and Beech, R.N. (2000) Glutamate modulates ivermectin binding to a glutamate-gated chloride channel subunit from *Haemonchus contortus*. *Joint Meeting of the American Society of Parasitologists and The Society of Protozoologists*, San Juan, PR, 24–28 June, Abstract 101.

Forrester, S.G., Prichard, R.K. and Beech, R.N. (2002) A glutamate-gated chloride channel subunit from *Haemonchus contortus*. Expression in a mammalian cell line, ligand binding and modulation of anthelmintic binding by glutamate. *Biochemical Pharmacology* 7144, 1–8.

Geary, T.G., Sims, S.M., Thomas, E.M., Vanover, L., Davis, J.P., Winterrowd, C.A., Klein, R.D., Ho, N.F.H. and Thompson, D.P. (1993) *Haemonchus contortus*: ivermectin-induced paralysis of the pharynx. *Experimental Parasitology* 77, 88–96.

Gill, J.H. and Lacey, E. (1998) Avermectin resistance in trichostrongylid nematodes. *International Journal for Parasitology* 28, 863–877.

Gill, J.H., Redwin, J.M., van Wyk, J.A. and Lacey, E. (1995) Avermectin inhibition of larval development in *Haemonchus contortus* – effects of ivermectin resistance. *International Journal for Parasitology* 25, 463–470.

Gill, J.H., Kerr, C.A., Shoop, W.L. and Lacey, E. (1998) Evidence of multiple mechanisms of avermectin resistance in *Haemonchus contortus*–comparison of selection protocols. *International Journal for Parasitology* 28, 783–789.

Gopal, R.M., Pomroy, W.E. and West, D.M. (1999) Resistance of field isolates of *Trichostrongylus colubriformis* and *Ostertagia circumcincta* to ivermectin. *International Journal for Parasitolology* 29, 781–786

Grant, W.N. (1994) Genetic variation in parasitic nematodes and its implications. *International Journal for Parasitology* 24, 821–830.

Grant, W. (2000) What is the real target for ivermectin resistance selection in *Onchocerca volvulus*? *Parasitology Today* 16, 458–459.

Hejmadi, M.V., Jagannathan, S., Delany, N.S., Coles, G.C. and Wolstenholme, A.J. (2000) L-glutamate binding sites of parasitic nematodes: an association with ivermectin resistance? *Parasitology* 120, 535–545.

Hoekstra, R., Otsen, M., Lenstra, J.A. and Roos, M.H. (1999) Characterisation of a polymorphic Tc1-like transposable element of the parasitic nematode *Haemonchus contortus*. *Molecular and Biochemical Parasitology* 102, 157–166.

Holden-Dye, L. and Walker, R.J. (1990) Avermectin and avermectin derivatives are antagonists at the 4-aminobutyric acid (GABA) receptor on the somatic muscle cells of *Ascaris*; is this the site of anthelmintic action? *Parasitology* 101, 265–271.

Holden-Dye, L., Hewitt, G.M., Wann, K.T., Krogsgaard-Larsen, P. and Walker, R.J. (1988) Studies involving avermectin and the 4-aminobutyric acid (GABA) receptor of *Ascaris suum* muscle. *Pesticide Science* 24, 231–245.

Hooke, F., Clement, P., Dell'Osa, D., Porter, R., MacColl, D. and Rew, R. (1997) Therapeutic and protective efficacy of doramectin injectable against gastro-intestinal nematodes in cattle in New Zealand: a comparison with moxidectin and ivermectin pour-on formulations. *Veterinary Parasitology* 72, 43–51.

Hugall, A., Stanton, J. and Moritz, C. (1997) Evolution of the AT-rich mito-chondrial DNA of the root knot nematode, *Meloidogyne hapla*. *Molecular Biology and Evolution* 14, 40–48.

Ihler, C.F. (1995) A field survey on anthelmintic resistance in equine small strongyles in Norway. *Acta Veterinaria Scandinavica* 36, 135–143.

Jackson, F. and Coop, R.L. (2000) The development of anthelmintic resistance in sheep nematodes. *Parasitology* 120, S95-S107.

Jagannathan, S., Laughton, D.L., Critten, C.L., Skinner, T.M., Horoszok, L. and Wolstenholme, A.J. (1999) Ligand-gated chloride channel subunits encoded by the *Haemonchus contortus* and *Ascaris suum* orthologues of the *Caenorhabditis elegans gbr-2* (*avr-14*) gene. *Molecular and Biochemical Parasitology* 103, 129–140.

Kass, I.S. (1984) Effects of avermectin and drugs related to acetylcholine and 4-aminobutyric acid on neurotransmission in *Ascaris suum*. *Molecular and Biochemical Parasitology* 13, 213–225

Keddie, E.M., Higazi, T. and Unnasch, T.R. (1998) The mitochondrial genome of *Onchocerca volvulus*: sequence, structure and phylogenetic analysis. *Molecular and Biochemical Parasitology* 95, 111–127.

Kotze, A.C. (1998) Effect of macrocyclic lactones on ingestion in susceptible and resistant *Haemonchus contortus* larvae. *Journal of Parasitology* 84, 631–635.

Kwa, M.S.G., Okoli, M.N., Schulz-Key, H., Okongkwo, P.O. and Roos, M.H. (1998) Use of P-glycoprotein gene probes to investigate anthelmintic resistance in *Haemonchus contortus* and comparison with *Onchocerca volvulus*. *International Journal for Parasitology* 28, 1235–1240.

Laughton, D.L., Amar, M., Thomas, P., Towner, P., Harris, P., Lunt, G.G. and Wolstenholme, A.J. (1994) Cloning of a putative inhibitory amino acid receptor subunit from the parasitic nematode *Haemonchus contortus*. *Receptors and Channels* 2, 155–163.

Laughton, D.L., Lunt, G.G. and Wolstenholme, A.J. (1997) Reporter gene constructs suggest that the *Caenorhabditis elegans* avermectin receptor beta-subunit is expressed solely in the pharynx. *Journal of Experimental Biology* 200, 1509–1514.

Le Jambre, L.F., Lenane, I.J. and Wardrop, A.J. (1999) A hybridisation technique to identify anthelmintic resistance genes in *Haemonchus*. *International Journal for Parasitology* 29, 1979–1985.

Le Jambre, L.F., Gill, J.H., Lenane, I.J. and Baker, P. (2000) Inheritance of avermectin resistance in *Haemonchus contortus*. *International Journal for Parasitology* 30, 105–111.

Lincke, C.R., The, I., van Groenigen, M. and Borst, P. (1992) The P-glycoprotein gene family of *Caenorhabditis elegans*. Cloning and characterization of genomic and complementary DNA sequences. *Journal of Molecular Biology* 228, 701–711.

Love, S., Murphy, D. and Mellor, D. (1999) Pathogenicity of cyathostome infection. *Veterinary Parasitology* 85, 113–121.

Ludmerer, S.W., Warren, V.A., Williams, B.S., Zhang, Y., Hunt, D.C., Ayer, M.B., Wallace, M.A., Chaudhary, A.G., Egan, M.A., Meinke, P.T., Dean, D.C., Garcia, M.L., Cully, D.F. and Smith, M.H. (2002) Ivermectin and nodulisporic acid receptors in *Drosphila melanogaster* contain both γ-amino butyril acid-gated Rdl and glutamate-gated Glu clα chloride channel subunits. *Biochemistry* 41, 6548–6560.

Molento, M.B. (2000) Effect of multidrug resistance modulators on activity against *Haemonchus contortus* and pharmacokinetics of ivermectin and moxidectin in sheep. PhD thesis, McGill University, Montreal.

Molento, M.B. and Prichard, R.K. (1999) Effects of the multidrug-resistance-reversing agents verapamil and CL 347,099 on the efficacy of ivermectin or moxidectin against unselected and drug-selected strains of *Haemonchus contortus* in jirds (*Meriones unguiculatus*). *Parasitology Research* 85, 1007–1011.

Okimoto, R., Macfarlane, J.L., Clary, D.O. and Wolstenholme, D.R. (1992) The mitochondrial genomes of two nematodes, *Caenorhabditis elegans* and *Ascaris suum*. *Genetics* 130, 471–498.

Paiement, J.P., Leger, C., Ribeiro, P. and Prichard, R.K. (1999a) *Haemonchus contortus*: effects of glutamate, ivermectin and moxidectin on pharyngeal pump activity in unselected and ivermectin-selected adults. *Experimental Parasitology* 92, 193–198.

Paiement, J.P., Prichard, R.K. and Ribeiro, P. (1999b) *Haemonchus contortus*: characterization of a glutamate binding site in unselected and ivermectin-selected larvae and adults. *Experimental Parasitology* 92, 32–39.

Paiva, F. and Menz, I. (2000) *In vitro* evaluation of ivermectin resistant field strains of *Haemonchus placei* and *Cooperia punctata* infective larvae. *21st World Buiatrics Congress*, Punta el Este, Uruguay, 4–8 December.

Pouliot, J.F., L'heureux, F., Liu, Z., Prichard, R.K. and Georges, E. (1997) Ivermectin: reversal of P-glycoprotein-associated multidrug resistance by ivermectin. *Biochemical Pharmacology* 53, 17–25.

Prichard, R.K. (1990) Anthelmintic resistance in nematodes: extent, recent understanding and future directions for control and research. *International Journal for Parasitolology* 20, 515–523.

Prichard, R.K. (1994) Anthelmintic resistance. *Veterinary Parasitology* 54, 259–268.

Prichard, R. (2000) What is the real target for ivermectin resistance selection in *Onchocerca volvulus*? Reply. *Parasitology Today* 16, 501–502.

Prichard, R.K., Hall, C.A., Kelly, J.D., Martin, I.C.A. and Donald, A.D. (1980) The problem of anthelmintic resistance in nematodes. *Australian Veteterinary Journal* 56, 239–251.

Rohrer, S.P., Birzin, E.T., Eary, C.H., Schaeffer, J.M. and Shoop, W.L. (1994) Ivermectin binding sites in sensitive and resistant *Haemonchus contortus*. *Journal of Parasitology* 80, 493–497.

Sangster, N.C. (1999) Anthelmintic resistance: past, present and future. *International Journal for Parasitology* 29, 115–124.

Sangster, N.C. and Gill, J. (1999) Pharmacology of anthelmintic resistance. *Parasitology Today* 15, 141–146.

Sangster, N.C., Bannan, S.C., Weiss, A.S., Nulf, S.C., Klein, R.D. and Geary, T.G. (1999) *Haemonchus contortus*: sequence heterogeneity of internucleotide binding domains from P-glycoproteins. *Experimental Parasitology* 91, 250–257.

Shoop, W.L. (1993) Ivermectin resistance. *Parasitology Today* 9, 154–159.

Skinner, T.M., Bascal, Z.A., Holden-Dye, L., Lunt, G.G. and Wolstenholme, A.J. (1998) Immunocytochemical localization of a putative inhibitory amino acid receptor subunit in the parasitic nematodes *Haemonchus contortus* and *Ascaris suum*. *Parasitology* 117, 89–96.

Slocombe, J. (1991) Anthelmintic resistance in strongyles of equids. *Equine Infectious Diseases – 6th International Conference*. R and W Publications, Cambridge.

van Wyk, J.A., Malan, F.S. and Bath, G.F. (1997a) Rampant anthelmintic resistance in sheep in South Africa – what are the options? In: Van Wyk, J. and van Schalkwyk, P.C. (eds) Workshop at the *16th International Conference of the World Association for the Advancement of Veterinary Parasitology*, pp. 51–63.

van Wyk, J.A., Malan, F.S. and Randles, J.L. (1997b) How long before resistance makes it impossible to control some field strains of *Haemonchus contortus* in South Africa with any of the modern anthelmintics? *Veterinary Parasitology* 70, 111–122.

Várady, M., Bjorn, H., Craven, J. and Nansen, P. (1997) *In vitro* characterization of lines of *Oesophagostomum dentatum* selected or not selected for resistance to pyrantel, levamisole and ivermectin. *International Journal for Parasitology* 27, 77–81.

Vassilatis, D.K., Arena, J.P., Plasterk, R.H., Wilkinson, H.A., Schaeffer, J.M., Cully, D.F. and Van der Ploeg, L.H. (1997) Genetic and biochemical evidence for a novel avermectin-sensitive chloride channel in *Caenorhabditis elegans*: isolation and characterization. *Journal of Biological Chemistry* 272, 33167–33174.

Vermunt, J.J., West, D.M. and Pomroy, W.E. (1995) Multiple resistance to ivermectin and oxfendazole in *Cooperia* species of cattle in New Zealand. *Veterinary Record* 137, 43–45.

Vermunt, J.J., West, D.M. and Pomroy, W.E. (1996) Inefficacy of moxidectin and doramectin against ivermectin-resistant *Cooperia* spp. of cattle in New Zealand. *New Zealand Veterinary Journal* 44, 188–193.

Waller, P. (1997) Anthelmintic resistance. *Veterinary Parasitology* 72, 391–412.

Waller, P.J., Echevarria, F., Eddi, C., Maciel, S., Nari, A. and Hansen, J.W. (1996) The prevalence of anthelmintic resistance in nematode parasites of sheep in southern Latin America: general overview. *Veterinary Parasitology* 62, 181–187.

Wang, G.T., Ranjan, S., Hirschlein, C. and Simkins, K. (1999) Experimental inducement of *H. contortus* resistance to macrocyclic lactones. *World Association for the Advancement of Veterinary Parasitology, 17th International Conference*, Copenhagen, 15–19 August, Abstract b.1.04.

Xu, M., Molento, M., Blackhall, W., Ribeiro, P., Beech, R. and Prichard, R. (1998) Ivermectin resistance in nematodes may be caused by alteration of P-glycoprotein homolog. *Molecular and Biochemical Parasitology* 91, 327–335.

Chapter 6

The Use of Macrocyclic Lactones to Control Parasites of Cattle

Introduction

With the worldwide availability and success of ivermectin, other macrocyclic lactones (MLs) have been discovered, registered and investigated extensively. Similarly to other animal host species, the control of parasites in cattle has come to rely heavily on the use of MLs. For the purpose of this chapter, the following will be discussed: (i) general characteristics of the therapeutic and prophylactic efficacy of the MLs used in cattle; (ii) the use of MLs to control parasites in Europe; (iii) the use of MLs to control parasites in North America; (iv) the use of MLs to control parasites in the temperate and (sub)tropical regions of South America; and (v) the use of of MLs to control parasites in Australia/New Zealand. Table 6.1 shows

Table 6.1. The MLs most widely available for cattle.

Drug	Formulation	Recommended dose rate ($\mu g\ kg^{-1}$)
Abamectin	s.c.	200
Doramectin	s.c.	200
	i.m.	200
	Pour-on	500
Eprinomectin	Pour-on	500
Ivermectin	s.c.	200
	Pour-on	500
	Long-acting, s.c.	600
	Oral	200
	Bolus	40 day^{-1}
Moxidectin	s.c.	200
	i.m.	250
	Pour-on	500
	Oral	200/400

the MLs available for cattle and the commercialized formulations. In the text, therapeutic dosages are not repeated further, except when they are different from the recommended therapeutic dose presented in Table 6.1.

Chapter 6.1

General Efficacy of the Macrocyclic Lactones to Control Parasites of Cattle

J. Vercruysse and R. Rew

The Efficacy of MLs Against Cattle Nematodes

The efficacy of the different macrocyclic lactones (MLs) (Table 6.1) against bovine nematodes has been studied extensively worldwide, by both the pharmaceutical companies and independent research laboratories. In the present review, only published studies were considered. Results published in abstracts of conferences are, with a few exceptions, not included as they do not show detailed data and they are often published later as a full paper. In most publications, efficacy is presented as the percentage reduction in worm counts comparing medicated and non-medicated animals and using analysis of geometric mean counts. Only results of worm counts were considered, as it is generally accepted that they represent a better and more precise parameter of efficacy than reductions in faecal egg counts.

A distinctive characteristic for most MLs is their very high efficacy against the major nematodes, often approaching 100%. In the present review, only three levels of efficacy are reported: (i) very high efficacy, efficacy is approaching 100%; (ii) high efficacy, efficacy is more variable but always higher than 90%; and (iii) insufficient efficacy, efficacy is below 90%. The (small) differences in percentage efficacy observed between studies (for the same product and parasite) are not discussed in detail as they are considered: (i) to have limited clinical relevance and (ii) to be due to changes in experimental design, parasite levels, age of animals, etc. It is also difficult to ascertain exactly which of the MLs, used in cattle, has the highest efficacy against a particular parasite because of: (i) differences in efficacy observed between studies; (ii) differences in infection levels between studies; (iii) the possibility that particular strains have a different susceptibility; and (iv) the lack of comparative studies and, if they exist, they are mostly limited to comparing only two MLs.

Consistency of efficacy in cattle would be expected to be greater with injectable formulations of MLs than with pour-on formulations. This would be a consequence of the maximum concentration (C_{max}) of pour-on MLs in plasma being considerably lower for doramectin and ivermectin

(Gayrard *et al.*, 1999) than for their injectable counterparts (Toutain *et al.*, 1997) and because variability around the C_{max} is greater for pour-on MLs. All pour-on MLs are considered 'rain fast', that is their efficacy is not reduced by 1 inch of rain/30 min prior to or following treatment. Data to support these claims were published for doramectin (Skogerboe *et al.*, 1999), eprinomectin (Gogolewski *et al.*, 1997a) and ivermectin (Rehbein *et al.*, 1999).

Other formulations and modes of administration of the MLs, presently commercialized, have been reported to be highly efficient, for example oral and intramuscular ivermectin (Egerton *et al.*, 1981), or bioequivalent for subcutaneous and intramuscular for doramectin (Nowakowski *et al.*, 1995); however, they are not discussed in the present review.

The efficacy of the MLs is presented according to the different nematode genera; references at the end of the section are grouped for abamection, doramectin, eprinomectin, ivermectin and moxidectin, respectively.

Abomasal nematodes

Ostertagia *spp.*

Infections with *Ostertagia* spp., especially *O. ostertagi*, are very common in temperate and subtropical regions and are considered to be the main pathogenic nematodes of cattle. Therefore, it is difficult to conceive that an ML would be commercialized without a very high efficacy against this parasite. Many studies have been performed with all commercialized MLs, and efficacies against adults, L_4 larvae and inhibited L_4 larvae were always very high, often approaching a 100% efficacy. One advantage of MLs compared with previously marketed anthelmintics is the consistent activity at any phase of inhibition of L_4 larvae (Miller, 1994).

Haemonchus *spp.*

In temperate regions, *Haemonchus* spp. may be present; however, it is mainly in (sub)tropical regions that this parasite is considered as a main pathogen. Some confusion on the species of *Haemonchus* occurring in cattle exists, that is *H. contortus*, *H. placei* and *H. similis* as species are not easily identified. All MLs have a very high efficacy against larval and adult stages, and again a 100% efficacy is often observed. Against inhibited L_4 stages, efficacy was very high for doramectin and ivermectin.

Mecistocirrus digitatus

In certain tropical countries, *M. digitatus* may be an important pathogen. Although it is likely that most MLs have a good activity, only one study and one abstract have been published in which it was found that both ivermectin and doramectin induced a 100% reduction.

Trichostrongylus axei

T. axei is a worldwide and commonly occurring abomasal parasite in ruminants, mostly of minor pathological importance. All MLs have a very high efficacy against the adult and larval stages; often a 100% efficacy is observed. Doramectin was, in one study, highly effective against inhibited stages of *T. axei*.

Intestinal nematodes

Trichostrongylus colubriformis

T. colubriformis, an intestinal nematode mainly of small ruminants, can also infect cattle and may be of importance in certain regions. Efficacy of abamectin, doramectin, eprinomectin and ivermectin has been reported to be very high against adult and larval stages; often a 100% efficacy is observed, and a very high efficacy of moxidectin is reported for adults.

Cooperia spp.

Cooperia spp. are the most common nematodes of cattle worldwide; their pathogenic role is rather moderate. *C. oncophora*, *C. pectinata*, *C. punctata*, *C. surnabada* and *C. spatula* are the species found in cattle. *C. oncophora* is mainly a parasite of the temperate regions; the others have a rather worldwide distribution. Dose determination studies identified this genus to be the dose-limiting species for doramectin, ivermectin and moxidectin. The reason for their relative insensivity has not been elucidated. *Cooperia* spp. have been shown to develop something resembling tachyphylaxis in treated animals (Bogan and McKellar, 1988). However, at the recommended therapeutic dose, efficacy against the adult and larval stages of *Cooperia* spp. is high to very high for all MLs.

Nematodirus spp.

Nematodes of the genus *Nematodirus* – in cattle, mainly *N. helvetianus* and to a lesser extent *N. spathiger* – are common parasites of young cattle in temperate regions. The genus is, as for *Cooperia*, also considered to be a dose-limiting parasite, and it is apparent that *Nematodirus* remains singular in cattle infections as the only nematode parasite not consistently controlled by ML treatment, except for eprinomectin and moxidectin.

Efficacy of abamectin against *N. helvetianus* was not uniform and ranged from insufficient to very high against adult and immature stages.

Mean efficacy of injectable doramectin was insufficient against the adult and larval stages (L_4) of *N. helvetianus* (induced infections). In contrast, in most studies with natural infections, a high to very high efficacy of doramectin (s.c.) was reported against adult worms and immature stages (Ranjan *et al.*, 1997). Eddi *et al.* (1993a) suggested that efficacy against field populations is more complete than for laboratory isolates because drug action and immune response from natural field infection may contribute to greater efficacy than is found with single dose-induced infection in the laboratory. Efficacy was insufficient and high against adults and larval stages of *N. helvetianus*, respectively, in studies where the pour-on formulation of doramectin was evaluated. Efficacy of doramectin (s.c.) against the adult stages of *N. spathiger* was high to very high according to two studies.

Ivermectin showed insufficient efficacy against adult and larval stages of *N. helvetianus* in most studies.

Eprinomectin and moxidectin have been shown to have a very high efficacy against L_4 and adult stages of *N. helvetianus*.

The consequence of this shortcoming, shared mainly by abamectin, doramectin and ivermectin (Flochlay and Deroover, 1997), may be viewed as minor, since *Nematodirus* infections are for the most part negligible or usually transient, albeit at times serious. Yazwinski *et al.* (1994c) and Eysker (1986) suggest from their findings, that is a reduction in worm counts without a corresponding reduction in egg counts, two possible scenarios: that survivors increased egg production, or that adulticide activity of the avermectins is directed against immature (non-fecund) adults, thereby not affecting post-treatment faecal egg count levels.

Capillaria *spp.*

Capillaria spp. are nematodes of little to no pathogenic importance, and no systematic efficacy studies have been performed. A high efficacy of doramectin has been observed against adults, but other studies with other MLs also suggest good efficacy.

Strongyloides papillosus

S. papillosus has a worldwide distribution and may be of pathological importance in young calves. Only a limited number of studies have been published on the efficacy against this parasite. Doramectin and ivermectin have a very high efficacy against adult female worms.

Toxocara vitulorum

Only a limited number of studies have been published on the efficacy of MLs against *Toxocara* infections in calves. The reasons may be that: (i) this parasite is rather a problem in tropical (poor) countries in calves and especially buffalo calves; (ii) pyrantel and levamisole, both cheap drugs, have a very high efficacy; and (iii) the efficacy of MLs is not superior to that of other anthelmintics. Gill *et al.* (1989) observed, based on egg counts, a high efficacy of ivermectin against adult stages; however, according to Hassanain and Degheidy (1990), ivermectin was not found to be effective.

Bunostomum phlebotomum

B. phlebotomum is an important pathogenic species of cattle occurring worldwide; however, its greatest importance is in (sub)tropical regions. Abamectin, doramectin, ivermectin, eprinomectin and moxidectin showed a very high efficacy against adult and immature stages.

Oesophagostomum radiatum

O. radiatum is present worldwide and is considered to be of moderate importance. Efficacy against adult stages is high to very high for doramectin, and very high for abamectin, eprinomectin and moxidectin. All MLs have a very high efficacy against immature stages, and one study reports on the very high efficacy of doramectin against inhibited stages.

Trichuris spp.

Infections with *Trichuris* spp. have been reported worldwide. Most infections are, however, light and of little pathological importance. Several field studies report on the efficacy of MLs against adult *Trichuris*, and the efficacy is from insufficient to high to very high. However, these results should be interpreted cautiously, as in most studies the number of animals infected and numbers of worms present were low.

Nematodes of other organs

Dictyocaulus viviparus

D. viviparus is an important pathogen in temperate regions. All MLs have a very high efficacy against adult and larval stages.

Thelazia spp.

The eyeworm, *Thelazia* spp., has a worldwide distribution, but its pathological importance is minor. Studies has been performed with doramectin and ivermectin, and both products had a very high efficacy against adult stages.

Parafilaria bovicola

P. bovicola is a filarial nematode, which causes cutaneous bleeding in cattle and bruise-like lesions on the subcutaneous and intramuscular surfaces of the carcasses of affected animals. Significantly fewer worms were recovered from ivermectin-treated cattle slaughtered 50 or 70 days after treatment than from controls. Reductions in the mean number and surface area of lesions, and in the weight of tissue trimmed, were statistically significant for the ivermectin-treated group slaughtered 70 days after treatment.

References on the efficacy of MLs against nematodes

1. Abamectin: Williams *et al.* (1992a); Heinze-Mutz *et al.* (1993b); Kaplan *et al.* (1994).
2. Doramectin: Eddi *et al.* (1993a); Jones *et al.* (1993); Kennedy and Phillips (1993); Yazwinski *et al.* (1994c, 1997a); Moreno *et al.* (1995); Watson *et al.* (1995a,b); Whelan *et al.* (1995); Couvillion *et al.* (1997); Flochlay and Deroover (1997); Marley *et al.* (1999).
3. Eprinomectin: Shoop *et al.* (1996); Gogolewski *et al.* (1997a,b); Pitt *et al.* (1997); Williams *et al.* (1997a); Yazwinski *et al.* (1997b)
4. Ivermectin: Egerton *et al.* (1979, 1981); Armour *et al.* (1980); Lyons *et al.* (1981); Williams *et al.* (1981, 1997b); Yazwinski *et al.* (1981, 1986, 1997a); Dorchies *et al.* (1982); Kerboeuf *et al.* (1985); Alva-Valdes *et al.* (1986); Ogunsusi *et al.* (1986); Pouplard *et al.* (1986); Barth (1987); Mendoza-de Gives *et al.* (1987); Robin and Razafindrakoto (1987); Yazwinski (1988); Gill *et al.* (1989); Hassanain and Degheidy (1990); Taylor *et al.* (1990); Soll *et al.* (1991, 1992a); Swan *et al.* (1991); Kennedy (1992); Williams (1992); Flochlay and Deroover (1997).
5. Moxidectin: Eysker and Boersema (1992); Ranjan *et al.* (1992); Samson *et al.* (1992); Williams *et al.* (1992b,c, 1996); Zimmerman *et al.* (1992); Hubert *et al.* (1995); Morin *et al.* (1996); Flochlay and Deroover (1997); Yazwinski *et al.* (1999).

The Persistent Efficacy of the MLs Against Nematodes

Factors interfering with persistent efficacy

Introduction

Sustained-release drugs have been known for a long time, mainly for antibiotics, insecticides and antiprotozoal drugs. With the advent of the avermectins, studies have demonstrated that a single therapeutic dose can persist in concentrations sufficient to be effective against incumbent nematode infections for prolonged periods after treatment.

The clinical significance of a prolonged efficacy is great. The sustained-release property may protect animals from reinfection by some nematode (and arthropod) species for several weeks, and will be of great value in the control of livestock pests under intermittent or constant challenge. The timing of an anthelmintic treatment may become less critical and the interval between treatments may be extended. It is, however, difficult to determine which is the optimal level and duration of persistence for an anthelmintic. The essential feature by which helminths differ from viruses, bacteria and protozoa is that the course of infections in the host does not conform to a standard pattern. Populations of helminths cannot increase in the host alone; each individual must undergo development outside the host before it can parasitize the host. The consequence is that the course and magnitude of worm populations are infinitely variable and determined not only by the host response, but also by climate and management (Michel, 1985). Furthermore, the maximum/optimal level of parasite reduction which could still ensure the success of preventive control schemes remains to some extent undefined. Therefore, different levels and lengths of persistence can be of interest depending on the required control.

The evaluation of persistence (nematodes)

The World Association of the Advancement of Veterinary Parasitology (WAAVP) guidelines (Wood *et al.*, 1995) include two test designs to determine the persistent efficacy against helminths. In the first test design, there are weekly treatment intervals prior to a single parasite challenge. An alternative test protocol involves a daily intake of larvae. Infections are given beginning on the day of treatment and continue daily for periods up to the designed extent of protection to be studied. The second test can also be done using animals to graze heavily contaminated pastures for various periods after treatment. Although this protocol has been used on a more limited scale, it reflects more closely the practical situation where animals, after treatment, graze on contaminated pastures. VICH[a] recommends a study design using multiple daily challenges, as this most closely mimics what occurs in nature (Vercruysse *et al.*, 2001).

Large variations in the persistent efficacy of a particular ML against a particular worm species have been reported. Potential reasons for these variable results include study design, and host- and parasite-related factors. The impact of trial design on persistent efficacy has been reviewed by Deroover *et al.* (1997). It was concluded that the efficacy values obtained

[a] The International Cooperation on Harmonisation of Technical Requirements for Registration of Veterinary Medicinal Products (VICH) is an international programme of cooperation between regulatory authorities and the animal health industries of the European Union, Japan and the USA which aims to harmonize the technical requirements for the registration of veterinary medicinal products. Australia and New Zealand participate as active observers.

from different types of studies used to evaluate persistent efficacy are not always comparable, and the study type itself may influence the apparent end point of the persistent efficacy of an anthelmintic. The influence of host-related factors, such as breed, nutritional status and amount of body fat, remains largely unstudied (Armour *et al.*, 1987). Persistent efficacy may also be influenced by parasite-related factors, such as inhibition proneness of the strain (Yazwinski *et al.*, 1994a) or the level of infection. Vercruysse *et al.* (1998) reported that the persistent efficacy of doramectin injectable against *C. oncophora* appeared to be shorter when treated animals were challenged with a high infection level (10,000 L_3 day^{-1}), compared with a lower infection level (1000 L_3 day^{-1}). A possible explanation for the effect of the infection level on the duration of persistent activity is that the establishment, maturation and survival of the worms in the untreated animals are density dependent. Differences in larval intake can influence worm mortality and the course of an infection in parasite-naïve calves (Michel, 1985). However, recently, Vercruysse *et al.* (2000) found that the duration of persistent efficacy of ivermectin against *Ostertagia* was, at a moderate infection level, somewhat shorter compared with a high dose level, indicating that other, unknown factors also may affect the duration of persistent activity.

According to the WAAVP and VICH guidelines (Wood *et al.*, 1995; Vercruysse *et al.*, 2001), persistence claims can only be determined on the basis of actual worm counts and not on number of eggs per gram of faeces. One could argue that worm counts indicate better the direct effects on the host, while faecal egg counts give a better view of the extent of pasture contamination. However, to measure the persistent effect of an anthelmintic by using faecal egg counts may be misleading. The relationship between worm counts and faecal egg counts is not necessarily linear and/or will vary substantially according to many, often unknown and/or uncontrolled, circumstances, for example the age of animal, parasite species, level of infection, reinfection pattern, inhibition and immunity development. The shortcomings of using faecal egg counts to assess persistent efficacy was illustrated by the observations done by Eddi *et al.* (1997). In his experiment, average faecal egg counts on day 56 post-treatment were either zero in doramectin-treated or very low in controls (46), ivermectin- (4) or fenbendazole-treated (25) groups. At the same time, however, mean total parasite counts, of which *O. ostertagi* was the predominant species, were: 250, 12,750, 2900 and 5600 parasites per animal, for the same groups, respectively. Ranjan *et al.* (1997) showed that although there was no difference in persistent efficacy between ivermectin and doramectin, based on faecal egg counts, animals receiving doramectin had 40% lower worm burdens 56 days after treatment. Entrocasso *et al.* (1996) and Meeus *et al.* (1997) could not detect significant differences between the persistent efficacy of the different MLs, based on faecal egg counts; this is in contrast to the many studies in which clear differences in duration of persistence were observed, based on worm counts. Also, results from some field evaluations of persistent activity based on post-treatment faecal egg counts (Entrocasso *et al.*, 1996; Eysker *et al.*, 1996; Meeus *et al.*, 1997; Talty *et al.*, 1998) suggest a longer persistent efficacy than those using worm counts. There is a suggestion that the MLs could have an effect on worm fecundity and/or immunity of the host (Meeus *et al.*, 1997). Finally, the expression of group faecal egg counts may mask the real differences in infection prevalence. Therefore, in the following review on duration of persistent efficacy of the MLs, only results based on worm counts were considered. Reductions in faecal egg counts may have effects on the extent of pasture contamination and would be important in infections over a grazing season.

Duration of persistent efficacy of the MLs

Persistent efficacy has been mainly investigated against the three major cattle nematodes, *O. ostertagi*, *C. oncophora* and *D. viviparus*, and to a lesser extent other nematodes. In the present review, persistent efficacy was only considered when published data were available of studies with worm counts and where at least a 90% reduction (geometric means) was considered to be the cut-off level. Although several papers compare the duration of persistent efficacy, these comparisons are mostly limited to two products. The few trials with cattle in which comparative persistent efficacy of several ML compounds was studied (Entrocasso *et al.*, 1996; Meeus *et al.*, 1997; Talty *et al.*, 1998; Williams *et al.*, 1999) used post-treatment faecal egg counts and are considered to be of limited interest. Therefore, and considering the variations in the persistent efficacy of a particular ML against a particular worm species, the duration of persistent efficacy of the different MLs is presented according to a specific ML, and only a few comparisons between MLs were made.

Abamectin injectable (Table 6.1.1)

Several studies are available on the duration of persistent efficacy of abamectin (Eagleson *et al.*, 1992; Yazwinski *et al.*, 1994b; Barth *et al.*, 1997b; Rolfe *et al.*, 1997). Abamectin had a 10–14 day residual activity against *Cooperia*, a 14–21 day residual activity against *Ostertagia* and a 28-day residual activity against *Dictyocaulus* (Table 6.1.1). Abamectin also had 10–14 day residual activity against *Trichostrongylus* spp. and *H. placei*, and residual activity against *O. radiatum* varied between 10 and 21 days.

Doramectin injectable (Table 6.1.2)

Ostertagia: doramectin injectable provided virtually total protection (100%) for 21 days after treatment in all studies, except one where efficacy was 94% (Houffschmitt *et al.*, 1999). At day 28, efficiency was less than 90% in three studies (Conder *et al.*, 1997; Rolfe *et al.*, 1997; Houffschmitt *et al.*, 1999) but, in four other studies, efficacy was higher than 90% (Weatherley *et al.*, 1993; Conder *et al.*, 1997; Eddi *et al.*, 1997; Vercruysse *et al.*, 1998). Vercruysse *et al.* (1998, 2000) observed in three different studies a significant reduction in infection (>90%) up to 35 days after treatment, and Eddi *et al.* (1997) observed a reduction up to 56 days.

Cooperia: doramectin injectable provided a very high protection (>99%) for 14 days after treatment (Weatherley *et al.*, 1993; Rolfe *et al.*, 1997; Stromberg *et al.*, 1999b). At day 21, protection was still high (>90%) (Weatherley *et al.*, 1993; Rolfe *et al.*, 1997; Stromberg *et al.*, 1999b), except in one study where almost no reduction in worm counts was observed (Houffschmitt *et al.*, 1999). Other studies reported high persistent efficacy at days 28 and 56 (Eddi *et al.*, 1997, Vercruysse *et al.*, 1998, 2000).

Table 6.1.1. Persistent efficacy of abamectin injectable.

Duration (days)	*O. ostertagi* % efficacy	Ref.	*C. oncophora* % efficacy	Ref.	*D. viviparus* % efficacy	Ref.
10	>99	2	>99	2		
14/15	88.8–99.4[a]	2, 4	63.1–100	1, 2, 3, 4		
21/22	93 to >99	1, 3	14	3	100	4
28	54	3			>99	1

[a]Arithmetic mean.
References: 1, Barth *et al.* (1997b); 2, Eagleson *et al.* (1992); 3, Rolfe *et al.* (1997); 4, Yazwinski *et al.* (1994b).

Table 6.1.2. Persistent efficacy of doramectin injectable.

Duration (days)	*O. ostertagi* % efficacy	Ref.	*C. oncophora* % efficacy	Ref.	*D. viviparus* % efficacy	Ref.
14/15	99.9–100	3, 11	99–100	6, 8, 11	99.3–100	1, 8
21/22	94–99.9	3, 5, 6, 11	54.2–99.9	5, 6, 8, 9	99.3–100	1, 8, 11
28	55–100	3, 4, 5, 6, 9, 10, 11	84.4–100[a]	4, 8, 9, 10	94.1–100	1, 8, 11
35	99.5–99.9	9, 10			96.4–100	2, 7
42					88.2–100	2, 7
49					100	7
56	100[b]	4	100	4		

[a]Moderate infection level.
[b]Adult worms.
References: 1, Barton *et al.* (1995); 2, Burden and Ellis (1997); 3, Conder *et al.* (1997); 4, Eddi *et al.* (1997); 5, Houffschmitt *et al.* (1999); 6, Rolfe *et al.* (1997); 7, Stromberg *et al.* (1999a); 8, Stromberg *et al.* (1999b); 9, Vercruysse *et al.* (1998); 10, Vercruysse *et al.* (2000); 11, Weatherley *et al.* (1993).

Dictyocaulus: doramectin injectable provided virtually total protection (100%) for 28 days after treatment; protection was still high (>90%) at day 35 and efficacy was 88.2% in one experiment at day 42. However, in other studies, efficacy was still 100% at day 49 (Weatherley *et al.*, 1993; Barton *et al.*, 1995; Burden and Ellis, 1997; Stromberg *et al.*, 1999a,b).

Others: doramectin was ≥96.9 and 100% efficacious in reducing infection with *H. placei* when challenged daily for 14–28 days (Ballweber *et al.*, 1999) and up to 56 days after treatment (Eddi *et al.*, 1997), respectively. At 56 days after treatment, worm burdens of *T. axei* and *T. colubriformis* were reduced by 100% in doramectin-treated animals (Eddi *et al.*, 1997).

Doramectin pour-on (Table 6.1.3)

Doramectin pour-on provided persistent efficacy against challenge infections of *O. ostertagia*, *C. oncophora*, *C. punctata*, *H. placei*, *H. contortus*, *O. radiatum* and *D. viviparus* for up to 35 days (Molento *et al.*, 1999; Stromberg *et al.*, 1999a,b). At 49 days, efficacy against *D. viviparus* was 81.5% (Stromberg *et al.*, 1999a). Dorny *et al.* (2000) showed that doramectin pour-on has a persistent efficacy against *C. oncophora* and *O. ostertagi* of at least 35 and 42 days, respectively.

Eprinomectin pour-on

Six studies were conducted to assess persistent efficacy of eprinomectin (Reid *et al.*, 1997), however without detailed results on percentage reductions. Eprinomectin controls infections of *H. placei* and *Trichostrongylus* spp. for 21 days after treatment and *Ostertagia* spp. (including *O. ostertagi*, *O. lyrata* and *O. leptospicularis*), *Cooperia* spp. (including *C. oncophora*, *C. punctata* and *C. surnabada*), *Nematodirus helvetianus*, *O. radiatum* and *D. viviparus* for 28 days after treatment. Reductions greater than 90% were also obtained 35 days after treatment for *N. helvetianus* (94.7%) and *D. viviparus* (96.9%), although they were tested in one study only for this time point. Dorny *et al.* (2000) could not determine the length of persistent efficacy for eprinomectin due to large individual variations in worm

Table 6.1.3. Persistent efficacy of doramectin pour-on.

Duration	*O. ostertagi* % efficacy	Ref.	*C. oncophora* % efficacy	Ref.	*D. viviparus* % efficacy	Ref.
21/22	100	2	99	2	100	2
28	99.9	2	98.7–100	1, 2	98.1–100	2, 3
35	99.8–99.9	1, 2	99.3–99.9	1, 2	93.1–98.7	2, 3
42	99.9	1			81.5	3

References: 1, Dorny *et al.* (2000); 2, Molento *et al.* (1999); 3, Stromberg *et al.* (1999a).

counts, but it was less than 28 days against *C. oncophora* and less than 35 days against *O. ostertagi*.

Ivermectin injectable (Table 6.1.4)

Ostertagia: many studies have been performed to determine the persistent efficacy of ivermectin against *Ostertagia*. The duration of persistent efficacy varied from 7 days up to 28 days; however, most studies suggest a duration of persistence of approximately 2 weeks. Williams and Broussard (1995) and Borgsteede and Hendricks (1986) concluded that a persistent effect of ivermectin was present for at least 1 week, but not after 2 and 3 weeks, respectively. Armour *et al.* (1985) and Barth (1983) observed that injectable ivermectin prevented reinfection for at least 14 days, but less than 3 weeks; in contrast, Barth *et al.* (1997b) still observed a greater than 99% efficacy at 3 weeks. A novel ivermectin formulation showed a persistent efficacy against *Ostertagia* up to 21 days (95%) (Houffschmitt *et al.*, 1999). Yazwinski *et al.* (1994a) and Eddi *et al.* (1997) observed that injectable ivermectin prevented reinfection for at least 28 days. Vercruysse *et al.* (2000) observed no persistent efficacy at day 35.

Cooperia: Bremner and Berrie (1983) and Swan and Harvey (1983) observed a persistent efficacy of higher than 95% after 7 days; after 9 days, efficacy was lower than 87.8%. Borgsteede and Hendriks (1986), Williams and Broussard (1995) and Armour *et al.* (1985) concluded that a persistent effect of ivermectin was present for at least 1 week, but not after 2–3 weeks. Yazwinski *et al.* (1994a), Barth (1983) and Barth *et al.* (1997b)

Table 6.1.4. Persistent efficacy of ivermectin injectable.

Duration (days)	*O. ostertagi* % efficacy	Ref.	*C. oncophora* % efficacy	Ref.	*D. viviparus* % efficacy	Ref.
7	>99–100	1, 2, 4, 7, 9, 10	92.4–100	1, 2, 9, 10	>99–100	1,4
9/10	95.8 to >99	1, 7	84	1	100	1
14	45–99	1, 2, 9, 10	0–99.6	1, 2, 9, 10, 11	98–100	1
21	0–99.7	1, 2, 4, 6, 11	0–96.2	1, 2, 4, 6, 11	85.3–100	1, 4, 11
28	0–100[a]	5, 6, 11	0–97.5	5, 6, 8	96.7	3, 11
35	69.4	8				
42					77.0	4
56	78.6[a]	5	100	5		

[a]Adult worms.

References: 1, Armour *et al.* (1985); 2, Barth (1983); 3, Barth *et al.* (1997b); 4, Borgsteede and Hendriks (1983); 5, Eddi *et al.* (1997); 6, Houffschmitt *et al.* (1999); 7, Swan and Harvey (1983); 8, Vercruysse *et al.* (2000); 9, Williams and Broussard (1995); 10, Williams (1992); 11, Yazwinski *et al.* (1994a).

observed that injectable ivermectin prevented reinfection for at least 14 days, but less than 3 weeks. Eddi *et al.* (1997) observed that injectable ivermectin prevented reinfection for more than 28 and less than 56 days. However, Vercruysse *et al.* (2000) did not observe any efficacy at day 28.

Dictyocaulus: Borgsteede and Hendriks (1986) concluded that a persistent effect of ivermectin was present for at least 1 week, but not after 3 weeks. Armour *et al.* (1985), Barth *et al.* (1997b) and Yazwinski *et al.* (1994a) observed that injectable ivermectin prevented reinfection for at least 14, 28 and 28 days, respectively.

Others: Williams and Broussard (1995) could not observe a persistent efficacy against *H. placei*. Yazwinski *et al.* (1994a) observed that the injectable formulation of ivermectin prevented reinfection of *T. axei, H. placei* and *O. radiatum* for at least 14, 21 and 21 days, respectively. According to Bremner and Berrie (1983), persistent activity against *C. pectinata, O. radiatum* and *H. placei* was 1 week. Swan and Harvey (1983) showed that the effect of ivermectin was virtually undiminished 9 days after administration against infections with *B. phlebotomum*. Eddi *et al.* (1997) still observed high efficacy at day 56 for *T. axei, H. placei* and *T. colubriformis*.

Ivermectin pour-on (Table 6.1.5)

Ostertagia: McKenna (1989) and Williams and Broussard (1995) concluded that a persistent effect of ivermectin pour-on against *O. ostertagia* was present for at least 2 weeks. Yazwinski *et al.* (1994a) observed that the pour-on formulation of ivermectin prevented reinfection for at least 21 days, also confirmed by Barth *et al.* (1997b), but not for 28 days. In contrast, Hong *et al.* (1995) still observed a high efficacy of ivermectin (>99%) 28 days after treatment.

Cooperia: Williams and Broussard (1995), Barth *et al.* (1997b) and Yazwinski *et al.* (1994a) observed that the pour-on formulation of ivermectin prevented reinfection against *Cooperia* for at least 7, 14 and 15 days, respectively.

Dictyocaulus: Ivermectin administered by topical application provides virtually complete protection against reinfection with *D. viviparus* for at

Table 6.1.5. Persistent efficacy of ivermectin pour-on.

Duration (days)	*O. ostertagi* % efficacy	Ref.	*C. oncophora* % efficacy	Ref.	*D. viviparus* % efficacy	Ref.
7	100	4, 5	99.8–100	4, 5		
14/15	>99–100	2, 3, 4, 5	89.1–100	1, 4, 5, 6	100	2, 3
21/22	96.8–100	1, 2, 6	60.4	6	96.7	6
28	72.7 to >89.1	3, 6			90–100	1, 2, 3, 6

References: 1, Barth *et al.* (1997b); 2, Hong *et al.* (1995); 3, McKenna (1989); 4, Williams (1992); 5, Williams and Broussard (1995); 6, Yazwinski *et al.* (1994a).

least 28 days after treatment (McKenna, 1989; Yazwinski *et al.*, 1994a; Hong *et al.*, 1995).

Others: Yazwinski *et al.* (1994a) observed that the pour-on formulation of ivermectin prevented reinfection of *T. axei*, *H. placei* and *O. radiatum* for at least 14, 21 and 21 days, respectively. McKenna (1989) found a high persistent efficacy against *T. axei* for 14 days. Barth *et al.* (1997b) found a high persistent efficacy against *O. radiatum* up to 22 days. Williams and Broussard (1995) observed a persistent efficacy against *H. placei* for at least 7 days, but less than 10 days.

Moxidectin injectable (aqueous formulation) (Table 6.1.6)

Against *O. ostertagia*, the efficacy of injectable moxidectin 28 days after treatment was only 86% (Hong *et al.*, 1995); however, according to Vercruysse *et al.* (1997), duration of activity was up to 5 weeks. In contrast, Eddi *et al.* (1993b) observed a duration of persistence of 3 weeks and less than 4 weeks. Moxidectin had no detectable persistent activity against *Cooperia*. In one study (Hong *et al.*, 1995), moxidectin provided virtually complete protection against reinfection with *D. viviparus* for at least 28 days after treatment, Vercruysse *et al.* (1997) showed a 42 day persistence.

Moxidectin pour-on (Table 6.1.7)

Moxidectin pour-on had a 99.9% efficacy for more than 5 weeks against *O. ostertagi* (Eysker and Eilers, 1995; Hubert *et al.*, 1997), or 6 weeks according to Vercruysse *et al.* (1997) and Hubert *et al.* (1997). Moxidectin pour-on also had a limited persistent efficacy against *Cooperia*. Moxidectin pour-on had a 100% efficacy for more than 5 weeks against *D. viviparus* (Eysker and Eilers, 1995), and for up to 6 weeks (Hubert *et al.*, 1997; Vercruysse *et al.*, 1997).

Against the abomasal species *T. axei*, moxidectin was virtually 100% effective for 4 weeks, in the same range as against *Ostertagia* species.

Table 6.1.6. Persistent efficacy of moxidectin injectable.

Duration (days)	*O. ostertagi* % efficacy	Ref.	*D. viviparus* % efficacy	Ref.
7	100	1	100	1
14	100	1	100	1
28	86	1	96	1
0–30			100	2
0–33	94.2	2		
0–42			99.8	2
0–45	78.1	2		

References: 1, Hong *et al.* (1995); 2, Vercruysse *et al.* (1997).

Table 6.1.7. Persistent efficacy of moxidectin pour-on.

Duration (days)	*O. ostertagi* % efficacy	Ref.	*C. oncophora* % efficacy	Ref.	*D. viviparus* % efficacy	Ref.
14	>99 to >99.9	1, 2, 3	89.7	2	100	2, 3
21	99.9–100	1, 2, 3	1	2	99.1–100	2, 3
28	99.7–100	1, 2, 3	22.2	2	99.1–100	2, 3
0–30					99.8	4
0–33	96.6	4				
35	99.8–100	2, 3	27.7	2	98.5–99.1	2, 3
0–42					100	4
42	97.5	3			94.8–99.1	2, 3
0–45	97.9	4				

References: 1, Chick *et al.* (1993b); 2, Eysker and Eilers (1995); 3, Hubert *et al.* (1997); 4, Vercruysse *et al.* (1997).

However, its effect persisted for only 2 weeks against the small intestinal sheep nematode *T. vitrinus*, which was similar to the effect observed for *Cooperia* species. This may suggest that the effectiveness of moxidectin varied with the location of the nematode in the host, the drug being more effective in the abomasum than in the small intestine (Eysker and Eilers, 1995). Chick *et al.* (1993b) observed a minimum of 28 days protection against reinfection by *T. axei* and *H. placei*.

The (Persistent) Efficacy of MLs Against Cattle Ectoparasites

Until the introduction of the avermectins, no single product could simultaneously provide control of mange, lice and other important ectoparasites as well as the important endoparasites of cattle. Moreover, the introduction of the MLs was an important improvement in the control of cattle obligate ectoparasites, as previous acaricides and insecticides rarely approached a 100% efficacy and repeated treatments were required. It must be stressed that, and in contrast to infections with nematodes, an adequate control of infestations with ectoparasites requires almost a 100% efficacy against the stages on the cattle.

Mange mites

Primarily, three genera of mange mites infect cattle: *Sarcoptes scabiei* var. *bovis*, *Psoroptes ovis* and *Chorioptes bovis*. The female *S. scabiei* var. *bovis* mites burrow in the superficial layers of the skin and evidently consume body fluids directly. *P. bovis* mites do not burrow in the skin but live at the base of hairs and cause irritation by piercing the skin. The mites are believed to feed on the superficial lipid emulsion of lymph, skin cells and skin secretions. *C. bovis* is a non-invasive mite, it is free-roaming over the skin and is thought to feed on dead tissue. It is evident that the living and feeding habits of the mange mites will impact the amount of drug that will be absorbed and the efficacy of the different formulations of the MLs.

When considering the efficacy of the different MLs against the mange mites, it is important to bear in mind that self-cure is not rare in *P. ovis*- (Pouplard *et al.*, 1990) and *C. bovis*-infected cattle (Losson, personal communication). Consequently, results must be interpreted with caution in studies without control animals. Moreover, the evaluation of therapeutic efficacy often relies only on the demonstration or absence of mites. This can have major drawbacks: (i) dead mites can be found several weeks after an effective treatment, making interpretation of data sometimes difficult; (ii) the repeated collection of infected material from lightly infected animals may induce self-cure; and (iii) there is a weak correlation between mite counts and clinical evaluation (Lonneux *et al.*, 1997). Only the injectable and pour-on formulations of the MLs seem to be effective against mange mites. Complete efficacy is lacking against mange mites following oral administration of ivermectin to cattle, presumably because less ivermectin is available to the site of infestation and it persists for a shorter duration compared with the subcutaneous route (Meleney, 1982; Benz *et al.*, 1989).

Sarcoptes scabiei *var.* bovis

Among the mange mites, *S. scabiei* var. *bovis* is the most sensitive to all registered MLs, including both injectable and pour-on formulations. Infestations with *S. scabiei* were completely eliminated (i.e. an efficacy of 100%) after abamectin treatment (injectable) (Heinze-Mutz *et al.*, 1993a), doramectin injectable (Logan *et al.*, 1993) and topical application (Rooney *et al.*, 1999), eprinomectin pour-on (Barth *et al.*, 1997a; Thompson *et al.*, 1997), ivermectin injectable and applied topically (Soll *et al.*, 1987, 1992b; Barth and Preston, 1988), and finally moxidectin injectable and pour-on (Losson and Lonneux, 1993a; Matthes *et al.*, 1993).

Barth and Preston (1988) concluded from the results of three trials that it is apparent that the topical formulation of ivermectin administered to healthy skin is fully effective against sarcoptic mange, but that its efficacy was impaired when the drug was applied over severe lesions caused by the parasite. This presumably results from reduced absorption of ivermectin at these thickened and encrusted areas, resulting in lower plasma levels of ivermectin, which would also reduce the efficacy of the drug against internal parasites.

Psoroptes ovis

An efficacy of more than 99% was recorded against *P. ovis* mange mites after abamectin treatment (Heinze-Mutz *et al.*, 1993a). Logan *et al.* (1993), Hendrickx *et al.* (1995) and Losson *et al.* (1996) showed that doramectin injectable completely eliminated a *P. ovis* infestation. Many studies described the excellent efficacy of injectable ivermectin against *P. ovis* in cattle (Guillot and Meleney, 1982; Meleney, 1982; Wright and Guillot, 1984; Soll *et al.*, 1987; Benz *et al.*, 1989; Lonneux *et al.*, 1997). Topical and injectable formulations of moxidectin gave similar parasitological and clinical results with full efficacy (Lonneux and Losson, 1992; Lonneux *et al.*, 1997). The topical formulations of doramectin, eprinomectin and ivermectin (Benz *et al.*, 1989) did not always completely eliminate *P. ovis* infections and are therefore not recommended for the treatment of *P. ovis* infections. However, Lonneux *et al.* (1997) showed that all cattle treated with ivermectin pour-on were parasitologically negative 28 and 42 days after treatment.

Although many studies reported that MLs completely eliminated mite populations, the clinical efficacy seems less consistent. In practice, often more than one treatment may be needed to obtain clinical cure, especially when the animals are suffering from the hyperkeratotic form of the disease (Lonneux and Losson, 1992, unpublished studies). A therapeutic failure, that is survival of a few mites that remained undetected, may result from difficulties of the drug reaching the mites in the hyperkeratotic lesions or reinfection from the environment.

The use of systematic endectocides such as MLs necessitates a period of quarantine after treatment before the treated animals are reintroduced to a healthy herd (Meleney and Christy, 1978; Guillot and Meleney, 1982; Guillot *et al.*, 1986; Strickland and Gerrish, 1987; Losson *et al.*, 1996). The recommended isolation period varies from 7 to 21 days.

Chorioptes bovis

Scheffler (1995) and Losson *et al.* (1998) showed a high efficacy of doramectin injectable against *C. bovis* in naturally infected animals. Doramectin pour-on had a 100% efficacy against *C. bovis* in naturally infested cattle, and efficacy was greater than 99% in artificially infested cattle (Rooney *et al.*, 1999). Eprinomectin is highly effective (≥95%) against *C. bovis* (Shoop *et al.*, 1996; Barth *et al.*, 1997a; Eagleson *et al.*, 1997b). *C. bovis* is controlled adequately by the pour-on formulation of ivermectin (Barth and Preston, 1988). Although the dose rate of ivermectin in the topical formulation is 2.5 times greater than in the subcutaneous injection, it has been suggested that the improved efficacy of the pour-on formulation of ivermectin (compared with the injectable formulation) against *C. bovis* was due to the presence of the drug on the skin and hair of the animal since, in the trials with chorioptic mange, none of the lesions was located close to the area of application. No data are available on the distribution of ivermectin applied topically over the skin (stratum corneum) and hair coat, but products such as the synthetic pyrethroids are known to move extensively over the surface of the animal from sites of 'point' or 'strip' application (Jenkinson *et al.*, 1986). Moxidectin injectable resulted in low to moderate reductions against *C. bovis* mites with large individual intervals in efficay (9–100%) (Losson and Lonneux, 1993a; Scheffler, 1995). This is in contrast to the moxidectin pour-on formulation that is fully effective against *C. bovis* within 14 days (Losson and Lonneux, 1996). Losson and Lonneux (1996) also assessed the activity of moxidectin cattle pour-on against *C. bovis* at a dosage of 0.25 mg kg^{-1}, half the recommended therapeutic dosage. Although a marked effect on mite population and clinical conditions could be observed, activity was incomplete.

Mange mites – persistent efficacy

Only a limited number of studies have been published on the persistent efficacy of the MLs against mange mites. Possible reasons may be the difficulties in setting up adequate protocols, the choice of challenge (natural or experimental) and/or interpretation of the results. Meleny *et al.* (1982) showed that an intramuscular ivermectin treatment protects cattle against *P. bovis* infestation for 3 weeks. In their study, they used direct challenge of stanchioned animals every 3 days as the infestation method. Clymer *et al.* (1997) demonstrated that a single treatment with doramectin injectable provided complete protection against infestation for 3 weeks and partial protection for an additional 2 weeks. The challenge infestation procedure used in their study, which involved the direct transfer of mite-infested material to animals prevented from self-grooming, represents a severe test of the protective capacity of the drug. In the face of a more natural challenge of exposure to infested animals, the period of complete protection conferred by doramectin was extended to 5 weeks. This was compared with a protection period of 4 weeks conferred by ivermectin in their study. It is interesting to observe that both ivermectin and doramectin demonstrate a longer period of residual protection with natural challenge than with artificial challenge, suggesting that the former is the better technique to use for predicting the protective efficacy of MLs under field use. Clymer *et al.* (1997) also designed a study to investigate the effects of grooming behaviour on the observed efficacy of the test drugs. It is not unusual for grooming activities that occur within groups of animals penned together to elicit a self-cure of scabies mite infestations. Comparison of the results from stanchioned animals where grooming was prevented and penned animals where normal grooming activities were permitted showed no differences in the period of complete protection conferred by either doramectin (5 weeks) or ivermectin (4 weeks). However, there was a difference between the two agents with respect to how soon, thereafter, significant mite infestations established. In the case of ivermectin, mite counts significantly greater than zero (complete protection) were observed 1 week later (at 5 weeks), whereas doramectin prevented establishment of infestations significantly greater than zero through to the last observation point (at 7 weeks).

Ticks

Single host ticks

Only a limited numbers of studies have been performed, and they were mostly limited to the southern cattle (single host) tick, *Boophilus microplus*.

Based on the number of engorged female *B. microplus* collected following treatment, overall efficacy of ivermectin applied topically or subcutaneously was 50 and 80%, respectively (Cramer *et al.*, 1988a). The index of reproduction for ivermectin given topically was reduced by 84% and that of ivermectin given subcutaneously by 94%. The persistence of ivermectin activity against *B. microplus* was clearly demonstrated within the first week after treatment on calves experimentally infected with about 5000 unfed larvae (Cramer *et al.*, 1988b).

The therapeutic and persistent efficacy of doramectin with *B. microplus* was described by Gonzales *et al.* (1993) using induced infections and by Muniz *et al.* (1995c) using natural infections. Results of both studies indicate that doramectin is highly efficacious in removing established tick populations and controlling re-establishment of *B. microplus* in grazing cattle for at least 28 days after treatment.

Moxidectin pour-on provided greater than 95% control of *B. microplus* (naturally) infestations from day 7 to 21 post-treatment (Remington *et al.*, 1994; Sibson, 1994; Guglielmone *et al.*, 2000). The moxidectin injectable formulation may provide effective control of *B. microplus* when applied at 4-weeks intervals (Sibson, 1994).

Multihost ticks

Treatment with an ML for multihost ticks is complex to evaluate, since several stages may be on the host at the time of treatment, each with a different sensitivity; stage of engorgement will affect sensitivity; and rapid reinfection from non-cattle hosts in the field will impact evaluation. However, treatment with an ML will often reduce adult tick counts temporarily, and several reproductive parameters of the tick may be affected.

Calves infected with (mainly) *Dermacentor* spp. and treated with injectable moxidectin showed a 100% reduction of engorged ticks during the first 2 weeks after treatment (Levasseur, 1997). Gray *et al.* (1996) demonstrated that adultical and impaired reproductive parameters (oviposition, egg mass weight and larval hatch) reduced larval production by 88% for doramectin against *Ixodes ricinus*. Soll *et al.* (1989) demonstrated that sustained-release ivermectin reduced survival and reproduction of multihost ticks feeding on cattle.

Lice

The four economically important lice species are *Haematopinus eurysternus* (short-nosed sucking louse), *Linognathus vituli* (long-nosed sucking louse), *Solenopotes capillatus* (small blue cattle louse) and *Bovicola* (*Damalinia*) *bovis* (biting louse). In addition, the tail head louse, *H. quadripertusus*, is often found during warm periods in subtropical areas of North America and Australia. In general, the injectable formulations of the MLs are recommended to treat mainly sucking lice (Table 6.1.8). The lower efficacy of the injectable MLs against *B. bovis* can be attributed to the lower exposure of the biting lice to body fluids containing the drug owing to the fact that it feeds on epithelial debris of the host. The pour-on formulations control both sucking and biting lice, although for certain of the MLs no data are available (Table 6.1.8). The complete efficacy of the pour-on formulations,

Table 6.1.8. Efficacy[a] of the MLs against biting and sucking lice.

Product	*B. bovis*	*L. vituli*	*H. eurysternus*	*S. capillatus*	References
Abamectin s.c.	–/±	+	+	?	Titchener *et al.* (1994) Heinze-Mutz *et al.* (1993a)
Doramectin s.c.	–/±	+	+	+	Lloyd *et al.* (1996) Logan *et al.* (1993) Watson *et al.* (1996)
Doramectin pour-on	+	+	+	+	Rooney *et al.* (1999)
Eprinomectin pour-on	+	+	+	+	Eagleson *et al.* (1997a) Shoop *et al.* (1996) Holste *et al.* (1997)
Ivermectin s.c.	±	+	+	+	Benz *et al.* (1989)
Ivermectin pour-on	+	+	?	?	Titchener *et al.* (1994)
Moxidectin s.c.	–	+	?	+	Titchener *et al.* (1994) Chick *et al.* (1993a) Webb *et al.* (1991)
Moxidectin pour-on	±/+	+	?	?	Losson and Lonneux (1996) Chick *et al.* (1993a)

[a]+, 100%; ±, aid-to-control 85–95% (average, in certain animals it may be lower); –, ineffective; ?, no references found.

as opposed to the injectable formulations, probably can be attributed to the presence of a high concentration of the active ingredient on the skin surface, killing lice by both contact and ingestion (Titchener *et al.*, 1994). It is important to stress that lice populations may decline spontaneously; this is generally associated with seasonality. Increasing numbers of lice in autumn and early winter are followed by decreasing numbers in spring (Villeneuve and Daigneault, 1997). Therefore, efficacy trials without adequate control animals have limited value.

Titchener *et al.* (1994) demonstrated that abamectin (injectable) was highly effective (100%) against *L. vituli*; against *B. bovis*, a 97% reduction in the count was seen. Heinze-Mutz *et al.* (1993a) demonstrated that abamectin was fully effective against *H. eurysternus* and *L. vituli*, but not against *B. bovis*.

Logan *et al.* (1993), Lloyd *et al.* (1996), Phillips *et al.* (1996) and Watson *et al.* (1996) reported the spectrum of activity of doramectin injectable. The three common species of sucking lice encountered were controlled completely. Efficacy against the biting louse was between 58 and 98%. Rooney *et al.* (1999) demonstrated that the efficacy of doramectin pour-on against the four common lice was 100% by 28 days after treatment.

Eprinomectin demonstrated complete efficacy against all four common species of cattle lice (Shoop *et al.*, 1996; Eagleson *et al.*, 1997a; Holste *et al.*, 1997).

Benz *et al.* (1989) confirmed, based on data from many studies, the high efficacy of ivermectin injectable against sucking lice; against biting lice an aid-in-control may be considered. Titchener *et al.* (1994) and Polley *et al.* (1998) demonstrated that ivermectin pour-on was highly effective (100%) against *L. vituli* and *B. bovis*.

Titchener *et al.* (1994) demonstrated that moxidectin (injectable) was completely effective against *L. vituli*; however, it was ineffective against *B. bovis*. Chick *et al.* (1993a) also showed that moxidectin injection did not eliminate *B. bovis* consistently (efficacy ranged from 0 to 85%). In contrast, moxidectin pour-on provided a consistently high efficacy (84–100%) against *B. bovis* at three trial sites. Topical moxidectin greatly reduced *B. bovis* populations; however, small numbers of live lice persisted for at least 8 weeks (Polley *et al.*, 1998). A high degree of control of *L. vituli* was achieved with both the injection (96.7 and 100%) and the pour-on formulations (94.6 and 100%) of moxidectin (Chick *et al.*, 1993a; Losson and Lonneux, 1996).

Lice – persistent efficacy

Surprisingly, very few studies have been performed to evaluate the preventive efficacy of MLs against lice and to determine the approximate period of protection. Villeneuve and Daigneault (1997) evaluated the protective efficacy of injectable doramectin against sucking lice under natural challenge conditions. Acquisition of infestation with *L. vituli* and *S. capillatus* was delayed by a mean period of 25.6 days in doramectin-treated animals compared with controls, the difference between the two groups being highly significant ($P \leq 0.001$). A repeated-exposure challenge model was used to evaluate the pour-on formulation of doramectin in preventing the establishment of *L. vituli* and *B. bovis* infestations in cattle (Skogerboe *et al.*, 2000). The acquisition of *L. vituli* and *B. bovis* infestations was delayed for 77 and 105 days, respectively. In a similar type of study evaluating the persistent efficacy of ivermectin pour-on against *B. bovis* infestations, it was concluded that ivermectin was 100% effective in preventing infestations for a period of 35–49 days after treatment (Clymer *et al.*, 1998). These studies confirmed the protective efficacy of ivermectin and doramectin against lice. The optimum programme for treatment and control of louse infestations is to treat all cattle in a herd at the same time to eliminate the opportunity for untreated cattle to reinfest the herd. Cattle management practices, however, often result in the commingling of treated and untreated cattle after the autumn and/or winter processing and, therefore, the use of doramectin and ivermectin would minimize the risk of reinfestation.

*Warble fly larvae and screw worms (***Hypoderma** *spp.,* **Dermatobia hominis, Cochliomyia hominovorax** *and* **Chrysomyia bezziana***)*

Warble flies (*H. bovis* and *H. lineatum*) are common and economically important parasites of cattle in the northern hemisphere between 25 and 60° of latitude. In most countries of Western Europe, measures were and are taken to eradicate foci.

Hendrickx *et al.* (1993) and Lloyd *et al.* (1996) reported that doramectin injectable is 100% efficacious at all stages of development. The pour-on formulation of doramectin is also highly effective. In grub studies, 107 of 136 control cattle had warbles, whereas two of 136 doramectin-treated cattle had one warble each, which represents a cure rate of 98.5% (Rooney *et al.*, 1999).

Ivermectin injectable and pour-on formulations were found to be, in most instances, 100% effective against all three larval stages (Alva Valdez *et al.*, 1986; Benz *et al.*, 1989). The use of microdoses of ivermectin injectable (1/60–1/100 of the therapeutic dosage of 0.2 mg kg^{-1}) have been suggested for the treatment of LI stages of *Hypoderma* (Argente and Hillion, 1984; Alvineri *et al.*, 1994). Although a high efficacy at this low dose is observed, the possibility of inducing resistance against gastrointestinal nematodes cannot be excluded.

Eprinomectin efficacy was complete against LI larvae. *Hypoderma* LII/LIII eradication approached 100% efficacy (one live larva was recorded) (Holste *et al.*, 1998).

The activity of moxidectin as an injectable and pour-on formulation was assessed in several studies (Scholl *et al.*, 1992; Losson and Lonneux, 1993b; Lonneux and Losson, 1994; Le Stang and Cardinaud, 1995; Boulard *et al.*, 1998). A 100% efficacy of both formulations was demonstrated against first-instar *H. bovis* and *H. lineatum* larvae. Efficacy of moxidectin against the other stages has not been reported.

The tropical warble fly, *Dermatobia hominis*, is a widespread and critical problem for the cattle industry in the tropical and subtropical regions of Latin America, up to an altitude of 5000 ft above sea level (Moya-Borja *et al.*, 1993a). Abamectin was 100% effective in the treatment of *D. hominis*. Reinfestation was first detected on day 44 (Cruz *et al.*, 1993). In a therapeutic efficacy study, parasitic nodules in doramectin (injectable)-treated animals were reduced by 74% at 48 h post-treatment, and efficacy reached 100% at 6 days post-treatment (Moya-Borja *et al.*, 1993a). Ivermectin has also proven to be effective; 7 and 10 days after treatment the overall efficacy against the three larval stages was 97 and 99%, respectively (Roncalli and Benitez Usher, 1988).

The New World screw worm, *Cochliomyia hominovorax*, is an obligatory myiasis-producing parasite that can infest all warm-blooded animals, including humans. With the inception of the sterile male release programme in the USA in 1958 and a few years later in Mexico, *C. hominovorax* was successfully eradicated from the USA and northern Mexico. Screw worm infestations are still important from southern Mexico to northern Argentina, and if the control programmes in Mexico were to be discontinued, would rapidly reappear in the southern USA.

Medication with doramectin was 100% effective in protecting treated calves against induced screw worm infestation (Moya-Borja *et al.*, 1993b). Results of other studies conducted in Latin America (Moya-Borja *et al.*, 1997) demonstrated that doramectin gave 100% protection against *C. hominivorax* induced 2 h after treatment, while ivermectin protected only 50% of the animals exposed to the same challenge.

Chrysomyia bezziana, the Old World screw worm, is controlled by ivermectin (Benz *et al.*, 1989).

Persistence against screw worms

According to Cruz *et al.* (1993), animals treated with abamectin were significantly protected against reinfection up to 79 days after treatment. In a trial conducted in Brazil, 21, 28 and 35 days after treatment with doramectin, calves were seeded with 25 first-instar *D. hominis* larvae. The persistent efficacy of a single injection of doramectin extended beyond 35 days, and no parasitic nodules developed in the treated calves at any time (Moya-Borja *et al.*, 1993a).

Moya-Borja *et al.* (1993b) demonstrated that the protection against *C. hominovorax* conferred by a single injection of doramectin lasted for a minimum of 14 days (the final end point of activity was not determined). Muniz *et al.* (1995a,b) confirmed from studies conducted in Latin America that a single injection with doramectin was 100% effective over the 12-day duration of the studies in the prevention and control of screw worm strikes in newborn calves, post-parturient cows and castrated cattle exposed to continuous field infestations of *C. hominovorax*. More recently, results of studies conducted in Brazil (Moya-Borja *et al.*, 1997) demonstrated that doramectin provided complete protection against *C. hominovorax* for 21 days, and partial protection (56%) at 28 days post-infection. Results of the study designed to compare the efficacy of doramectin and ivermectin against induced challenge of *C. hominivorax* substantiate initial reports of incomplete efficacy of ivermectin against the screw worm (Benz *et al.*, 1989; Baez Kohn *et al.*, 1995). This study was conducted under exacting conditions to detect efficacy of treatment. However, even under these conditions, 29% of the incisions became infested in ivermectin-treated animals and larvae completed development to adult flies. When calves, previously treated with doramectin or ivermectin at recommended use levels, were castrated 7 and 15 days later and exposed to natural challenge, the incidence of screw worm myiasis was 60% in control and ivermectin groups and 0% in doramectin-treated animals up to 12 days. Ivermectin injected subcutaneously within 24 h of birth or at the time of castration was highly effective in preventing navel and scrotal myiasis due to *C. hominivorax*, respectively. Against navel and scrotal myiasis, only two of the 26 and two of the 99 treated calves, respectively, developed myiasis that required therapy. Most calves of the untreated groups became infected (Benitez Usher *et al.*, 1997). Recently, Anziani *et al.* (2000) conducted a study to evaluate the activity of a single administration of doramectin or ivermectin (both injectable formulations) against severe, induced infections of *C. hominivorax*. Doramectin provided reduction in myiasis of 90.9 and 83.3% at 12 and 15 days after treatment, respectively, compared with the saline treatment. In contrast, there were no significant differences in the number of calves with myiasis between those treated with either ivermectin or the saline control.

*Hornflies (***Haematobia irritans***)*

Haematobia irritans, the hornfly, is one of the most obvious blood-sucking parasites of cattle. It occurs almost worldwide and induces damage due to blood loss, annoyance and disease transmission. Accepted estimates of the economic threshold of hornflies on cattle range from 50 to 200 flies per animal, depending on geographic location (Andress *et al.*, 2000). Hornfly control involves mainly killing adult flies sucking blood on cattle, but inhibiting hornfly development in the manure may also be an action (Miller *et al.*, 1981; Lysyk and Colwell, 1996). Marley *et al.* (1993) showed that the subsequent increase in hornfly numbers on an ivermectin-treated herd was predicted by a developmental rate equation interpreted by the time of cessation of larval fly inhibition in field dung. Estimating efficacy against hornflies is not easy, and large differences, with the same ML drug, between studies may be found. A likely cause of variation among study sites may be to the number of hornflies, the size of herd and pastures, and differences in herd management factors that affect the number of immigrating flies.

Treatment with doramectin pour-on (Farkas *et al.*, 2000) resulted in a persistent efficacy of 98.9, 95.9 and 93.4% at 21, 28 and 35 days post-treatment, respectively. At the end of the study, on day 49, efficacy was still 75.1%. Acceptable hornfly control was obtained for 6–7 weeks with one treatment and 13 weeks by the use of two treatments of doramectin pour-on (Andress *et al.*, 2000).

Several studies evaluating the persistent efficacy of ivermectin pour-on against naturally occurring hornflies on cattle were reported. Lysyk and Colwell (1996) observed that ivermectin pour-on reduced adult populations of *H. irritans* by 90% for 8–16 days and by 50% for 18–26 days. Larval survival in manure was reduced by 90% for 8 and 15 days and by 50% for 19 and 24 days, consistent with the results presented for reduction in adult numbers. Marley *et al.* (1993) reported that late spring treatment of cattle with a single dose of pour-on ivermectin resulted in reduced hornfly populations for approximately 6 weeks, with percentage efficacy exceeding 80% for at least 26 days post-treatment. Other studies generally demonstrated longer duration effects on adult populations. Lancaster *et al.* (1991) reported a greater than 50% adult reduction for 35–42 days following spring application of ivermectin, and also reported increased levels of control with repeated applications of ivermectin in isolated treatment herds. Finally, Uzuka *et al.* (1999) found that topically applied ivermectin was 100% efficacious against hornflies for up to 35 days.

Mosquitoes

Adult *Culicoides brevitarsis*, an important vector of arboviruses affecting livestock in Australia, were fed on cattle which were treated with ivermectin injectable. The mean mortality of engorged insects 48 h after feeding was 99% for 10 days post-treatment (Standfast *et al.*, 1984).

References

Alva-Valdes, R., Wallace, D.H., Holste, J.E., Egerton, J.R., Cox, J.L., Wooden, J.W. and Barrick, R.A. (1986) Efficacy of ivermectin in a topical formulation against induced gastrointestinal and pulmonary nematode infections, and naturally acquired grubs and lice in cattle. *American Journal of Veterinary Research* 47, 2389–2392.

Alvineri, M., Sutra, J.F., Galtier, P. and Toutain, P.L. (1994) Microdose d'ivermectine chez la vache laitière: concentrations plasmatiques et résidus dans le lait. *Revue de Médecine Vétérinaire* 145, 761–764.

Andress, E.R., DeRouen, S.M. and Foil, L.D. (2000) Efficacy of doramectin 0.5% w/v pour-on for control of the horn fly, *Haematobia irritans. Veterinary Parasitology* 90, 327–331.

Anziani, O.S., Flores, S.G., Moltedo, H., Derozier, C., Guglielmone, A.A., Zimmermann, G.A. and Wanker, O. (2000) Persistent activity of doramectin and ivermectin in the prevention of cutaneous myiasis in cattle experimentally infested with *Cochliomyia hominivorax. Veterinary Parasitology* 87, 243–247.

Argenté, G. and Hillion, E. (1984) Utilisation de petites doses d'ivermectine pour le traitement préventif de l'hypodermose bovine. *Le Point Vétérinaire* 16, 614–620.

Armour, J., Bairden, K. and Preston, J.M. (1980) Anthelmintic efficiency of ivermectin against naturally acquired bovine gastrointestinal nematodes. *Veterinary Record* 107, 226–227.

Armour, J., Bairden, K., Batty, A.F., Davison, C.C. and Ross, D.B. (1985) Persistent anthelmintic activity of ivermectin in cattle. *Veterinary Record* 116, 151–153.

Armour, J., Bairden, K., Opirie, H.M. and Ryan, W.G. (1987) Control of parasitic bronchitis and gastroenteritis in grazing cattle by strategic prophylaxis with ivermectin. *Veterinary Record* 121, 5–8.

Baez Kohn, A.R., Bresanovich, C.A. and Jara, J. (1995) Efecto preventivo de ivermectina y doramectina contra bicheras de castracion en toritos. *Therios* 24, 247–250.

Ballweber, L.R., Siefker, C., Engelken, T., Walstrom, D.J. and Skogerboe, T. (1999) Persistent activity of doramectin injectable formulation against experimental challenge with *Haemonchus placei* in cattle. *Veterinary Parasitology* 86, 1–4.

Barth, D. (1983) Persistent anthelmintic effect of ivermectin in cattle. *Veterinary Record* 113, 300.

Barth, D. (1987) Treatment of inhibited *Dictyocaulus viviparus* in cattle with ivermectin. *Veterinary Parasitology* 25, 61–66.

Barth, D. and Preston, J.M. (1988) Efficacy of topically administered ivermectin against chorioptic and sarcoptic mange of cattle. *Veterinary Record* 123, 101–104.

Barth, D., Hair, J.A., Kunkle, B.C., Langholff, W.K., Löwenstein, M., Rehbein, S., Smith, L.L., Eagleson, J.S. and Kutzer, E. (1997a) Efficacy of eprinomectin against mange mites in cattle. *American Journal of Veterinary Research* 58, 1257–1259.

Barth, D., Ericsson, G.F., Kunkle, B.N., Rehbein, S., Ryan, W.G. and Walace, D.H. (1997b) Evaluation of the persistence of the effect of ivermectin and abamectin against gastrointestinal and pulmonary nematodes in cattle. *Veterinary Record* 140, 278–279.

Barton, N.J., Mitchell, P.J., Hooke, F.G. and Reynolds, J. (1995) The therapeutic efficacy and prophylactic activity of doramectin against *Dictyocaulus viviparus* in cattle. *Australian Veterinary Journal* 72, 349–351.

Benitez Usher, C., Cruz, J., Carvalho, L.A.F., Bridi, A.A., Farrington, D., Barrick, R.A. and Eagleson, J. (1997) Prophylactic use of ivermectin against cattle myiasis caused by *Cochliomyia hominivorax* (Coquerel, 1858). *Veterinary Parasitology* 72, 215–220.

Benz, G.W., Roncalli, R.A. and Gross, S.J. (1989) Use of ivermectin in cattle, sheep, goats, and swine. In: Campbell, W.C. (ed.) *Ivermectin and Abamectin.* Springer-Verlag, New York, pp. 215–229.

Bogan, J.A. and McKellar, Q.A. (1988) The pharmacodynamics of ivermectin in sheep and cattle. *Journal of Pharmacology and Therapeutics* 11, 260–268.

Borgsteede, F.H.M. and Hendriks, J. (1986) The residual effect of treatment with ivermectin after experimental reinfection with nematodes in calves. *Veterinary Quarterly* 8, 98–104.

Boulard, C., Banting, A. de L. and Cardinaud, B. (1998) Activity of moxidectin 1% injectable solution against first instar *Hypoderma* spp. in cattle and effects on antibody kinetics. *Veterinary Parasitology* 77, 205–210.

Bremner, K.C. and Berrie, D.A. (1983) Persistence of the anthelmintic activity of ivermectin in calves. *Veterinary Record* 113, 569.

Burden, D.J. and Ellis, R.N.W. (1997) Use of doramectin against experimental infections of cattle with *Dictyocaulus viviparus. Veterinary Record* 141, 393.

Chick, B., McDonald, D., Cobb, R., Kieran, P.J. and Wood, I. (1993a) The efficacy of injectable and pour-on formulations of moxidectin against lice on cattle. *Australian Veterinary Journal* 70, 212–213.

Chick, B.F., Cobb, R., Kieran, P. and Fraser, G. (1993b) Efficacy and persistent effect features of moxidectin pour-on when used for parasite control in cattle. *Proceedings of the 10th Seminar for The Society of Dairy Cattle Veterinarians of the New Zealand Veterinary Association*, Waitangi, Bay of Islands. New Zealand, 24–28 June, pp. 177–185.

Clymer, B.C., Janes, T.H. and McKenzie, M.E. (1997) Evaluation of the therapeutic and protective efficacy of doramectin against psoroptic scabies in cattle. *Veterinary Parasitology* 72, 79–89.

Clymer, B., Newcombe, K.M., Ryan, W.G. and Soll, M.D. (1998) Persistence of the activity of topical ivermectin against biting lice (*Bovicola bovis*). *Veterinary Record* 143, 193–195.

Conder, G.A., Cruthers, L.R., Slone, R.L., Johnson, E.G., Zimmerman, G.L., Zimmerman, L.A., Shively, J.E., Logan, N.B. and Weatherley, A.J. (1997) Persistent efficacy of doramectin against experimental challenge with *Ostertagia ostertagi* in cattle. *Veterinary Parasitology* 72, 9–13.

Couvillion, C.E., Pote, L.M.W., Siefker, C. and Logan, N.B. (1997) Efficacy of doramectin for treatment of experimentally induced infection with gastrointestinal nematodes in calves. *American Journal of Veterinary Research* 58, 282–285.

Cramer, L.G., Carvalho, L.A.F., Bridi, A.A., Amaral, N.K. and Barrick, R.A. (1988a) Efficacy of topically applied ivermectin against *Boophilus microplus* (Canestrini, 1887) in cattle. *Veterinary Parasitology* 29, 341–349.

Cramer, L.G., Bridi, A.A., Amaral, N.K. and Gross, S.J. (1988b) Persistent activity of injectable ivermectin in the control of the cattle tick *Boophilus microplus. Veterinary Record* 122, 611–612.

Cruz, J.B., Benitez-Usher, C., Cramer, L.G., Gross, S.J. and Kohn, A.B. (1993) Efficacy of abamectin injection against *Dermatobia hominis* in cattle. *Parasitology Research* 79, 183–185.

Deroover, E., Cobb, R., Rock, D.W. and Guerino, F. (1997) Persistent efficacy, importance and impact of trial design. *Veterinary Parasitology* 73, 365–371.

Dorchies, P., Franc, M. and Ducos de Lahitte, J. (1982) Le traitement antiparasitaire des bovins par l'ivermectine. *Revue de Médecine Vétérinaire* 133, 709–713.

Dorny, P., Demeulenaere, D., Smets, K., Claerebout, E. and Vercruysse, J. (2000) Persistent efficacy of topical doramectin and eprinomectin against *Ostertagia ostertagi* and *Cooperia oncophora* infections in cattle. *Veterinary Record* 147, 139–140.

Eagleson, J.S., Scott, P.G. and Gross, S.J. (1992) Persistent anthelmintic activity of abamectin injection in cattle. *New Zealand Veterinary Journal* 40, 21–23.

Eagleson, J.S., Holste, J.E. and Pollmeier, M. (1997a) Efficacy of topically applied eprinomectin against the biting louse *Damalinia (Bovicola) bovis. Proceedings of the 16th International Conference of the World Association for the Advancement of Veterinary Parasitology*, Sun City, South Africa, 10–15 August.

Eagleson, J.S., Holste, J.E., Kunkle, B.N. and Barth, D. (1997b) The efficacy of topically applied eprinomectin for treatment of *Chorioptes bovis* infestations. *Proceedings of the 16th International Conference of the World Association for the Advancement of Veterinary Parasitology*, Sun City, South Africa, 10–15 August.

Eddi, C., Bianchin, I., Honer, M.R., Muniz, R.A., Caracostantogolo, J. and do Nascimento, Y.A. (1993a) Efficacy of doramectin aginst field nematode infections of cattle in Latin America. *Veterinary Parasitology* 49, 39–44.

Eddi, C.S., Caracostantogolo, J., Marangunich, L., Bulman, G.M., Morley, M.E. and Ambrustolo, R.R. (1993b) Moxidectin: persistence of activity against cattle nematodes in Argentina. *Proceedings of the 14th International Conference of the World Association for the Advancement of Veterinary Parasitology*, Cambridge, UK, 8–13 August, p. 130.

Eddi, C., Muniz, R.A., Caracostantogolo, J., Errecalde, J.O., Rew, R.S., Michener, S.L. and McKenzie, M.E. (1997) Comparative persistent efficacy of doramectin, ivermectin and febendazole against natural nematode infections in cattle. *Veterinary Parasitology* 72, 33–41.

Egerton, T.R., Ostlind, D.A., Blair, L.S., Eary, C.H., Suhayda, D., Cifelli, S., Riek, R.F. and Campbell, W.C. (1979) Avermectins, a new family of potent anthelmintic agents: efficacy of the B1a component. *Antimicrobial Agents and Chemotherapy* 15, 372–378.

Egerton, J.R., Eary, C.H. and Suhayda, D. (1981) The anthelmintic efficacy of ivermectin in experimentally infected cattle. *Veterinary Parasitology* 8, 59–70.

Entrocasso, C., Parra, D., Vottero, D., Farias, M., Uribe, L.F. and Ryan, W.G. (1996) Comparison of the persistent activity of ivermectin, abamectin, doramectin and moxidectin in cattle. *Veterinary Record* 138, 91–92.

Eysker, M. (1986) The prophylactic effect of ivermectin treatment of calves, three weeks after turn-out, on gastro-intestinal helminthiasis. *Veterinary Parasitology* 22, 95–103.

Eysker, M. and Boersema, J.H. (1992) The efficacy of moxidectin against *Dictyocaulus viviparus* in cattle. *Veterinary Quarterly* 14, 79–80.

Eysker, M. and Eilers, C. (1995) Persistence of the effect of a moxidectin pour-on against naturally acquired cattle nematodes. *Veterinary Record* 137, 457–460.

Eysker, M., Boersema, J.H., Cornelissen, J.B.W.J., Kloosterman, A. and Kooyman, F.N.J. (1996) Residual effect of injectable moxidectin against lungworm and gastrointestinal nematodes in calves exposed to high pasture infectivity levels in the Netherlands. *Veterinary Parasitology* 61, 61–71.

Farkas, R., Gräfner, G. and Hendrickx, M.O. (2000) Persistent efficacy of doramectin pour-on against *Haematobia irritans* in cattle. *Veterinary Record* 146, 378–380.

Flochlay, A. and Deroover, E. (1997) Efficacy of four endectocides against *Nematodirus helvetianus* in cattle: moxidectin, ivermectin, doramectin 1% injectable solutions and moxidectin 0.5% pour-on solution. *Revue de Médecine Vétérinaire* 148, 229–234.

Gayrard, V., Alvinerie, M. and Toutain, P.L. (1999) Comparison of pharmacokinetic profiles of doramectin and ivermectin pour-on formulations in cattle. *Veterinary Parasitology* 81, 47–55.

Gill, B.S., Singh, J., Gill, B.S., Singh, A., Khehra, S.S., Rai, A. and Hussain, O. (1989) Efficacy of ivermectin against mange and gastrointestinal nematodes of buffalo (*Bubalus bubalis*). *Veterinary Parasitology* 31, 141–147.

Gogolewski, R.P., Allerton, G.R., Pitt, S.R., Thompson, D.R., Langholff, W.K., Hair, J.A., Fulton, R.K. and Eagleson, J.S. (1997a) Effect of simulated rain, coat length and exposure to natural climatic conditions on the efficacy of a topical formulation of eprinomectin against endoparasites of cattle. *Veterinary Parasitology* 69, 95–102.

Gogolewski, R.P., Slacek, B., Familton, A.S., Paterson, B., Langholff, W.K., Allerton, G.R., McAnulty, R. and Eagleson, J.S. (1997b) Efficacy of a topical formulation of eprinomectin against endoparasites of cattle in New Zealand. *New Zealand Veterinary Journal* 45, 1–3.

Gonzales, J.C., Muniz, R.A., Farias, A., Goncalves, L.C.B. and Rew, R.S. (1993) Therapeutic and persistent efficacy of doramectin against *Boophilus microplus* in cattle. *Veterinary Parasitology* 49, 107–119.

Gray, J.S., Ellis, R.N.W. and Purnell, R.E. (1996) The effects of doramectin on induced infestations of *Ixodes ricinus* on cattle. *Irish Veterinary Journal* 49, 609–610.

Guglielmone, A.A., Mangold, A.J., Munoz Cobenas, M.E., Scherling, N., Garcia Posse, F., Anziani, O.S. and Ioppolo, M. (2000) Moxidectin pour-on for control of natural populations of the cattle tick *Boophilus microplus* (Acarina: Ixodidae). *Veterinary Parasitology* 87, 237–241.

Guillot, F.S. and Meleney, W.P. (1982) The infectivity of surviving *Psoroptes ovis* (Hering) on cattle treated with ivermectin. *Veterinary Parasitology* 10, 73–78.

Guillot, F.S., Wright, F.C. and Oehler, D. (1986) Concentration of ivermectin in bovine serum and its effect on the fecundity of psoroptic mange mites. *American Journal of Veterinary Research* 47, 525–527.

Hassanain, M.A. and Degheidy, N.S. (1990) Studies on the efficiency of some anthelmintics in treatment of *Toxocara vitulorum* in buffalo-calves. *Egyptian Journal of Comparative Pathology and Clinical Pathology* 3, 249–256.

Heinze-Mutz, E.M., Barth, D., Cramer, L.G., Gross, S.J. and Visser, M. (1993a) Efficacy of abamectin against ectoparasites of cattle. *Veterinary Record* 132, 455–457.

Heinze-Mutz, E.M., Pitt, S.R., Bairden, K., Baggott, D.G., Armour, J., Barth, D. and Cramer, L.G. (1993b) Efficacy of abamectin against nematodes in cattle. *Veterinary Record* 132, 35–37.

Hendrickx, M.O., Anderson, L., Boulard, C., Smith, D.J. and Weatherley, A.J. (1993) Efficacy of doramectin against warble fly larva (*Hypoderma bovis*). *Veterinary Parasitology* 49, 75–84.

Hendrickx, M.O., Levasseur, G., Saget, Y., Gevrey, J. and Shanks, D.-J. (1995) Etude de l'efficacité de la doramectine dans le traitement de la gale psoroptique des bovins. *Bulletin des GTV* 2-B-513, 53–57.

Holste, J.E., Smith, L.L., Hair, J.A., Lancaster, J.L., Lloyd, J.E., Langholff, W.K., Barrick, R.A. and Eagleson, J.S. (1997) Eprinomectin: a novel avermectin for control of lice in all classes of cattle. *Veterinary Parasitology* 73, 153–161.

Holste, J.E., Colwell, D.D., Kumar, R., Lloyd, J.E., Pinkall, N.P.M., Sierra, M.A., Waggoner, J.W., Langholff, W.K., Barrick, R.A. and Eagleson, J.S. (1998) Efficacy of eprinomectin against *Hypoderma* spp. in cattle. *American Journal of Veterinary Research* 59, 56–58.

Hong, C., Harris, T.J., Grimshaw, W.T.R. and Newcomb, K.M. (1995) Persistent activity of ivermectin topical and moxidectin injection against *Ostertagia ostertagi* and *Dictyocaulus viviparus* in calves. *Veterinary Record* 137, 640–641.

Houffschmitt, P., Martin, P. and Chick, B. (1999) Persistence of the efficacy of a novel injectable ivermectin formulation and injectable doramectin against gastrointestinal nematodes in cattle. *Veterinary Record* 144, 564–565.

Hubert, J., Kerboeuf, D., Le Stang, J.P., Cardinaud, B. and Blond, F. (1995) Efficacy of moxidectin pour-on against nematode infections in cattle. *Veterinary Record* 136, 632–634.

Hubert, J., Kerboeuf, D., Cardinaud, B., Blond-Riou, F. and Fournier, R. (1997) Persistent efficacy of topical moxidectin against *Dictyocaulus viviparus* and *Ostertagia ostertagi*. *Veterinary Parasitology* 68, 187–190.

Jenkinson, D., Hutchison, G., Jackson, D. and McQueen, L. (1986) Route of passage of cypermethrin across the surface of sheep skin. *Research in Veterinary Science* 41, 237–241.

Jones, R.M., Logan, N.B., Weatherley, A.J., Little, A.S. and Smothers, C.D. (1993) Activity of doramectin against nematode endoparasites of cattle. *Veterinary Parasitology* 49, 27–37.

Kaplan, R.M., Courtney, C.H., Kunkle, W.E., Zeng, Q.Y., Jernigan, A.D. and Eagleson, J.S. (1994) Efficacy of injectable abamectin against gastrointestinal tract nematodes and lungworms of cattle. *Journal of American Veterinary Research* 55, 353–357.

Kennedy, M.J. (1992) The efficacy of ivermectin against the eyeworm, *Thelazia skrjabini*, in experimentally infected cattle. *Veterinary Parasitology* 45, 127–131.

Kennedy, M.J. and Phillips, F.E. (1993) Efficacy of doramectin against eyeworms (*Thelazia* spp.) in naturally and experimentally infected cattle. *Veterinary Parasitology* 49, 61–66.

Kerboeuf, D., Hubert, J. and Le Stang, J.-P. (1985) Efficacité de l'ivermectine dans le traitement de l'ostertagiose bovine de prétype II. Essai en Normandie. *Recueil de Médecine Vétérinaire* 161, 773–777.

Lancaster, J.L. Jr, Kilgore, R.L., Simco, J.S., Parham, R.W., Hubbel, D. and Cox, J.L. (1991) Efficacy of a topical ivermectin formulation against naturally occurring adult horn flies on cattle. *Southwest Entomology* 16, 339–345.

Le Stang, J.-P. and Cardinaud, B. (1995) Efficacité de la moxidectine 1% injectable dans le traitement de l'hypodermose bovine. *Bulletin des GTV* 1-B-505, 55–57.

Levasseur, G. (1997) Les tiques dans le sud du charolais: contrôle de l'infestation avec la moxidectine 1% injectable. *Bulletin des GTV* 1-B-538, 38–47.

Lloyd, J.E., Kumar, R., Waggoner, J.W., Phillips, F.E. and Kumar, R. (1996) Doramectin systemic activity against cattle grubs, *Hypoderma lineatum* and

H. bovis (Diptera: Oestridae), and cattle lice, *Bovicola bovis* (Mallophaga: Trichodectidae), *Linognathus vituli* and *Solenopotes capillatus* (Anoplura: Linognathidae), and *Haematopinus eurysternus* (Anoplura: Haematopinidae), in Wyoming. *Veterinary Parasitology* 63, 307–317.

Logan, N.B., Weatherley, A.J., Phillips, F.E., Wilkins, C.P. and Shanks, D.J. (1993) Spectrum of activity of doramectin against cattle mites and lice. *Veterinary Parasitology* 49, 67–73.

Lonneux, J.-F. and Losson, B.J. (1992) Field efficacy on injectable and pour-on moxidectin in cattle naturally infested with *Psoroptes ovis* (Acarina: Psoroptidae). *Veterinary Parasitology* 45, 147–152.

Lonneux, J.-F. and Losson, B.J. (1994) The efficacy of moxidectin 0.5% pour-on against *Hypoderma bovis* in naturally infested cattle: parasitologial and serological data. *Veterinary Parasitology* 52, 313–320.

Lonneux, J.-F., Nguyen, T.Q. and Losson, B.J (1997) Efficacy of pour-on and injectable formulations of moxidectin and ivermectin in cattle naturally infected with *Psoroptes ovis*: parasitological, clinical and serological data. *Veterinary Parasitology* 69, 319–330.

Losson, B. and Lonneux, J.-F. (1993a) Field efficacy on injectable moxidectin in cattle naturally infested with *Chorioptes bovis* and *Sarcoptes scabiei*. *Veterinary Parasitology* 51, 113–121.

Losson, B. and Lonneux, J.-F. (1993b) Une estimation de l'activité rémanente de la moxidectine 1% injectable chez le bétail infesté par le premier stade larvaire d'*Hypoderma* sp. *Annales de Médecine Vétérinaire* 137, 105–108.

Losson, B. and Lonneux, J.-F. (1996) Field efficacy of moxidectin 0.5% pour-on against *Chorioptes bovis*, *Damalinia bovis*, *Linognathus vituli* and *Psoroptes ovis* in naturally infected cattle. *Veterinary Parasitology* 63, 119–130.

Losson, B., Lonneux, J.-F., Roelants, B. and De Keulenaer, K.I.T. (1996) Activité de la doramectine, une nouvelle avermectine, sur *Psoroptes ovis* chez le bovin lors d'infestations naturelles. *Annales de Médecine Vétérinaire* 140, 439–444.

Losson, B.J., Mignon, B., Bossaert, K., Leclipteux, T. and Lonneux, J.-F. (1998) Field efficacy on injectable doramectin against *Chorioptes bovis* in naturally infected cattle. *Veterinary Record* 142, 18–19.

Lyons, E.T., Tolliver, S.C., Drudge, J.H. and LaBore, D.E. (1981) Ivermectin: controlled test of anthelmintic activity in dairy calves with emphasis on *Dictyocaulus viviparus*. *American Journal of Veterinary Research* 42, 1225–1227.

Lysyk, T.J. and Colwell, D.D. (1996) Duration of efficacy of diazinon ear tags and ivermectin pour-on for control of horn fly (Diptera: Muscidae). *Journal of Economic Entomology* 89, 1513–1520.

Marley, S.E., Hall, R.D. and Corwin, R.M. (1993) Ivermectin cattle pour-on: duration of a single late spring treatment against horn flies, *Haematobia irritans* (L) (Diptera: Muscidae) in Missouri, USA. *Veterinary Parasitology* 51, 167–172.

Marley, S.E., Illyes, E.F., Keller, D.S., Meinert, T.R., Logan, N.B., Hendrickx, M.-O. and Conder, G.A. (1999) Efficacy of topically administered doramectin against eyeworms, lungworms, and gastrointestinal nematodes of cattle. *American Journal of Veterinary Research* 60, 665–668.

Matthes, H.F., Busch, A., Dröbigk, U. and Hiepe, T. (1993) Efficacy of moxidectin against *Sarcoptes bovis* in experimental infected calves. *Proceedings of the 14th International Conference of the World Association for the Advancement of Veterinary Parasitology*, Cambridge, UK, 8–13 August, p. 264.

McKenna, P.B. (1989) Persistent anthelmintic activity of topically administered ivermectin in cattle. *New Zealand Veterinary Journal* 37, 146–147.

Meeus, P.F.M., De Bont, J. and Vercruysse, J. (1997) Comparison of the persistent activity of ivermectin, abamectin, doramectin and moxidectin in cattle in Zambia. *Veterinary Parasitology* 70, 219–224.

Meleney, W.P. (1982) Control of psoroptic scabies on calves with ivermectin. *American Journal of Veterinary Research* 43, 329–331.

Meleney, W.P. and Christy, J.E. (1978) Factors complicating the control of psoroptic scabies of cattle. *Journal of the American Medical Association* 173, 1473–1478.

Meleney, W.P., Wright, F.C. and Guillot, F.S. (1982) Residual protection against cattle scabies offered by ivermectin. *American Journal of Veterinary Research* 43, 1767–1769.

Mendoza-de Gives, P., Fuentes, R.N., Herrera-Rodriguez, D. and Roncalli, R.A. (1987) Efficacy of ivermectin against the abomasal nematode *Mecistocirrus digitatus* in naturally infected zebu calves. *American Journal of Veterinary Research* 48, 1611–1612.

Michel, J.F. (1985) Epidemiology and control of gastrointestinal helminths in domestic animals. In: Vanden Bossche, H., Thienpont, D. and Janssens, P.G. (eds) *Chemotherapy of Gastro-intestinal Helminths.* Springer-Verlag, Berlin, pp. 68–123.

Miller, J.A., Kunz, S.E., Oehler, D.D. and Miller, R.W. (1981) Larvicidal activity of Merck ML-933, an avermectin, against the horn fly, stable fly, face fly, and house fly. *Journal of Economic Entomology* 74, 608–611.

Miller, J.E. (1994) Variable efficacy of benzimidazole anthelmintics against inhibited *Ostertagia ostertagi*. *Bovine Proceedings* 26, 150–153.

Molento, M.B., Trudeau, C., Prichard, R.K., Zimmerman, G.L., Johnson, E.G., Marley, S. and Conder, G.A. (1999) Persistent efficacy of doramectin pour-on against artificially induced infections of nematodes in cattle. *Veterinary Parasitology* 82, 297–303.

Moreno, L.G., Rew, R.S., Muniz, R.A. and Moreno, J. (1995). Efficacy of doramectin against *Mecistocirrus digitatus* and other abomasal parasites of cattle in Venzuela. *Proceedings of the 15th International Conference of the World Association for the Advancement of Veterinary Parasitology*, Yokohama, 30 August–2 September, p. 72, A10.

Morin, D., Valdez, R., Lichtensteiger, C., Paul, A., DiPietro, J. and Guerino, F. (1996) Efficacy of moxidectin 0.5% pour-on against naturally acquired nematode infections in cattle. *Veterinary Parasitology* 65, 75–81.

Moya-Borja, G.E., Muniz, R.A., Sanavria, A., Goncalves, L.C.B. and Rew, R.S. (1993a) Therapeutic and persistent efficacy of doramectin against *Dermatobia hominis* in cattle. *Veterinary Parasitology* 49, 85–93.

Moya-Borja, G.E., Muniz, R.A., Sanavria, A., Goncalves, L.C.B. and Rew, R.S. (1993b) Prophylactic and persistent efficacy of doramectin against *Cochliomyia hominivorax* in cattle. *Veterinary Parasitology* 49, 95–105.

Moya-Borja, G.E., Muniz, R.A., Umehara, O., Goncalves, L.C.B., Silva, D.S.F. and McKenzie, M.E. (1997) Protective efficacy of doramectin and ivermectin against *Cochliomyia hominivorax*. *Veterinary Parasitology* 72, 101–109.

Muniz, R.A., Anziani, O.S., Ordónez, J., Errecalde, J., Moreno, J. and Rew, R.S. (1995a) Efficacy of doramectin in the protection of neonatal calves and post-parturient cows against field strikes of *Cochlyomyia hominivorax*. *Veterinary Parasitology* 58, 155–161.

Muniz, R.A., Coronado, A., Anziani, O.S., Sanavria, A., Moreno, J., Errecalde, J. and Goncalves, L.C.B. (1995b) Efficacy of injectable doramectin in the protection of castrated cattle against field infestations of *Cochlyomyia hominivorax*. *Veterinary Parasitology* 58, 327–333.

Muniz, R.A., Hernandez, F., Lombardero, O., Leite, R.C., Moreno, J., Errecalde, J. and Goncalves, L.C.B. (1995c) Efficacy of injectable doramectin against natural *Boophilus microplus* infestations in cattle. *American Journal of Veterinary Research* 56, 460–463.

Nowakowski, M.A., Lynch, M.J., Smith, D.G., Logan, N.B., Mouzin, D.E., Lukaszewicz, J., Ryan, N.I., Hunter, R.P. and Jones, R.M. (1995) Pharmacokinetics and bioequivalence of parenterally administered doramectin in cattle. *Journal of Veterinary Pharmacology and Therapeutics* 18, 290–298.

Ogunsusi, R.A., Ajanusi, O.J. and Ogunkoya, Y.O. (1986) The effect of ivermectin against nematode parasites of white fulani calves. *Veterinary Parasitology* 19, 333–335.

Phillips, F.E., Logan, N.B. and Jones, R.M. (1996) Field evaluation of doramectin for treatment of gastrointestinal nematode infections and louse infestations of cattle. *American Journal of Veterinary Research* 57, 1468–1471.

Pitt, S.R., Langholff, W.K., Eagleson, J.S. and Rehbein, S. (1997) The efficacy of eprinomectin against induced infections of immature (fourth larval stage) and adult nematode parasites in cattle. *Veterinary Parasitology* 73, 119–128.

Polley, L.R., Wagner, B.A., Ward, T.I. and Campbell, J.R. (1998) Effect of topical ivermectin and moxidectin for naturally acquired *Damalinia bovis* infestations in cattle treated under winter conditions in Canada. *Veterinary Record* 143, 80–81.

Pouplard, L., Lekeux, P. and Detry, M. (1986) Efficacy of ivermectin and levamisole against immature *Dictyocaulus viviparus* in cattle. *Veterinary Record* 118, 557–559.

Pouplard, L., Losson, B., Detry, M. and Hollanders, W. (1990) Les gales bovines. *Annales de Médecine Vétérinaire* 134, 531–539.

Ranjan, S., Trudeau, C., Prichard, R.K., von Kutzleben, R. and Carrier, D. (1992) Efficacy of moxidectin against naturally acquired nematode infections in cattle. *Veterinary Parasitology* 41, 227–231.

Ranjan, S., Trudeau, C., Prichard, R.K., Daigneault, J. and Rew, R.S. (1997) Nematode reinfection following treatment of cattle with doramectin and ivermectin. *Veterinary Parasitology* 72, 25–31.

Rehbein, S., Barrick, R.A., Battu, A.F., Drag, M.D., Rolfe, P.F. and Cox, J.L. (1999) Evaluation of the effect of simulated rainfall on the efficacy of ivomec pour-on against *Cooperia* spp. infection in cattle. *Parasitology Research* 85, 783–786.

Reid, J.F.S., Eagleson, J.S. and Langholff, W.K. (1997) Persistent efficacy of eprinomectin pour-on against gastrointestinal and pulmonary nematodes in cattle. *Proceedings of the 16th International Conference of the World Association for the Advancement of Veterinary Parasitology*, Sun City, South Africa, 10–15 August.

Remington, B., Kieran, P., Cobb, R. and Bodero, D. (1994) The application of moxidectin formulations for cattle tick (*Boophilus microplus*) control under Queensland field conditions. *Proceedings of the Australian Association of Cattle Veterinarians*, Canberra, Australia, 6–11 March, pp. 111–114.

Robin, B. and Razafindrakoto, C. (1987) Efficacité de l'ivermectine dans le traitement des principaux nématodes du zébu à Madagascar. *Revue de Médecine Vétérinaire* 138, 51–54.

Rolfe, P.F., Dawson, K.L., Soll, M.D., Nichols, G.K. and Ryan, W. (1997) Persistent efficacy of abamectin and doramectin against gastrointestinal nematodes of cattle. *Australian Veterinary Journal* 75, 33–35.

Roncalli, R.A. and Benitez Usher, C. (1988) Efficacy of ivermectin against *Dermatobia hominis* in cattle. *Veterinary Parasitology* 28, 343–346.

Rooney, K.A., Illyes, E.F., Sunderland, S.J., Sarasola, P., Hendrickx, M.O., Keller, D.S., Meinert, T.R., Logan, N.B., Weatherley, A.J. and Conder, G.A. (1999) Efficacy of a pour-on formulation of doramectin against lice, mites, and grubs of cattle. *American Journal of Veterinary Research* 60, 402–404.

Samson, D., Charleston, W.A.G., Pomroy, W.E. and Alexander, A.M. (1992) Evaluation of moxidectin for the treatment of internal parasites of cattle. *New Zealand Veterinary Journal* 40, 15–17.

Scheffler, C. (1995) Ergebnisse der einmaligen subkutanen Behandlung der *Chorioptes* – Räude in einem Mastrinderbestand mit Doramectin und Moxidectin. *Tierärztliche Umschau* 50, 713–718.

Scholl, P.J., Guillot, F.S. and Wang, G.T. (1992) Moxidectin: systemic activity against common cattle grubs (*Hypoderma lineatum*) (Diptera: Oestridae) and trichostrongyle nematode in cattle. *Veterinary Parasitology* 41, 203–209.

Shoop, W.L., Egerton, J.R., Eary, C.H., Haines, H.W., Michael, B.F., Mrozik, H., Eskola, P., Fisher, M.H., Slayton, L., Ostlind, D.A., Skelly, B.J., Fulton, R.K., Barth, D., Costa, S., Gregory, L.M., Campbell, W.C., Seward, R.L. and Turner, M.J. (1996) Eprinomectin: a novel avermectin for use as a topical endectocide for cattle. *International Journal for Parasitology* 26, 1237–1242.

Sibson, G.J. (1994) The effects of moxidectin against natural infestations of the cattle tick (*Boophilus microplus*). *Australian Veterinary Journal* 71, 381–382.

Skogerboe, T.L, Cracknell, V.C., Walstrom, D.J., Ritzhaupt, L. and Karle, V.K. (1999) The effect of simulated rainfall on the efficacy of doramectin pour-on against nematode parasites of cattle. *Veterinary Parasitology* 86, 229–234.

Skogerboe, T.L., Smith, L.L., Karle, V.K. and Derozier, C.L. (2000) The persistent efficacy of doramectin pour-on against biting and sucking louse infestations of cattle. *Veterinary Parasitology* 87, 183–192.

Soll, M.D., Carmichael, I.H., Swan, G.E. and Scherer, H. (1987) Control of cattle mange in Southern Africa using ivermectin. *Tropical Animal Health and Production* 19, 93–102.

Soll, M.D., Carmichael, I.H., Swan, G.E. and Gross, S.J. (1989) Control of induced infestations of three African multihost tick species with sustained-release ivermectin. *Experimental and Applied Acarology* 7, 121–130.

Soll, M.D., Carmichael, I.H. and Barrick, R.A. (1991) Ivermectin treatment of feedlot cattle for *Parafilaria bovicola*. *Preventive Veterinary Medicine* 10, 251–256.

Soll, M.D., Carmichael, I.H., Scherer, H.R. and Gross, S.J. (1992a) The efficacy of ivermectin against *Thelazia rhodesii* (Desmarest, 1828) in the eyes of cattle. *Veterinary Parasitology* 42, 67–71.

Soll, M.D., d'Assonville, J.A. and Smith, C.J.Z. (1992b) Efficacy of topically applied ivermectin against sarcoptic mange (*Sarcoptes scabiei* var. *bovis*) of cattle. *Parasitology Research* 78, 120–122.

Standfast, H.A., Muller, M.J. and Wilson, D.D. (1984) Mortality of *Culicoides brevitarsis* (Diptera: Ceratopogonidae) fed on cattle treated with ivermectin. *Journal of Economic Entomology* 77, 419–421.

Strickland, R.K. and Gerrish, R.R. (1987) Infestivity of *Psoroptes ovis* on ivermectin-treated cattle. *American Journal of Veterinary Research* 48, 342–344.

Stromberg, B.E., Averbeck, G.A., Anderson, J.F., Woodward, B.W., Cunningham, J., Brake, A. and Skogerboe, T. (1999a) Comparison of the persistent efficacy of the injectable and pour-on formulations of doramectin against artificially-induced infection with *Dictyocaulus viviparus* in cattle. *Veterinary Parasitology* 87, 45–50.

Stromberg, B.E., Woodward, B.W., Courtney, C.H., Kunkle, W.E., Johnson, E.G., Zimmerman, G.L., Zimmerman, L.A., Marley, S.E., Keller, D.S. and Conder, G.A. (1999b) Persistent efficacy of doramectin injectable against artificially infections with *Cooperia punctata* and *Dictyocaulus viviparus* in cattle. *Veterinary Parasitology* 83, 49–54.

Swan, G.E. and Harvey, R.G. (1983) Persistent anthelmintic effect of ivermectin in cattle. *Journal of the South African Veterinary Association* 54, 249–250.

Swan, G.E., Soll, M.D. and Gross, S.J. (1991) Efficacy of ivermectin against *Parafilaria bovicola* and lesion resolution in cattle. *Veterinary Parasitology* 40, 267–272.

Talty, P.J., Grimshaw, W.T.R. and Ryan, W.G. (1998) Persistent effect of injectable abamectin, doramectin, moxidectin and ivermectin on faecal egg counts of calves. *Irish Veterinary Journal* 51, 251–253.

Taylor, S.M., Mallon, T.R. and Green, W.P. (1990) Comparison of the efficacy of dermal formulations of ivermectin and levamisole for the treatment and prevention of *Dictyocaulus viviparus* infection in cattle. *Veterinary Record* 126, 357–359.

Thompson, D.R., Rehbein, S., Loewenstein, M., Villeneuve, A., Bowman, D. and Eagleson, J.S. (1997) Efficacy of eprinomectin against *Sarcoptes scabiei* infestation in cattle. *Proceedings of the 16th International Conference of the World Association for the Advancement of Veterinary Parasitology*, Sun City, South Africa, 10–15 August.

Titchener, N.R., Parry, J.M. and Grimshaw, W.T.R. (1994) Efficacy of formulations of abamectin, ivermectin and moxidectin against sucking and biting lice of cattle. *Veterinary Record* 134, 452–453.

Toutain, P.L., Upson, D.W., Terhune, T.N. and McKenzie, M.E. (1997) Comparative pharmacokinetics of doramectin and ivermectin in cattle. *Veterinary Parasitology* 72, 3–8.

Uzuka, Y., Yoshioka, T., Tanabe, S., Kinoshita, G., Nagata, T., Yagi, K., Funaki, H., Hanyu, H. and Sarashina, T. (1999) Chemical control of *Haematobia irritans* with 0.5% topical ivermectin solution in cattle. *Journal of Veterinary Medical Science* 61, 287–289.

Vercruysse, J., Claerebout, E., Dorny, P., Demeulenaere, D. and Deroover, E. (1997) Persistence of the efficacy of pour-on and injectable moxidectin against *Ostertagia ostertagi* and *Dictyocaulus viviparus* in experimentally infected cattle. *Veterinary Record* 140, 64–66.

Vercruysse, J., Claerebout, E., Dorny, P., Demeulenaere, D., Agneessens, J. and Smets, K. (1998) Persistent efficacy of doramectin against *Ostertagia osteragi* and *Cooperia oncophora* in cattle. *Veterinary Record* 143, 443–446.

Vercruysse, J., Dorny, P., Claerebout, E., Demeulenaere, E., Smets, K. and Agneessens, J. (2000) Evaluation of the persistent efficacy of doramectin and ivermectin injectable against *Ostertagia ostertagi* and *Cooperia oncophora* in cattle. *Veterinary Parasitology* 89, 63–69.

Vercruysse, J., Holdsworth, P., Letonja, T., Barth, D., Conder, G., Hamamoto, K. and Okano, K. (2001) International harmonisation of anthelmintic guidelines. *Veterinary Parasitology* 96, 171–193.

Villeneuve, A. and Daigneault, J. (1997) Evaluation of the protective efficacy of doramectin against sucking lice of cattle. *Veterinary Parasitology* 72, 91–99.

Watson, T.G., Hosking, B.C. and Hooke, F.G. (1995a) Efficacy of doramectin against naturally acquired adult and inhibited larval infections by some nematode parasites in cattle in new Zealand. *New Zealand Veterinary Journal* 43, 64–66.

Watson, T.G., Hosking, B.C. and Hooke, F.G. (1995b) Efficacy of doramectin against experimental infections by some nematode parasites in cattle in New Zealand. *New Zealand Veterinary Journal* 43, 67–69.

Watson, T.G., Bishop, D.M., Hooke, F.G., Heath, A.C.G. and Cole, D.J.W. (1996) Efficacy of injectable doramectin against natually acquired louse infestations on cattle. *New Zealand Journal of Agricultural Research* 39, 401–404.

Weatherley, A.J., Hong, C., Harris, T.J., Smith, D.G. and Hammet, N.C. (1993) Persistent efficacy of doramectin against experimental nematode infections in calves. *Veterinary Parasitology* 49, 45–50.

Webb, J.D., Burg, J.G. and Knapp, F.W. (1991) Moxidectin evaluation against *Solenoptes capillatus* (Anoplura: Linognathidae), *Bovicola bovis* (Mallophaga: Trichodectidae), and *Musca autumnalis* (Diptera: Muscidae) on cattle. *Journal of Economic Entomology* 84, 1266–1269.

Whelan, N.C., Charleston, W.A.G., Pomroy, W.E. and Alexander, A.M. (1995) Evaluation of the efficacy of doramectin against an artificial infection of *Dictyocaulus viviparus* in calves. *New Zealand Veterinary Journal* 43, 21–22.

Williams, J.C. (1992) A comparison of anthelmintic activity of injectable and pour-on ivermectin. *Louisiana Agriculture* 35, 15–18.

Williams, J.C. and Broussard, S.D. (1995) Persistent anthelmintic activity of ivermectin against gastrointestinal nematodes of cattle. *American Journal of Veterinary Research* 56, 1169–1175.

Williams, J.C., Knox, J.W., Baumann, B.A., Snider, T.G., Kimball, M.G. and Hoerner, T.J. (1981) Efficacy of ivermectin against inhibited larvae of *Ostertagia ostertagi*. *American Journal of Veterinary Research* 42, 2077–2080.

Williams, J.C., Loyacano, A.F., Nault, C., Ramsey, R.T. and Plue, R.E. (1992a) Efficacy of abamectin against natural infections of gastrointestinal nematodes and lungworm of cattle with special emphasis on inhibited early fourth stage larvae of *Ostertagia ostertagi*. *Veterinary Parasitology* 41, 77–84.

Williams, J.C., Barras, S.A. and Wang, G.T. (1992b) Efficacy of moxidectin against gastrointestinal nematodes of cattle. *Veterinary Record* 131, 345–347.

Williams, J.C., Nault, C., Ramsey, R.T. and Wang, G.T. (1992c) Efficacy of Cydectin® moxidectin 1% injectable against experimental infections of *Dictyocaulus viviparus* and *Bunostomum phlebotomum* superimposed on natural gastrointestinal infections in calves. *Veterinary Parasitology* 43, 293–299.

Williams, J.C., Broussard, S.D. and Wang, G.T. (1996) Efficacy of moxidectin pour-on against gastrointestinal nematodes and *Dictyocaulus viviparus* in cattle. *Veterinary Parasitology* 64, 277–283.

Williams, J.C., Stuedemann, J.A., Bairden, K., Kerboeuf, D., Ciordia, H., Hubert, J., Broussard, S.D., Plue, R.E., Alva-Valdes, R., Baggott, D.G., Pinkall, N. and Eagleson, J.S. (1997a) Efficacy of a pour-on formulation of eprinomectin (MK-297) against nematode parasites of cattle, with emphasis on inhibited early fourth-stage larvae of *Ostertagia* spp. *American Journal of Veterinary Research* 58, 379–383.

Williams, J.C., DeRosa, A., Nakamura, Y. and Loyacano, A.F. (1997b) Comparative efficacy of ivermectin pour-on, albendazole, oxfendazole and febendazole against *Ostertagia ostertagi* inhibited larvae, other gastrointestinal nematodes and lungworm of cattle. *Veterinary Parasitology* 73, 73–82.

Williams, J.C., Loyacano, A.F., DeRosa, A., Gurie, J., Clymer, B.C. and Guernio, F. (1999) A comparison of persistent anthelmintic efficacy of topical formulations of doramectin, ivermectin, eprinomectin and moxidectin against naturally acquired nematode infections of beef calves. *Veterinary Parasitology* 85, 277–288.

Wood, I.B., Amaral, N.K., Bairden, K., Duncan, J.L., Kassai, T., Malone, J.B., Pankavich, S.A., Reinecke, R.K., Slocombe, O., Taylor, S.M. and Vercruysse, J. (1995) World Association of the Advancement of Veterinary Parasitology (WAAVP) second edition of guidelines for evaluating the efficacy of anthelmintics in ruminants (bovine, ovine, caprine). *Veterinary Parasitology* 58, 181–213.

Wright, F.C. and Guillot, F.S. (1984) Effect of ivermectin in heifers on mortality and egg production of *Psoroptes ovis*. *American Journal of Veterinary Research* 45, 2132–2134.

Yazwinski, T.A. (1988) Use of febantel or ivermectin for treatment of calves with experimentally induced *Bunostomum phlebotomum* infection. *American Journal of Veterinary Research* 49, 1407–1408.

Yazwinski, T.A., Williams, M., Greenway, T. and Tilley, W. (1981) Anthelmintic activities of ivermectin against gastrointestinal nematodes of cattle. *American Journal of Veterinary Research* 42, 481–482.

Yazwinski, T.A., Wood, N., Holtzen, H.M., Presson, B.L. and Kilgore, R. (1986) Measuring the anthelmintic efficacy of a new formulation of ivermectin. *Veterinary Medicine* 81, 1175–1177.

Yazwinski, T.A., Featherston, H.E., Tucker, C.A. and Johnson, Z. (1994a) Residual nematocidal effectiveness of ivermectin in cattle. *American Journal of Veterinary Research* 55, 1416–1419.

Yazwinski, T.A., Tucker, C.A. and Featherston, H.E. (1994b) Residual anthelmintic activity of abamectin in artificially infected calves. *Veterinary Record* 134, 195.

Yazwinski, T.A., Tucker, C.A. and Featherston, H.E. (1994c) Efficacy of doramectin against naturally acquired gastrointestinal nematode infections in cattle. *Veterinary Record* 135, 91–92.

Yazwinski, T.A., Tucker, C.A., Featherston, H.E. and Walstrom, D.J. (1997a) Comparative therapeutic efficacy of doramectin and ivermectin against naturally acquired nematode infections in cattle. *Veterinary Record* 140, 343–344.

Yazwinski, T.A., Johnson, E.G., Thompson, D.R., Drag, M.D., Zimmerman, G.L., Langholff, W.K., Holste, J.E. and Eagleson, J.S. (1997b) Nematocidal efficacy of eprinomectin, delivered topically, in naturally infected cattle. *American Journal of Veterinary Research* 58, 612–614.

Yazwinski, T.A., Tucker, C., Copeland, S., Yazwinski, T. and Guerino, F. (1999) Dose confirmation of moxidectin pour-on against natural nematode infections in lactating dairy cows. *Veterinary Parasitology* 86, 223–228.

Zimmerman, G.L., Hoberg, E.P. and Pankavich, J.A. (1992) Efficacy of orally administered moxidectin against naturally acquired gastrointestinal nematodes in cattle. *American Journal of Veterinary Research* 53, 1409–1410.

Chapter 6.2

Use of Macrocyclic Lactones to Control Cattle Parasites in Europe

J. Vercruysse and R. Rew

Importance and Epidemiology of Cattle Parasites in Europe

Gastrointestinal (GI) nematodes, lungworms and ectoparasites (lice, warbles, hornflies and mites) are, as in many other temperate regions of the world, the most commonly occurring and important parasites of cattle in Europe.

Due to demographic and environmental constraints, space for grazing cattle has become more and more reduced in Western Europe. Many farmers have replaced pastures with crops such as maize for animal feeding. The direct consequences of less pasture availability are that pasture management practices for nematode control have been abandoned in most places and that stocking density tends to be too high. Therefore, use of anthelmintics, is now the cornerstone of anthelmintic control in (Western) Europe. The macrocyclic lactones (MLs) are considered the product of choice not only because of their high efficacy against the adult and larval (and inhibited) stages of the common nematodes, but also because of their persistent efficacy and activity against arthropods.

Gastrointestinal nematodes

GI nematode infections of cattle involving the abomasal-dwelling *Ostertagia ostertagi*, *Trichostrongylus axei* and *Haemonchus placei*, and the intestinal-dwelling *Cooperia oncophora* and *Nematodirus helvetianus* are very common in temperate regions of Europe (Armour, 1989; Shaw *et al.*, 1998a,b). Of these, *O. ostertagi* and *C. oncophora* are the most common, with the former being the major cause of pathological damage. The relative importance of the different GI nematodes differs with host age because of acquired immunity. The acquisition of immunity depends on the cumulative exposure to infection; however, normally, a protective host resistance develops against *Nematodirus* and *Cooperia* within 1 year. *Ostertagia* and *Trichostrongylus* engender a protective immune response more slowly and are therefore also present in older cattle (Armour, 1989).

From the many epidemiological studies conducted in Europe, the following important facts about first grazing season (FSG) calves have emerged:

1. A considerable number of third-stage larvae (L_3s) can survive the winter on pasture. Sometimes the number are sufficient to precipitate disease in calves 3–4 weeks after (early) turnout on pasture. However, this is unusual, and the role of the surviving L_3s is rather to infect calves at a level which produces patent subclinical infections and ensures contamination of the pasture for the rest of the grazing season.
2. A high mortality of overwintered L_3s on the pastures occurs in spring, and only negligible numbers can usually be detected in June. This mortality combined with the dilution effect of the rapidly growing herbage renders most pastures safe until mid-July.
3. The eggs deposited in spring develop slowly; most reach the infective stage from mid-July onwards. If sufficient numbers of L_3s are ingested, parasitic gastroenteritis (PGE) can occur any time from July until October.
4. As autumn progresses and temperatures fall, an increasing proportion of the ingested L_3s – mainly *Ostertagia* – become inhibited (early L_4). Type II disease may occur when larvae mature synchronously in early spring; however, due to frequent housing treatment, this is becoming rare.

During the second and further grazing seasons, a strong acquired immunity develops and adult stock in endemic areas are highly immune to reinfection, being of comparatively little significance in the epidemiology.

Since FGS calves are the most susceptible to infections with GI nematodes, clinical and severe PGE is mainly seen in this age class. Shaw *et al.* (1998a,b) analysed 85 studies, covering a 26-year period and carried out in 13 countries in Western Europe, on GI nematode infections in FGS calves and identified a number of parameters that were associated with outbreaks of PGE in FGS calves in Western Europe. The mean initial age (and weight) at turnout of calves not receiving any chemoprophylaxis, was significantly associated with PGE outbreaks: 82% of calf herds ≤6 months of age had outbreaks of PGE, compared with only 33% of control calf herds more than 6 months of age. In 92% of trials where the geometric mean faecal egg count was ≥200 eggs per gram (EPG) on day 56, PGE outbreaks were observed, but where it was less than 200 EPG only 29% had PGE. The use of these two factors in assessing likelihood of PGE outbreaks in untreated calf groups in future FGS is therefore proposed. Higher stocking densities were significantly associated with higher pasture contamination; however, a significant association of PGE outbreaks with stocking density could not be confirmed despite the higher stocking densities (>10 ha^{-1}) being exclusively associated with outbreaks of PGE.

Although primarily a disease of FGS calves, ostertagiosis can nevertheless affect second grazing season animals, particularly, if they had little previous exposure to the parasite, since there is no significant age immunity to infections.

Lungworms

Dictyocaulus viviparus infections are very common in most parts of northern Western Europe; much less in southern regions around the Mediterranean Sea. Although infections have a high prevalence in certain regions, lungworm disease (parasitic bronchitis (PB) or husk) in cattle is much less common. Probably *D. viviparus* infections occur on many farms without causing serious clinical signs and result in the build-up of an adequate degree of immunity. In the epidemiology of lungworm disease, the source of primary infections of susceptible animals is a very important factor. The main sources for primary infections are overwintered larvae on pasture or contamination of pasture by carrier animals (Michel, 1969; Eysker *et al.*, 1993, 1994). A second important phenomenon is the very rapid translation of infective larvae to pasture throughout the grazing season. This and the pre-patent period of 3–4 weeks imply that the intervals between subsequent generations will be approximately 1 month. Disease may occur in the first generation when infection levels are high. Alternatively, a second generation can cause disease when primary infection is low. When the development of immunity is hampered due to very low primary or secondary infections, a third generation may cause disease. The seeming unpredictability of lungworm infections is due rather to the fact that the start of the initial infection is unknown; once this is known, the pattern becomes predictable (Ploeger and Eysker, 2000). Primary infection may occur in cattle of all age classes and during the whole pasture season. Traditionally, outbreaks of PB are seen mainly in FSG calves. In several European countries (e.g. the UK, Belgium and The Netherlands), there seems to be a significant increase in the number of incidents of PB. The Veterinary Investigation Diagnosis Analysis of the UK reported 226 and 520 incidents year^{-1} of PB in 1993 and 1997, respectively. The remarkable feature of this increase has been the involvement of adult cattle (>18 months) as the predominant age group. The risk factors thought to be involved in the appearance of a high level of disease in adult cattle were significant reductions in the level of vaccination, increased use of powerful anthelmintic strategies in successive grazing seasons, creating possible pharmacological barriers to the maintenance of immunity within herds, and favourable climatic factors.

Ectoparasites

Parasitic skin diseases present in Europe are economically important and include lice, mange and warble infestations (Fadok, 1984; Lonneux and Losson, 1996).

Lice are highly host-specific parasites that spend their complete life cycle on the host. The lice are spread by direct contact or contact with bedding or other inanimate objects that an infected animal has rubbed to relieve the itching sensation. Lice are a greater problem in winter (stabling period) when nutrition may be poor, crowding occurs, which facilitates spread of infection, and animals have long hair coats, which provide a good environment for louse reproduction. Exposure to cold and debilitating disease increases the likelihood of heavy infestation. Some animals may remain infected all year round and infect other animals the next winter. When this happens, the infected animals should be treated for lice when housed or in the autumn.

Psoroptic mange is caused by *Psoroptes ovis* and is the most common and important mange type in Western Europe. The condition is more common in winter than in summer months, and beef cattle (Charolais, Belgian White Blue) are more susceptible. Sarcoptic mange is becoming rare in Western Europe. Chorioptic mange (*Chorioptes bovis*) is most evident in the winter in stabled cattle, particularly dairy animals. Spread of mange occurs by contact as well as by inanimate objects, and is a greater problem in winter (stabling period).

Warbles in Europe are mainly larvae of the fly *Hypoderma bovis* and, to a lesser extent, *H. lineatum*. Animals are infected in autumn. During winter, larvae (*H. bovis*) migrate in the connective tissue and fat surrounding the spinal cord. The larvae migrate to the subcutaneous region over the back, usually in early spring.

Control of Gastrointestinal Nematodes

The control procedures against GI nematodes will be discussed according to age and/or type of breeding: FSG calves, heifers and young bulls, adult cows and cow–calf pairs.

First grazing season calves

Prevention of nematodes by chemoprophylaxis fits in the general concept of modern farm management where intensive farming is combined with minimal labour. A wide range of different effective chemoprophylactic systems has been developed to prevent outbreaks of PGE and control infections (for a review, see McKellar, 1994) in FGS calves. Suppressive anthelmintic medication during the first half of the grazing season, using various forms of intraruminal boli or carefully timed administration of anthelmintics, has proven to be highly effective in Western Europe for the control of GI nematodes of grazing calves during their first year (Shaw *et al.*, 1998a,b). The objective of this chemoprophylaxis is to limit the faecal egg output during the first half of the grazing season, thereby reducing the build-up of large numbers of larvae on pasture in late summer. A comprehensive review on GI nematode infections of FSG calves in Western Europe and the effect of chemoprophylaxis was given by Shaw *et al.* (1998a,b). Generally, all chemoprophylactic systems were found to give satisfactory results. No symptoms of PGE were observed in any of the chemoprophylactically treated groups, while these symptoms occurred in the control groups of 53 of the 85 studies. In addition, chemoprophylaxis resulted in a significant increase in weight gain. An interesting observation was that in studies with an outbreak of PGE in the untreated control group, the chemoprophylactically treated group had significantly higher faecal egg and pasture larval counts than treated groups in 'subclinical' studies. The overall weight gain in chemoprophylactically treated calves in 'clinical' studies was also significantly lower than the chemoprophylactically treated calves in 'subclinical' studies. 'Subclinical' control groups gain significantly more weight during the grazing season than 'clinical' control groups ($P < 0.001$), with the former gaining on average 85 kg (540 g day^{-1}), and the latter only 60 kg (370 g day^{-1}). 'Subclinical' chemoprophylactically treated groups also gained significantly more weight than 'clinical' chemoprophylactic groups ($P < 0.005$). On average, the former groups gained 110 kg (690 g day^{-1}), compared with 95 kg (600 g day^{-1}) in the latter. These results indicate that on heavily infected pastures, chemoprophylaxis will prevent PGE, but calves will still suffer some production losses. Other observations were made with regard to the type and especially the duration afforded by a particular chemoprophylactic system: there was a highly significant negative relationship between maximum faecal egg counts in the chemoprophylactically treated calves and the proportion of a trial covered by the different chemoprophylactic systems. A highly significant positive relationship between the weight gained in the chemoprophylactically treated groups and the estimated duration of the various chemoprophylactic systems was also found.

Although chemoprophylaxis theoretically can be achieved with all anthelmintics (Shaw *et al.*, 1998a,b), MLs are the preferred choice, owing to their prolonged effect, which reduces the establishment of incoming larvae over several weeks. The following paragraph describes the results from studies in which MLs were used to prevent infections with GI nematodes. Tables are used to summarize results.

The early concept of chemoprophylaxis was to reduce, as much as possible, the excretion of eggs for at least 3 months, that is the interval between treatments was scheduled considering the pre-patent period of

GI nematodes (3 weeks) and the duration of persistence of the ML. As an example, for ivermectin, the interval between treatments was scheduled to be 5 weeks: a 3-week pre-patent period and the period of high persistence, which was considered to be 2 weeks; the two installed ivermectin programmes were treatments at weeks 3, 8 and 13 after turnout, and at weeks 3 and 8 after turnout, which were designed to suppress egg excretion during 18 and 13 weeks, respectively. At a later stage, programmes were less suppressive, that is a certain numbers of eggs were allowed to develop, not to the extent that these pasture infections would be likely to cause disease, but sufficient that the small number of GI parasites which do reach maturity might be able to stimulate immunity in the animals. An example here is the doramectin programme of treatment 0 and 8 weeks after turnout. Between the two treatments, animals remain unprotected for approximately 2 weeks.

Due to differences in management, pasture infectivity, climate, etc., it is difficult to corroborate which programme is to be preferred. It must be clear that any chemoprophylactic programme has beneficial effects as long as it is adapted to the epidemiological situation on the farm.

Ivermectin (abamectin)

The first chemoprophylactic programmes developed with MLs were using ivermectin. In the 1980s, mainly the 3–8–13 and 3–8 programmes were advocated. Although highly efficient, the programmes were labour intensive and, therefore, in practice, difficult to implement. Many studies suggested that they were too suppressive and could interfere with normal immunity developments (see pages 239–240). Therefore, from the 1990s, a less rigid programme of 0 and 6 weeks after turnout was promoted. At the beginning of the 1990s, an indwelling rumino-reticular ivermectin sustained-release bolus was commercialized.

THE 3–8–(13) IVERMECTIN PROGRAMME. Many published studies have demonstrated the success of the 3–8–(13) programme in controlling PGE in FSG calves (Tables 6.2.1and 6.2.2). The necessity of three treatments (weeks 0, 8 and 13) instead of two (weeks 3 and 8) is mainly in the event that the pasture challenge is heavy and the grazing season is longer (Armour *et al.*, 1987). The first studies (Taylor *et al.*, 1985, 1986, 1988; Armour *et al.*, 1987, 1988) assessed the efficacy of the programme in controlling both PGE and PB, with focus on control or eradication of *D. viviparus* or for farmers who cannot use an oral lungworm vaccine (see pages 240–241). The more recent studies focused more on the use of the programme in preventing PGE (Hollanders *et al.*, 1992; Taylor *et al.*, 1995a,b; Vercruysse *et al.*, 1995a). In the first studies (Taylor *et al.*, 1985), questions about second season immunity and grazing season duration became a concern, particularly in view of the finding of Armour *et al.* (1988) of high burdens of inhibited

Table 6.2.1. GI nematodes: chemoprophylaxis by 3–8–13 scheme with ivermectin.

Location	Length of pasture season (days)	Maximum faecal egg counts		Reduction[a] of pasture infection (%) (at housing)	Weight gain (kg)		References
		Control	Treated		Control	Treated	
Belgium	154	700[b]	260	L 88	14	109	Vercruysse *et al.* (1995a)
Belgium	163	1500[b]	<50	L >99	14	64	Hollanders *et al.* (1992)
Denmark	170	115	<50	L 39	60	93	Satrija *et al.* (1996)
England	145	700	<50	ND	50	70	Jacobs *et al.* (1989)
Ireland (Northern)	162	1400	<200	L >90	65	83	Taylor *et al.* (1995a)
Scotland	127	>300[b]	80	ND	ND	ND	Armour *et al.* (1987)

[a]L, based on pasture larval counts.
[b]Salvage treatment performed.
ND, not determined.

Table 6.2.2. GI nematodes: chemoprophylaxis by 3–8 scheme with ivermectin.

Location	Length of pasture season (days)	Maximum faecal egg counts		Reduction[a] of pasture infection (%) (at housing)	Weight gain (kg)		References
		Control	Treated		Control	Treated	
England	160 (1983)	500	50	L 80–90	ND	ND	Jacobs *et al.* (1987)
	152 (1984)	>1000[b]	100	L 70–90	40–60	80–90	
The Netherlands	163	1400	±70	ND	ND	ND	Eysker *et al.* (1988)
Scotland	127[b]	>300	>800	ND	ND	ND	Armour *et al.* (1987)

[a]L, based on pasture larval counts.
[b]Salvage treatment performed.
ND, not determined.

larvae in yearlings that had received a 3–8–13 programme as calves (see pages 239–240).

THE 0–6 IVERMECTIN/ABAMECTIN PROGRAMME. Ivermectin application at turnout and 6 weeks later would virtually prevent establishment of *O. ostertagi* infections during the first 9 weeks after turnout. Clear and substantial differences between a treated and a non-treated group in parasitological and serological parameters and weight gain have been observed (Table 6.2.3). The 0–6 ivermectin/abamectin scheme has to be considered among the least suppressive nematode anthelmintic treatment schemes because it will not suppress nematode infection for much longer than 2 months. Both Pollmeier *et al.* (1992) and Eysker *et al.* (1998) observed a build-up of nematode infections at the end of the grazing season in treated animals and concluded that anthelmintic treatment at housing is essential when this system is used. In conclusion, this scheme may be relevant when turnout is late, as seen in The Netherlands where, on the majority of the farms, calves are not turned out before June (Eysker *et al.*, 1998). Consequently, initial infections are usually low and the first grazing season will be short. Thus systems with a long suppression of nematode infections are not required. They may even be counterproductive because it has been demonstrated that overprotection of calves against nematodes in the first grazing season may result in economic losses on nematodosis in the second grazing season (Eysker *et al.*, 1998).

THE IVOMEC SR BOLUS. In 1985, Pope *et al.* developed an indwelling rumino-reticular ivermectin sustained-release bolus (I-SRB). Studies with different prototype boluses (Egerton *et al.*, 1986; Alva-Vades *et al.*, 1988; Zimmerman *et al.*, 1991; Williams and Plue, 1992; Baggot *et al.*, 1994) have indicated that ivermectin delivered via an I-SRB effectively removed existing infections and prevented establishment of infections with common GI and pulmonary nematodes, including inhibited *O. ostertagi*. The I-SRB commercialized in Western Europe is designed to release ivermectin at 12 mg day^{-1} for 135 days and is recommended for use in cattle weighing 100–300 kg at the time of treatment. Several studies have been conducted to evaluate the prophylactic efficacy of the I-SRB, given at turnout, in preventing PGE in FSG calves under European conditions (Table 6.2.4). The I-SRB was highly effective in suppressing faecal nematode egg output and preventing PGE and has to be considered among the highest suppressive nematode anthelmintic treatment schemes because it will suppress nematode infection for as long as 4 months.

LATE-SEASON TACTICAL TREATMENTS WITH IVERMECTIN. Steffan and Nansen (1990) and Satrija *et al.* (1996) found that late-season tactical treatment (metaphylaxis) with ivermectin given at 4-weekly intervals can significantly alleviate the pathogenic effects of GI nematodes and improve

Table 6.2.3. GI nematodes: chemoprophylaxis by abamectin/ivermectin 0–6 programme.

Location	Length of pasture season (days)	Maximum faecal egg counts		Reduction[a] of pasture infection (%) (at housing)	Weight gain (kg)		References
		Control	Treated		Control	Treated	
Belgium[b]	126	110	99	L 80	94	90	Vercruysse *et al.* (1996)
England	166	160	65	L >95 (*Ostertagia*)	58	80	Jacobs *et al.* (1995)
Germany	168	520	<50	L 68	86.7	105.8	Pollmeier *et al.* (1992)
	182	150	<50	L 33	85	100	Schnieder *et al.* (1996c)
The Netherlands	154	128	25	L 82	71.4	83.5	Eysker *et al.* (1998)

[a]L, reduction in pasture larval counts.
[b]Controls were treated with doramectin at turnout.

Table 6.2.4. GI nematodes: chemoprophylaxis with Ivomec bolus.

Location	Length of pasture season (days)	Maximum faecal egg counts		Reduction[a] of pasture infection (%) (at housing)	Weight gain (kg)		References
		Control	Treated		Control	Treated	
Belgium	140	288	0	T 100 (*Ostertagia* and *Cooperia*)	118	117	Claerebout *et al.* (1994)
	140 (1993)	195	<1	T 100 (*Ostertagia*)	112	135	Claerebout *et al.* (1997)
	154 (1994)	56	1	T 91	95	133	
England	142	376[b]	31	ND	80	121	Taylor *et al.* (1997)
Germany	154	140	< 50	L 20	116	142	Schnieder *et al.* (1996a)

[a]L, based on pasture larval counts; T, based on worm counts of tracers.
[b]Salvage treatment performed.

performance in calves exposed to a continuous high larval intake on a permanent pasture. Taylor *et al.* (1995a) did not observe significant performance differences between calves treated according to a 3–8–13 ivermectin programme and calves treated with ivermectin at 10, 15 and 20 weeks after turnout. Mid-season administration of an I-SRB has also been recommended (Pitt *et al.*, 1996; Claerebout, 1998). Because mid-season treatment does not interfere with host–parasite interactions during the first half of the grazing season, it has been hypothesized that metaphylaxis does not interfere with the build-up of acquired immunity (Steffan and Nansen, 1990). Moreover, since mid-season treatment allows for recycling of larval populations earlier in the grazing season, the animals are also exposed to infestation levels during the second half of the season. Claerebout (1998) found that mid-season administration with an I-SRB effectively suppressed GI nematode infections during the second half of the first grazing season but could not make firm conclusions on the effect of metaphylactic treatment on acquired immunity (see also pages 239–240). In general, when metaphylaxis and prophylaxis were compared within studies, weight gains were usually better in the prophylactically treated animals (Satrija *et al.*, 1996). Late-season tactical treatments may also be used where worm control for various reasons has not been planned in advance.

Doramectin

Because of the longer persistent efficacy compared with ivermectin, intervals between doramectin treatments can be extended. In contrast to ivermectin, the chemoprophylaxis programmes with doramectin had a less suppressive approach, and mainly the 0–8 programme was considered to be the programme of choice for doramectin. Treatment at turnout is also convenient to the farmers because the calves are treated at a time when animals are handled routinely; thus only the follow-up dose requires handling of the animals on pasture.

THE 0–8 (10) DORAMECTIN PROGRAMME. A dosing interval of 8 weeks was selected for doramectin to allow a certain number of eggs to develop, but not to the extent that these pasture infections would be likely to cause disease (Vercruysse *et al.*, 1993, 1995b). The small number of GI parasites that do reach maturity might be able to stimulate immunity in the animals. Table 6.2.5 summarizes the results of the different studies published in Western Europe (injectable doramectin). All studies clearly demonstrated that with the 0–8 week schedule of doramectin treatment, effective control of PGE was achieved: lower faecal egg counts and pasture infection levels resulted in higher weight gains. Only a limited number of studies were published on the 0–10 doramectin programme (Borgsteede *et al.*, 1995; Le Stang *et al.*, 1995; Satrija and Nansen, 1996).

Table 6.2.5. Control of GI nematodes by 0–8 scheme with doramectin.

Location	Length of pasture season (days)	Maximum faecal egg counts		Reduction[a] of pasture infection (%) (at housing)	Weight gain (kg)		References
		Control	Treated		Control	Treated	
Belgium	161	352	<50	T 94	97	130	Vercruysse *et al.* (1995b)
	154	700[b]	304	L 77.1	14	115	Vercruysse *et al.* (1995a)
England	145	676	291	ND	80	117.5	Taylor *et al.* (1997)
	154	250	<50	L 74	114	140	Parr *et al.* (1995)
	161	>900	180	L 88 (*Ostertagia*) L 13 (*Cooperia*) L 93 (*Nematodirus*)	25	53	Fisher *et al.* (1995)
France	Study (a) 190	260[b]	<50	L 76.6	65	130	Le Stang *et al.* (1995)
	Study (b) 144	180	50	L 57.1	15	95	
Ireland	179	—	—	L 0	150	171	Talty and McSweeney (1996)
The Netherlands	182	350	150	ND	60	100	Borgsteede *et al.* (1995)

[a]L, based on pasture larval counts; T, based on worm counts of tracers.
[b]Salvage treatment.

According to Le Stang *et al.* (1995) and Borgsteede *et al.* (1995), no significant differences were seen between the 0–8 and 0–10 schemes under moderate to high challenge conditions.

Moxidectin

Only a limited number of published studies are available on the effects of strategic early-season treatments with moxidectin on trichostrongylosis in FSG calves on a permanent pasture. One possible reason is the limited persistent efficacy of moxidectin against *Cooperia*, with the possible consequence that high levels of infections with this parasite may occur. Yang *et al.* (1996) showed that treatment with moxidectin at turnout and again 8 weeks later significantly suppressed faecal egg excretion in the early part of the season, leading to reduced herbage infectivity and parasitism over the rest of the season.

Eprinomectin

Eprinomectin has been developed recently and, until now, only a few studies (Epe *et al.*, 1999; Dorny *et al.*, 2000) on a 0–8 programme have been published. Similarly to a 0–8 doramectin programme, the eprinomectin pour-on treatments at turnout and again after 8 weeks on pasture controlled GI strongyle infections in FGS calves during the entire grazing season and kept pasture larval counts relatively low until housing. This was associated in one study (Dorny *et al.*, 2000) with a significantly better weight gain of 20 kg when compared with a non-chemoprophylactically treated control group. No differences in weight gain were seen in another study (Epe *et al.*, 1999).

Miscellaneous

SINGLE TREATMENT COMBINED OR NOT WITH LATE TURNOUT OR EARLY HOUSING. A single ivermectin treatment 3 weeks after early turnout did not prevent heavy helminth infections (Eysker, 1986). However, such single treatments of calves 3 weeks after a late turnout on mown pasture appears to be a promising way of preventing heavy infections (Eysker, 1986; Eysker *et al.*, 1988). Yang *et al.* (1996) found that a single moxidectin treatment at turnout conferred a significant lowering of infection levels. The discrepancy between the results from Eysker (1986) and Yang *et al.* (1996) may be attributed to differences in numbers of overwintered larval populations and to climatic conditions, rather than to differences in persistent efficacy between the two MLs. Doramectin given as a single dose at turnout, at the start of a short grazing season (126 days), controlled parasite infections sufficiently in FGS calves to match weight gains when compared with an effective 0–6 ivermectin programme with excellent parasite control (Vercruysse *et al.*, 1996).

HOUSING TREATMENT. Only a few published studies are available on the effects of a housing treatment. Gosselin *et al.* (1995) observed that heifers, treated at housing with doramectin, gained during the subsequent stabling period 23 kg more than untreated heifers. This difference in weight gains was due to the efficacy of doramectin against both GI nematodes and lice. A problem with housing treatment may be the assessment of the parasite level at that moment of the host–parasite contact that has been experienced during the grazing season (with and without treatment). According to Dorny *et al.* (1999), pepsinogen levels determined at housing gave the clearest division between calves who do not need treatment at housing and calves who have had too much exposure during their FGS (pepsinogen level >3.5 U tyr), and presumably treatment would be a well-considered choice.

USE OF MLS UNDER EXTENSIVE GRAZING CONDITIONS IN THE ALPINE REGION. Hertzberg and Eckert (1996) and Hertzberg *et al.* (1998) studied the necessity of protecting FSG calves against GI nematodes (and lungworms) under the extensive grazing conditions in the alpine region. They investigated, among others, the metaphylactic effect of a single injection of moxidectin administered before transfer to the alpine pastures in June and the effect of doramectin pour-on administered before and after the alpine grazing season. Both control systems provided good protection; however, infection levels with GI nematodes may be considered to be of minor relevance under the extensive grazing conditions on high alpine pastures. Strategic treatments in these regions may be more important to prevent outbreaks of PB (Hertzberg *et al.*, 1998)

Heifers and young bulls – second grazing season

The requirement for control strategies in the second grazing season (SGS) depends on the immunity acquired by the host in the first grazing season. Whether the level of immunity acquired will afford sufficient protection to second year grazing animals depends on the level of first year exposure and the challenge they encounter during their SGS. Some of the anthelmintic strategies could readily be adapted to second year animals, if this is considered necessary. Ploeger *et al.* (2000) used a questionnaire (86 farms randomly distributed) to enquire about use of anthelmintic drugs in SGS cattle in The Netherlands. On 60.2% of the farms, SGS cattle were treated at least once with an anthelmintic drug, and on 13.4% at least twice. The percentage for the SGS animals indicates that the use of anthelmintic drugs in those animals has increased over the last 10–15 years. On 36.1% of the farms ($n = 30$), SGS animals received a preventive treatment, for example (for MLs) on four farms an I-SRB, and on three and one farms a doramectin and ivermectin programme, respectively. In SGS animals the interval between two treatments was 9–11 weeks, that is longer than in FSG calves. Considering the intensive use of MLs in SGS animals, surprisingly few data are available on the (economic) consequences of such treatments.

Dorchies and Gosselin (1996) and Gosselin *et al.* (1998) investigated the efficacy of a single application of doramectin pour-on, at turnout, in controlling PGE in SGS cattle. Treated animals were compared with control animals grazed on separate but similar pastures. Control animals remained untreated (Dorchies and Gosselin, 1996) or were treated, according to normal farm practice, at mid-season with levamisole or fenbendazole (Gosselin *et al.*, 1998). In the four studies, doramectin-treated cattle showed weight gain advantages of 11, 35, 38 and 44 kg. The weight gain improvement was observed despite little measurable effects of treatment on the level of infection (egg counts). Mage *et al.* (1998) compared three treatment programmes with moxidectin, that is a 0–10 programme, treatment at turnout and treatment 16 weeks after turnout on animals in their SGS. Infection levels were low and small differences in faecal egg counts and pasture larval counts between treated and control groups were observed. However, the 0–10 programme and treatment 16 weeks after turnout resulted in substantially higher weight gains.

Until now, it seems that there are insufficient data available uniformly to recommend systematic treatment of animals during their SGS.

Cows

For dairy cows, the degree to which subclinical losses as measured by milk production due to GI nematodes can be prevented has been debated and investigated over more than four decades. Recently, Gross *et al.* (1999) reviewed the results of more than 80 studies on GI parasitism and the impact of anthelmintic treatment on milk production in dairy cattle. In 70 of 87 experiments, there was an increase in milk production after anthelmintic treatment, with a median increase of 0.63 kg day^{-1}. All anthelmintics were included in the survey and no consistent and significant differences were seen between anthelmintics; however, MLs (e.g. Ploeger *et al.*, 1989; Walsh *et al.*, 1995; Kloosterman *et al.*, 1996) seem to provide a more consistent positive response in a greater proportion of studies. In the selection of the anthelmintic, and in the development of herd parasite control programmes, consideration should be given to the milk withdrawal times of the products. In Europe, the only registered ML without a withdrawal time is eprinomectin (Alvinerie *et al.*, 1999). Other MLs have withdrawal times of 28 days and longer, or are not recommended to be used in dairy cattle. However, many problems and questions still need to be solved before uniform recommendations can be given: for example, whether the intensity of infection with GI nematodes can be diagnosed by available tests and when treatment should be given (mid-lactation, pre-calving or early lactation). Until now, no published studies are available on the utilization of eprinomectin in dairy cows.

Treatment of cows may also reduce calving intervals, increase calf weight, etc.; however, only results from small-scale studies are available.

Cow–calf pairs

Cow–calf beef production systems in Europe consist of late autumn- up to early spring-born calves remaining with their dams on pasture until weaned in summer or late autumn. The presence of the dams alongside their calves is unlikely to result in pasture contamination, which is dangerous to the calves (Laiblin *et al.*, 1996; Agneessens *et al.*, 1997). The effect of strategic anthelmintic treatments of beef cattle and cow–calf pairs on body weight gain is well established and recently has been underlined by studies in North America (see pages 251–255). However, in Europe, the need for a strategic anthelmintic control of strongyle infections in cow–calf systems has been a matter of discussion for some time (Wacker *et al.*, 1999) and further studies are needed. Taylor *et al.* (1995b) stress that to give advice on the merits of anthelmintic treatment of suckling calves in their second season at grass, it would be desirable to obtain accurate birth dates, an estimate of the degree of larval contamination on their grazing during the previous year and use of any anthelmintic prophylaxis that might have reduced the animals' exposure to infective larvae. This information would make it possible to assess more certainly the risk of a clinical or substantial subclinical parasitic infections developing during their second year. In the absence of such information, two treatments with ivermection at 3 and 8 weeks after turnout has been shown to give adequate prophylaxis in the face of a severe parasitic challenge (Taylor *et al.*, 1995b).

The Effect of Chemoprophylaxis on Acquired Immunity to Gastrointestinal Nematodes in Cattle

Chemoprophylaxis, especially when MLs are used, strongly reduces the number of GI nematodes in FGS calves and prevents the build-up of a high pasture infection during the second part of the grazing season. Because acquisition of immunity depends on an animal's cumulative experience of infection, a strong reduction of exposure to GI nematodes results in a reduced level of acquired immunity at the end of the first grazing season. A negative effect of chemoprophylaxis on the acquired resistance of the animals to an artificial challenge infection has been clearly demonstrated (Claerebout *et al.*, 1996, 1998a,b). Moreover, the level of acquired resistance was negatively related to the degree of suppression of host–parasite contact (Fisher and Jacobs, 1995; Claerebout *et al.*, 1996, 1998a). Under natural conditions, studies on the effect of chemoprophylaxis on acquired resistance to reinfection have given conflicting results. Claerebout *et al.* (1998a) found that resistance to a natural challenge infection was diminished in animals receiving an I-SRB at turnout and to a lesser extent in animals treated with doramectin at turnout and 8 weeks later, especially early in the SGS, when a large overwintered population of larvae was present. Similar observations were made with anthelmintics other than MLs (Herbert and Probert, 1987). In contrast, in a number of other studies, only minor differences in worm counts and/or egg counts were found, if any, between previously treated and untreated yearlings during a natural SGS (Jacobs *et al.*, 1989; Taylor *et al.*, 1995a; Satrija *et al.*, 1996; Schnieder *et al.*, 1996a,b; Claerebout *et al.*, 1997). The low levels of natural challenge infections early in the SGS, together with the timing of the assessment of the acquired resistance of the animals may explain why in most natural SGS only limited differences in acquired resistance between previously treated and untreated yearlings are observed. Thus when yearlings that have received prophylactic treatments in their FGS are turned out on a pasture with a low infestation level in the second year, since a low challenge infection is unlikely to unveil an incomplete resistance/resilience, it may boost further development of acquired immunity.

Control of Lungworms

In many studies, primarily designed to study the chemoprophylaxis of GI nematodes, infections with lungworms were also present and efficiently controlled (Taylor *et al.*, 1985, 1988; Armour *et al.*, 1987; Parr *et al.*, 1995; Vercruysse *et al.*, 1998). Other studies report on efficacy of the same programmes with MLs on farms where mainly PB was considered to be important (Taylor *et al.*, 1986; Talty *et al.*, 1996; Vercruysse *et al.*, 1998). The control of lungworms through chemoprophylaxis is a matter of debate, and it is important to sound three notes of caution.

1. Due to the speed and unpredictability of lungworm flare ups in the field, it is important to realize that cattle under any early season prophylactic programme may be at risk in a late autumn lungworm challenge (Vercruysse *et al.*, 1987).
2. There has been considerable concern that effective lungworm prophylaxis in FGS calves may so suppress lungworm infections that a natural immunity fails to develop. This can mean that adult cattle, and particularly those in dairy herds, can suffer outbreaks of PB, which may be expensive in terms of lost milk production. However, it was noted that,

if infection with *D. viviparus* took place during the FGS when the ML regimes were applied, that cattle in their second season were mostly functionally immune to reinfection (Taylor *et al.*, 1988, 1997, 2000; Jacobs *et al.*, 1989, 1995; Schnieder *et al.*, 1996b; Taltry *et al.*, 1996; Vercruysse *et al.*, 1998). Taylor *et al.* (2000) found that in situations of heavy (experimental) lungworm infection and simultaneous systemic doramectin treatment, significant immunity can be stimulated without further development of the larvae. This must have taken place after the early death of the larvae and systematic antigenic release, both from the larval sheath and somatic antigens. However, it may well be that the degree of immunity is dependent on the degree of larval intake and that when the larval intake is low (as often in natural conditions), protection may still be insufficient. Studies indicated that where it is considered necessary to use lungworm vaccination in addition to an I-SRB or doramectin 0–8 week treatment programme (Grimshaw *et al.*, 1996; Jacobs *et al.*, 1996), the compatibility of these programmes with lungworm vaccination will allow development of a protective level of immunity to *D. viviparus*.

3. It is important to stress that infection with *D. viviparus* may well not take place during the FGS and animals will remain fully susceptible to PB in their subsequent grazing years. Because of this, many studies on chemoprophylaxis against PB utilized experimental infections with *D. viviparus*, to be sure that infections with lungworm larvae would occur, however biasing the natural situation!

Control of Ectoparasites

In Europe, specific programmes to control ectoparasites are not commonly practised, except for hypodermosis. Trials for the control of cattle grub are facilitated by the high efficacy of the MLs against all infesting larvae of *Hypoderma* (Boulard, 1999). Another advantage of the MLs is the fact that they induce a slow death of the larvae of the first stage, limiting risks of secondary hypersensitivity reactions (Boulard *et al.*, 1991). Also, the use of microdoses of ivermectin (see before) contributed largely to the intensive use of the drug. Eradication programmes have been initiated successfully in the UK, (ex) Czechoslovakia, Ireland, France and Switzerland (Boulard, 1999). Eradication can only be achieved following a strict organization of the treatment courses and a follow-up of the efficacy of elimination. All national programmes have stressed a second phase of surveillance against the resurgence of the disease.

During the pasture season, it is known that programmes with MLs used to control endoparasites will eventually also control ectoparasites; however, published data on efficacy and economic benefits are scarce. The I-SRB controlled *P. ovis* mites and prevented the establishment of *P. ovis* mites on cattle weighing more than 300 kg at the time of treatment (Rehbein *et al.*, 1997).

Often, treatments with MLs are recommended at housing to control both endo- and ectoparasites; however, again few (published) data are available to confirm the efficiency of such treatments

References

Agneessens, J., Dorny, P., Hollanders, W., Claerebout, E. and Vercruysse, J. (1997) Epidemiological observations on gastrointestinal nematode infections in grazing cow–calf pairs in Belgium. *Veterinary Parasitology* 69, 65–75.

Alva-Valdes, R., Wallace, D.H., Egerton, J.R., Benz, G.W., Gross, S.J., Wooden, J.W. and Reuter, V.E. (1988) Prophylactic efficacy of an ivermectin sustained-release bolus against challenge exposure with gastrointestinal and pulmonary nematode infective larvae in calves. *American Journal of Veterinary Research* 40, 1726–1728.

Alvinerie, M., Sutra, J.F., Galtier, P. and Mage, C. (1999) Pharmacokinetics of eprinomectin in plasma and milk following topical administration to lactating dairy cattle. *Research in Veterinary Science* 67, 229–232.

Armour, J. (1989) The influence of host immunity on the epidemiology of strongyle infections in cattle. *Veterinary Parasitology* 32, 5–19.

Armour, J., Bairden, K., Pirie, H.M. and Ryan, W.G. (1987) Control of parasitic bronchitis and gastroenteritis in grazing cattle by strategic prophylaxis with ivermectin. *Veterinary Record* 121, 5–8.

Armour, J., Bairden, K. and Ryan, W.G. (1988) Immunity of ivermectin treated cattle to challenge from helminth parasites in the following season. *Veterinary Record* 122, 223–225.

Baggott, D.G., Ross, D.B., Preston, J.M. and Gross, S.J. (1994) Nematode burdens and productivity of grazing cattle treated with a prototype sustained-release bolus containing ivermectin. *Veterinary Record* 135, 503–506.

Borgsteede, F.H.M., van Walbeek, M.G., Gaasenbeek, C.P.H. and van der Linden, J.N. (1995) Doramectin zur Verhinderung des Auftretens der Trichostrongylose bei Kälbern. *Der Praktische Tierarzt* 3, 217–220.

Boulard, C. (1999) La lutte contre l'hypodermose en Europe. *Le Point Vétérinaire* 30, 301–307.

Boulard, C., Argenté, G. and Hillion, E. (1991) Effets indésirables des antiparasitaires (hypersensibilité). *Receuil de Médecine Vétérinaire* 167, 1127–1132.

Claerebout, E. (1998) The effect of chemoprophylaxis on acquired immunity to gastrointestinal nematodes in cattle. PhD thesis, Faculty of Veterinary Medicine, Ghent University, Belgium.

Claerebout, E., Hollanders, W., De Cock, H., Vercruysse, J. and Hilderson, H. (1994) A field study of the ivermectin sustained-release bolus in the seasonal control of gastrointestinal nematode parasitism in first season grazing calves. *Journal of Veterinary Pharmacology and Therapeutics* 17, 232–236.

Claerebout, E., Hilderson, H., Meeus, P., De Marez, T., Behnke, J., Huntley, J. and Vercruysse, J. (1996) The effect of truncated infections with *Ostertagia ostertagi* on the immune response in calves. *Veterinary Parasitology* 66, 225–239.

Claerebout, E., Hollanders, W., Dorny, P. and Vercruysse, J. (1997) Effect of chemoprophylaxis with an ivermectin sustained-release bolus on acquired resistance to gastrointestinal nematodes in cattle. *Veterinary Record* 141, 441–445.

Claerebout, E., Dorny, P., Vercruysse, J., Agneessens, J. and Demeulenaere, D. (1998a) Effects of preventive anthelmintic treatment on acquired resistance to gastrointestinal nematodes in naturally infected cattle. *Veterinary Parasitology* 76, 287–303.

Claerebout, E., Vercruysse, J., Dorny, P., Demeulenaere, D. and Dereu, A. (1998b) The effects of different levels of exposure on acquired resistance to gastrointestinal nematodes in artificially infected cattle. *Veterinary Parasitology* 75, 153–167.

Dorchies, P. and Gossellin, J. (1996) Intérêt d'une injection unique de doramectine administrée à la mise à l'herbe dans le contrôle des strongles gastrointestinaux chez les bovis en seconde saison de pâturage. Résultats de deux essais conduits en France. *Revue de Médecine Vétérinaire* 147, 145–150.

Dorny, P., Shaw, D.J. and Vercruysse, J. (1999) The determination at housing of exposure to gastrointestinal nematode infections in first-grazing season calves. *Veterinary Parasitology* 80, 325–340.

Dorny, P., Demeulenaere, D., Smets, K. and Vercruysse, J. (2000) Control of gastrointestinal nematodes in first season grazing calves by two strategic treatments with eprinomectin. *Veterinary Parasitology* 89, 277–286.

Egerton, J.R., Suhayda, D. and Eary, C.H. (1986) Prophylaxis of nematode infections in cattle with an indwelling rumino-reticular ivermectin sustained release bolus. *Veterinary Parasitology* 22, 67–75.

Epe, C., Woidtke, S., Pape, M., Heise, M., Kraemer, F., Kohlmetz, C. and Schnieder, T. (1999) Strategic control of gastrointestinal nematode and lungworm infections with eprinomectin at turnout and eight weeks later. *Veterinary Record* 144, 380–382.

Eysker, M. (1986) The prophylactic effect of ivermectin treatment of calves, three weeks after turnout, on gastro-intestinal helminthiasis. *Veterinary Parasitology* 22, 95–103.

Eysker, M., Kooyman, F.N.J. and Wemmenhove, R. (1988) The prophylactic effect of ivermectin treatments on gastrointestinal helminthiasis of calves turned out early on pasture or late on mown pasture. *Veterinary Parasitology* 27, 345–352.

Eysker, M., Saatkamp, H.W. and Kloosterman, A. (1993) Infection build-up and development of immunity in calves following primary *Dictyocaulus viviparus* infections of different levels at the beginning or in the middle of the grazing season. *Veterinary Parasitology* 49, 243–254.

Eysker, M., Boersema, J.H., Cornelissen, J.B.W.J., Kooyman, F.N.J., De Leeuw, W.A. and Saatkamp, H.W. (1994) An experimental field study on the build up of lungworm infections in cattle. *Veterinary Quarterly* 16, 144–147.

Eysker, M., Boersema, J.H., Githiori, J.B. and Kooyman F.N.J. (1998) Evaluation of the effect of ivermectin administered topically at zero and six weeks after turnout on gastrointestinal nematode infection in first season grazing cattle. *Veterinary Parasitology* 78, 277–286.

Fadok, V.A. (1984) Parasitic skin diseases of large animals. *Veterinary Clinics of North America: Large Animal Practice* 6, 3–26.

Fisher, M.A. and Jacobs, D.E. (1995) Influence of chemoprophylaxis on protective immunity to nematodes in cattle: a two-year study comparing four control strategies. *Veterinary Record* 137, 581–585.

Fisher, M.A., Jacobs, D.E., Hutchinson, M.J. and Simon, A.J. (1995) Evaluation of doramectin in a programme for season-long control of parasitic gastroenteritis in calves. *Veterinary Record* 137, 281–284.

Gossellin, J., Couquet, C., Joly, B. and Franqueville, P. (1995) Efficacité de la doramectine en traitement de rentrée à l'étable contre les strongles gastro-intestinaux et les ectoparasites des bovins. *Bulletin de la Société Vétérinaire Pratique de France* 79, 381–391.

Gossellin, J., Dorchies, P., Mage, C. and Vercruysse, J. (1998) Control of gastro-intestinal trichostrongyles with a single application of doramectin pour-on at turn-out in second-season beef cattle. *Revue de Médecine Vétérinaire* 149, 331–338.

Grimshaw,W.T.R., Hong, C., Webster, R. and Hunt, K.R. (1996) Development of immunity to lungworm in vaccinated calves treated with an ivermectin sustained released bolus or an oxfendazole pulse release bolus at turnout. *Veterinary Parasitology* 62, 119–124.

Gross, S.J., Ryan, W.G. and Ploeger, H.W. (1999) Anthelmintic treatment of dairy cows and its effect on milk production. *Veterinary Record* 144, 581–587.

Herbert, I.V. and Probert, A.J. (1987) Use of an oxfendazole pulse release bolus in calves exposed to natural subclinical infection with gastrointestinal nematodes. *Veterinary Record* 121, 536–540.

Hertzberg, H. and Eckert, J. (1996) Epidemiologie und Prophylaxe des des Magen-Darm-und Lungenwurmbefalls bei erstsömmrigen Rindern unter alpinen Weidenbedingungen. *Wiener Tierärztliche Monatschrift* 83, 202–209.

Hertzberg, H., Ochs, H., Perl, R. and Tschopp, A. (1998) Prophylaxe des Magen-Darm- und Lungenwurmbefalls bei gealpten Jungrindern: Einsatz von Doramectin pour-on bei Alpauftrieb und Alpabtrieb. *Schweizer Archiv für Tierheilkunde* 140, 419–426.

Hollanders, W., Berghen, P., Dorny, P., Hilderson, H., Vercruysse, J. and Ryan, W.G. (1992) Prevention of parasitic gastroenteritis and parasitic bronchitis in first and second season grazing cattle. *Veterinary Record* 130, 355–356.

Jacobs, D.E., Fox, M.T. and Ryan, W.G. (1987) Early season parasitic gastroenteritis in calves and its prevention with ivermectin. *Veterinary Record* 120, 29–31.

Jacobs, D.E., Foster, J., Gowling, G., Pilkington, J.G., Fox, M.T. and Ryan, W.G. (1989) Comparative study of early-season prophylaxis using ivermectin with lungworm vaccination in the control of parasitic bronchitis and gastroenteritis in cattle. *Veterinary Parasitology* 34, 45–56.

Jacobs, D.E., Fisher, M.A., Hutchinson, M.J., Bartram, D.J. and Veys, P. (1995) An evaluation of abamectin given at turnout and six weeks after turnout for the control of nematode infections in calves. *Veterinary Record* 136, 386–389.

Jacobs, D.E., Hutchinson, M.J. and Burr-Nyberg, E. (1996) Compatibility of the programmed use of doramectin with lungworm vaccination in calves. *Veterinary Record* 139, 191–192.

Kloosterman, A., Ploeger, H.W., Pieke, E.J., Lam, T.J.G.M. and Verhoef, J. (1996) The value of bulk milk ELISA *Ostertagia* antibody titres as indicator of milk production response to anthelmintic treatment in the dry period. *Veterinary Parasitology* 64, 197–205.

Laiblin, C., Ilchmann, G. and Metzner, M. (1996) Management in Mutterkuhherden unter besonderer Berücksichtigung parasitologischer Fragestellungen. *Der Praktische Tierarzt* 6, 538–543.

Le Stang, J.P., Gossellin, J. and Herout, C. (1995) Efficacité de deux programmes de traitement utilisant la doramectine dans le controle des strongyloses gastro-intestinales des jeunes bovins au pâturage: résultats de quatre essais conduits en France. *Revue de Médecine Vétérinaire* 146, 93–102.

Lonneux, J.-F. and Losson, B. (1996) Epidémiologie des gales bovines. *Annales de Médecine Vétérinaire* 140, 317–327.

Mage, C., Reynal, P.-H. and Bourguignon, L. (1998) Contrôle de l'infestation par des strongles gastro-intestinaux chez des génisses limousines de 2e année d'herbe selon trois programmes de traitement avec la moxidectine. *Bulletin des GTV* 596, 33–39.

McKellar, Q.A. (1994) Chemotherapy and delivery systems – helminths. *Veterinary Parasitology* 54, 249–258.

Michel, J.F. (1969) The epidemiology of some nematode infections in calves. *Veterinary Record* 85, 323–326.

Parr, S.L., Gray, J.S., Sheehan, P. and Simon, A.J. (1995) Effect of doramectin on the performance of cattle exposed to gastrointestinal worms and lungworms in Ireland. *Veterinary Record* 137, 617–618.

Pitt, S., Baggott, D. and Forbes, A.B. (1996) Use of the Ivomec® sustained-release bolus midway through the grazing season. Its effect on the productivity of parasite naïve cattle and their subsequent development of immunity to parasitic nematodes. BCVA, Edinburgh, pp. 61–62.

Ploeger, H.W. and Eysker, M. (2000) Simulating *Dictyocaulus viviparus* infection in calves: the parasitic phase. *Parasitology* 120, S3–S15.

Ploeger, H.W., Schoenmaker, G.J.W., Kloosterman, A. and Borgsteede, F.H.M. (1989) Effect of anthelmintic treatment of dairy cattle on milk production related to some parameters estimating nematode infection. *Veterinary Parasitology* 34, 239–253.

Ploeger, H.W., Borgsteede, F.H.M., Sol, J., Mirck, M.H., Huyben, M.W.C., Kooyman, F.N.J. and Eyker, M. (2000) Cross-sectional serological survey on gastrointestinal and lung nematode infections in first and second-year replacement stock in The Netherlands: relation with management practices and use of anthelmintics. *Veterinary Parasitology* 90, 285–304.

Pollmeier, M., Schaper, R., Schillinger, D. and Hörchner, F. (1992) Effizienzuntersuchung einer Austriebsbehandlung mit IVOMEC(R) Pour-On bei erstsömmrigen Rindern. *Deutschen Veterinärmedizinschen Gesellschaft*, Husum, 6–7 April, pp. 83–94.

Pope, D.G., Wilkinson, P.K., Egerton, J.R. and Conroy, J. (1985) Oral controlled-release delivery of ivermectin in cattle via an osmotic pump delivery device. *Journal of Pharmaceutical Sciences* 74, 1108–1110.

Rehbein, S., Pitt, S.R., Langholff, K., Barth, D. and Eagleson, J.S. (1997) Therapeutic and prophylactic efficacy of the Ivomec SR bolus against nematodes and *Psoroptes ovis* in cattle weighing more than 300 kg at the time of treatment. *Parasitology Research* 83, 722–726.

Satrija, F. and Nansen, P. (1996) The effects of early-season treatments with doramectin on set-stocked calves naturally exposed to trichostrongyles. *Veterinary Research Communications* 20, 31–39.

Satrija, F., Nansen, P., Jørgensen, R.J., Monrad, J. and Esfandiari, A. (1996) The effects of first-season strategic and tactical ivermectin treatments on trichostrongylosis in the first- and second-season grazing. *Veterinary Parasitology* 64, 219–237.

Schnieder, T., Epe, C., von Samson-Himmelstjerna, G. and Kohlmetz, C. (1996a) Use of an ivermectin bolus against gastrointestinal nematode and lungworm infections in first-year grazing calves. *Applied Parasitology* 37, 38–44.

Schnieder, T., Epe, C., von Samson-Himmerlstjerna, G. and Kohlmetz, C. (1996b) The development of protective immunity against gastrointestinal nematode and lungworm infections after use of an ivermectin bolus in first-year grazing calves. *Veterinary Parasitology* 64, 239–250.

Schnieder, T., Epe, C., von Samson-Himmerlstjerna, G., Kohlmetz, C. and Woidtke, S. (1996c) Strategische Bekämpfung von Magen-Darmwürmen. Vergleichende Applikation von Ivermectin beim Austrieb und 6 oder 8 Wochen danach. *Der Praktische Tierarzt* 77, 529–536.

Shaw, D.J., Vercruysse, J., Claerebout, E. and Dorny, P. (1998a) Gastrointestinal nematode infections of first-grazing season calves in Western Europe: general patterns and the effects of chemoprophylaxis. *Veterinary Parasitology* 75, 115–131.

Shaw, D.J., Vercruysse, J., Claerebout, E. and Dorny, P. (1998b) Gastrointestinal nematode infections of first-grazing season calves in Western Europe: associations between parasitological, physiological and physical factors. *Veterinary Parasitology* 75, 133–151.

Steffan, P.E. and Nansen, P. (1990) Effects of tactical late-season treatments with ivermectin on calves naturally exposed to trichostrongyles. *Veterinary Parasitology* 37, 121–131.

Talty, P.J. and McSweeney, C. (1996) Use of a prophylactic doramectin programme to control subclinical parasitic gastroenteritis in calves in County Clare. *Irish Veterinary Journal* 49, 596–600.

Talty, P.J., McSweeney, C. and Simon, A.J. (1996) Control of lungworms in dairy calves on a farm in County Clare using doramectin in a treatment programme. *Irish Veterinary Journal* 49, 661–663.

Taylor, S.M., Mallon, T.R. and Kenny, J. (1985) Comparison of early season suppressive anthelmintic prophylactic methods for parasitic gastroenteritis and bronchitis in calves. *Veterinary Record* 117, 521–524.

Taylor, S.M., Mallon, T.R. and Green, W.P. (1986) Comparison of vaccination and ivermectin treatment in the prevention of bovine lungworm. *Veterinary Record* 119, 370–372.

Taylor, S.M., Mallon, T.R., Green, W.P., McLoughlin, M.F. and Bryson, D.G. (1988) Immunity to parasitic bronchitis of yearling cattle treated with ivermectin during their first grazing season. *Veterinary Record* 123, 391–395.

Taylor, S.M., Mallon, T.R., Kenny, J. and Edgar, H. (1995a) A comparison of early and mid grazing season suppressive anthelmintic treatments for first year grazing calves and their effects on natural and experimental infection during the second year. *Veterinary Parasitology* 56, 75–90.

Taylor, S.M., McMullin, P.F., Mallon, T.R., Kelly, A. and Grimshaw, W.T.R. (1995b) Effects of treatment with topical ivermectin three and eight weeks after turnout on nematode control and the performance of second-season beef suckler cattle. *Veterinary Record* 136, 558–561.

Taylor, S.M., Kenny, J., Edgar, H.W. and Whyte, M. (1997) Protection against *Dictyocaulus viviparus* in second year cattle after first year treatment with doramectin or an ivermectin bolus. *Veterinary Record* 141, 593–597.

Taylor, S.M., Kenny, J., Edgar, H.W., Mallon, T.R. and Canavan, A. (2000) Induction of protective immunity to *Dictyocaulus viviparus* in calves while under treatment with endectocides. *Veterinary Parasitology* 88, 219–228.

Vercruysse, J., Dorny, P., Berghen, P. and Frankena, P. (1987) The use of an oxfendazole pulse release bolus in the control of parasitic gastroenteritis and parasitic bronchitis in first-season grazing calves. *Veterinary Record* 121, 297–300.

Vercruysse, J., Dorny, P., Hong, C., Harris, T.J., Hammet, N.C., Smith, D.G. and Weatherley, A.J. (1993) The efficacy of doramectin in the prevention of gastrointestinal nematode infections in grazing cattle. *Veterinary Parasitology* 49, 51–59.

Vercruysse, J., Hilderson, H. and Claerebout, E. (1995a) Effect of chemoprophylaxis with avermectins on the immune response to gastro-intestinal nematodes in first-season grazing calves. *Veterinary Parasitology* 58, 35–48.

Vercruysse, J., Hilderson, H., Claerebout, E. and Roelants, B. (1995b) Control of gastrointestinal nematodes in first-season grazing calves by two-strategic treatments with doramectin. *Veterinary Parasitology* 58, 27–34.

Vercruysse, J., Claerebout, E., Dereu, A. and Lonneux, J.F. (1996) Control of gastrointestinal nematodes in beef calves by prophylactic treatments with doramectin and ivermectin. *Veterinary Record* 139, 547–548.

Vercruysse, J., Dorny, P., Claerebout, E. and Weatherley, A.J. (1998) Field evaluation of a topical doramectin evaluation for the chemoprophylaxis of parasitic bronchitis in calves. *Veterinary Parasitology* 75, 169–179.

Wacker, K., Roffeis, M. and Conraths, F.J. (1999) Cow–calf herds in Eastern Germany: status quo of some parasite species and pasture management in the control of gastrointestinal nematodes. *Journal of Veterinary Medicine B* 46, 475–483.

Walsh, T.A., Younis, P.J. and Morton, J.M. (1995) The effect of ivermectin treatment of late pregnant dairy cows in South-West Victoria on subsequent milk production and reproductive performance. *Australian Veterinary Journal* 72, 201–207.

Williams, J.C. and Plue, R.E. (1992) Efficacy of ivermectin delivered from a sustained-release bolus against inhibited early fourth stage larvae of *Ostertagia ostertagi* and other nematodes in cattle. *American Journal of Veterinary Research* 53, 793–795.

Yang, X., Satrija, F. and Nansen, P. (1996) Strategic effects of early season treatments with moxidectin on trichostrongylosis in young calves. *Applied Parasitology* 37, 8–16.

Zimmerman, G.L., Mulrooney, D.M. and Wallace, D.H. (1991) Efficacy of ivermectin administered via sustained-release bolus against gastrointestinal nematodes in cattle. *American Journal of Veterinary Research* 52, 62–63.

Chapter 6.3

Use of Macrocyclic Lactones to Control Cattle Parasites in the USA and Canada

R. Rew and J. Vercruysse

Epidemiology of Cattle Parasites and Cattle Management in North America

Parasite epidemiology in North America is based upon regional weather pattern differences between the northern USA/Canada and the southern USA, age of the cattle and management type (grazing, feedlot, drylot, etc.). Calves are born primarily in late winter/early spring (though autumn calving is becoming more popular) on pastures, drylots or crop residues. Beef steer calves are grazed with the cow for 6–8 months, then weaned and either backgrounded on pasture or in a drylot or sold directly to the feedlot where they are placed on high-energy, high-protein rations. Beef heifer calves may be treated similarly or go into the reproductive herd after weaning. Internal parasites will begin infecting calves after 2–3 months of grazing as they rely more on grazing and less on suckling.

Ostertagia and *Cooperia* are generally the two most predominant genera of internal nematodes throughout the animal's life span (Gibbs and Herd, 1986). Less predominant genera such as lungworm (*Dictyocaulus viviparus*), abomasal worms (*Haemonchus placei* and *Trichostrongylus axei*), intestinal worms (*T. colubriformis, Oesophagostomum radiatum, Strongyloides papillosus, Trichuris* sp. and *Nematodirus helvetianus*) can become problematic at times, but except for *Ostertagia* and *T. axei* are immunologically controlled by cattle with healthy immune systems (Armour, 1989).

Two of the most economically important external parasite groups are sucking (*Linognathus vituli, Solenopotes capillatus, Hematopinus eurysternus* and *H. quadripertusus*) and biting or chewing lice (*Bovicola* (*Damalinia*) *bovis*), and hornflies (*Haematobia irritans*) (Drummond, 1987). Cattle grubs and sarcoptic mange have been all but eliminated from cattle due to widespread macrocyclic lactone (ML) usage. Psorptic and chorioptic mange are seen occasionally, usually in confined operations. The USDA quarantine programme at the southern border of the USA continues to keep the single host tick *Boophilus microplus* from re-entering from Mexico. Several species of multihost ticks are present that cause economic losses in cattle, especially *Bos taurus*. Multihost ticks and stable flies (*Stomoxys calcitrans*) are two ectoparasites of cattle that are not controlled by MLs that do have extensive economic impact in the USA.

Though the same parasites are common throughout North America, the timing of their prevalence is a function of a northern temperate climate versus a southern subtropical climate.

Generally, parasitic infections in North American cattle are subclinical apart from the exceptional situation of high stocking densities in the southeastern USA in late winter or early spring. However, productivity responses to ML treatment can be shown in rapidly growing animals on pasture or on drylots even when faecal egg counts are relatively low. Inconsistent productivity responses are seen after ML treatment of older cows on low stocking densities with adequate forage, but higher stocking densities and/or poor forage result in productivity benefits for treatment (cow body conditioning scores, calf birth weights and calf weaning weights) even in adult cows (Reinemeyer, 1992). Some research has been conducted on the impact of deworming on immune responses in young calves (Klesius *et al.*, 1984; Yang *et al.*, 1993; Gasbarre, 1994, 1997; Almeria *et al.*, 1998). This research would support an enhanced immune response by reduction in *Ostertagia* burdens. This area of research may provide useful guidelines for more effective vaccination of cattle for other diseases as well as point us to a better understanding of the role and limitations of the immune system and MLs in controlling parasites.

A recent survey by Bowman *et al.* (1999) demonstrated a surprisingly high percentage of dairy cows shedding nematode faecal eggs even when the operator indicated that they were total confinement operations. This survey in Pennsylvania, Wisconsin, New York and Ohio showed that 77% of the 141 farms had positive first and second calf heifers for nematode eggs of primarily *Ostertagia*, *Cooperia* and *Haemonchus*. Of the 84 farms with grazing heifers, 86% were positive, while of the 66 farms that did not graze, 50% were positive. On an animal basis, 46% of the 674 heifers on grass were infected, while 37% of the 496 heifers confined were positive.

Southern USA

Cows and calves are generally on grass (permanent pastures) 12 months of the year. Grass growth begins in autumn, reaches a maximum in late winter or early spring, and dries out in early summer. Nematode pasture infectivity peaks in April, a time when spring-born calves are just beginning to ingest grass. Lice and, to a lesser extent, psoroptic and chorioptic mites will peak in February, though compared with the northern USA and Canada, lice infestations are much less important. Hornflies are the most economically important ectoparasite (Byford *et al.*, 1992), and populations will begin to increase in June and will peak in August. Inhibited *Ostertagia* will increase from June to August. One geographical exception to this pattern is that *Ostertagia* seems to be of lesser importance in Florida, especially in dairy heifers, where *Haemonchus* is more dominant (Courtney and Zeng, 1994).

A study examining nematode species in beef cow/calf herds has shown *O. ostertagia* and *Cooperia* spp. as the major parasites in calves, and *O. ostertagi* and *T. axei* predominate in cows in Mississippi (Couvillion *et al.*, 1996). Inhibited L_4 *Ostertagia* build up from early summer to autumn. Williams *et al.* (1977, 1979, 1981) and Williams (1991) reported mean values of 3700–107,000 inhibited *Ostertagia* per head in Louisiana, and Miller (1994) reported mean values of 340,000 in Louisiana in summer. *Haemonchus* and *Oesophagostomum* may build up in warmer weather. Low numbers of *Dictyocaulus, Bunostomum, Strongyloides, Trichuris* and *Nematodirus* are also seen. Hornfly is the predominant ectoparasite in summer, and the economic threshold has been determined to be 100 flies per side. This response appears to be correlated with increased heart and respiration rate induced by 100 flies per side in cattle (Schwinghammer *et al.*, 1986). Fly numbers higher than 100 per side will cause a 0.1–0.2 lb day^{-1} weight loss (Drummond, 1987).

As calves mature and summer arrives, adult *Cooperia* spp. (*pectinata* and *punctata*) predominate. In the autumn and winter, adult *O. ostertagi* predominate (Williams, 1990). Many autumn-weaned calves will move onto 'clean' wheat pasture in Texas, Oklahoma, Kansas, etc. during the winter, if they do not go to the feedlots. Stocking densities are higher in the southeastern states on non-irrigated land than in other parts of the country, since the growing season for the grass and rainfall are conducive to excellent pastures for about 9 months per year (September–June). Pasture nematode larval survival declines significantly in June–September (Williams, 1990).

Northern USA and Canada

In the northern USA and Canada, lice (both biting and sucking) are the predominant winter external parasite in cattle. Losses due to lice infestations are dependent upon number of lice, but 0.1 lb day^{-1} losses due to sucking lice frequently have been reported (Drummond, 1987). Again, *Ostertagia ostertagi* and *Cooperia* spp. (*oncophora*) are the dominant internal parasites (Leland *et al.*, 1973; Gibbs and Herd, 1986). Inhibited *Ostertagia* is the predominant winter internal parasite. Malczewski *et al.* (1996) reported an average of 20,000 inhibited larvae per animal in Wyoming cattle (2–5 years old) in January; Kistner *et al.* (1979) reported mean values of nearly 200,000 inhibited *Ostertagia* larvae in cattle in Oregon in winter; and Westcott (1984) reported mean values of nearly 1000 inhibited *Ostertagia* in cattle in Washington in winter. Oregon and

Washington are in the transition zone of inhibition (Williams, 1986), such that large variations in percentage inhibition in winter may be seen from year to year.

Cattle are usually put on to grass in late spring in the northern USA and Canada, and moved to drylot, wheat stubble or cornstalks in late autumn. Low stocking densities (extensive grazing) are the norm in cow–calf areas. Grass quality and quantity increase from April to August in the northern USA and Canada. Internal parasite numbers in the northern USA and Canada begin to build in cattle in the spring. In young calves, *Strongyloides* may be seen even before turnout on to grass due to direct skin penetration or transmammary transmission. *D. viviparus* and *N. helvetianus* become the first species that may be seen in late spring. If exposure to these three genera is not seen in spring, an opportunity for infection of these parasite-specific naïve calves exists in the summer or autumn. *O. ostertagi* and *C. oncophora* are the primary internal parasites of cattle in summer and autumn, similar to northern Europe. Inhibited *Ostertagia* begin to appear in the tissues in late autumn and peak in winter (Malczewski *et al.*, 1996).

The primary external parasite in summer would be hornflies. The primary winter ectoparasites are lice (both biting and sucking) and mites (psoroptic and chorioptic) as hair length on the cattle increases and UV light decreases. Nematode faecal egg counts in adult cows will rise quickly after calving, and will remain relatively high for 4–6 weeks as a result of relaxation of the immune system (Stromberg *et al.*, 1991). This impact allows for the presence of fresh larvae for calves as they begin to ingest grass at 2–3 months of age.

Control of Nematodes and Arthropods

The timing of treatment of cattle with MLs in North America tends to be driven by other management activities. This is more predominant in North America than Europe or the subtropical climates because internal parasite infections tend to be mostly subclinical in nature as a result of low stocking densities and interruptions of transmission periods by very cold or very hot, dry weather in North America. Though the timing of treatment may not be optimal for parasite control, treatment is given when cattle are shipped, cows are pregnancy tested, calves are weaned, ear-tagged for flies or castrated, or stocker cattle enter a backgrounding operation or feedlot.

Cow—calf

Traditionally, a single treatment of cows in the autumn has been the majority of use of MLs in cows in the USA. More recently, an increasing number of producers are seeing the value of adding a spring deworming with the MLs because of the persistent efficacy characteristics of these molecules for infection prevention and pasture larval clean-up. Calf weaning weight responses to treatment with an ML in cow–calf systems have been much more variable than in stocker or feedlot trials (Reinemeyer, 1992). Trials in beef breed cows demonstrate that the impact of treatment of the cow has two effects: first, increased milk production (Myers, 1988) and, secondly, a reduction in pasture nematode egg contamination as a benefit for calf infection (Stromberg *et al.*, 1991). The result of cow treatment can be measured in the calf as a decrease in egg counts in the calf and an increased weaning weight that ranges from 0.02 to 0.11 lb day^{-1} (Reinemeyer, 1992). One would expect that the older the cow, the more effective its immune response (as measured by faecal egg counts), and thus the less impact that treatment would have. However, other factors, such as poor nutrition, abnormal physiological conditions or infectious diseases, have such an overriding impact that the age of the cow is not a good indicator of the impact of ML treatment. ML treatment timing to address body conditioning scores (which are predictive for calving rates and weights) has not been tested adequately, but may be a very important parameter that could be significantly improved by MLs.

Northern USA and Canada

The primary reasons for treatment of cows in the northern USA and Canada in autumn with MLs are lice control and internal parasite clean-up (especially inhibited *Ostertagia*). Cows generally go on to poor quality feed during the winter, but need to maintain good body condition for spring calving; therefore removing parasites is important during this period. Calves are weaned in the autumn, vaccinated and dewormed prior to going to a backgrounding operation or a feedlot.

The primary reason for cow–calf treatments in spring in these northern areas is prevention of early season pasture contamination by nematodes and an adjunct to hornfly control. Treatment of calves less than 3 months of age probably has little, if any, impact. At 4–6 months of age, calf treatment is useful in controlling parasites and increasing weaning weight.

Wohlgemuth and Melancon (1988) demonstrated over 2 years of trials that calves of cows in North Dakota treated with ivermectin injectable only in the spring, in spring and autumn or only in the autumn were 15.5 lb ($P < 0.05$) heavier than calves from untreated cows. Ballweber *et al.* (2000a) demonstrated that spring treatment of cow and calf with doramectin pour-on in Idaho resulted in calves 21.5 lb heavier at weaning than untreated controls. In addition, with the presence of a high proportion of pyrethroid-resistant hornflies, having an ML as part of a control programme is becoming more and more attractive.

Southern USA

The primary reason for treatment of cows in autumn in the southern areas is prevention of internal parasite infection (similar to spring deworming in the north). Higher stocking densities, more moisture and longer grazing seasons make internal parasite challenge more severe in the southeast USA. Thus productivity responses to treatment tend to be greater in the southeast than in other regions. Ballweber *et al.* (2000a) reported a 38 lb advantage in weaning weight of calves when cows and autumn-born calves were treated with doramectin pour-on versus untreated controls.

The primary reason for spring treatment in southern areas is a clean-up of the maximum challenge in March, but potentially this treatment could initiate hornfly control as well. Ciordia *et al.* (1984) demonstrated a mean calf weaning weight gain advantage versus untreated controls of 35 lb ($P < 0.05$) following spring treatment of cows and calves and early summer treatment of calves with ivermectin in Georgia. Andress *et al.* (2000) demonstrated that a single doramectin pour-on treatment in March controlled hornfly populations below economic levels for 5–9 weeks. In a second season, two treatments 8 weeks apart (May and July) kept hornfly numbers below 50 per side for 15 of the 16 weeks. Foil *et al.* (1998) demonstrated similar data with ivermectin pour-on over two seasons. In the first season with a single treatment, ivermectin pour-on controlled hornflies for 6–8 weeks when applied in spring and 4–5 weeks when applied in midsummer. In a second season, a programme of ivermectin pour-on and permethrin ear tags provided 7–11 weeks of control.

Stocker

The primary reason for treating stockers with MLs is to improve growth rate during a period of rapid growth. The most obvious response to treatment is in this management group, since they are only partially immune and they are growing rapidly, usually at the time that maximum parasite numbers are building up on the pasture.

Response to treatment is based on a number of variables. Rew (1999) reviewed a series of four trials with doramectin and ivermectin injectables being compared with untreated controls on separate pastures in the USA with a single spring/summer treatment for weaned calves or stockers (Ballweber *et al.*, 1997), one trial with two treatments (Williams *et al.*, 1997) and one trial with a similar protocol except on the same pasture. A single treatment with an ML in the USA yielded a significant weight gain response for a single treatment of approximately 0.3–0.4 lb day^{-1} for treatment. These trials ranged from 60 to 161 days of grazing and were under conditions of low to high challenge, poor to good nutrition, and presence or absence of growth implants in the northern USA (Wisconsin) and the southeastern USA (Alabama, Louisiana, Mississippi and Georgia). Though the absolute rate of growth was quite different from condition to condition (see Table 6.3.1) due to food availability, growth implants or parasite challenge, the difference in increased growth rate for treatment was quite consistent from trial to trial. Skogerboe *et al.* (2000) demonstrated a 0.12–0.46 lb day^{-1} increase in weight gain over 140 days of grazing following a single treatment of steers with doramectin pour-on versus untreated controls on separate pastures in northern USA (Wisconsin) and the southeastern USA (Tennessee and Louisiana). The authors demonstrated that the majority of gain difference was over the first 56 days of each trial. Williams *et al.* (1999) compared doramectin, ivermectin, moxidectin and eprinomectin pour-ons with untreated controls on the same pasture for 112 days in Louisiana. Parasite control and weight gain were significantly improved by doramectin, moxidectin, and eprinomectin versus untreated controls by 0.4–0.5 lb day^{-1}. Loyacano *et al.* (2001) compared doramectin injectable with ivermectin/clorsulon injectable on separate pastures for 140 days in Louisiana. Parasite control as measured by nematode faecal egg counts was significantly reduced for doramectin on days 7–56 versus untreated controls (UC) and days 14–56 versus ivermectin/clorsulon (I/C). I/C faecal egg counts were significantly reduced versus UC on days 7–49. Average daily weight gain was significantly improved over the untreated controls (1.37 lb day^{-1}) by both products and was significantly improved for doramectin (1.74 lb day^{-1}) versus ivermectin/clorsulon (1.56 lb day^{-1}) on day 140.

Table 6.3.1. Average daily gains (ADGs) of yearling grazing calves in US trials (from Rew, 1999).

Location (state)	Breed/ sex	Days on pasture	Weather/ feed/implant status	No. of ML doses			ADG (lb day^{-1})		
				Cont.	DOR	IVE	Cont.	DOR	IVE
Wisconsin	Cross/F	140	N	0	1	1	1.69^{a}	2.05^{b}	2.02^{b}; 1.98^{b}
Arkansas	Cross/S	84	N/s	0	1	1	1.41^{a}	2.00^{b}	1.94^{b}
Georgia	Angus/S	112	D	0	1	1	0.44^{a}	0.86^{b}	0.75$^{a/b}$
Mississippi	Cross/F	112	N/s	0	1	1	1.76^{a}	2.09^{a}	1.98^{a}
Louisiana	Cross/S	161	N	0	2	2	1.32^{a}	2.16^{b}	2.00^{b}; 2.02^{b}
Louisiana	Cross/S	140	N	0	1	1	1.36^{a}	1.74^{b}	1.56^{c}
Alabama	Cross/F	60	N/s/i	0	1	1	1.98^{a}	2.46^{b}	2.29^{b}

F, Female; S, steer; N, normal; D, drought; s, supplement feed; ^{a,b,c}, differing letters in a row are significantly different ($P < 0.05$); Cont., control; DOR, doramectin; IVE, ivermectin; i, implant.

Programme use of MLs is an uncommon practice in North America unless one calls two treatments per year, one in spring and one in autumn, a programme. However, use of programmes for stockers in the southeastern USA has been investigated. Williams *et al.* (1990) compared three or eight treatments of ivermectin over an 11-month period with untreated controls and a positive control. There was no difference in gain from November to March among the three groups. From March to October, there was a significant increase in gain of 180 lb for the eight treatments of ivermectin programme versus untreated controls. Also, Williams *et al.* (1997) showed a significant increase in productivity of 0.68 and 0.84 lb day^{-1} versus untreated controls with two doses of ivermectin or doramectin, respectively, to grazing stockers at an 84-day interval over a 161-day grazing period.

Sustained-release bolus of ivermectin is another form of programme treatment. Sixteen 6-month-old stocker calves (407 lb) were grazed for 135 days on eight separate Arkansas pastures (two per pasture) along with one tracer per pasture. At the end of the winter/spring grazing, the principal treated calves gained 110 lb more than their control principal counterparts (Yazwinski *et al.*, 1995).

Most of the 35 milllion acres (14 million ha) of tall fescue in the USA are infested with an endophytic fungus that is associated with a toxic syndrome including poor productivity. Bransby (1999) demonstrated that treatment of cattle with ivermectin at turnout and 56 days later grazed on highly infected endophyte pastures enhanced productivity by 0.66 lb day^{-1} over 140 days compared with untreated cattle, whereas ivermectin treatment on low endophyte infected pastures enhanced productivity by 0.22 lb day^{-1} (Alabama). This impact may be related to the short-term reduction observed by pre-treatment of rats with ivermectin in the thermoregulatory response to ergovaline, the primary toxin in endophyte-infested tall fescue (Spiers *et al.*, 1997).

Feedlot

Feeders in a feedlot will show similar enhanced gains over the first 30–60 days, as do grazing stockers. However, since they are no longer challenged with incoming larvae and untreated cattle will lose a major portion of their infection during this time of high nutrition, there is much less advantage to treatment after 60 days. Since *Ostertagia* is the most important internal parasite in animals over 6 months of age, the majority of the productivity response to treatment (70%) is due to enhanced appetite (Fox *et al.*, 1989). Response for treatment with MLs in the feedlot have given us additional information on the quantitative benefits of ML treatment in cattle. Grimson *et al.* (1987) demonstrated a significant increase in average daily gain following treatment with ivermectin injectable versus untreated or levamisole/fenthion-treated steers of 0.33 and 0.20 lb day^{-1}, respectively. In addition, feed intake was increased by 0.37 lb day^{-1} versus untreated, and feed conversion was significantly enhanced by 0.54 and 0.38 versus untreated and levamisole/fenthion, respectively. Johnson *et al.* (1998) demonstrated in two research sized feedlot trials (180 head) that treatments with doramectin or ivermectin at entry increased feed consumption and gain. In one trial, where pre-treatment nematode counts were only 400 per head (eggs per gram (EPG) 35), there were 22 and 8 lb increases in gain at 225 days versus untreated controls for doramectin and ivermectin, respectively. Feed intake increases mirrored the gain: 0.097 and 0.035 lb day^{-1} increase in daily feed consumption for doramectin and ivermectin. In the second trial, where pre-treatment nematode counts were nearly 5000 per head (EPG 75), there were 52 and 48 lb significant increases in gain versus untreated controls at day 168 for doramectin and ivermectin, respectively. Again, daily feed intake mirrored the gain: 0.310 and 0.286 lb day^{-1} increase in feed intake for doramectin and ivermectin, respectively.

Guichon *et al.* (2000) in a commercial sized feedlot trial (14,000 head) showed a significant body weight gain following treatment with ivermectin of 0.11 lb day^{-1} with a pre-treatment EPG of 4–5 versus fenbendazole/fenthion/permethrin. MacGregor *et al.* (2001) in commercial sized feedlot trials (6000 head) showed a significant body weight gain following treatment with doramectin of 0.18 lb day^{-1} with a pre-treatment EPG of 9–10 versus untreated controls. If we look at intervals post-treatment, these increases in weight gain were 0.25 lb day^{-1} for the first 60 days and 0.04 lb day^{-1} for the next 60 days. In addition, they have shown that the feed intake is increased significantly (0.48 lb day^{-1}), which would be predicted by the appetite suppression mechanism demonstrated by Fox *et al.* (1989). Along with this increased feed intake and gain in all of these trials are a significant increase for MLs in feed efficiency and a significant improvement in carcass quality. There are approximately 6–9% more choice carcasses (a grading system used by the USDA inspector at the packing plant) and subsequently a 6–9% decrease in select carcasses compared with treated or untreated controls in the two commercial sized trials. This derived benefit represents a mechanism of impact not well understood, but highly valuable to producers in North America who are paid at the slaughterhouse on a combination of weight and carcass characteristics. Interestingly, when a half recommended dose of doramectin injectable was compared with a full dose of doramectin injectable at entry into the feedlot, Edwards *et al.* (2001) demonstrated that though faecal egg counts at day 75 post-treatment were similar between fenthion-treated, half-dose, and full dose doramectin-treated steers, significant average daily gains of 0.14 and 0.17 lb day^{-1} and carcass quality (15% choice) improvements were only seen with a full dose of doramectin versus half-dose doramectin or fenthion-treated steers.

Ballweber *et al.* (2000b) examined the impact of a sustained-release bolus (SR-bolus) on 550 lb stockers grazing for 125 days and then in the feedlot for 124 days versus untreated or benzimidazole (bz)-treated steers treated at grazing and at feedlot entry. They demonstrated an additional 60 lb gain in the grazing period for SR-bolus treatment versus bz or untreated. During the 124-day feedlot period, the SR-bolus treated steers gained an additional 30 lb versus untreated and gained the same as the bz-treated steers.

Dairy

The majority of the dairy cattle in the USA are raised and maintained in confinement. Calves are raised in hutches or barns and male calves are used primarily for veal. Though some more traditional dairy operations are age integrated and graze their animals, many operations in the USA (especially larger operations) send female calves to heifer raisers where they may be raised on drylots. Then first calf heifers are sent to the dairy and cows are maintained in confinement for the remainder of their lives.

Two MLs are approved for lactating dairy cows with a zero milk withdrawal in the USA, eprinomectin and moxidectin. The primary usage for these MLs in lactating herds is lice and mange control. If the faecal egg counts of confined operations are meaningful or if the operation is not a totally confined operation, MLs are an excellent choice for control and prevention of nematode infections, especially at freshening. A study with eprinomectin treatment of grazing dairy cows in Canada demonstrated a 2.07 lb day^{-1} increase in milk production over the 200 days. The study included 28 herds (14 treated and 14 control) with 954 adult cows (Nodvedt *et al.*, 2001). All MLs are approved for use in replacement heifers. This group of animals, like beef stockers, may benefit the most from excellent therapeutic and persistent control of nematodes. The rate at which these replacement heifers reach breeding weight controls the lifetime milk production of the cow. The sooner breeding weight can be achieved, the sooner that animal becomes an economic benefit to the owner. Like beef stockers, 0.2–0.4 lb day^{-1} of increased gain might be expected with MLs over untreated controls in a similar grazing operation. If stocking densities are high and the grazing period is over 120 days, a second dose would be recommended to obtain maximum growth from these heifers. Several trials have been conducted to examine the production impact of the SR-bolus (ivermectin) on dairy heifer productivity. Smith (1994) demonstrated a 0.24 lb day^{-1} increase in average daily gain (ADG) in SR-bolus (ivermectin) treated 318 lb dairy heifers versus untreated over 175 days of grazing in Wisconsin. Caldwell (1998) demonstrated a 0.18 lb day^{-1} increase in ADG in SR-bolus-treated 893 lb dairy heifers versus untreated over 160 days of grazing in Quebec. The impact of MLs in drylot on growth of dairy heifers or lactation of cows has not been reported in North America. This area of research will need to be investigated in relation to ectoparasites as well as low-level transmission of internal nematodes.

References

Almeria, S., Canals, A., Gomez-Munoz, M.T., Zarlenga, D.S. and Gasbarre, L.C. (1998) Characterization of protective immune responses in local lymphoid tissues after drug-attentuated infections with *Ostertagia ostertagi* in calves. *Veterinary Parasitology* 80, 53–64.

Andress, E.R., DeRouen, S.M. and Foil, L.D. (2000) Efficacy of doramectin 0.5% w/v pour-on for control of the horn fly *Haematobia irritans*. *Veterinary Parasitology* 90, 327–331.

Armour, J. (1989) The influence of host immunity on the epidemiology of trichostrongyle infections in cattle. *Veterinary Parasitology* 31, 5–19.

Ballweber, L.R., Stuedemann, J.A., Smith, L.L., Yazwinski, T.A. and Skogerboe, T.L. (1997) The effectiveness of a single treatment with doramectin or ivermectin in the control of gastrointestinal nematodes in grazing yearling stocker cattle. *Veterinary Parasitology* 72, 53–68.

Ballweber, L.R., Evans, R.R., Siefker, C., Johnson, E.G., Rowland, W.K., Zimmerman, G.L., Thompson, L., Walstrom, D.J., Skogerboe, T.L., Brake, A.C. and Karle, V.K. (2000a) The effectiveness of doramectin pour-on in the control of gastrointestinal nematode infections in cow-calf herds. *Veterinary Parasitology* 90, 93–102.

Ballweber, L.R., Brown, J., Hawkins, J.A., Bechtol, D.T., Black, S., Alva, R. and Plue, R.E. (2000b) Comparison of ivermectin SR bolus, benzimidazole anthelmintics, and topical fenthion on productivity of stocker cattle from grazing through feedlot. *Veterinary Therapeutics* 1, 192–198.

Bowman, D., Smith, G., Smith, L., Stromberg, B.E., Melancon, J.J., Maye, D., Robertson-Plouch, C. and Guerrero, J. (1999) Prevalence of internal parasites among dairy herds in the northern United States. *American Association of Veterinary Parasitologists*. New Orleans, Louisiana, Abstract #86.

Bransby, D.L. (1999) Steer weight gain responses to ivermectin when grazing fescue. *Large Animal Practice* 18, 16–19.

Byford, R.L., Craig, M.E. and Crosby, B.L. (1992) A review of ectoparasites and their effect on cattle production. *Journal of Animal Science* 70, 597–602.

Caldwell, V., DesCoteaux, L. and Doucet, M. (1998) Impact of a sustained-release ivermectin bolus on weight gain in breeding age Holstein heifers under commercial pasture conditions in southern Quebec. *Canadian Veterinary Journal* 39, 701–705.

Ciordia, H., McCampbell, H.C., Calvert, G.V. and Plue, R.E. (1984) Effect of ivermectin on performance of beef cattle on Georgia pastures. *American Journal of Veterinary Research* 45, 2455–2457.

Courtney, C.H. and Zeng, Q.-Y. (1994) Dairy replacement heifers in Florida harbor depauperate populations of gastrointestinal nematodes. *American Association of Veterinay Parasitologists*. San Francisco, California, Abstract #26.

Couvillion, C.E., Seifker, C. and Evans, R.R. (1996) Epidemiological study of nematode infections in a grazing beef cow–calf herd in Mississippi. *Veterinary Parasitology* 64, 207–218.

Drummond, R.O. (1987) Economic aspects of ectoparasites of cattle in North America. In: Leaning, W.H.D. and Guerrero, J. (eds) *Proceedings of the MSD AGVET Symposium*, pp. 9–24.

Edwards, T.A., MacGregor, S., Lechtenberg, K.F. and Hunsaker, B.D. (2001) Effects of single administration of half-dose injectable doramectin, full-dose injectable doramectin, or fenthion on fecal egg counts, growth performance and carcass characteristics in cattle. *Bovine Practitioner* 35, 85–89.

Foil, L.D., Strother, G.R., Hawkins, J.A., Gross, S.J., Coombs, D.F., Derouen, S.M., Wyatt, W.E., Kuykendall, L.K. and Spears, B.G. Jr (1998) The use of Ivomec (ivermectin) pour-on and permethrin ear tags for horn fly control. *Southwestern Entomologist* 23, 317–323.

Fox, M.T., Gerrelli, S.R., Pitt, D. and Jacobs, D.E. (1989) *Ostertagia ostertagi* infection in the calf: effects of a trickle challenge on appetite, digestibility, rate of passage of digesta and liveweight gain. *Research in Veterinary Science* 47, 294–298.

Gasbarre, L.C. (1994) *Ostertagia ostertagi*: changes in lymphoid populations in the local lymphoid tissues after primary and secondary infection. *Veterinary Parasitology* 55, 105–114.

Gasbarre, L.C. (1997) Effects of gastrointestinal nematode infection on the ruminant immune system. *Veterinary Parasitology* 72, 327–343.

Gibbs, H.C. and Herd, R.P. (1986) Nematodiasis in cattle: importance, species involved, immunity and resistance. *Veterinary Clinics of North American: Food Animal Practice* 2, 211–224.

Grimson, R.E., Stilborn, R.P., Gummeson, P.K., Leaning, W.H.D., Guerrero, J. and Newcomb, K.M. (1987) Effect of antiparasitic treatments on performance and profitablity in feedlot steers. *Modern Veterinary Practice* 68, 361–364.

Guichon, P.T., Jim, G.K., Booker, C., Schunicht, O.C. and Brown, J.R. (2000) Relative cost-effectiveness of treatment of feedlot calves with ivermectin versus treatment with a combination of fenbendazole, permethrin, and fenthion. *Journal of the American Veterinary Medical Association* 216, 1965–1969.

Johnson, E.G., Rowland, W.K., Zimmerman, G.L., Walstrom, D.J. and Skogerboe, T.L. (1998) Performance of feedlot cattle with parasite burdens treated with anthelmintics. *Compendium for Continuing Education for Veterinarians*, S116–S123.

Kistner, T.R., Wyse, D. and Averkin, E. (1979) Efficacy of oxfendazole against inhibited *Ostertagia ostertagi* in natural infected cattle. *Australian Veterinary Journal* 55, 232–235.

Klesius, P.H., Washburn, S.M., Ciordia, H., Haynes, T.B. and Snider, T.G. (1984) Lymphocyte reactivity to *Ostertagia ostertagi* antigen in type I ostertagiasis. *American Journal for Veterinary Research* 45, 230–233.

Leland, S.E., Caley, H.K. and Ridley, R.K. (1973) Incidence of gastrointestinal nematodes in Kansas cattle. *American Journal for Veterinary Research* 34, 581–585.

Loyacano, A.F., Skogerboe, T.L., Williams, J.C., DeRosa, A.A., Gurie, J.A. and Shostrum, V.K. (2001) Effects of parenteral administration of doramectin or a combination of ivermectin and clorsulon on control of gastrointestinal nematode and liver fluke infections and on growth performance in cattle. *Journal of the American Veterinary Medical Association* 218, 1465–1468.

MacGregor, D.S., Yoder, D. and Rew, R.S. (2001) Impact of doramectin treatment of feedlot entry on the productivity of steers with natural nematode infections. *American Journal for Veterinary Research* 62, 622–624.

Malczewski, A., Jolley, W.R. and Woodard, L.F. (1996) Prevalence and epidemiology of trichostrongylids in Wyoming cattle with consideration of the inhibited development of *Ostertagia ostertagi*. *Veterinary Parasitology* 64, 285–297.

Miller, J.E. (1994) Variable efficacy of benzimidazole anthelmintics against inhibited larvae of *Ostertagia ostertagi*. *Bovine Proceedings* 26, 150–153.

Myers, G.H. (1988) Strategies to control internal parasites in cattle and swine. *Journal of Animal Science* 66, 1555–1564.

Nodvedt, A., Gonboy, G., Dohoo, I., Sanchez, J., Keefe, G., Descoteaux, L., Leslie, K. and Campbell, J. (2001) The effect of Ivomec-Eprinex on milk production in pastured dairy cattle. *American Association for Veterinary Parasitologists*. Boston, Massachusetts, Abstract #81.

Reinemeyer, C.R. (1992) The effect of anthelmintic treatment of beef cows on parasitology and performance parameters. *Compendium for Continuing Education for the Practicing Veterinarian* 14, 678–687.

Rew, R.S. (1999) Production-based control of parasitic nematodes of cattle. *International Journal of Parasitology* 29, 177–182.

Schwinghammer, K.A., Knapp, F.W., Boling, J.A. and Schillo, K.K. (1986) Physiological and nutritional response of beef steers to infestations of the horn fly (Diptera: Muscidae). *Journal of Economic Entomology* 79, 1010–1015.

Skogerboe, T.L., Thompson, L., Cunningham, J.M., Brake, A.C. and Karle, V.K. (2000) The effectiveness of a single dose of doramectin pour-on in the control of gastrointestinal nematodes in yearling stocker cattle. *Veterinary Parasitology* 87, 173–181.

Smith, L.L. (1994) Evaluation of the impact of the ivermectin sustained release (SR) bolus on weight gain and parasite control in dairy heifers during their first grazing season. *Bovine Proceedings* 26, 154–156.

Spiers, D.E., Snyder, B.L., Eichen, P.A., Rottinghaus, G.E. and Garner, G.B. (1997) Potential benefit of an anthelmintic in reducing hyperthermia associated with fescue toxicosis. *Journal of Animal Science* 75 (Supplement 1), 212.

Stromberg, B.E., Schlotthauer, J.C., Haggard, D.L., Vatthauer, R.J., Hanke, H.E. and Myers, G.H. (1991) The role of epidemiology in the management of growing cattle. *International Journal for Parasitology* 29, 33–39.

Westcott, R.B. (1984) Bovine parasites of particular interest in the northwest. *Proceedings of the American Association of Bovine Practitioners* 17, 60–63.

Williams, J.C. (1986) Epidemiologic patterns of nematodiasis in cattle. *Veterinary Clinics of North America: Food Animal Practice* 2, 235–246.

Williams, J.C. (1990) Epidemiology of bovine nematode parasites in warm temperate regions of the United States. In: Guerrero, J. and Leaning, W.H.D. (eds) *Epidemiology of Bovine Nematode Parasites in the Americas*. Veterinary Learning Systems, New Jersey, pp. 73–82.

Williams, J.C. (1991) Efficacy of albendazole, levamisole, and fenbendazole against gastrointestinal nematodes of cattle with emphasis on inhibited early fourth stage *Ostertagia ostertagi* larvae. *Veterinary Parasitology* 40, 59–71.

Williams, J.C., Knox, J.W., Sheehan, D. and Fuselier, R.H. (1977) Efficacy of albendazole against inhibited early fourth stage larvae of *Ostertagia ostertagia*. *Veterinary Record* 101, 484–486.

Williams, J.C., Knox, J.W., Sheehan, D. and Fuselier, R.H. (1979) Further evaluation of the activity of albendazole against inhibited larvae of *Ostertagia ostertagi*. *Veterinary Record* 105, 98–100.

Williams, J.C., Knox, J.W., Baumann, B.A., Snider, T.G. and Hoerner T.J. (1981) Anthelmintic efficacy of albendazole against inhibited larvae of *Ostertagia ostertagi*. *American Journal for Veterinary Research* 42, 318–321.

Williams, J.C., Knox, J.W., Barras, S.A. and Hawkins, J.A. (1990) Effects of ivermectin and fenbendazole in strategic treatment of gastrointestinal nematode infections in cattle. *American Journal for Veterinary Research* 51, 2034–2043.

Williams, J.C., Loyacano, A.F., DeRosa, A., Gurie, J., Coombs, D.F. and Skogerboe, T.L. (1997) A comparison of the efficacy of the two treatments of doramectin injectable, ivermectin injectable and ivermectin pour-on against naturally acquired gastrointestinal nematode infections of cattle during winter-spring grazing season. *Veterinary Parasitology* 72, 69–77.

Williams, J.C., Loyacano, A.F., DeRosa, A., Gurie, A., Clymer, B.C. and Guerino, F. (1999) A comparison of persistent anthelmintic efficacy of topical formulations of doramectin, ivermectin, eprinomectin, and moxidectin against naturally acquired nematode infections of beef calves. *Veterinary Parasitology* 85, 277–288.

Wohlgemuth, K. and Melancon, J.J. (1988) Relationship between weaning weights of North Dakota beef calves and treatment of their dams with ivermectin. *Agri-Practice* 9, 23–26.

Yang, C., Gibbs, H.C. and Xioa, L. (1993) Immunological changes in *Ostertagia ostertagi* infected calves treated strategically with an anthelmintic. *American Journal of Veterinary Research* 54, 1074–1083.

Yazwinski, T.A., Featherston, H. and Tucker, C. (1995) Effectiveness of the ivermectin sustained-release bolus in the control of bovine nematodiasis. *American Journal of Veterinary Research* 56, 1599–1602.

Chapter 6.4

Use of Macrocyclic Lactones to Control Cattle Parasites in South America

C. Eddi, A. Nari and J. Caracostantogolo

Introduction

The geographic and climatic diversity of the South American territory (12°00′ S–56°00′ S), as well as its history of socio-economic development, have resulted in a variety of livestock raising and marketing systems. These systems have tried to adapt parasite control to local ecological conditions and to ever-changing economic and political situations in different South American countries (Nari, 1992). Macrocyclic lactones (MLs) are used extensively in South America to control nematode infections and ectoparasites in cattle.

Importance and Epidemiology of Nematodes in South America

South America has a cattle population in excess of 200 million head: 130 million are found in Brazil, 50 million in Argentina and 30 million in the remaining South American countries. Climatically, southern Brazil, Uruguay, Argentina and Chile would be considered temperate, whereas the remainder of South America would be considered subtropical or tropical.

Although parasitic diseases are distributed throughout the continent, they have different economic impacts according to production system, management and geo-climatic conditions. In intensive production systems, the economic impact is greatest due both to loss of productivity and to an increase in the related cost of control. In mixed farming systems and extensive grazing conditions in the tropics and subtropics, the environment is usually very suitable for parasite development. The variety and prevalence of parasitic diseases are much greater than in temperate climates. The result is often reduced productivity expressed in the form of reduced weight gain, delayed and weak oestrus and lower calving rates.

Gastrointestinal (GI) nematode infections in cattle may be attributed to a number of species such as *Ostertagia ostertagi*, *Trichostrongylus axei*,

Haemonchus placei, *Cooperia* spp., *Bunostomum phlebotomum*, *Nematodirus* spp., *T. colubriformis* and *Oesophagostomum* spp. From this list of species, *O. ostertagi*, is well recognized as the most important GI helminth of cattle in temperate areas of the world (Levine, 1968; Anderson *et al.*, 1969; Armour, 1970; Bayly, 1970; Armour *et al.*, 1973). This nematode parasite list and the importance of *Ostertagia* is also consistent with temperate areas of South America.

Although attempts to assess economic losses due to GI parasites in South American cattle have been made, it is not easy to establish accurately the impact of nematodes on animal husbandry. In general terms and based on productivity trials where a non-treated control group is compared with treated groups, weight losses due to nematode infections range between 10 and 50 kg per animal over a year (Eddi *et al.*, 1985a,b, 1993b) according to levels of infectivity and availability of pasture. In more extensive production systems in Uruguay, these differences ranged between 44 and 109 kg per animal from weaning (7 months) to the age of 18–19 months (Nari and Fiel, 1994). Entrocasso (1987) studied the general impact of GI nematodes in Argentina, showing that approximately $200 millon were lost yearly. The main economic impact was due to weight loss ($160 millon), while expenses in antiparasitic drugs accounted for $65 million and losses due to mortality were estimated at $22 million. Carcasses of cattle with light or heavy parasite infections were studied for tissue quality. In non-treated controls, muscles with high or moderately high growth patterns were negatively affected (Garris *et al.*, 1987). In relation to milk production, even though it was difficult to observe statistical results between animals exposed to suppressive anthelmintic treatment and non-treated controls, a reduction of approximately 5% in milk production was reported (Biondani and Steffan, 1988).

Temperate climates

The epidemiology of GI nematodes in South America depends on the area in consideration. In the humid pampas of Argentina, one of the largest beef production regions, temperatures and moistures are highly favourable during the entire year for parasite development. Average temperatures range between 12 and 19°C, while average rainfall is between 800 and 1500 mm. The main nematode species found in the humid pampas during the autumn–winter period are *O. ostertagi*, *T. axei*, *H. placei*, *C. oncophora* and *O. radiatum* (Entrocasso, 1988). *O. ostertagi* arrested development takes place during spring–summer, and the inhibited stages average between 70 and 90% of the parasite population (Muñoz Cobeñas *et al.*, 1993).

The species pattern of the epidemiology of cattle parasites in Uruguay is quite similar to that observed in Argentina, with up to 97%

of inhibited stages of *O. ostertagi* found during the spring–summer period and relatively high percentages of *Cooperia* spp. (Berdie *et al.*, 1988; Nari and Fiel, 1994).

Tropical and subtropical climates

In Colombia and Venezuela, *Cooperia* spp. and *Haemonchus* spp. were found to be the most prevalent species observed at the beginning of the rainy season, from May through June (Rivera *et al.*, 1985). Moreover, as secondary parasites, it must be mentioned that *Mecistocirrus digitatus* and *Agriostomum vryburgii* are commonly found in cattle in this area (Rivera and Hurtado, 1983; Parra and Uribe, 1990; Moreno and Gomez, 1991).

Temperatures range from 18 to 33°C in Brazil, while rainfall reaches more than 500 mm during the rainy season. Brazil is considered to have the third largest national herd in the world, with approximately 130 million head. Brazil is a very large country with many climatic zones, thus the epidemiology of parasites varies regionally. Campo Grande, State of Mato Grosso do Sul is one of the most important areas for cattle production. It has well-defined wet and dry seasons and an average temperature range from 18 to 25° C, while rainfall ranges from 200 to 500 mm from November to April. Although *Cooperia* spp. are the most prevalent parasites, accounting for approximately 80% of the helminths observed, there is also a common presence of *H. contortus, H. similis, T. axei, T. longispicularis, O. radiatum, Trichuris discolor* and *Bunostomum phlebotomum*. High burdens of parasites are found in the state of Mato Grosso during the dry season, decreasing in the rainy season in the summer (Bianchin and Honer, 1985; Bianchin *et al.*, 1990). In southern Brazil, the state of Minas Gerais has approximately 20 million head of cattle, half of these for beef production. Temperatures in general remain constant during the year at 20°C, decreasing only during wintertime to 14°C. Rainfall is considered the most important limiting factor to parasites. Precipitation below 50 mm per month is registered from May to September. These minimal rainfall amounts during some months of the dry season limit the development and migration of larvae from the faeces to the pasture. However, most of the time, the humidity in the faecal pad is enough for larval development. As was mentioned previously, *Cooperia* is the most prevalent parasite during the year (Furlong *et al.*, 1985). In the southeastern region of Rio Grande do Sul, Brazil, maximum temperatures range from 29 to 35°C during the summer. Minimum temperatures reach 1–5°C during winter. Rainfall is very close to or above 50 mm for the full year. Temperatures and rainfall are generally favourable throughout the year for survival and migration of nematode larvae on the pasture. *Haemonchus, Trichostrongylus* and *Ostertagia* are present in most months of the year. According to Rassier *et al.* (1990) inhibition of *O. ostertagi* may occur from early spring to early summer in order to maintain the life cycle through the hot and dry months of the year.

Control of Nematodes in South America

The MLs are used extensively in South America. In general, two different programmes are in use: tactical treatments and strategic treatments.

Tactical treatments

Tactical treatments are used by veterinarians and farmers to solve specific situations such as: clinical symptoms like diarrhoea, rough hair coat, weight loss; introduction of new cattle to a farm; after heavy rains to prevent infections from newly activated larvae; stress (nutritional, transport, etc.); coprological analysis; and at weaning.

When tactical treatments are implemented, although the immediate situation apparently is solved, the pasture in which the treated animals are grazing remains contaminated. Therapeutic tactical treatment has a beneficial impact on the worm burden infecting the cattle, but has no impact on the total population of worms in the pasture. In this situation, reinfection occurs in a few weeks, since the pasture is not affected by the treatments.

Strategic treatments

Strategic treatments are being used in large farms with good management, not only as a therapeutic tool to clean the animals of worms, but also as a prophylactic measure to clean-up or reduce parasite populations on existing pasture to avoid future reinfections. The productivity trials carried out since the introduction of ivermectin in the area, plus those carried out with moxidectin, abamectin and doramectin, provided the vast majority of the basic epidemiological information in the region (Eddi *et al.*, 1985c).

Many farmers are using suppressive treatments given on a monthly basis or every 2 months to keep the livestock clean of parasites and, consequently, to decrease levels of pasture contamination. This is a method implemented by many farmers in Argentina to avoid nematode infections, in Venezuela to deal with ticks and in Brazil to deal with ticks on dairy cattle. The frequent and irrational administration of drugs to cattle caused several important problems related to drug resistance, residues and environmental contamination, and should not be recommended (Echevarria, 1996).

Relationship between management and use of MLs

In South America, animals are kept mainly on natural pasture under extensive conditions, being exposed to reinfection continuously. Because of this, farmers need to treat animals regularly to avoid helminthiasis. At weaning, which is carried out during autumn, calves traditionally are treated. Since MLs have become competitvely priced, farmers are increasing the use of endectocides at this period. When MLs are used, a second treatment is recommended at 60 days post-weaning. Steers are treated at the end of the spring mainly to avoid arrested *Ostertagia* (tropical and subtropical areas) and, according to levels of pasture contamination, animals receive treatment every 2–3 months. Heifers are treated for GI worms before the mating period (15 months old) and at the end of the spring to avoid inhibition of *O. ostertagi* (tropical and subtropical areas). At parturition, in tropical and subtropical regions, due to the presence of adult screw worms in the warm weather, cows are treated with MLs to avoid vaginal myiasis, while calves are treated to prevent umbilical myiasis. In Table 6.4.1 the different treatments being used in the region are presented.

Therapeutic and persistent activity of MLs

MLs have been used extensively in South America since the introduction in the 1980s of ivermectin. During nearly a decade, this was the only ML available. Since 1989, moxidectin, abamectin and doramectin have been introduced in South America. A large number of clinical and productivity trials have been carried out by the pharmaceutical industry in joint programmes with South American research groups. MLs were very well evaluated regarding not only their therapeutic and persistent efficacy, but also their productivity responses.

Table 6.4.1. Livestock categories, time of treatment with MLs and purpose.

Categories	Time of treatment and purpose
Newborn calves	1 dose to avoid umbilical myiasis
Calves at calving	1 dose at 3–4 months old
Castrated calves	1 dose to prevent myiasis
Weaning and heifers	1 dose at weaning time 1 dose at 60 days after weaning to treat GI parasites and to prevent animals from psoroptic mange (temperate areas)
Steers and heifers (1 year)	1 dose at the end of spring to avoid inhibition of *Ostertagia* and latent psoroptic mange
Cows at parturition	1 dose to avoid vaginal myiasis

Therapeutic activity

Niec *et al.* (1982) presented the first report of the anthelmintic action of ivermectin in the region, observing that ivermectin injectable was highly effective against the most common gastrointestinal parasites in the region. Eddi *et al.* (1993b) reported two studies, one in Argentina and one in Brazil, to evaluate the therapeutic efficacy of doramectin administered subcutaneously at a dose rate of 200 μg kg^{-1} (1 ml per 50 kg) to cattle harbouring mixed field infections of GI nematodes. The efficacy of doramectin was at least 99.9% against adult stages of *O. ostertagi, H. placei, H. similis, T. axei, T. longispicularis, C. oncophora, C. punctata, C. pectinata, C. spatulata, C. surnabada* (*C. mcmasteri*), *O. radiatum* and *Dictyocaulus viviparus*. Efficacy against *Nematodirus helvetianus* was 97.9%, while efficacy against *T. discolor* was 92.3%. Activity against inhibited larval forms of *O. ostertagi, H. placei, O. radiatum* and *T. axei* was 99.9%.

Steffan and Castro (1990) reported 100% efficacy with ivermectin pour-on against *H. placei, O. radiatum, O. venulosum* and *D. viviparus*, over 99% efficacy against *O. ostertagi* and *T. axei*, and 89.7% efficacy against *C. oncophora* and *C. punctata* in cattle.

Persistent efficacy

After moxidectin, abamectin and doramectin were launched in the South American market, the target in clinical trials with MLs was not only directed to study the therapeutic efficacy of those drugs, but also the persistent therapeutic effects. The experimental design to study the persistent effect varies depending on whether trials were carried out under pen conditions using experimental infections or under natural infection in the field.

Using experimental infections and working under housing conditions, Eddi *et al.* (1996a) observed a persistence of activity of non-aqueous injectable moxidectin of 28–35 days post-treatment. The main parasites in these trials, were *Ostertagia, Haemonchus, Cooperia, Trichostrongylus* and *Oesophagostomum*.

In Colombia, and based on eggs per grain (EPG) counts in animals exposed to field conditions with natural nematode infections, Entrocasso *et al.* (1996) could not observe differences in the persistent activity between the MLs in reducing EPG values.

However, to evaluate the persistent efficacy, Eddi *et al.* (1997) compared the persistent efficacy of doramectin, ivermectin and fenbendazole against natural nematode infections in cattle using egg counts and worm burdens. Although at day 28, the differences in parasite counts between the necropsied animals in the different groups were not statistically significant, at day 56 post-treatment all groups differed significantly ($P < 0.05$), with the control group having the most parasites, followed by the fenbendazole and ivermectin groups, and the smallest number in the doramectin group. The authors concluded that the duration of protection conferred by a single injection of doramectin was longer than that for ivermectin or fenbendazole, and that the duration of activity of the anthelmintics was demonstrated more accurately by parasite counts than by faecal egg counts.

In Venezuela, Sogbe *et al.* (1998a,b) using EPG counts observed that ivermectin and abamectin at 1% were highly effective to control nematode populations (96 and 92% reductions at 30 days post-treatment, respectively).

Productivity

The positive therapeutic effect of the MLs in a basic research trial was observed by Mejia *et al.* (1999). These authors showed that weaning calves treated with ivermectin grew faster than untreated calves, and differences in body weight became significant at 6 weeks of age. Ivermectin-treated heifers reached puberty 3 weeks earlier than infected, untreated heifers as assessed by serum progesterone concentrations. In addition, the pelvic area at 39 weeks and at 15 months of age was increased in treated heifers (8 and 11%, respectively) compared with parasitized animals. No differences in the wither heights were observed. It was concluded by Mejia *et al.* (1999) that ivermectin treatment in dairy heifers may increase growth rate during development, advance the onset of ovarian function and positively affect yearling pelvic area.

Moxidectin was used to study the variation in the response of Shorthorn bulls to GI parasites by Costa *et al.* (1997, personal communication). Shorthorn bull-calves were weaned and treated with moxidectin. The authors observed that liveweight gain was significantly higher in the treated animals. The most common genera of nematodes present were *Cooperia*, *Ostertagia* and *Haemonchus*.

Similar results were observed by Alvarez *et al.* (1997) using ivermectin in northeast Santa Fe Province, Argentina. Animals treated with ivermectin and non-treated controls were kept in separate pastures. Ivermectin-treated animals gained over 20 kg more than the controls.

Programmes of use with MLs

Bulman *et al.* (1990) studied the productivity of 285 calves, divided into three groups of 95 animals each treated strategically with ivermectin (in May, July, September, November and February) or oxfendazole (in May, June, July, September, October, December and again in July) or not treated (control group) grazing in the same paddock. Comparisons were based on the number of steers reaching a minimum weight and finish for slaughter when approximately 2 years old (585 days, 19.5 months of trial).

The helminths present were *Haemonchus*, *Cooperia*, *Oesophagostomum*, *Trichostrongylus*, *Ostertagia* and *Bunostomum*. The authors reported that in the first 525 days (6 May 1986 to 13 October 1987), the groups treated with ivermectin and oxfendazole had gained 23.5 and 19.2% more weight respectively, than in the controls.

The effect of previous pasture management and ivermectin treatment on GI parasitism, weight gain and carcass composition of steers was studied in the semiarid Pampeana region of Argentina (Suarez *et al.*, 1991). These authors observed that the liveweight gain responses of animals treated strategically with ivermectin during the autumn–winter period and those treated monthly with ivermectin were significantly greater than those of non-treated controls, during autumn, winter and early spring. At the end of the study when cattle reached market condition, the liveweight gains of both groups receiving ivermectin were 74.1 (strategic) and 81.9 kg (monthly), respectively, greater than non-treated controls. Carcass analyses showed significantly greater weight and dressing percentages in both groups treated with ivermectin than in controls. Reduced total bone, muscle and fat weights were observed in control animals.

Using strategic control of GI helminths in Aberdeen Angus heifers at autumn–winter time, Steffan *et al.* (1995) observed, in a group of 156 calves divided into three groups, that ivermectin gave the best results in terms of reduced worm burdens, weight gain and conception rate when compared with oxfendazole-treated or non-treated control groups.

Control programmes carried out in Brazil using MLs also confirmed the positive economic impact of ML treatment on GI parasitism in cattle. Working with dairy cattle, Lima *et al.* (1997) reported that ivermectin-treated calves (monthly doses of 200 $\mu g\ kg^{-1}$) weighed on average 53.8 kg more than untreated calves by the end of the study period. Vieira-Bressan *et al.* (1998), also in Brazil, observed that animals that received two and three doses of doramectin showed higher mean weight gains of 14.89 and 48.85 kg, respectively, than the mean weight gain of routinely treated animals after a period of 365 days.

Additional South American ML issues

Generics

As soon as the international patent of ivermectin expired, several generic ivermectins and abamectins appeared in the market in South America. Some of these products have not only the same efficacy as ivermectin, but also good persistent therapeutic properties. Nevertheless, it is during the registration process that the government authority approves the sale and use of an antiparasitic agent after having evaluated whether the product is effective and that its use involves no risk for animals, public health or the environment. This process is often difficult for many developing countries of South America because it requires a sophisticated infrastructure and specialized personnel to carry out the tests (a common issue with MLs). Notwithstanding this, some countries have made important advances both nationally and regionally. Another obligation that has proven more complicated to implement and for which the poorer countries are on their own is the continuous monitoring of the quality of antiparasitic agents in order to prevent abuses, including adulteration, the sale of substandard preparations and drug combinations of dubious stability (Nari and Hansen, 1999).

Ivermectin 1% long-acting injectable

Due to the epidemiological impact that can be achieved using persistent drugs, the persistent activity (PA) of a new injectable long-acting ivermectin 1% (w/w) formulation was studied by Eddi *et al.* (1998) using 36 male 6- to 9-month-old calves against the principal GI nematodes in Argentina. The calves were allocated to groups of six animals each: groups 1–5 were treated on days –35, –28, –21, –14 and –7, respectively; the sixth group served as untreated controls. On day 0, all the calves were challenged orally with more than 60,000 infective larvae (L_3) from a mixed culture. At necropsy time, on days +26 and +27 post-infection, comparing parasite counts showed a significant ($P < 0.05$) PA of 28 days against adult and immature *O. ostertagi* and *Oesophagostomum* spp., and a PA of 35 days against *Haemonchus* spp., *Trichostrongylus* spp. and *Cooperia* spp. The results showed that the ivermectin formulation has a high efficacy and persistency against the principal GI nematodes of cattle. These results were confirmed by Steffan *et al.* (2000) in a study trial to measure economic impact on grazing beef cattle, at the end of a 6-month period, using natural nematode infections, and by Mercier *et al.* (2000) in a persistent efficacy trial using experimental nematode infecions in calves.

Ivermectin 3.15% long-acting injectable

Since the main problem in parasite control is the reinfection of animals under natural grazing conditions, an ivermectin 3.15% long-acting injectable formulation was launched recently in the South American market. The long-acting ivermectin has persistent effect against endo- and ectoparasites (Carvalho *et al.*, 2000). Levels of protection vary according to parasite species, being at least 6 weeks against GI nematodes, more than 100 days against the cattle grub *Dermatobia hominis*, more than 70 days against *Boophilus microplus* and more than 50 days against *Psoroptes ovis*. In comparative efficacy trials to control GI nematodes (Eddi *et al.*, 2000) and *B. microplus* (Grisi *et al.*, 2000), no significant differences ($P < 0.05$) were observed among doramectin and the long-acting ivermectin 3.15%.

Problems with the frequent use of MLs

MLs have been used extensively in South America during the last 15 years. However, up to the present no problems have been reported due to the frequent use of those drugs. Even though anthelmintic resistance to MLs was reported in sheep parasites (Echevarria *et al.*, 1996; Eddi *et al.*, 1996b; Maciel *et al.*, 1996a; Nari *et al.*, 1996) in several South American countries, there are only two reports of reduced efficacy to MLs in cattle. These studies used the faecal egg reduction test to evaluate efficacy. Ivermectin and doramectin showed 75% efficacy in EPG reduction, while moxidectin showed 90% efficacy in cattle harbouring *Cooperia* spp. (Anziani *et al.*, 2000) The other study showed resistance of *Cooperia* spp. and *Trichostrongylus* spp. to ivermectin and doramectin, which had 65 and 85% efficacy, respectively, while fenbendazole and moxidectin provided EPG reductions of 100 and 95%, respectively (Fiel *et al.*, 2000).

Dung degradation/beetle colonization

In studies examining the environmental impact of ML faecal excretion, no differences in beetle colonization of faecal samples were observed between those faecal pads from treated animals and those of non-treated controls even though hornfly development was reduced (Mariategui *et al.*, 1998, 2000).

Use of MLs Against Ectoparasites in South America

The economically important cattle ectoparasites of South America are multi-host ticks, single host tick (*B. microplus*), tropical warble (*D. hominis*), and screw worms (*Cochliomyia hominovorax*) in the tropical and subtropical regions and mange (*P. ovis*) in the temperate regions. Recently, hornflies (*Haematobia irritans*) have migrated into South America and are currently considered an important problem. MLs have become an important class of compounds in controlling all of these parasites except multihost ticks (found in northern South America only), partly because of their excellent therapeutic and persistent activities, but also because many of these organisms have developed resistance to previously used classes of compounds such as pyrethroids and organophosphates. In addition, a number of additional advantages for the use of MLs are also readily apparent due to convenience, product stability and environmental contamination.

The cattle tick **Boophilus microplus**

B. microplus ticks are a constraint on America's cattle industry from Mexico to Argentina (Guglielmone *et al.*, 2000). Annual losses due to this tick were calculated as $150 million in Argentina, $800 million in Brazil and $40 million in Uruguay (Horn and Arteche, 1985; Spath *et al.*, 1994; Rimbaud Giambruno, 1999). The plunge vat has been the most effective mean of applying acaricides but, in many regions, the existence of small cattle populations reduces the cost/efficiency of this method. Furthermore, in some areas, it is difficult to find a readily accessible and sufficient water source to charge and replenish the vats (Nari, 1990). Resistance of *B. microplus* to conventional acaricides (organophosphates, pyrethroids and formamidines) is also a relevant problem in the region (Cardozo, 1995; Caracostantogolo *et al.*, 1996; Martins, 1996). In those regions with widespread resistance to these compounds, their lower cost in relation to MLs is no longer an advantage for treatment. When tick susceptibility to conventional acaricides is lost, control measures become more expensive. MLs are excellent tools in these conditions. In order to optimize the use of MLs, it is necesary to do epidemiological studies in each region to determine the optimum time to apply treatments according to the status of the non-parasitic tick population, tick burdens in cattle and persistent efficacy of each ML (Martins *et al.*, 1996).

In Table 6.4.2, the comparative efficacy of the differents MLs available in the South American's veterinary market for the treatment of *B. microplus* is presented. Since the present review, no new MLs have been developed for the control of *B. microplus*.

The results in Table 6.4.2 show that cattle can be treated successfully against *B. microplus* using ivermectin, moxidectin and doramectin. The percentage efficacy in the different study trials ranges from 84.7 to 99%, at least during 21 days. However, in some trials, excellent efficacy was observed for even more than 35 days post-treatment.

Myiasis caused by Cochliomyia hominivorax

Screw worm, *C. hominivorax*, is an obligate, myiasis-causing, parasitic dipteran. The female fly deposits 200–400 eggs on the edge of a wound, the navels of newborns and other sites on the lacerated skin of any warm-blooded animal. The screw worm fly larvae then hatch and burrow into the flesh and feed on tissue, causing extensive damage and infection until they complete development and drop on the soil to pupate and eclose to become adults. Because infested wounds (myiases) attract other gravid screw worm females and secondary infestations, untreated wounds frequently become self-perpetuating, often with fatal results (Moya-Borja *et al.*, 1993). *C. hominivorax* is endemic in the tropical and subtropical areas of South America (FAO, 1993).

Perhaps the more widespread application of MLs for all countries in the tropical and subtropical areas of South America is the treatment against these screw worm fly larvae. Before the advent of MLs, there was no secure way to avoid or prevent navel myiasis of newborn calves, vaginal myiasis in post-parturient cows, scrotal myiasis post-castration of male calves or myiasis developed on any surgical or accidental wound. Therefore, calving, dehorning and castration were avoided when adult flies were present. With the exception of moxidectin, the different current formulations of MLs are highly efficient in preventing the development of

Table 6.4.2. Study trials carried out in South America against *Boophilus microplus*, using MLs.

Formulation	Dose	Country[a]	Results	Reference
Ivermectin 1% injectable	200 μg kg^{-1} at days 0 and 9	Argentina, EI	>99% efficacy between days 7 and 39	Perez Arrieta *et al.* (1982)
Ivermectin 1% injectable	200 μg kg^{-1} at days 0–9 and 18	Argentina, EI	>99% efficacy between days 12 and 43	Bulman *et al.* (1982)
Ivermectin 1% injectable	200 μg kg^{-1} at day 0	Argentina, EI	98.6% efficacy between days 0 and 23	D'Agostino *et al.* (1997)
Ivermectin 1% injectable	200 μg kg^{-1} at day 0	Brazil, NI	>99% efficacy between days 0 and 23	Marques *et al.* (1995)
Moxidectin 1% injectable	200 μg kg^{-1} at day 0	Argentina, NI	84.7% efficacy on adult female ticks; 94.2% efficacy to reduce egg laying between days 1 and 35	Eddi *et al.* (1994)
Moxidectin 1% injectable	200 μg kg^{-1} at day 0	Uruguay, EI	>95% efficacy between days 1 and 30	Larrosa *et al.* (1994)
Moxidectin 1% injectable	200 μg kg^{-1} at day 0	Argentina, EI	97.9% efficacy to reduce number of ticks; 70.4% efficacy to reduce engorgement	Eddi *et al.* (1996c)
Moxidectin 0.5% pour-on	500 μg kg^{-1} at day 0	Argentina, NI	95% efficacy to reduce number of ticks between days 7 and 21	Guglielmone *et al.* (2000)
Doramectin 1% injectable	200 μg kg^{-1} at day 0	Brazil, EI	>99% efficacy between days 4 and 21; 20 days of persistent efficacy	Gonzales *et al.* (1993)
Doramectin 1% injectable	200 μg kg^{-1} at day 0	Argentina, NI	98–100% efficacy to reduce number of ticks between days 8 and 28	Lombardero *et al.* (1995)
Doramectin 1% injectable	200 μg kg^{-1} at day 0	Brazil, NI	>98.6% efficacy between days 8 and 28	Leite *et al.* (1995)
Doramectin 1% injectable	200 μg kg^{-1} at day 0	Brazil, NI	91–99% efficacy between days 8 and 28	Muñiz *et al.* (1995b)

[a]NI, natural infestation; EI, experimental infestation.

myiasis when they are applied within a few hours or days of the presence of the occurrence of a susceptible area on the host.

The use of MLs in this regard, can be summarized as follows.

- To prevent navel myiasis, newborn calves are treated with 1 ml of ivermectin, or doramectin injectable formulation within 3 days of birth.
- To prevent scrotal myiasis animals are treated at the time of castration with 200 μg kg^{-1} of abamectin, doramectin or ivermectin injectable formulations.
- Animals treated with ivermectin slow-release bolus 15 days before castration are also protected from scrotal myiasis.

The results in Table 6.4.3 show that the MLs can be used successfully to cure and/or prevent myasis caused by *C. hominivorax*. The percentage efficacy in the different study trials ranges from 96.5 to 100%.

The warble fly, Dermatobia hominis

The warble fly, *D. hominis,* is widespread in the tropical and subtropical regions of South America, with the exception of Chile (Moya-Borja *et al.*, 1993). The female uses other insects as carriers of its eggs. Infective larvae emerge from the eggs after 4–9 days when the carrier insect visits a warm-blooded host. Larvae penetrate the unbroken skin and develop in the subcutaneous tissue where they produce a parasitic nodule or warble. Any warm-blooded animal, including man and birds, can be affected. *D. hominis* larvae require 4–18 weeks to complete their cycle at the original site of penetration, and then drop to pupate (Kettle, 1995). The activity of adult female D. *hominis* is confined to specific times of the year and depends on rainfall (Lombardero and Fontana, 1968). The incidence of nodules tends to decrease and disappear during the dry and cool seasons. *D. hominis* does not cause mortality in adult cattle, but young animals are severely affected and appear to lose condition (Moya-Borja *et al.*, 1993). The hides of affected animals have low value due to multiple orifices, causing economic losses to the leather industry. In Brazil, only 15% of hides are of good quality (Oliveira, 1988). Annual losses due to the effect of *D. hominis* on meat and milk production and on leather industries in South America were estimated at $260 millon (Pinheiro *et al.*, 1999)

MLs have high levels of efficacy against *D. hominis* larvae infestations. Ivermectin, abamectin and doramectin applied at 200 μg kg^{-1} by the subcutaneous route kill larvae, promote the healing of lesions and can prevent new infestations. Similar effects are obtained with the ivermectin bolus and with eprinomectin or abamectin pour-on formulations at a dose rate of 500 μg kg^{-1}.

Table 6.4.3. Study trials to evaluate the efficacy of the MLs in curing or preventing *Cochliomyia hominivorax*.

Formulation	Dose	Country[a]	Result	Reference
Abamectin 1% injectable	200 μg kg^{-1}	Argentina, NI	96.5% prophylactic control of scrotal myiasis	Anziani *et al.* (1995a)
Ivermectin 1% injectable	200 μg kg^{-1}	Argentina, NI	100% efficacy using wounded calves, castrated calves and newborns (four trials)	Anziani and Loreficce (1993)
Ivermectin 1% injectable	200 μg kg^{-1}	Argentina, NI	100% efficacy in navel myiasis when treated within 4 days of birth. 100% prophylactic control of scrotal myiasis	Lombardero *et al.* (1999a,b)
Ivermectin SR bolus (oral)	One bolus delivered at 12 mg day^{-1} for 135 days	Brazil, NI	100% prophylactic control of scrotal myiasis when administered 15 days before castration	Lessa and Lessa (1999)
Doramectin 1%	200 μg kg^{-1} injectable	Brazil, EI	100% prophylactic control of myiasis in wounds	Oliveira *et al.* (1993)
Doramectin 1%	200 μg kg^{-1}	Brazil, EI	100% prophylactic control of myiasis in wounds for 21 days	Moya-Borja *et al.* (1997)
Doramectin 1% injectable	200 μg kg^{-1}	Argentina, NI	100% efficacy in navel myiasis when treated within 4 days of birth. 100% prophylactic control of scrotal myiasis	Lombardero *et al.* (1999a,b)
Doramectin 1% injectable	200 μg kg^{-1}	Argentina, Brazil, Venezuela, NI	100% prophylactic control of scrotal myiasis	Muñiz *et al.* (1995c)
Doramectin 1% injectable	200 μg kg^{-1}	Argentina, NI	100% prophylactic control of scrotal myiasis. 100% control of navel myiasis when treated within 24 h of birth	Anziani *et al.* (1995b)
Doramectin 1% injectable	1 ml to each calf	Argentina, Venezuela, NI	100% prophylactic control of screw worm strikes in newborn calves and post-parturient cows	Muñiz *et al.* (1995a)

[a]NI, natural infestation; EI, experimental infestation.

Numerous trials have been carried out in South America in order to determine the efficacy of MLs against *D. hominis* larvae. Most of them were field trials in which naturally infested cattle were allocated to a non-treated control group and to a treated group that received the test formulation. After treatment, animals of each group were examined repeatedly at intervals for the presence of parasitic nodules. The published results of study trials to evaluate the efficacy of the MLs in curing or preventing *D. hominis* are presented in Table 6.4.4.

The results in Table 6.4.4 show that the MLs can be used successfully to cure and/or prevent *D. hominis* infestations. The protection attained with the different MLs varies in relation to formulations and route of administration. Oral bolus provided the longest period of protection.

The hornfly, Haematobia irritans

Hornflies (mainly pyrethroid-resistant strains) have moved from North America through Venezuela and Brazil into Argentina in the last few years. A limited economic impact has been seen in *Bos indicus* cattle in the tropical regions of South America apparently due to the poor fly larval development conditions of the faecal pats of these cattle (Moya-Borja, personal communication). However, as the flies move into the *Bos taurus* herds of southern Brazil and Argentina, economic damage by the flies is expected to become substantial (Eddi *et al.*, 1993a).

Several trials were carried out in South America to determine the efficacy of MLs in reducing and controlling the number of *H. irritans* on cattle. Use of MLs exclusively to control hornflies where pyrethroids are effective would not be cost effective. However, as part of an overall control programme that included hornflies with other susceptible parasites or where pyrethroid resistance is present, MLs would be an excellent choice for hornfly control on cattle.

Pour-on MLs are very effective against adult flies, and both injectable and pour-on MLs are active against developing larvae. MLs help to maintain *H. irritans* populations below the threshold of economic damage. The application of abamectin 0.5% pour-on at 500 μg kg^{-1} has been shown to be effective when applied to *H. irritans*-infested cattle. Ivermectin or doramectin 1% (w/w) injectables are useful in the control of *H. irritans* as well as against endoparasites and other ectoparasites such as *B. microplus*, *C. hominivorax*, warbles and mites. In Table 6.4.5 the results in the reduction of hornfly populations, using injectable and pour-on MLs are presented.

MLs provided approximately 14–20 days of protection against the hornfly *H. irritans*. Both routes of administration reported (injectable and pour-on) gave excellent therapeutic efficacy.

Mange

Since 1980, the MLs have been considered the most effective treatment against psoroptic mange. The disease is endemic in southern South America, and particularly severe in some areas of Argentina. Up to 1985, the disease was widely spread in all the Humid Pampa. Economic losses

Table 6.4.4. Study trials to evaluate the efficacy of the MLs to cure or prevent *Dermatobia hominis.*

Formulation	Dose	Country	Result	References
Abamectin 1% injectable	200 μg kg^{-1}	Argentina and Brazil	Reinfestation on day 44 p.t.	Coumendouros *et al.* (1993)
Ivermectin 1% injectable	200 μg kg^{-1}	Brazil	Free of nodules until day 31 p.t.	Oliveira *et al.* (1997)
Ivermectin 1% (w/w)	1 ml per 50 kg	Argentina	Free of nodules until day 35 p.t.	Quiroz *et al.* (1998)
Ivermectin 1% injectable	200 μg kg^{-1}	Brazil	Free of nodules from day 21 to 63 p.t.	Silva *et al.* (1995)
Ivermectin 1%	200 μg kg^{-1}	Venezuela	Free of nodules from day 14 to 21 p.t.	Garcia *et al.* (1998)
Ivermectin SR bolus (oral)	One bolus delivered at 12 mg day^{-1} for 135 days	Brazil	Free of nodules from day 30 to 148 p.t.	Maciel *et al.* (1996b)
Ivermectin SR bolus (oral)	One bolus delivered at 12 mg day^{-1} for 135 days	Argentina Brazil Colombia Uruguay	Free of nodules from day 14 to 135 p.t.	Eagleson *et al.* (1996)
Eprinomectin 0.5% pour-on	500 μg kg^{-1}	Brazil	81.8–100% efficacy at day 35 p.t. (three trials)	Coumendouros *et al.* (1999)
Abamectin 0.5% pour-on	500 μg kg^{-1}	Brazil	Free of nodules until day 14 p.t.	Souza *et al.* (1999)
Doramectin 1% injectable	200 μg kg^{-1}	Argentina Brazil Venezuela	100% efficacy from day 7 to 30 p.t.	Muñiz *et al.* (1997)
Doramectin 1% injectable	200 μg kg^{-1}	Brazil (induced infestation)	100% efficacy at day 6 p.t. Persistent efficacy 35 days p.t.	Moya-Borja *et al.* (1993)
Doramectin 1% injectable	200 μg kg^{-1}	Brazil	Free of nodules until day 45 p.t.	Oliveira *et al.* (1997)

p.t., post-treatment.

Table 6.4.5. Study trials to evaluate the efficacy of MLs against the hornfly *Haematobia irritans*.

Formulation	Dose	Country	Result	References
Abamectin 1% injectable	200 μg kg^{-1}	Argentina	73% reduction 14 days p.t. Immature stages did not develop until 7 days p.t.	Guglielmone *et al.* (1999)
Abamectin 1% pour-on	500 μg kg^{-1}	Brazil	Reduction ranged from >99 to 68% between days 1 and 14 p.t.	Sabatini *et al.* (1999)
Abamectin 1% pour-on	1 ml per 20 kg	Argentina	75% reduction 14 days p.t.; 52% reduction 21 days p.t.	Eddi and Caracostan-togolo (2000)
Ivermectin 1% (w/w) injectable	1 ml per 50 kg	Argentina	>75% reduction over the first 2 weeks p.t.	Guglielmone *et al.* (1998)
Eprinomectin 0.5% pour-on	500 μg kg^{-1}	Brazil	100% reduction over the first 3 weeks p.t.	Scott *et al.* (1999)
Doramectin 1% injectable	200 μg kg^{-1}	Argentina	Almost 81% reduction over the first 2 weeks p.t., followed by a slow decrease to 40% on day 49	Anziani *et al.* (1999)

p.t., post-treatment.

due to psoroptic cattle scab were calculated by Nuñez and Moltedo (1985) as nearly $150 million per year. The frequent use of MLs in Argentina significantly decreased the number of mange outbreaks. Exacerbating factors exist in the development of the disease: high concentration of cattle; intensive animal movement; cool and windy winters; low levels of feed; and internal parasitosis.

MLs have overcome many of the drawbacks of dipping baths or sprays with products that required two applications 10–12 days apart. Even though no product is effective against mite eggs, the MLs are still present when mite eggs hatch; therefore, the larvae are killed following their initial exposure. The poor maintenance of vats, defective hand sprayers and lack of compliance with interval applications that led to incomplete efficacy in the past (Giudici, 1985) can be overcome with the formulations and efficacy of the MLs.

A relatively small number of trials were published in South America on the efficacy of MLs against mange mites in cattle. *P. ovis* var. *ovis* infections were eliminated with a single injection of 0.2 mg kg^{-1} of ivermectin 1% (w/v) or (w/w) or doramectin 1% (Benitez Usher, 1982; Errecalde,

1993; Marguerritte *et al.*, 1997). However, the use of MLs to treat specifically psoroptic mange cases, added to control of GI nematodes in autumn–winter and late spring throughout the late 1980s has been very important in reducing the incidence of the disease. The continued use of MLs over approximately 15 years, treating not only the apparent cases of cattle scab but also young susceptible animals that do not yet show any lesions, in each treatment against GI nematodes, reduced the incidence of cattle mange. In 1999, the National Service of Animal Health found only six farms with psoroptic cattle scab in a survey that involved 23,539 farms (SENASA, 1999).

References

Alvarez, J.D., Anziani, O.S., Vottero, D.A. and Peruchena, C. (1997) Effect of the control of gastrointestinal nematodes on weight gain in heifers in the northeast of the Santa Fe Province, Argentina. *Revista de Medicina Veterinaria, Buenos Aires* 78, 9–13.

Anderson, N., Armour, J., Jennings, W.F., Ritchie, J.S.D. and Urquhart, G.M. (1969) The sequential development of naturally occurring ostertagiasis in calves. *Research in Veterinary Science* 10, 18–28.

Anziani, O.S. and Loreficce, C. (1993) Prevention of cutaneous myiasis caused by screw worm larvae (*Cochliomyia hominivorax*) using ivermectin. *Journal of Veterinary Medicine* Series B 40, 287–290.

Anziani, O.S., Guglielmone, A.A. and Aguirre, D.H. (1995a) Abamectin in the prevention of post castration myiasis (*Cochliomyia hominivorax*) in cattle. *Veterinaria Argentina* 114, 233–236.

Anziani, O.S., Errecalde, J.O. and Muniz, R.A. (1995b) Efficacy of doramectin in the prevention of myiasis caused by *Cochliomyia hominivorax*. *Revista de Medicina Veterinaria, Buenos Aires* 76, 433–436.

Anziani, O.S., Guglielmone, A.A., Flores, S.G. and Moltedo, H. (1999) Evaluation of injectable doramectin to control natural infestations of *Haematobia irritans* (Diptera: Muscidae) in cattle. *Veterinaria Argentina* 157, 501–505.

Anziani, O.S., Zimmermann, G., Guglielmone, A.A., Vasquez, R. and Suarez, V. (2000) Resistance to avermectins in cattle harboring *Cooperia* spp. Preliminary communication. *Veterinaria Argentina* 164, 280–281.

Armour, J. (1970) Bovine ostertagiasis: a review. *Veterinary Record* 86, 184–189.

Armour, J., Jennings, F.W., Murray, M. and Selman, I. (1973) Bovine ostertagiasis. In: Urquhart, G.M. and Armour, J. (eds) *Helminth Parasites of Cattle, Sheep, and Horses in Europe*. Oxford University Press, Oxford, pp. 11–23.

Bayly, W.M. (1970) Type II bovine ostertagiasis – a review. *Southwestern Veterinarian* 30, 171–175.

Benitez Usher, C. (1982) Efficacy of Ivermectin against cattle scab. *IV Congreso Argentino de Ciencias Veterinarias*, La Plata, Buenos Aires. Hemisferio Sur, Buenos Aires, Argentina, November, p. 14.

Berdie, J., Genovese, J., Zunini, C., Molinavi, C., Chevlone, A., Castro, E. and Duncan, J.L. (1988) Epidemiological study on gastrointestinal parasites of beef cattle in Uruguay. In: *Nuclear Techniques in the Study and Control of Parasitic Diseases of Livestock*. IAEA, Vienna, pp. 95–105.

Bianchin, I. and Honer, R. (1985) Helminth parasites of beef cattle in the Cerrados of Brazil. *Tropical Animal Health and Production* 19, 39–45.

Bianchin, I., Honer, M.R. and Nascimento, Y.A. (1990) The epidemiology of Helminths in nellore beef cattle in the Cerrados of Brazil. In: Guerrero, J. and Leaning, W. (eds) *Epidemiology of Bovine Nematode Parasites in the Americas*. Salvador, Bahia, Brazil, pp.

Biondani, C. and Steffan, P. (1988) Effect of gastrointestinal parasitism in milk production in dairy herds. *Veterinaria Argentina* 42, 116–127.

Bulman, G.M., D'Agostino, B.I., Monzon, C.M. and Brunal, C.M. (1982) Ivermectin: eficacia de este nuevo sistemico para el control de la garrapata comun del vacuno *Boophilus microplus* (Can) su evaluacion en la zona subtropical de la Republica Argentina. *IV Congreso Argentino de Ciencias Veterinarias*. La Plata, Buenos Aires, November, 7.

Bulman, G.M., Ingouville, E., Fiel, C.A. and Ambrustolo, R.R. (1990) Productivity in Brahman/Hereford steers from weaning to slaughter weight on improved pastures in the central mesopotamia area (Entre Rios, Argentina), comparing different parasite control strategies. *Veterinaria Argentina* 65, 301–305.

Caracostantogolo, J., Muñoz Cobeñas, M.E., Eddi, C., Ambrústolo, R.R., Bulman, G.M. and Marangunich, L. (1996) Primera determinación en la República Argentina de una población de *Boophilus microplus* (Can.) resistente al piretroide sintético alfacipermetrina caracterizada mediante pruebas preliminares. *Veterinaria Argentina* 13, 575–582.

Cardozo, H. (1995) Situación de resistencia del *Boophilus microplus* en Uruguay. Medidas para controlarla. *Mem. III Sem. Int. Parasitol. Anim. 'Resistencia y control en garrapatas y moscas de importancia veterinaria'*. Acapulco, México, 11–12 October, pp. 30–38.

Carvalho, L.A., Bianchin, I., Bridi, A.A., Maciel, A.E., Malacco, M.A., Santos, C.A., Cruz, J.B., Cox, J.L. and Cramer, L. (2000) Field evaluation of the effect of Ivomec Gold on weight gain on beef cattle. *XXI World Buiatric Congress*, Punta del Este, Uruguay, 4–8 December.

Coumendouros, K., Scott, F.B., Sant'Anna, F.B., Dabes, C.E., Cruz, J.B., Benitez Usher, C., Cramer, L.G., Gross, S.J. and Kohn, A.B. (1993) Efficacy of abamectin injection against *Dermatobia hominis* in cattle. *Parasitology Research* 79, 183–185.

Coumendouros, K., Scott, F.B., Sant'Anna, F.B., Dabes, C.E. and Fernandez, M.M.M. (1999) Eficacia vernicida a nivel de campo de uma formulaçao pour on contendo 0.5% de eprinomectin em bovinos. *XI Seminario Brazileiro de Parasitologia Veterinaria*, Salvador, Bahia, Brazil, October.

D'Agostino, B.I., Citroni, D., Lamberti, J.C. and Bulman, G.M. (1997) Efficacy of a new injectable formulation of ivermectin 1% w/w against the common cattle tick *Boophilus microplus* (Can.) in experimentally infected housed heifers. *Veterinaria Argentina* 138, 518–524.

Eagleson, J., Cramer, L., Uribe, L., Maciel, E.Y. and Cruz, J. (1996) Field efficacy of Ivomec SR bolus against *Boophilus microplus* and *Dermatobia hominis*. *XV Congresso Panamericano de Ciencias Veterinarias*. Campo Grande, MS, Brazil, October, p. 319.

Echevarria, F. (1996) Anthelmintic resistance. In: *Controle dos Nematodes Gastrintestinais em Ruminantes*. EMBRAPA, Terezinha Padilla, Brazil.

Echevarria, F., Borba, M.F.S., Pinheiro, A.C., Waller, P.J. and Hansen, J.W. (1996) The prevalence of anthelmintic resistance in nematode parasites of sheep in Southern Latin America: Brazil. *Veterinary Parasitology* 62, 199–206.

Eddi, C. and Caracostantogolo, J. (2000) Efficacy of a pour on formulation against the horn fly *Haematobia irritans*. Technical Report, Rosenbusch Laboratories, March.

Eddi, C., McLoughlin, R., Niec, R., Dughetti, R. and Benitez Usher, C. (1985a) Evaluación de una estrategia antiparasitaria en bovinos de invernada. *Acintacnia (Buenos Aires)* 17, 58–62.

Eddi, C., Niec, R., McLoughlin, R., Carcagno, M., Dughetti, R., Corbellini, C. and Benitz, A. (1985b) Efficacy of Tobimin (Sch 32481) in cattle in Argentina. *11th International Conference of the World Association for the Advancement of Veterinary Parasitology (WAAVP).* Río de Janeiro, Brazil, 5–9 August.

Eddi, C., McLaughlin, R., Niec, R., Dughetti, R. and Benitez Usher, C. (1985c) Evaluation of an antiparasitic strategy in weaner cattle. *ACINTACNIA*, II, No. 17, pp. 58–62.

Eddi, C., Caracostantogolo, J., Aufranc, C., Peralta, J., Marangunich, L., Noaco, A. and Balbiani, G. (1993a) *Haematobia irritans*: epidemiology and economical impact assay in Argentina. *14th International Conference of the World Association for the Advancement of Veterinary Parasitology (WAAVP)*, Cambridge, 8–13 August.

Eddi, C., Bianchin, I., Honer, R., Muñiz, R., Caracostantogolo, J. and doNascimento, Y.O. (1993b) Efficacy of doramectin against field nematode infections of cattle in Latin America. *Veterinary Parasitology* 49, 39–44.

Eddi, C., Caracostantogolo, J., Bulman, G.M., Ledesma, C. and Balbi, A. (1994) Eficacia de moxidectin administrado a diferentes dosis contra *Boophilus microplus*. *VII Congreso Argentino de Ciencias Veterinarias*, Buenos Aires, November, p. 259.

Eddi, C., Caracostantogolo, J., Bulman, G.M., Schapiro, J., Muñoz Cobeñas, M., Marangunich, L. and Ambrustolo, R.R. (1996a) Persistency of activity of moxidectin 1% non-aqueous formulation against gastrointestinal nematodes in cattle. *Proceedngs of the XV Panamerican Congress of Veterinary Science*, Campo Grande, Brazil, October, p. 333.

Eddi, C., Caracostantogolo, J., Peña, M., Schapiro, J., Marangunich, L., Waller, P.J. and Hansen, J.W. (1996b) The prevalence of anthelmintic resistance in nematode parasites of sheep in Southern Latin America: Argentina. *Veterinary Parasitology* 62, 189–197.

Eddi, C.S., Caracostantogolo, J., De Montigny, L.S., Bulman, G.M. and Delay, R.L. (1996c) Comparative efficacy of two injectable formulations of moxidectin against *Boophilus microplus* in experimentally infested cattle. *XV Congreso Panamericano de Ciencias Veterinarias*, Campo Grande, MG-Brasil, October, p. 306.

Eddi, C., Muñiz, R., Caracostantogolo, J., Errecalde, J., Rew, R., Michener, S. and McKenzie, M.E. (1997) Comparative persistent efficacy of doramectin, ivermectin and fenbendazole against naturally acquired nematode infections in cattle. *Veterinary Parasitology* 72, 33–41.

Eddi, C., Caracostantogolo, J., Lamberti, J., Bulman, G.M. and Schapiro, J. (1998) Persistence of an injectable formulation of 1% ivermectin against cattle gastrointestinal nematodes. *Veterinaria Argentina* 146, 408–414.

Eddi, C., Caracostantogolo, J., Moltedo, H., Derozier, C., Cutulle, C. and Schapiro, J. (2000) Efficacy of doramectin (200 mcg/kg) and ivermectin (630 mcg/kg) against gastro-intestinal nematode infections in cattle. *XXI World Buiatric Congress*, Punta del Este, Uruguay, 4–8 December.

Entrocasso, C. (1987) Economic impact of gastrointestinal verminosis in the temperate climate areas of South America, with special reference to Argentina. In: Leaning, B. and Guerrero, J. (eds) *The Economic Impact of Parasitism in Cattle. Proceedings of the MSD AGVET Symposium, XXIII World Veterinary Congress*, Montreal, Quebec, Canada.

Entrocasso, C. (1988) Epidemiology and control of bovine osteragiasis in South America. *Veterinary Parasitology* 27, 59–65.

Entrocasso, C., Parra, D., Vottero, D., Farias, M., Uribe, L.F. and Ryan, W.G. (1996) Comparison of the persistent activity of ivermectin, abamectin, doramectin and moxidectin in cattle. *Veterinary Record* 138, 91–92.

Errecalde, J.O. (1993) Efficacy of doramectin against field infestations of *Psoroptes bovis*. *Pfizer Symposium*, Iguaçu Falls, Brazil, June.

FAO (1993) *Manual for the Control of the Screwworm fly* Cochliomyia hominivorax *(Coquerel)*. Vol. 2.

Fiel, C.A., Saumell, C.A., Steffan, P.E., Rodrigues, E.M. and Salaberry, G. (2000) Resistance of *Cooperia* and *Trichostrongylus* species to ivermectin treatments in grazing cattle of the Humid Pampa-Argentina. *Revista de Medecina Veterinaria (Buenos Aires)* 81, 310–315.

Furlong, J., Abreu, H.G.L. and Verneque, R.S. (1985) Parasitoses dos bovinos na regiao da Zona da Mata de Minas Gerais: I. Comportamento estacional de nematodeos gastrintestinais. *Pesquisa Agropecuaria Brasileira* 20, 143–153.

Garcia, L., Gutierrez, J. and Angulo-Cubillan, F.J. (1998) Efficacy of treatment with ivermectin 1% on bovine dermatobiosis. *Revista Cientifica, Facultad de Ciencias Veterinarias, Universidad del Zulia* 8 (Supplement 1), 117–118.

Garris, C., Touraille, C., Steffan, P., Fiel, C., Ambrustolo, R., Biondani, C., Zamorano, M. and Bulman, G. (1987) Gastrointestinal parasitism: its effects on muscle, fat and bone composition of the carcass and organoleptic characteristics of meat. In: Leaning, B. and Guerrero, J. (eds) *The Economic Impact of Parasitism in Cattle. Proceedings of the MSD AGVET Symposium, XXIII World Veterinary Congress*, Montreal, Quebec, Canada.

Giudici, C. (1985) Estudio de la prevalencia de Sarna Ovina y caracterizacion del manejo sanitario en 22 partidos de la Provincia de Buenos Aires. *V Congreso Argentino de Ciencias Veterinarias*. Buenos Aires, Argentina, September, p. 153.

Gonzales, J.C., Muñiz, R.A., Farias, A., Goncalves, L.C.B. and Rew, R.S. (1993) Therapeutic and persistent efficacy of doramectin against *Boophilus microplus* in cattle. *Veterinary Parasitology* 49, 107–119.

Grisi, L., Umehara, O., Caproni, L., Gonçalves, L.C.B., Derozier, C. and Amarantes, E.E.V.G. (2000) Comparative efficacy of doramectin (200 mcg/kg) and Ivermectin (630 mcg/kg) against induced infestations of the cattle tick *Boophilus microplus*. *XXI World Buiatric Congress*, Punta del Este, Uruguay, 4–8 December.

Guglielmone, A.A., Volpogni, M.M., Flores, S.G., Bulman, G.M., Anziani, O.S., Lamberti, J.C. and Mancebo, O.A. (1998) Evaluation of a 1% w/w injectable formulation of ivermectin in the control of natural infestations of cattle by *Haematobia irritans* (Linnaeus, 1758) (Diptera: Muscidae) in Santa Fe (Argentina). *Veterinaria Argentina* 143, 170–175.

Guglielmone, A.A., Volpogni, M.M., Anziani, O.S. and Flores, S.G. (1999) Evaluation of injectable abamectin to control natural infestations of *Haematobia irritans* (Diptera: Muscidae) in cattle. *Journal of Medical Entomology* 36, 325–328.

Guglielmone, A.A., Mangold, A.J., Muñoz Cobeñas, M.E, Scherling, N., Garcia Posse, F., Anziani, O.S. and Ioppolo, M. (2000) Moxidectin pour-on for control of natural populations of the cattle tick *Boophilus microplus* (Acarina: Ixodidae). *Veterinary Parasitology* 87, 237–241.

Horn, S.C. and Arteche, C.C.P. (1985) Situaçao parasitaria da pecuaria no Brazil. *A Hora Veterinaria* 23, 12–32.

Kettle, D.S. (1995) *Medical and Veterinary Entomology*, 2nd edn. CAB International, Wallingford, UK.

Larrosa, A., Petraccia, C., Cardozo, H. and Nari, A. (1994) Estudio de la eficacia del Moxidectin como garrapaticida en bovinos. *VII Congreso Argentino de Ciencias Veterinarias*, Buenos Aires, November, p. 260.

Leite, R.C., Muñiz, R.A., Oliveira, P.R., Gonçalves, L.C.B. and Rew, R.S. (1995) Efficacy of doramectin against natural infestations of *B. microplus* (Canestrini, 1887) (Acari: Ixodidae) in cattle. *Revista Brasileira de Parasitologia Veterinaria* 4, 53–56.

Lessa, C.S.S. and Lessa, D.A.B. (1999) Prophylactic effect of ivermectin sustained release on bovine castration myiasis. *Revista Brazileira de Medicina Veterinaria* 21, 147–148.

Levine, N.D. (1968) *Veterinary Parasitology*. Burgess Publishing Co., Minneapolis, Minnesota, pp. 93–100.

Lima, W.S., Fakuri, E., Guimaraes, M.P. and Malacco, M.A. (1997) Population dynamics of helminths in dairy cattle in the mining area of Minas Gerais. *Revista Brasileira de Parasitologia Veterinaria* 6, 97–103.

Lombardero, O.J. and Fontana, B.A. (1968) La 'Ura' (*Dermatobia hominis*) en la provincia de Formosa. *Gaceta Veterinaria* 30, 297–306.

Lombardero, O.J., Moriena, R.A., Racioppi, O. and Errecalde, J.O. (1995) Efficacy of doramectin against natural *Boophilus microplus* infestations in cattle in the province of Corrientes, Argentina. *Veterinaria Argentina* 115, 318–324.

Lombardero, O.J., Moriena, R.A., Racioppi, O., Billaudots, A. and Maliandi, E.S. (1999a) A comparative efficacy evaluation of ivermectin and doramectin in the control and prevention of navel myasis in calves with natural infestation, in Corrientes (Argentina). *Veterinaria Argentina* 158, 588–591.

Lombardero, O.J., Moriena, R.A., Racioppi, O., Billaudots, A. and Maliandi, F.S. (1999b) Comparison of the action of ivermectin and doramectin on post-castration myiasis in calves in Corrientes, Argentina. *Veterinaria Argentina* 155, 340–343.

Maciel, S., Giménez, A.M., Gaona, C., Waller, P.J. and Hansen, J.W. (1996a) The prevalence of anthelmintic resistance in nematode parasites of sheep in Southern Latin America: Paraguay. *Veterinary Parasitology* 62, 207–212.

Maciel, A.E.B., Bordin, E.L. and Cruz, J.B. (1996b) Ivomec SR bolus for the control of *Dermatobia hominis*. *XV Congresso Panamericano de Ciencias Veterinarias*, Campo Grande, Ms, Brazil, October, p. 326.

Mariategui, G., Speicys, C., Urretabizkaya, N., Caracostantogolo, J., Eddi, C. and Pazzos, L. (1998) Impact of the endectocides on faecal pads. *Brazilian Journal of Veterinary Parasitology* 8.

Mariategui, G., Speicys, C., Urretabizkaya, N., Caracostantogolo, J. and Eddi, C. (2000) Activity of the macrocyclic lactones on the dung fauna in cattle faecal pads. *XVI Reunión Latinoamericana de Producción Animal*. Montevideo, R.O. del Uruguay, 28 March.

Marguerite, J., Lamberti, J., Filippi, J., Elordi, L. and Bescos, J. (1997) Testing a new injectable, long-acting 1% formulation of ivermectin against psoroptic mange in calves in Argentina. *Veterinaria Argentina* 134, 251–255.

Marques, A.O., Arantes, G.J. and Silva, C.R. (1995) Evaluation of the efficacy of ivermectin of 1% (injectable solution) in the treatment of cattle naturally infested with the tick *Boophilus microplus* (Canestrini, 1887) (Acari: Ixodidae) and maintained on pasture. *Revista Brasileira de Parasitologia Veterinaria* 4, 117–119.

Martins, J.R. (1996) Resistència a carrapaticidas no estado do Rio Grande do Sul: Relato de situaçào. *Veterinaria (Montevideo)* 32, 12–14.

Martins, J.R., Cordoves, C.O.C., Correa, B.L. and Cereser, V.H. (1996) Control of the cattle tick *Boophilus microplus* with strateglc application of avermectin in Rio Grande do Sul-Brazil. *XV Congresso Panamericano de Ciencias Veterinarias*.

Mejia, M., Gonzalez Iglesias, A., Diaz Torga, G.S., Villafañe, P., Formia, N., Libertun, C., Becu Villalobos, D. and Lacau Mengido, I.M. (1999) Effects of continuous ivermectin treatment from birth to puberty on growth and reproduction in dairy heifers. *Journal of Animal Science* 77, 1329–1334.

Mercier, P., White, C.R., Eddi, C. and Caracostantogolo, J. (2000) Persistent efficacy of 3 endectocides after experimental nematode infection in calves. *XXI World Buiatric Congress*. Punta del Este, Uruguay, 4–8 December.

Moreno, L. and Gomez, E.A. (1991) Gastrointestinal and lung parasites in cattle in the state of Bolivar. *Veterinaria Tropical* 16, 55–67.

Moya-Borja, G.E., Muniz, R.A., Sanavria, A., Goncalves, L.C.B. and Rew, R.S. (1993) Therapeutic and persistent efficacy of doramectin against *Dermatobia hominis* in cattle. *Veterinary Parasitology* 49, 85–93.

Moya-Borja, G.E., Muñiz, R.A., Umehara, O., Gonçalves, L.C.B., Silva, D.S.F. and McKenzie, M.E. (1997) Protective efficacy of doramectin and ivermectin against *Cochliomyia hominivorax*. *Veterinary Parasitology* 72, 101–109.

Muñiz, R.A., Anziani, O.S., Ordoñez, J., Errecalde, J., Moreno, J. and Rew, R.S. (1995a) Efficacy of doramectin in the protection of neonatal calves and post-parturient cows against field strikes of *Cochliomyia hominivorax*. *Veterinary Parasitology* 58, 155–161.

Muñiz, R.A., Hernández, F., Lombardero, O., Leite, R.C., Moreno, J., Errecalde, J. and Gonçalves, L.C.B. (1995b) Efficacy of injectable doramectin against natural *Boophilus microplus* infestations in cattle. *American Journal of Veterinary Research* 56, 460–463.

Muñiz, R.A., Coronado, A., Anziani, O.S., Sanavria, A., Moreno, J., Errecalde, J. and Goncalves, L.C.B. (1995c) Efficacy of injectable doramectin in the protection of castrated cattle against field infestations of *Cochliomyia hominivorax*. *Veterinary Parasitology* 58, 327–333.

Muñiz, R.A., Cerqueira Leite, R., Coronado, A., Soraci, O., Umehara, O., de Oliveira, C.B., Muñiz, R.A., Umehara, O., Caproni, L. Jr, Oliveira, L.O., Rodrigues, V.J.P. and De Oliveira, C.B. (1997) Comparative persistence of doramectin and ivermectin against natural infestations of *Dermatobia hominis* in cattle in Rio Grande do Sul. *Revista Brasileira de Parasitologia Veterinaria* 6, 169–173.

Muñoz Cobeñas, M.E., Eddi, C., Caracostantogolo, J., Nolazco, J., Gross, S., Guerrero, J. and Mascotena, A. (1993) Study of the epidemiology and control of Ostertagiasis in the northern humid pampa of Argentina. *14th International Conference of the World Association for the Advancement of Veterinary Parasitology (WAAVP)*, Cambridge, UK, 8–13 August.

Nari, A. (1990) Methods currently used for the control of one-host ticks: their validity and proposals for future control strategies. *Parassitologia* 32, 12–143.

Nari, A. (1992) *Control y Prevención de Enfermedades Parasitarias de los Bovinos en el Trópico Americano*. Fernández Vaca, FAO, Santiago de Chile, Chile.

Nari, A. and Fiel, C. (1994) *Enfermedades Parasitarias de Importancia Economica en Bovinos. Bases Epidemiológicas para su Prevención y Control*. Hemisferio Sur, Montevideo, Uruguay.

Nari, A. and Hansen, J.W. (1999) Resistance of ecto- and endo-parasites: current and future solutions. *67th General Session, International Commitee, OIE*, Paris, 17–21 May, 22 pp.

Nari, A., Salles, J., Gil, A., Waller, P.J. and Hansen, J.W. (1996) The prevalence of anthelmintic resistance in nematode parasites of sheep in Southern Latin America: Uruguay. *Veterinary Parasitology* 62, 213–222.

Niec, R., Eddi, C. and Gomez, B. (1982) Anthelmintic action of ivermectin in cattle. *Revista de Medicina Veterinaria (Buenos Aires)* 63, 456–458.

Nuñez, J.L. and Moltedo, H.L. (1985) *Sarna Psoroptica en Ovinos y Bovinos*. Hemisferio Sur, Buenos Aires, Argentina.

Oliveira, C.M.B., Muñiz, R.A., Gonçalves, L.C.B. and Oliveira, L.O. (1993) The efficacy of doramectin against infestations caused by *Cochliomyia hominivorax* (Coquerel, 1858) in cattle in Rio Grande do Sul, Brazil. *Revista Brasileira de Parasitologia Veterinaria* 2, 7–10.

Oliveira, C.M.B., Muñiz, R.A., Umehara, O., Caproni, L., Oliveira, L.O., Rodrigues, V.J.P. and De Oliveira, C.B. (1997) Comparative persistence of Doramectin and Ivermectin against natural infestations of *Dermatobia hominis* in cattle in Rio Grande do Sul. *Revista Brazileira de Parasitologia Veterinaria* 6, 169–173.

Oliveira, G.P. (1988) Fatores que prejudicam economicamente a qualidade do couro dos bovinos In: *Conferencia Anual da Sociedade Paulista de Medicina Veterinaria*, Campinas, Sao Paulo.

Parra, D. and Uribe, L.F. (1990) Epidemiology of bovine nematode parasites in the Eastern plains of Colombia. In: Guerrero, J. and Leaning, B. (eds) *Epidemiology of Bovine Nematode Parasites in the Americas. Proceedings of the MSD AGVET Symposium, XVI World Buiatrics Congress and VI Latin American Buiatric Congress*, Salvador, Bahia, Brazil, 14 August.

Perez Arrieta, A., Marti Vidal, J. and Benitez Usher, C. (1982) Efficacy of ivermectin against *Boophilus microplus*. *IV Congreso Argentino de Ciencias Veterinarias*, La Plata, Buenos Aires, November, p. 8.

Pinheiro, A.C., Alves-Branco, E.P.J. and Sapper, M.E.M. (1999) Impacto economico das parasitoses nos paises do mercosul. *Xl Seminario Brasileiro de Parasitologla Veterinaria*, Salvador, Bahia, Brazil, October, p. 59.

Quiroz, R.G., Bulman, G.M., Lamberti, J.C., Marguerítte, J.A., Filippi, J.L. and Elordi, L.C. (1998) Efficacy of a new injectable formulation of 1% w/w ivermectin against *Dermatobia hominis* (Linnaeus, 1781) in calves with natural infestations in two field trials in the Argentine Mesopotamia. *Veterinaria Argentina* 144, 257–264.

Rassier, D., Farias, M.T., Bordin, E. and Newcomb, K.M. (1990) Epidemiological study of parasites in young weaned beef cattle in the temperate zone of Brazil. In: Guerrero, J. and Leaning, B. (eds) *Epidemiology of Bovine Nematode Parasites in the Americas. Proceedings of the MSD AGVET Symposium, XVI World Buiatrics Congress and VI Latin American Buiatric Congress*, Salvador, Bahia, Brazil, 14 August.

Rimbaud Giambruno, E. (1999) Impacto de las ecto endoparasitosis en el sistema productivo pecuario del Uruguay. *XI Seminario Brasileiro de Parasitologia Veterinaria*, Salvador, Bahia, Brazil, October.

Rivera, M.A. and Hurtado, E. (1983) *Mecistocirrus digitatus* (Von Linstow, 1906) Railliet y Henry, 1912 (Nematoda: Trhichostrongylidae) in cattle in Apure state, Venezuela. *Revista de la Faculdad de Ciencias Veterinarias, UCV* 30, 75–80.

Rivera, M.A., Garcia, F.A. and Sabate, C. (1985) Gastrointestinal parasitism in young cattle at the experimental station of 'La Antonia', San Felipe, Yaracuy, Venezuela. *Revista de la Faculdad de Cienciass Veterinarias, Universidad* 32, 37–45.

Sabatini, G.A., Costa, A.J., Scarpelli, L.C., Soares, V.E., Souza, L.M. and Henrique, C.H. (1999) Eficacia terapeutica do abamectin a 0,5%, via dorso lombar ('Pour-On'), contra la *Haematobia irritans*, em bovinos criados a campo. *XI Seminario Brasileiro de Parasitologia Veterinaria*, Salvador, Bahia, Brazil, October.

Scott, E.B., Coumendouros, K., Sant'Anna, E.B., Dabes, C.E., Fernandes, M.M. and Tancredi, I.P. (1999) Eficacia mosquicida a nivel de campo de uma formulacao pour-on contendo 0,5% de eprinomectin em bovinos. *XI Seminario Brasileiro de Parasitologia Veterinaria*. Salvador, Bahia, Brazil, October.

SENASA (1999) Servicio Nacional de Sanidad y Calidad Agroalimentaria, Direccion Nacional de Sanidad Animal – DNSA – Informacion basica del pais por provincia, partidos y/o departamentos a diciembre de 1999. Available at: senasa.mecon.gov.ar

Silva, C.R., Arantes, G.J. and Marques, A.O. (1995) Evaluation of the efficacy of 1% ivermectin (injectable solution) in the treatment of cattle maintained on pasture and naturally parasitized with larvae of the fly *Dermatobia hominis* (Linnaeus Jr., 1781) (Diptera: Cuterebridae). *Revista Brasileira de Parasitologia Veterinaria* 4, 121–123.

Sogbe, E., Ascanio, E., Vega, F., Diaz, C.T., Ascanio, M. and Gustavo, A. (1998a) Efficacy of broad spectrum 1% ivermectin for the control of gastrointestinal nematodes in cattle, and its action against *Boophilus microplus* and *Haematobia irritans* alone and associated with ethion/cypermethrin (84:4). *Revista Cientifica, Facultad de Ciencias Veterinarias, Universidad del Zulia* 8 (Supplement 1), 131–133.

Sogbe, E., Ascanio, E., Vega, F., Diaz, C.T., Ascanio, M. and Gustavo, A. (1998b) Efficacy of broad spectrum 1% ivermectin (Virbamec) and broad spectrum 1% abamectin (Virbamex) for the control of gastrointestinal nematodes in cattle. Action against *Boophilus* microplus alone and associated with 10% cypermethrin. *Revista Cientifica, Facultad de Ciencias Veterinarias, Universidad del Zulia* 8 (Supplement 1), 134–136.

Souza, L.M., Pelo, M.A.A., Soares, V.E., Costa, A.J. and Rocha, U.E. (1999) Ensaio de eficacia terapeutica do abamectin a 0,5%, via dorso-lombar (pour-on), contra larvas de *Dermatobia hominis* (Linnaeus, 1781) em bovinos naturalmente infestados. *XI Seminario Brasileiro de Parasitologia Veterinaria*, Salvador, Bahia, Brazil, October.

Spath, E.J.A., Guglielmone, A.A., Signorini, A.R. and Mangold, A.J. (1994) Estimacion de las perdidas economicas directas producidas por la garrapata *Boophilus microplus* y las enfermedades asociadas en la Argentina. 1ª parte *Therios* 23, 341–360; 2ª parte *Therios* 23, 389–396; 3ª parte *Therios* 23, 454–468; 4ª parte *Therios* 23, 524–553.

Steffan, P.E. and Castro, T.E. (1990) Efficacy of ivermectin in pour-on formulation against gastrointestinal and pulmonary nematodes in naturally infected cattle. *Veterinaria Argentina* 70, 668, 670–674.

Steffan, P.E., Entrocasso, C.M., Almada, A.A., Buck, S. and Arosteguy, J. (1995) Strategic control of gastrointestinal nematodes in Aberdeen angus heifers: effect on weight gain and reproductive function. *Revista Argentina de Produccion Animal* 15, 3–4, 823–826.

Steffan, P.E., Mercier, P. and White, C.R. (2000) Economic impacts on grazing beef cattle treated with injectable endectocides. *XXI World Buiatric Congress*. Punta del Este, Uruguay, 4–8 December.

Suarez, V.H., Bedotti, D.O., Larrea, S., Busetti, M.R. and Garriz, C.A. (1991) Effects of an integrated control programme with ivermectin on growth, carcass composition and nematode infection of beef cattle in Argentina's western pampas. *Research in Veterinary Science* 50, 195–199.

Vieira Bressan, M.C.R., Gennari, S.M., Caproni, L. Jr, Goncalves, L.C.B. and Umehara, O. (1998) Comparative efficacy of two doramectin treatment schemes with a conventional treatment program for the control of endo and ectoparasites of crossbred zebu cattle. *Revista Brasileira de Parasitologia Veterinaria* 7, 63–68.

Chapter 6.5

Use of Macrocyclic Lactones to Control Cattle Parasites in Australia and New Zealand

P.A. Holdsworth

Introduction

In Australia and New Zealand (ANZ), development of preventive parasite management programmes is the most effective way to control parasites. These programmes involve adopting a strategic approach to parasite control and treating early in the season to reduce the first generations of worms, ticks and flies. Commonly, these programmes will require treating when only a few parasites are present and the cattle do not have clinical signs of parasitism.

Recognition that parasite control is a herd health problem rather than an individual animal problem and that the main aim is to reduce pasture contamination, particularly that grazed by young susceptible animals, are important considerations in the formulation of control programmes. The main aim of helminth control programmes for cattle in ANZ should be to provide safe pasture, particularly for weaners, to avoid build-up of worm populations in this susceptible stock. Integrated control programmes involving effective anthelmintic and pasture/stock management strategies are the only effective ways to maintain low level pasture contamination.

Australia

Control of gastrointestinal parasites

In Australia, two distinct climatic zones, the temperate and the tropical/subtropical, delineate different gastrointestinal helminth species of cattle based on the epidemiology of their infection and their economic importance (Steele, 1998). In each zone, the prevalence of parasitism in cattle is determined largely by rainfall. The temperate zone can be subdivided further on the basis of summer or winter rainfall dominance.

Throughout the temperate (southern New South Wales (NSW), Victoria, South Australia and the southwest of Western Australia) zone, *Ostertagia ostertagi*, *Trichostrongylus axei* and *Cooperia oncophora* are the main species contributing to parasitic gastroenteritis in cattle (Anderson *et al.*, 1983), but *O. ostertagi* is the most important of the three species. Recommended treatment programmes are directed principally at preventive control of this parasite in young cattle (Steele, 1998). It had been generally accepted that only young cattle up to 2 years of age needed to be treated on a regular basis, and both macrocyclic lactones (MLs) and benzimidazole carbamates were effective anthelmintics and were recommended for this purpose (Hall, 1990). More recently, however, this focus just on young cattle has been challenged with research demonstrating the value of worm control in adult dairy cattle (Spence *et al.*, 1992; Walsh *et al.*, 1995; Spence *et al.*, 1996; Gross *et al.*, 1999).

Throughout the southern, or winter rainfall, region of the temperate zone where autumn calving is practised, a drench at weaning in December–January is recommended, with follow-up treatments in March–April and, on a minority of properties, in July–August (Anderson *et al.*, 1983). Further treatment of 2-year-old weaners the following January is designed to prevent type II ostertagiasis caused by the emergence of inhibited larvae which accumulate over the late spring and summer. In southern NSW, where spring calving is preferred, a weaning drench is recommended in March–April with follow-up treatment in May–June and, again, on a minority of properties, in July–August (Hall, 1990). On the Northern Tablelands of NSW in the summer rainfall zone, a single drench at weaning provides adequate control with little or no production response to follow-up treatments (Steele, 1998). Since healthy, non-reproducing cattle have developed a strong immunity to *Ostertagia* by 2 years of age, routine use of anthelmintics to prevent clinical disease may not be necessary in mature animals except in the few individuals that show signs of the disease (Steele, 1998). Bulls should be treated prior to and at the end of the joining period (Rolfe, 1998). MLs are largely involved in the anthelmintic treatment programmes because of their high efficacy and persistent efficacy.

In the tropical/subtropical zone of Australia, the helminth parasites of greatest economic significance are *Haemonchus placei*, *Oesophagostomum radiatum*, *Cooperia pectinata* and *Cooperia punctata* (Winks *et al.*, 1983). *Bunostomum phlebotomum* can also reach significant numbers in dairy calves in coastal Queensland. Infections with *O. ostertagi*, *C. oncophora* and *T. axei*, which are more common in the temperate regions of Australia, can also be a problem in southeast Queensland and on the north coast of NSW. Clinical disease is usually seen in dairy calves (4–12 months old) and in beef calves during the first year after weaning (Steele, 1998). Adult cattle are generally less susceptible to infection, and clinical signs of parasitism are unusual in animals more than 2 years old (Winks *et al.*, 1983).

For these reasons, anthelmintic treatment at weaning and for 9 months thereafter at intervals of no longer than 6 weeks is usually recommended for preventive control of gastrointestinal parasites in the tropical/subtropical zone (Steele, 1998). Rolfe (1998) however recommended treatment at weaning followed by another 2 months later, while M. Murphy (1998) advocated that 2- to 10-month-old dairy calves should be monitored at 4 or 8 week time periods depending on which ML was used. She added that stock 12–20 months of age should be monitored every 8 weeks (M. Murphy, 1998) depending on the ML used.

In both climatic zones, effectiveness of parasite control can be enhanced and treatment frequency reduced by moving treated animals to pastures of low infectivity immediately following anthelmintic therapy. Widespread adoption of these general recommendations would be expected to confine usage of MLs largely to cattle aged between 9 months (weaning) and 18–24 months (Steele, 1998). In Northern Australia, however, the scope of beef cattle production is such that stock may only be available once per year for management activities such as drenching. The value of ML residual efficacy is limited if only one treatment is given each year to adult cattle. However, adult cattle do suffer from type II ostertagiasis, cows do suffer a periparturient rise in worm egg production and a single anthelmintic treatment may result in higher milk production and weaning weight (Walsh *et al.*, 1995; Gross *et al.*, 1999). Such benefits may be realized if the timing of cattle mustering can coincide with strategic imperatives.

Non-combination MLs (abamectin, ivermectin, moxidectin, doramectin and eprinomectin) are available for cattle in Australia as either subcutaneous injection (SCI) or pour-on (PO) formulations for use against cattle parasites. All SCI formulations are administered at a rate of 200 μg kg^{-1} body weight (bw) while all PO formulations are administered at a rate of 500 μg kg^{-1} body weight. These products are the most expensive option in chemical parasite control, especially the PO formulations; however, they have significant benefits in convenience of application as well as spectrum of activity and have variable residual efficacy after treatment, depending on the drug and its formulation. A significant factor in dairy operations with the PO formulations of moxidectin and eprinomectin along with two ivermectin PO formulations is their availability to be used in lactation with convenience of administration and long residual efficacy. Ivermectin (SCI) and abamectin (SCI and PO) formulations in Australia have minor market share and have diminishing significance in parasite management programmes.

No ML controls liver fluke (*Fasciola hepatica*) infections; however, two formulations, one oral product delivering 0.2 mg kg^{-1} of ivermectin and 12 mg kg^{-1} of triclabendazole and the other an SCI delivering 0.2 mg kg^{-1} ivermectin and 2 mg kg^{-1} clorsulon, have been approved in Australia with a liver fluke claim. The approved label claim for the oral formulation includes the treatment of early immature, immature and adult *F. hepatica*, while the SCI formulation includes the treatment of adult *F. hepatica*. These formulations can be used as part of a liver fluke strategic control programme.

Table 6.5.1 details a management programme for ML product use against gastrointestinal helminths of cattle under various Australian geographic locations.

Control of ectoparasites

Cattle tick (*Boophilus microplus*) and buffalo fly (*Haematobia irritans exigua*) are the predominant external parasites of the tropical/subtropical zone of Australia. *B. microplus* infection is associated with causing loss of condition, anaemia and death, hide damage and the transmission of babesiosis in cattle (Seddon, 1967a). *H. irritans exigua* infestations result in loss of cattle condition or delay in fattening of beef cattle, as well as adverse effects on milk production in dairy cattle (Seddon, 1967b). Such negative effects on meat and milk production and leather quality, along with animal welfare implications, justify graziers treating to manage these two ectoparasites.

When targeting *B. microplus* and *H. irritans exigua*, it is important to remember that whole herd treatment is required to help break the life cycle and reduce environmental contamination with eggs and larvae. All ivermectin (excluding the triclabendazole combination), moxidectin and doramectin formulations along with one abamectin PO formulation have the approved claim for control of *B. microplus*, while the abamectin SCI and most of the abamectin PO formulations along with the eprinomectin PO formulation have an approved claim for aiding in the control of *B. microplus*. Doramectin SCI has a registered protection period against reinfection with *B. microplus* and can be used at 28-day intervals. All abamectin formulations are not recommended for strategic dipping programmes for control of *B. microplus*. PO formulations of abamectin, ivermectin, doramectin and eprinomectin also have an approved claim for the control of *H. irritans exigua* and, in some fly seasons, the use of these product types may be all that is required to control this parasite on cattle.

In the summer rainfall and tropical/subtropical zone, *B. microplus* can be controlled by combining the tick vaccine (at present not being marketed) with early season (spring rise) broad spectrum ML parasiticide treatment and strategic plunge dipping (amidine or synthetic pyrethroids) in spring and early summer (O'Sullivan, 1998). This type of integrated programme reduces pastoral larval tick contamination levels throughout the season. Timing of treatment will be based on local property conditions and historical parasite patterns. Younger stock (<18 months old) are most at risk at this time because they will be most

Table 6.5.1. Recommended parasite control programme for use of MLs in cattle in Australia.

Class of cattle	Environment	Frequency of use
Dairy calves up to 12 months old	Tropical and subtropical environments	At 6–8 weeks of age, and then at intervals dependent on grazing management, season conditions and results of faecal egg monitoring
Dairy and beef calves up to 12 months old	Mild, temperate, coastal environments	At weaning, then at intervals dependent on grazing management, season conditions and results of faecal egg monitoring. The following times are minimal strategic treatments and vary with region: • Two treatments within the period of March–June to prevent autumn and winter contamination of pastures • Late winter to stop spring contamination
Dairy cattle	Tropical and subtropical coastal environments (*Haemonchus* areas)	Prior to calving and at 3 and 6 months after calving
	Mild, temperate coastal environments	After drying off and, where this coincides in seasonal milking herds with late autumn, a treatment at this time, and a possible retreatment in early winter, acts to reduce winter pasture contamination. Use of POs with a nil withholding period enables strategic drenching to occur irrespective of management systems in use
Beef calves	Cool, high rainfall areas of Southern Australia	Usually at weaning, then at intervals dependent on grazing management, season conditions and results of faecal egg monitoring. A strategic programme would include a treatment in late winter with a move to 'low-worm' pastures and/or a treatment in September and December for *Ostertagia*
Adult beef cattle	All environments	Depending on grazing and pasture management, adult cattle may not require regular treatment. If grazing management or seasonal conditions have predisposed to worm or fluke challenge, then tactical treatments may be required and faecal egg monitoring may assist in this decision

susceptible to both worms and ticks. In weaners and young cattle, MLs can be incorporated into an effective tick treatment and control strategy. Some of the MLs also kill worms, lice and buffalo fly for a period after use. Long-acting MLs available as PO or SCI formulations treat and protect animals and reduce pasture contamination for long periods as well as precluding the need for plunge dipping (O'Sullivan, 1998).

For *B. microplus* management alone, Cox and Watson (1998) have recommended two strategically timed doramectin SCI treatments 35 days apart, with the first at the time of the spring rise in tick larvae infestations from pasture as being likely to provide highly effective and sustained control in Queensland. Jonsson (1997) has also recommended that to reduce the risk of introducing cattle in Queensland with *B. microplus* resistant to synthetic pyrethroids and amitraz, the best strategy could be to treat all purchased cattle with moxidectin or ivermectin 4 days before transporting or on arrival, followed by a quarantine period of at least 4 days.

In temperate zones of Australia, biting (*Bovicola bovis*) and sucking (*Linognathus vituli*, *Haematopinus eurysternus* and *Solenopotes capillatus*) lice are the most common ectoparasites of cattle. *B. bovis* is found on cattle of all ages and, when numerous, is capable of causing considerable annoyance and irritation. *L. vituli* and *S. capillatus* are found mainly on dairy cattle, with *H. eurysternus* on calves. Lice irritate cattle while sucking blood or feeding on epithelial debris. Affected cattle relieve the irritation by rubbing or scratching. Affected animals become anaemic and unthrifty, and death may occur (Seddon, 1967b).

Farmers generally consider control of *B. bovis*, *L. vituli*, *H. eurysternus* and *S. capillatus* as being one of the routine management procedures in temperate Australia, particularly when nutritional intake is restricted (Freer and Gahan, 1968). Treatment of cattle with MLs for lice usually commences prior to winter when lice numbers increase. In addition, the annual treatment programmes for internal parasite control using ML PO formulations will also control lice species on cattle. All SCI formulations of MLs do not have an approved claim for control of *B. bovis*; however, the PO formulations do possess a full control claim. Full claim of PO is probably due to drug availability for biting lice both inside and outside the bloodstream. One ivermectin PO formulation has a registered persistency claim of up to 56 days against lice reinfestation.

For situations where all major cattle parasites may exist on a property, the following regimen gives an example of management practices that may be adopted. ML PO formulations can be applied as the first treatment in the spring to control worms and *B. microplus* and then follow-up with another treatment or conventional tickicide if *B. microplus* persist through the spring. When *H. irritans exigua* becomes a problem in early summer, ML PO formulations can be applied and followed-up with a conventional control option such as ear tags if the fly continues to be a problem. In late autumn or early winter when worms and lice become a problem again, the use of ML PO (*B. bovis, L.vituli, H. eurysternus* and *S. capillatus*) or ML SCI (*L. vituli, H. eurysternus* and *S. capillatus*) formulations are the best options.

New Zealand (NZ)

Control of gastrointestinal parasites

NZ cattle production is based around the dairy cow, the beef weaner and the intensive (dairy–beef/bull–beef) industries. In general, parasite control in NZ beef cattle is heavily dependent on use of anthelmintics (McPherson *et al.*, 1997). *O. ostertagia* is regarded as the most economically significant parasite (McPherson *et al.*, 1997), being generally recognized as pathogenic, with *T. axei* slightly less so and *Cooperia* spp. as relatively non-pathogenic, but strongly immunogenic, so infection is usually confined to calves (Bisset, 1994).

The effects of parasitism are likely to be most severe and of long-term significance in cattle less than 12 months old. The principal objective of NZ control strategies is to minimize cattle exposure to infective larvae on pasture. The aim is to control levels of parasitism in these calves because these animals are most vulnerable and they are potentially the main source of pasture contamination which has a direct bearing on the challenge to which older animals are exposed and the potential for type II ostertagiasis (Charleston, 1994).

The maximum number of *O. ostertagia* and *Cooperia* spp. in calves occurs in their first winter. By 12 months of age, calves usually develop a high level of immunity to *Cooperia* spp. By 12–15 months of age, *O. ostertagia* adults have reduced in numbers but inhibited larvae may be present (Charleston, 1994). However, peak numbers of *T. axei* (both adults in animals and larvae on pasture) occur later in about October though numbers decline shortly afterwards and the peak may be quite low (Brunsdon, 1971; Bisset and Marshall, 1987). It takes around 18–20 months for cattle to establish high levels of immunity to infections of all the commonly found parasites *O. ostertagia, T. axei* and *Cooperia* spp. (Charleston, 1994). Therefore, calves in the first year are the major source of pasture contamination; after about 12 months, faecal egg counts are generally low (<100 eggs per gram) although yearlings may contribute significant numbers of *T. axei* eggs to pasture in spring (Brunsdon, 1980; Bisset, 1994).

Given NZ's climatic conditions and year round pastoral farming systems, it is unlikely that under most farming conditions, adequate productivity and profitability can be achieved without the use of anthelmintics and, for practical purposes, in most circumstances this will entail a regular drenching programme of some kind (Charleston, 1994). MLs (abamectin, ivermectin, moxidectin, doramectin and eprinomectin) are available in

NZ as either SCI, sustained-release bolus or PO formulations for use against cattle parasites. All SCI formulations are administered at 200 μg kg^{-1} while all PO formulations are administered at 500 μg kg^{-1} and the sustained-release bolus delivers 30 μg day^{-1} of ivermectin for 400 kg maximum weight animals with a pay out period of 135 days. MLs most commonly used in NZ are PO formulations of eprinomectin, abamectin, ivermectin and moxidectin, as well as doramectin SCI.

In the NZ dairy cattle industry, there are two seasonal peaks of helminth larval availability on pasture grazed by dairy cattle, in (early) summer (December) and in May or June (Bisset, 1994). High, on-farm parasite challenge occurs where cows graze on pasture throughout the year and replacements also graze on the home farm. If cows graze the same pasture for only part of the year, parasite challenge is reduced. Similarly, if replacements graze off the farm, then the challenge to milking cows also tends to be reduced (Charleston, 1994).

Drenching at times of increased challenge is likely to give better results than drenching at other times (A.W. Murphy, 1998). MLs with persistent activity can be used strategically to cover the periods of peak larval availability. Nutrition is often more limiting in early lactation, and the effects of parasites may be more significant at this time. A strategic parasite control programme in a seasonal dairy herd in NZ would involve treatment on one or more occasions as detailed in Table 6.5.2. As in Australia, recent studies in NZ have demonstrated the value of worm control in adult cattle (A.W. Murphy, 1998; McPherson *et al.*, 1999).

In the beef weaner system, calves usually are weaned at 6–9 months of age and so are less dependent on grazing early on, compared with calves

Table 6.5.2. A strategic programme for use of MLs to control helminth parasites in a NZ seasonal dairy herd.

Time	Herd management/drenching
July–September	Calving – greatest response expected and benefits may continue throughout lactation
October–December	A second treatment within 100 days of calving may be beneficial
Summer	The smaller of the two larval peaks on pasture occurs at this time if calves are grazed on the same area as cows. In dry conditions, larval challenge will be reduced, but feed shortages may mean effects of parasite challenge are increased
April–May	Treatment prior to drying off. The larger larval peak occurs at this time and it also coincides with treatment for lice

Taken from A.W. Murphy (1998).

weaned earlier. These cattle are usually run in conjunction with sheep at lower stocking rates than in dairy and intensive beef enterprises. As such, the effects of gastrointestinal parasitism are less likely to be as dramatic as they can be in intensive systems, but they are still of considerable economic importance (Charleston, 1994).

Recommendations for the required number of anthelmintic treatments and the intervals between them vary considerably (Brunsdon and Adams, 1975; McMullan *et al.*, 1981). Table 6.5.3 details advice for an ML management programme for NZ beef weaners. This programme is based on trials reported by McPherson *et al.* (1989) where economic advantages were reported when six treatments of ivermectin SCI in beef weaners were given at 6-week intervals compared with four treatments at 6-week intervals and untreated control animals.

Weaning in the beef weaner system may start as early as February or as late as April/May, depending on the area and farm practice (McPherson *et al.*, 1989). Where weaning occurs as late as April/May, most of the season's contamination is likely to be already on pasture by the time the drenching programme starts. If young stock is grazed after weaning on ground where they have been in the previous 3 months, they will be exposed to significant challenge and the drenching programme becomes 'protective' rather than 'preventive'. In such circumstances, it is suggested, if possible, that starting the treatment programme before weaning should reduce levels of pasture contamination and give better pasture control (Charleston, 1994).

In the intensive beef system, calves are usually reared on milk replacer with or without concentrate supplements, weaned at around 8 weeks and then placed on pasture. Management systems vary, with some animals slaughtered at around 18 months of age and others not until 2–3 years old. Cattle may be grazed rotationally or set stocked according to feed availability. In stand-alone enterprises or where it is confined to a particular area of the farm, successive mobs of cattle are run on the same area until slaughter weights are reached or they are sold. This monoculture of predominantly young cattle means that a high level of pasture infectivity is inevitable without effective parasite control. There is little or

Table 6.5.3. Recommended parasite control programme for use of MLs in NZ beef weaner production systems.

Time	(Beef) Weaner management/drenching
February–April/May	Calves at weaning. If weaning occurs in April/May, it is recommended to also drench pre-weaning and continue at 6-weekly intervals
March/April–October	Five treatments at 6-weekly intervals

Taken from McPherson *et al.* (1989).

no scope for using grazing management to reduce it. This then becomes a critical issue and inevitably is dependent on regular anthelmintic treatments (Charleston, 1994).

Table 6.5.4 details an ML drenching programme for the dairy–beef/bull–beef production system commencing about November, or when the weaners are brought in, and continue at 6–8 week intervals until July. The problem is that there is no way of telling if all the drenches are necessary, but returns are almost certain to more than cover costs and the economic stakes are too high to risk stopping. Risk of developing helminth resistance is high (Charleston, 1994). Inefficacy of various ivermectin formulations against *Cooperia* has been detected in several cattle herds in NZ (Bisset *et al.*, 1990; Eagleson and Allerton, 1992; West *et al.*, 1994; McKenna, 1995; Vermunt *et al.*, 1995; Watson *et al.*, 1995). Vermunt *et al.* (1996) presented trial results that indicated that both moxidectin and doramectin might be ineffective against ivermectin-resistant *Cooperia* spp., as would any ML.

In the beef weaner systems, preventative measures against type II ostertagiasis must start with effective parasite control in young stock as they contribute most of the pasture contamination from which the larvae that become inhibited are derived. In some cases, it may be desirable to treat cows at the end of winter or in early spring, a few weeks before calving. In some situations, especially where parasite control in the previous year is less than good, drenching yearlings in the spring may also be advisable (Charleston, 1994).

In the intensive dairy–beef/bull–beef systems, the fact that the cattle are drenched almost continuously at least until the end of the first winter should mean that the type II ostertagiasis is very unlikely to occur subject to effective drenching.

Control of ectoparasites

Although the impact of lice on cattle productivity in NZ may be questionable, animals in poor condition may experience increased irritation resulting in intensive grooming and rubbing (Kettle, 1974; Chambers and Charleston, 1980). Such infestations may reduce feeding activities, affect

Table 6.5.4. Recommended parasite control programme for use of MLs in NZ intensive (dairy–beef/bull–beef) production systems.

Time	Bull–beef weaner management/drenching
November	Treat when weaners are brought in
November/December–July	Treat at 6- to 8-week intervals

Supplied by Merial New Zealand Ltd.

Box 6.5.1. ML products (by registered name) in Australia and New Zealand for use in/on cattle.

AUSTRALIA

Abamectin

- Avomec Antiparasitic Injection for Cattle
- Duotin Antiparasitic Injection for Cattle
- Virbamec Antiparasitic Injection for Cattle
- Virbamec Pour-on for Cattle
- Rycomectin Antiparasitic Cattle Injection
- Genesis Injection Abamectin Antiparasitic for Cattle and Sheep
- Paramectin Injection for Cattle
- Paramectin Pour-on for Cattle
- Genesis Abamectin Pour-on Endectocide for Cattle
- Virbac Virbamec Pour-on for Cattle
- Vetmec Antiparasitic Cattle Injection
- Farm Direct Injectable Endectocide for Cattle

Doramectin

- Dectomax Injectable Endectocide
- Dectomax Pour-on Endectocide

Eprinomectin

- Ivomec Eprinex (Eprinomectin) Pour-on for Beef and Dairy Cattle

Ivermectin

- Ivomec Plus (Ivermectin Plus Clorsulon) Broadspectrum Antiparasitic Injection for Cattle
- Ivomec (Ivermectin) Pour-on for Cattle
- Ivomec Antiparasitic Injection for Cattle
- Coopers Paramax Pour-on for Beef and Dairy Cattle
- Baymec Pour-on for Cattle
- Genesis Injection Ivermectin Antiparasitic for Cattle
- Genesis Pour-on Ivermectin Endectocide for Cattle
- Ivermectin Baymec Pour-on for Cattle
- Noromectin Pour-on for Cattle
- Noromectin Injectable for Cattle
- Farm Direct Ivermectin Cattle Pour-on
- Fasimec Cattle Oral Flukicide and Broadspectrum Drench
- Ecomectin Antiparasitic Injection for Cattle

Moxidectin

- Cydectin Injection for Cattle
- Cydectin Pour-on for Cattle and Red Deer
- Cydectin Injection for Cattle and Sheep

NEW ZEALAND

Abamectin

- Duotin Injection for Cattle
- Abamec Injection
- Genesis Pour-on
- Genesis Injection
- Paramectin
- Paramectin Pour-on for Cattle

Doramectin

- Dectomax
- Dectomax Pour-on Endectocide

Eprinomectin

- Ivomec Eprinex Pour-on for Cattle and Deer

Ivermectin

- Ivomec Oral Solution for Cattle
- Ivomec Injection for Cattle
- Ivomec Pour-on for Cattle and Deer
- Ivomec Plus Injection for Cattle
- Ivomec SR Bolus for Cattle
- Ivocare Injection
- Ivomec Injection for Cattle, Sheep and Pigs
- Ivocare Pour-on
- Bomectin Injection
- Bomectin Pour-on
- Noromectin Injection
- Noromectin Pour-on
- Iver Pour

Moxidectin

- Cydectin Injection
- Cydectin Pour-on
- Vetdectin Pour-on
- Vetdectin Injection

hide quality and damage farm fixtures (Watson *et al.*, 1996). In NZ, farmers generally treat stock for lice at least once annually, usually in the autumn to early winter period to minimize losses and reduce louse transmission between stock. All SCI formulations of MLs do not have an approved claim for control of *B. bovis*; however, the PO formulations do possess a full control claim for *B. bovis*, *L. vituli*, *S. capillatus* and *H. eurysternus*.

With ANZ being major exporters of beef and dairy products, cattle producers use MLs as one (albeit a very important one) of various production tools in quality assurance programmes to satisfy customer requirements. Adhering to ML product labelling, especially for use directions, withholding periods and export slaughter intervals (in Australia), will contribute to ensuring export market quality requirements are met. Box 6.5.1 details the registered ML products in ANZ that are approved for use in/on cattle.

Acknowledgements

The technical advice from the following companies and organizations is acknowledged: Merial Australia Pty Ltd; Merial New Zealand Pty Ltd; Fort Dodge Australia Pty Ltd; Pfizer Animal Health Pty Ltd; Schering Plough Animal Health Ltd; Virbac Australia Pty Ltd; Novartis Animal Health Australasia Pty Ltd; International Animal Health Pty Ltd; CSIRO; and Meat and Livestock Australia.

References

Anderson, N., Donald, A.D. and Waller, P.J. (1983) Epidemiology and control of parasitic gastroenteritis of cattle in the temperate climatic zone. In: Anderson, N. and Waller, P.J. (eds) *The Epidemiology and Control of Gastrointestinal Parasites of Cattle in Australia*. CSIRO, Australia, pp. 47–63

Bisset, S.A. (1994) Helminths of economic importance in cattle in New Zealand. *New Zealand Journal of Zoology* 21, 9–22.

Bisset, S.A. and Marshall, E.D. (1987) Dynamics of *Ostertagia* spp. and *Cooperia oncophora* in field grazed cattle from weaning to 2 years old in New Zealand, with particular reference to arrested development. *Veterinary Parasitology* 24, 103–116.

Bisset, S.A., Brunsdon, R.V. and Forbes, S. (1990) Efficacy of a topical formulation of ivermectin against naturally acquired gastrointestinal nematodes in weaner cattle. *New Zealand Veterinary Journal* 38, 4–6.

Brunsdon, R.V. (1971) Trichostrongyle worm infection in cattle: further studies on problems of diagnosis and on seasonal patterns of occurrence. *New Zealand Veterinary Journal* 19, 203–212.

Brunsdon, R.V. (1980) Bovine ostertagiosis in New Zealand. In: *Ostertagia Symposium Proceedings*, Roseworthy Agricultural College, May, pp. 79–100.

Brunsdon, R.V. and Adams, J.L. (eds) (1975) *Internal Parasites and Animal Production.* New Zealand Society of Animal Production, Occasional Publication No. 4.

Chambers, K. and Charleston, W.A.G. (1980) Cattle lice in New Zealand: effects on host liveweight gain and haematocrit levels. *New Zealand Veterinary Journal* 28, 235–237.

Charleston, T. (1994) Control of gastrointestinal parasites in beef production systems. *Proceedings of the 24th Seminar Sheep and Beef Society New Zealand Veterinary Society*, July, pp. 157–174.

Cox, J.W. and Watson, T.G. (1998) Doramectin injectable as a practical and effective management tool to control *Boophilus microplus* infestation on cattle in Queensland. *XX World Buiatrics Congress Proceedings*, Sydney, Australia, p. 193.

Eagleson, J.S. and Allerton, G.P. (1992) Efficacy and safety of ivermectin applied topically to cattle under field conditions in Australia. *Australia Veterinary Journal* 69, 133–134.

Freer, R.E. and Gahan, R.J. (1968) Controlling lice in beef herds – is it economic? *Agricultural Gazette of NSW* 79, 308.

Gross, S.J., Ryan, W.G. and Ploeger, H.W. (1999) Anthelmintic treatment of dairy cows and its effect on milk production. *Veterinary Record* 144, 581–587.

Hall, B. (1990) *Ostertagia* control in cattle. Elders Pastoral Technical Services Bulletin, Armidale.

Jonsson, N.N. (1997) Control of cattle ticks (*Boophilus microplus*) on Queensland dairy farms. *Australian Veterinary Journal* 75, 802–807.

Kettle, P.R. (1974) The influence of cattle lice (*Damalinia bovis* and *Linognathus vituli*) on weight gain in beef animals. *New Zealand Veterinary Journal* 22, 10–11.

McKenna, P.B. (1995) Topically applied ivermectin and *Cooperia* infections in cattle. *New Zealand Veterinary Journal* 43, 44.

McMullan, M.J., Leaning, W.H.D., Holmden, J. and Cairns, G.C. (1981) The effects of anthelmintic treatment on the growth rate of beef calves following weaning. *New Zealand Journal of Experimental Agriculture* 2, 129–134.

McPherson, W.B., Cairns, G.C., Leaning, W.H.D. and Newcomb, K.M. (1989) Effects of different ivermectin treatments on weight gains in beef weaners. *Proceedings of the New Zealand Society of Animal Production* 49, 307–311.

McPherson, W.B., Bowie, J.Y., Ryan, W.G., Gross, S.J. and Webster, M.C. (1997) Effects of topical ivermectin treatment on weight gains in beef weaners. *Proceedings of the New Zealand Society of Animal Production* 57, 199–203.

McPherson, W.B., Slacek, B., Familton, A., Gogolewski, R.P., Cramer, L.G. and Gross, S.J. (1999) The benefits of topical eprinomectin on milk production in dairy cattle. *Proceedings of the 17th International Conference of the World Association for the Advancement of Veterinary Parasitology*, Copenhagen, Denmark, Abstract b 5.04.

Murphy, A.W. (1998) The effects of treatment with moxidectin, a long acting endectocide, on milk production in lactating dairy cows. *XX World Buiatrics Congress Proceedings*, Sydney, Australia, pp. 463–470.

Murphy, M. (1998) Wormbusting Queensland's internal cattle parasites. A report by Australia's leading parasitologists and cattle veterinarians, commissioned by Pfizer Animal Health.

O'Sullivan, B. (1998) Queensland's cattle parasite problem. A report by Australia's leading parasitologists and cattle veterinarians, commissioned by Pfizer Animal Health.

Rolfe, P. (1998) Cattle parasites in Australia. In: *Cattle Parasites in Australia: are we Treating the Problem Effectively?* A report by Australia's leading parasitologists and cattle veterinarians, commissioned by Pfizer Animal Health.

Seddon, H.R. (1967a) *Diseases of Domestic Animals in Australia. Part 3. Arthropod Infestations (Ticks and Mites).* Commonwealth Department of Health.

Seddon, H.R. (1967b) *Diseases of Domestic Animals in Australia. Part 2. Arthropod Infestations (Flies, Lice and Fleas).* Commonwealth Department of Health.

Spence, S.A., Fraser, G.C., Dettmann, E.B. and Battese, D.F. (1992) Production responses to internal parasite control in dairy cattle. *Australian Veterinary Journal* 69, 217–220.

Spence, S.A., Fraser, G.C. and Chang, S. (1996) Responses in milk production to the control of gastrointestinal nematode and paramphistome parasites in dairy cattle. *Australian Veterinary Journal* 74, 456–459.

Steele, J.W. (1998) *Assessment of the Effects of the Macrocyclic Class of Chemicals on Dung Beetles and Dung Degradation in Australia in NRA Special Review of Macrocyclic Lactones.* NRA Special Review Series 98.3. NRA, Australia.

Vermunt, J.J., West, D.M. and Pomroy, W.E. (1995) Multiple resistance to ivermectin and oxfendazole in *Cooperia* species on cattle in New Zealand. *Veterinary Record* 137, 43–45.

Vermunt, J.J., West, D.M. and Pomroy, W.E. (1996) Inefficacy of moxidectin and doramectin against ivermectin-resistant *Cooperia* spp. of cattle in New Zealand. *New Zealand Veterinary Journal* 44, 188–193.

Walsh, T.A., Younis, P.J. and Morton, J.M. (1995) The effect of ivermectin treatment of late pregnant dairy cows in south-west Victoria on subsequent milk production and reproductive performance. *Australian Veterinary Journal* 72, 201–207.

Watson, T.G., Hosking, B.C. and McKee, P.F. (1995) Preliminary evidence of multiple anthelmintic resistance in Friesian bull beef weaners in New Zealand. *New Zealand Journal of Zoology* 22, 183–184.

Watson, T.G., Bishop, D.M., Hooke, F.G., Heath, A.C.G. and Cole, D.J.W. (1996) Efficacy of injectable doramectin against naturally acquired louse infestations on cattle. *New Zealand Journal of Agricultural Research* 3, 401–404.

West, D.M., Vermunt, J.J., Pomroy, W.E. and Bentall, H.P. (1994) Inefficacy of ivermectin against *Cooperia* spp. infection in cattle. *New Zealand Veterinary Journal* 42, 192–193.

Winks, R., Bremner, K.C. and Barger, I.A. (1983) Epidemiology and control of parasitic gastroenteritis of cattle in the tropical/subtropical zone. In: *The Epidemiology and Control of Gastrointestinal Parasites of Cattle in Australia.* CSIRO, Australia.

Chapter 7

The Use of Macrocyclic Lactones to Control Parasites of Sheep and Goats

R.L. Coop, I.A. Barger and F. Jackson

Introduction

Reductions in the performance and welfare of small ruminants are a common consequence of endo- and ectoparasitoses throughout the world. Current production systems rely largely upon chemotherapy and chemoprophylaxis for the treatment and control of these parasitoses. Although it is also possible for producers to use pasture management to help control nematode infections, the trend for intensification within agricultural production systems during the latter half of the 20th century led inevitably to a reduced reliance upon rotational and/or managemental strategies for controlling endoparasites. Although there can be no doubt of the short-term production benefits arising from the intensive use of chemoprophylaxis, the practice has led to the selection of anthelmintic resistant worm populations.

Because of the widespread prevalence of anthelmintic resistance in nematodes of small ruminants in both the southern (Waller, 1977; Waller *et al.*, 1996) and northern hemispheres (Jackson and Coop, 2000), macrocyclic lactones (MLs) occupy a position of special importance. The unique broad-spectrum efficacy against both endo- and ectoparasites and persistence characteristics of drugs within the ML family, coupled with the low prevalence of resistance against these drugs in many countries, gives them a crucial role in combating parasitic disease in small ruminants.

ML Anthelmintics' Spectrum of Activity in Small Ruminants

Spectrum of activity against endoparasites of small ruminants

Avermectin and milbemycin anthelmintics have been registered for use in small ruminants in different formulations that are administered either orally or in an injectable format. Since pharmaceutical companies register products on a country by country basis and tailor registration to suit the prevalent regional production systems, there is considerable between-country variation in product availability. Different ML compounds registered for use in small ruminants include ivermectin, abamectin, doramectin and moxidectin. At present, eprinomectin is registered only for use in cattle, but off licence use has been recorded in dairy goats in France. These MLs have a broad spectrum of efficacy against many genera of gastrointestinal and pulmonary nematodes of goats and sheep (Oakley, 1990). While it is possible to publish initial efficacy data for novel compounds, journals often are loathe to publish large numbers of efficacy studies on the grounds of novel data. For these reasons, much of the efficacy data are not in the public domain but are used by companies for registration purposes. Table 7.1 uses products currently available for small ruminants within the UK to illustrate their spectrum of activity based upon label claims (National Office of Animal Health, 2000). All of the products are administered at 0.2 mg kg^{-1} liveweight except for injectable doramectin treatments given for adult *Nematodirus battus*, which are given at 0.3 mg kg^{-1} liveweight.

Disruption of normal neuromuscular activity by MLs has been shown to affect not only motility but also feeding, by interfering with the pharyngeal pump (Geary *et al.*, 1993) and egg laying (McKellar *et al.*, 1988; Sutherland *et al.*, 1999). There appear to be marked differences between the relative potencies of the different ML compounds (Shoop *et al.*, 1996) and, as persistency (Bairden *et al.*, 1995) and dose titration (Shoop *et al.*, 1990) studies show, also between-species differences in susceptibility. There may also be between-species differences in mechanisms of removal *in vivo*. An ovine study examining parasite rejection post-ivermectin treatment showed that *Haemonchus contortus* and *Trichostrongylus colubriformis* were expelled rapidly following treatment but that *Teladorsagia* (*Ostertagia*) was lost more slowly (Gill *et al.*, 1998). The authors concluded that rapid expulsion was associated with effects upon motility but that the protracted rate of loss of *Teladorsagia* (*Ostertagia*) resulted from an energy imbalance through reduced feeding activity.

Differences in lipophilicity of the different compounds are important in explaining differences in persistent effects. Orally administered moxidectin has a persistent effect preventing reinfection in sheep with *T. circumcincta* and *H. contortus* for 5 weeks and with *Oesophagostomum columbianum* for 4 weeks (Peter *et al.*, 1994; Kerboeuf *et al.*, 1995). The

Table 7.1. UK efficacy claims for ML compounds against gastrointestinal and pulmonary nematodes of small ruminants.

Site	Species	COS	ODS	ODG	RDS	RDG	CIS	DICS	PICSP	RI
Abomasum	*Haemonchus contortus*	A/I/In	A/I/In	A/I	A/I/In	A/I	A	A/I	A/I/In	A/I/In
	Ostertagia circumcincta	A/I/In	A/I/In	A/I	A/I/In	A/I	A/I/In	A/I/In	A/I/In	A/I/In
	Ostertagia trifurcata	A/I							A/I	A/I
	Trichostrongylus axei	A/I/In	A/I		A/I		A	A/I	A	A
Small intestine	*Trichostrongylus colubriformis*	A/I	A/I	A/I	A/I	A/I	A/I	A/I	A/I	A/I
	Trichostrongylus vitrinus	A/I	A/I		A/I			A/I	A	A
	Nematodirus battus	A/I	A/I		A/I			I A*		
	Nematodirus spathiger	A/I	A/I	A/I	A/I	A/I	A			
	Nematodirus filicollis	A	A/I		A/I				A/I	A/I
	Strongyloides papillosus	I	A/I	A/I	A/I	A/I		A/I		
	Cooperia curticei	A	A/I		A/I		A	I	A/I	A/I
	Cooperia oncophora	A/I	A/I		A/I			A/I		
	Cooperia punctata		A/I		A/I		A			
Large intestine	*Gaigeria pachyscelis*						I	A/I		
	Oesophagostomum columbianum	A/I	A/I	A/I	A/I	A/I	I	A/I	A/I	A/I
	Oesophagostomum venulosum	A	A/I		A/I				A	A/I
	Chabertia ovina	A/I	A	A	A	A	A	A	A/I	A/I
	Trichuris ovis	A						A	A	A
Respiratory tract	*Dictyocaulus filaria*	A	A/I	A	A/I	A	A	A	A/I	A/I
	Protostrongylus rufescens								A	A

Oral MLs: COS, Cydectin 0.1% oral drench for sheep moxidectin; ODS, Oramec drench (sheep) ivermectin; ODG, Oramec drench (goats) ivermectin; RDS, Rycomec drench (sheep) ivermectin; RDG, Rycomec drench (goats) ivermectin.
Injectable MLs: CIS, Cydectin 1% injectable solution for sheep moxidectin; DICS, Dectomax injectable solution for cattle and sheep doramectin; PICSP, Panomec injection for cattle, sheep and pigs ivermectin; RI, Rycomec injection ivermectin.
A, adult worms; I, immature worms; In, inhibited stages; A*, removal of adults requires dosing at 300 μg kg^{-1} liveweight.

injectable formulation also has similar persistence claims together with a 4-week persistence against *Gaigeria pachyscelis* and a 2-week persistence claim against *T. colubriformis* (Peter *et al.*, 1994). Given the importance of *T. circumcincta* and *H. contortus* in the periparturient increase in faecal egg output, moxidectin is now being used for the treatment of periparturient ewes in many temperate production systems.

The recommended withdrawal periods for meat and milk products are also determined by national licensing authorities and may vary considerably between countries. In the UK, orally administered MLs have a meat withdrawal period of 14 days, whereas the more persistent injectable formulations have withdrawal periods of 42 days (ivermectin), 56 days (doramectin) and 70 days (moxidectin).

Macrocyclic lactone dose rates for goats

Although MLs have excellent efficacy against the nematode parasites of goats, which are essentially the same parasites common in sheep, their use in goats is more problematical. Goats metabolize anthelmintics faster than sheep, so, for the same dose rate, plasma levels of anthelmintic and persistence of effective plasma concentrations are substantially reduced in goats compared with sheep. This phenomenon is seen with benzimidazoles (Bogan *et al.*, 1987; Hennessy *et al.*, 1993) and with MLs including ivermectin (Scott *et al.*, 1990) and eprinomectin (Alvinerie *et al.*, 1999). Thus, worms in goats are exposed to lower concentrations of anthelmintic for shorter times than they would be in sheep given an identical dose of the same drug. Drug-resistant worms are therefore more likely to survive in treated goats than in treated sheep, and it is probably no coincidence that anthelmintic-resistant strains of nematodes seem routinely to be discovered first in goat herds. A major concern of sheep farmers in all countries is that resistant worms selected in goats may then be transmitted to sheep.

It is for this reason, among others, such as the small relative size of the goat market, that manufacturers of MLs often do not seek registration for the use of their products in goats. For example, in Australia, neither Merial nor Fort Dodge has registered ivermectin or moxidectin, respectively, for use in goats. Fort Dodge has followed a similar policy in New Zealand and South Africa, and Merial has not registered eprinomectin for use in goats in any market.

Orally administered MLs appear to work well when given to goats at the sheep dose rate of 0.2 mg kg^{-1}; however, an early pharmacokinetic study (Alvinerie *et al.*, 1993) suggested that the most appropriate dose for subcutaneously administered ivermectin in goats was 0.3 mg kg^{-1}. Evidence from a study using the topically applied ML eprinomectin recently registered for use in dairy cattle (Alvinerie *et al.*, 1999) suggests that this drug should be administered at higher dose rates to goats than those given to cattle (0.5 mg kg^{-1}).

Spectrum of activity against ectoparasites

Permanent ectoparasites (mange mites, lice and keds) complete their entire life cycle on the host, whereas semipermanent ectoparasites (nasal bot flies, head flies and blowflies) have one or more free-living stage off the host. Many of the ectoparasites of ruminants are host specific, and transmission occurs when animals are in close association. The relative importance of the endo- and ectoparasiticidal properties of the MLs varies widely between countries in both the northern and the southern hemisphere. While the anthelmintic activity of these compounds is of major importance in all markets, activity against arthropods is of minor importance only in Australia and New Zealand, where sheep scab (*Psoroptes communis ovis*) has long been eradicated. Specific programmes for control of ectoparasites using MLs are not common but are very important for the containment of psoroptic mange.

Use of MLs to Control Endo- and Ectoparasites of Small Ruminants

Endoparasites

World census data for the last 5 years show that for every small ruminant there were 0.76 cattle, and the most recent data, for the year 2000, suggest that for every sheep there were 0.67 goats (FAOSTAT, 2000). Given the variety of agroclimatic zones used in the production of small ruminants, it is hardly surprising that there are a range of different genera implicated in nematodoses; however, economically, the most important genera are *Haemonchus*, *Teladorsagia* (*Ostertagia*), *Trichostrongylus*, *Nematodirus*, *Cooperia* and *Oesophagostomum*. In animals that are not compromised nutritionally, disease attributable to these nematodes is generally restricted to young stock during their first grazing season or occasionally in second grazing season animals. Although non-reproducing adult sheep on an adequate plane of nutrition generally express an effective acquired immunity, the acquisition and expression of acquired immunity against gastrointestinal nematodes appears to be much more limited in goats (Le Jambre and Royal, 1976; Jallow *et al.*, 1994; Huntley *et al.*, 1995). In many intensive small ruminant production systems, chemoprophylactic control of gastrointestinal nematodoses is rarely attempted using prescriptive regimes of the type described for calves (McKellar, 1994). Small ruminant chemoprophylaxis is generally used to limit the extent of challenge from pasture in order to maintain the productivity and welfare of susceptible stock.

Epidemiological factors influencing treatment strategies

An understanding of the epidemiology of the various nematode species implicated in disease in small ruminants is a crucial prerequisite for anthelmintic control strategies. However, since anthelmintic strategies are required to cope with regional differences in production systems, climate and prevailing parasite species, they inevitably need to be site specific. However, there are some common strategic and epidemiological elements that influence treatment regimes. The principal strategies used in sheep and goat production systems are either to administer treatments during 'epidemiologically' sensitive periods when they are intended to have the greatest effect in reducing pasture contamination, or to use anthelmintics suppressively, within the pre-patent period of the target species, or at frequent intervals to achieve the same effect. Unfortunately, treatments administered suppressively and at these 'epidemiologically' sensitive periods may also select most strongly for anthelmintic resistance, since they both confer considerable advantage to worms carrying the resistance genes.

Temperate production systems

In many temperate production systems in both hemispheres, sheep and goat breeding is seasonal, which usually results in relatively repeatable patterns of pasture contamination and challenge. In those systems where rainfall is non-seasonal or has a summer prevalence, large numbers of infective larvae become available on pasture from mid-summer onwards, usually reaching peak numbers in autumn/early winter. In more arid regions and under drought conditions, these suprapopulation peaks may be shifted towards the wetter periods, often the autumn and/or winter months. In temperate production systems, pasture larval populations tend to decline over the winter months, when temperatures do not suit the development and translation of larvae, and usually reach their minimum values during spring.

Much small ruminant production is also carried on in the Mediterranean climatic zones of the northern hemisphere and in the corresponding climatic zones of southern Australia and southern South Africa. Here, winters are mild and wet, and followed by a reliably hot, dry summer, during which pastures become effectively sterile as far as free-living stages of nematode parasites are concerned. Following the first rains in autumn, eggs passed by adult worms surviving the summer in their hosts are able to hatch and develop to infective larvae, with numbers of infective larvae usually reaching their seasonal peak in late winter or early spring.

There are two main sources of infection for newly born sheep and goats, which are epidemiologically sensitive targets for anthelmintic treatments. The first source of infection derives from eggs and larvae that have survived overwinter on pasture, and the second source of infection derives from contamination deposited during the ewe or doe periparturient relaxation in immunity (PPRI). In fat lamb production systems, early season treatments are aimed at reducing contamination and delaying the incrementation of the pasture larval population. The periparturient increase in faecal egg count derives from an increase in fecundity of established worms, the maturation of previously inhibited populations and an increased susceptibility to new infection (Armour, 1980). It is now an almost universal practice in temperate production systems to include one or more periparturient ewe or doe treatments in order to minimize the epidemiological contribution made by periparturient animals, particularly for economically important species such as *Teladorsagia* (*Ostertagia*), *Trichostrongylus* and *Haemonchus*. Flushing (pre-tupping) treatments are administered routinely to adult females and rams mostly for therapeutic rather than prophylactic reasons, and serve to bring the animals into peak condition prior to mating.

The ewe appears to play a lesser or negligible role compared with the pasture in the epidemiology of some nematodes. In species of nematodes where this is true, the overwintered suprapopulation serves to transmit infection to newly born animals. *Nematodirus battus* provides an extreme example of an intergenerational lamb to lamb transmitted disease. In those countries where this species poses a problem, early season (spring) treatments are given to young stock to minimize the risk to the grazing animal and the infectivity of those pastures in the subsequent grazing season. For those temperate climate species where the ewe plays a lesser role in contaminating pastures, intragenerational transmission is important in generating the challenge that leads to disease. In permanently stocked fat lamb production systems, routine chemoprophylaxis is used to minimize the risks from intragenerational transmission and enable lambs to achieve target weights in the shortest possible time. The contribution made by young stock to pasture contamination is influenced by the number of parasite generations that contribute to intragenerational transmission and the minimum generation interval, which varies from species to species. In fat lamb production systems in temperate areas, where genera with relatively short minimum generation intervals such as *Teladorsagia*, *Haemonchus* and *Trichostrongylus* tend to predominate, acquired immunity generally restricts the number of generations that can occur during the grazing season (Waller and Thomas, 1978).

Although the patterns of infection and disease are essentially similar in temperate extensive and intensive production systems, treatments are often administered in the former systems whenever animals are gathered for routine management, that is at lambing, tail docking, weaning and mating.

Tropical/subtropical production systems

In many tropical/subtropical production systems, the population dynamics of the principal parasites and diseases attributable to them may be driven by environmental factors which affect the development, translation and longevity of the suprapopulation. Under these circumstances and particularly in those areas/systems where significant rainfall only occurs on occasions, disease and treatment patterns are often linked to seasonal rainfall patterns. In those production systems where seasonal droughts are a routine occurrence, then hypobiotic larvae and persistent adult stages provide an epidemiologically sensitive target and treatments may be administered at the onset of rainfall or during the middle of the dry season. Treatments administered at these times may be highly selective as far as anthelmintic resistance is concerned since they may 'screen' virtually the entire parasite population. As a consequence of their exposure to intense periodic challenge, indigenous ruminant breeds often display a superior ability to regulate their parasite populations even under the relatively poor nutritional environment that these systems often provide.

In those tropical/subtropical production systems where rainfall is not a limiting factor, ruminants are produced against a background of high continuous challenge and it is difficult to identify specific epidemiologically sensitive treatment times. Intensive large-scale production systems in these areas have often resorted to the use of suppressive (treatment frequency within the pre-patent period) or neo-suppressive (treatment frequency close to the pre-patent period) regimes which have resulted in high levels of anthelmintic resistance. Small-scale producers in many tropical and subtropical regions generally use anthelmintics for therapeutic rather than prophylactic purposes.

One possible consolation to small ruminant producers in wet tropical areas is the fact that, although egg hatching and larval development is rapid and continuous throughout the year, the resulting infective larvae on pasture have a very short life expectancy. Studies in Fiji and Tonga (Banks *et al.*, 1990; Barger *et al.*, 1994) showed that infective larvae of *Haemonchus*, *Trichostrongylus* and *Oesophagostomum* survived for only 3–7 weeks under these conditions, which contrasts strongly with their survival for at least as many months in temperate regions. This short survival of infective larvae can be exploited by simple rotational grazing or tethered or herding husbandry systems that can obviate the need for most and perhaps all anthelmintic treatment.

Persistent efficacy

The persistent efficacy of moxidectin, which varies from 2 to 5 weeks depending on species, formulation and regulatory requirements, gives it a special role in control of haemonchosis (Kerboeuf *et al.*, 1995). Persistent efficacy against *H. contortus* is always at the upper end of the range, and the ability thus to prevent *Haemonchus* egg output for several weeks at times of the year when these eggs would otherwise hatch and contribute to pasture infectivity greatly simplifies control of this species. Even longer persistent activity against all susceptible species has been obtained for ivermectin through its incorporation into a controlled-release capsule that lodges in the rumen and releases ivermectin (0.02–0.04 mg kg^{-1} day^{-1} for 100 days) over an extended period (Rehbein *et al.*, 2000a). The delivery rate of the ivermectin controlled-release capsules is 0.8 mg ivermectin day^{-1} for sheep 20–40 kg in weight and 1.6 mg day^{-1} for sheep weighing 41–80 kg. Hence, on a per kg liveweight basis, the daily dose is 0.02–0.04 mg. Ivermectin controlled-release capsules have been used by sheep producers in Australia and New Zealand as the extended activity offers the opportunity to use treated sheep to prepare clean pastures. Both forms of persistent MLs (moxidectin, and ivermectin capsules) are also used typically in particularly susceptible animals, such as lambing ewes or young animals at weaning (Taylor *et al.*, 1997) or as an alternative to frequent treatment during outbreaks of helminthosis when clean pastures are not available.

Goats

In young goats, the acquisition of immunity is more protracted and the maximum expression of immunity generally occurs during their second grazing season (Vlassoff *et al.*, 1999). The limited capacity of goats to acquire and express immunity against gastrointestinal nematodes has made the treatment of all age classes a common practice in goats used in intensive production systems where they are obliged to graze rather than browse. Intensively grazing goats has been an important factor in accounting for the relatively high incidence and prevalence of anthelmintic resistance in goats.

Control and treatment of ectoparasites

Ectoparasitoses

MITE INFESTATIONS. Mange is caused by a variety of mites that induce an allergic dermatitis and four genera of parasitic mites are associated with mange in sheep.

1. *Psoroptic mange* (sheep scab). Sheep scab is present in most of the sheep-rearing areas of the world apart from Australia, New Zealand, Canada and the USA. It results from an allergic skin dermatitis caused by the excretory products of the non-burrowing mite, *Psoroptes ovis*. Infestations can be acute or chronic in form and lead to high morbidity, extensive loss of wool and, in severe cases, poor body condition and death. Transmission of scab is via live mites through sheep to sheep contact or through contaminated equipment/clothing or tags of wool. Scab is the most contagious of mange infections.

2. *Sarcoptic mange* (head scabies). Head mange of sheep is caused by the mite *Sarcoptes scabiei* var. *ovis* and is cosmopolitan, although absent from the UK. The mites prefer regions without extensive wool cover and therefore they frequently are found on the ears, face and groin where they burrow into the epidermis. Irritation can lead to itching with exudation and crust formation, which can result in wool and hide damage.

3. *Chorioptic mange* (foot mange). The mites of *Chorioptes ovis* tend to infest areas of the body with a sparse covering of wool or hair, such as the lower legs and scrotum. The mites feed superficially on skin debris and induce an allergic dermatitis with fluid exudate, which causes marked irritation of affected areas. Frequent dipping for control of sheep scab has resulted in eradication of the chorioptes mite from parts of Europe and the UK.

4. *Psorobic mange* (itch mite). The itch mite, *Psorobia* (*Psorergates*) *ovis* mainly affects Merino sheep and is confined to Australia, New Zealand, South and North America, and South Africa. The mites burrow under the superficial layers of the skin causing dry flaky lesions and irritation with wool loss.

LICE INFESTATIONS. Lice infestations occur throughout the world but are only considered of economic significance where wool production is of primary importance. Lice do not leave the host during any phase of their life cycle. Three species of louse infest sheep. The face or blue body louse (*Linognathus ovillus*) is cosmopolitan and feeds by piercing the skin and sucking blood, and prefers the face, ears and neck areas. The foot louse (*Linognathus pedalis*) inhabits the skin between the hooves and the hocks and sometimes the crotch, scrotum and belly. It is also a blood feeder. Chewing body lice (*Bovicola ovis*, formally *Damalinia ovis*) feed on skin debris and wool fibres and often infest the whole body. Irritation causes rubbing and scratching, resulting in damage to the fleece or skin. Infestation with *B. ovis* is regarded very seriously in the wool-producing areas of Australia and New Zealand, and is a notifiable disease. Infestation is difficult to detect at low intensities, and most producers have opted for prophylactic control through strategic dipping programmes using organophosphate or synthetic pyrethroid insecticides, usually in ignorance of whether or not the parasite is actually present in their flock.

Insecticide residue problems when their wool is scoured probably pose a greater threat to sustainable wool production than does the louse itself, and these dipping programmes are being actively discouraged.

FLY INFESTATIONS. Flies may cause irritation to sheep and goats by their persistent activity or use the host to complete a stage of their life cyle.

1. Hippoboscid flies – sheep ked (*Melophagus ovinus*). This is a wingless blood-feeding fly, which completes its entire life cycle on the host. Severe infestation can cause anaemia and mechanical damage to the fleece through scratching and rubbing.
2. *Oestrus ovis* (nasal bot fly). *Oestrus ovis* is common in sheep throughout the world. Infection results from adult flies depositing first-stage larvae into the nasal cavity, which subsequently mature to the third-stage instar. The larval stages induce a nasal discharge, sneezing and in some cases head shaking.
3. Blowfly strike. The larvae of blowflies (Diptera), particularly *Lucilia cuprina*, *Lucilia sericata* (greenbottles) and some *Calliphora* spp. (bluebottles) and *Phormia* spp. (black blowfly) cause a primary cutaneous myiasis in sheep which can result in marked loss of wool production and is a serious welfare problem. The larvae remain on the surface of the skin, but the head can penetrate deep into the dermis. Secondary flies (many *Calliphora* spp. and *Chrysomyia* spp.) do not initiate strike but will attack an area already damaged. Cutaneous myiasis is widespread throughout sheep-rearing areas of the world.

Treatment of ectoparasitoses

It is impracticable to detail a comprehensive listing of the use of the various MLs and different formulations against a complete range of ectoparasites of sheep and goats. This is partly due to the differences in product registration between countries. For example, the avermectins (ivermectin and generic abamectin) carry label claims for activity against itch mite (*Psorergates ovis*) and larval stages of the nasal bot (*O. ovis*) in both Australia and New Zealand, as does moxidectin injectable in Australia. Oral moxidectin has a claim only against itch mite in Australia, while injectable moxidectin is registered for use only against nasal bot in New Zealand. There is also occasional off-label use of MLs against sarcoptic and chorioptic mange in both countries and in other areas of the world.

PSOROPTIC MANGE. Prior to the introduction of the MLs, prevention and treatment of psoroptic mange (sheep scab) was through plunge dipping in persistent organochlorines and, more recently, organophosphates (diazinon, propetamphos) or the synthetic pyrethroids (flumethrin, high *cis* cypermethrin), and this method is still recommended in many sheep-rearing areas. In the UK, sheep scab was almost eliminated through a compulsory dipping programme but, since deregulation in 1992, the disease has re-emerged and currently is widespread. In South Africa and most South American countries where scab is prevalent, use of MLs for control of scab probably ranks similarly in importance to their use in controlling helminths, with control of other ectoparasites such as mange and lice of minor importance.

It is important to emphasize the difference between treatment of scab infestations and prevention of infestation. Treatment is usually administered when clinical signs are present and often a welfare problem exists. Preventative therapy will ensure that sheep are protected for a limited period when exposed to other affected sheep. The systemic ML endectocides, ivermectin, doramectin and moxidectin, are effective in eliminating the *P. ovis* mite from sheep when given by injection. Ivermectin (0.2 mg kg^{-1}) is effective at treating and controlling the scab mite when subcutaneously injected twice at either 7-day (Soll *et al.*, 1992) or 10-day intervals (O'Brien *et al.*, 1993) as the ML has little residual activity. The reason for the 7–10 day interval is that some mites may lay eggs after initial treatment and these will then be able to survive before hatching. The second injection will kill mites emerging from eggs or from a moult. A long-acting formulation of ivermectin recently has been shown to be effective for treatment of natural *P. ovis* infestation of sheep given as either a single dose of 0.3 mg kg^{-1} or as two doses of 0.2 mg kg^{-1} 7 days apart (Gomez Blanco *et al.*, 1998). Doramectin given as a single injection at the higher dose rate of 0.3 mg kg^{-1} is also effective therapeutically against experimental infestations of *P. ovis* (Bates *et al.*, 1995). A review of therapeutic sheep scab trials in Great Britain, South Africa, Argentina and Uruguay with doramectin injection at 0.2 or 0.3 mg kg^{-1} demonstrated 97.7–100% efficacy (McKenzie, 1997). Eddi *et al.* (1999) demonstrated a persistent efficacy of doramectin at 0.2 mg kg^{-1} against *P. ovis* for at least 14 days.

Moxidectin administered subcutaneously at 0.2 mg kg^{-1} as either a single injection or two injections 10 days apart was effective for treatment of clinical psoroptic mange (O'Brien *et al.*, 1994). A single injection protected against experimental infestation for up to 35 days. Two large field trials (6500–10,000 sheep) were conducted in Ireland (O'Brien *et al.*, 1996) or in Wales (Williams and Parker, 1996), and these confirmed that a single injection with moxidectin at 0.2 mg kg^{-1} was highly effective as a prophylactic treatment against sheep scab when all sheep in the flock were treated. The studies by O'Brien *et al.* (1996), Ortega-Mora *et al.* (1998) and Parker *et al.* (1999) also showed that two injections of moxidectin given 10 days apart were effective in the treatment of outbreaks of psoroptic mange. Moxidectin was shown to prevent infection or re-infestation for at least 28 days following treatment. This means that treated sheep can be returned to infested pasture immediately, as the period of protection is longer than the 16 days in which mites can be viable off the host. It is worth noting that following treatment with injectable acaricides, animals can still have irritation from skin lesions in the absence of mites.

Recently, an ivermectin controlled-release (CR) intraruminal bolus has been developed for sheep, which delivers a minimum dose of 0.02 mg kg^{-1} day^{-1} of ivermectin for 100 days. This bolus showed high efficacy for both therapeutic treatment of *P. ovis*-infected sheep and an extended (up to 100 days) period of prophylaxis for the control of psoroptic mange (Bridi *et al.*, 1998; Forbes *et al.*, 1999; O'Brien *et al.*, 1999). Ivermectin bolus-treated sheep have been shown to exhibit greater liveweight gain, improved carcass yield and quality and have significantly increased clean fleece weight and yield compared with untreated controls (Rehbein *et al.*, 2000b,c). A further advantage of the bolus treatment is that sheep can be moved freely in the flock for 3 months without the problem of mite infestation. In addition, the ivermectin CR bolus will provide prolonged control of gastrointestinal nematodes that accumulate from summer grazing.

The introduction of the MLs has the advantage of reducing the problem of disposal of large volumes of acaricidal dip and the associated health risk of handling organophosphates and the environmental impact of synthetic pyrethroids (SPs) on fish and aquatic insects. Also no expensive dipping equipment is required. The main perceived disadvantage, raised in relation to the treatment of endoparasites, is that the excreted ML endectocides have been shown to have ecotoxicological effects on some arthropods in the environment, but this effect will be influenced by the rate of breakdown of faeces, timing of administration and agricultural practices (Taylor, 1999). Indeed, some of the justification raised by various conservation bodies for using MLs is to avoid the much larger environmental risks associated with the use of dips, particularly the SPs.

OTHER MANGE INFESTATIONS. A single injection (0.2 mg kg^{-1}) of either ivermectin, doramectin or moxidectin is effective in most cases for treatment of mild outbreaks of sarcoptic mange in sheep and goats (Manurung *et al.*, 1990; Corba *et al.*, 1995). Two injections of ivermectin or moxidectin (0.2 mg kg^{-1}) given 10–15 days apart have shown high efficacy against ovine (Papadopoulos and Fthenakis, 1999; Fthenakis *et al.*, 2000) or caprine (Ghosh and Nanda, 1997; Sengupta *et al.*, 1997) sarcoptic mange. Ivermectin injection (0.2 mg kg^{-1}) effectively removed mites from sheep infested with *Psorergates ovis* (Soll and Carmichael, 1988). The CR ivermectin bolus offers high efficacy against several mange infestations (Rehbein *et al.*, 1998).

LICE INFESTATIONS. The systemic MLs are effective against *Linognathus* spp. when administered by injection as a single dose but are virtually ineffective against biting/chewing lice (*Bovicola ovis*) and have no label claim against these species. It is for this reason that a differential diagnosis between sheep lice and sheep scab is so important with the MLs (Bates,

1999). Misdiagnosis of sheep scab has resulted in the incorrect use of MLs for lice, and unsatisfactory efficacy has resulted. Ivermectin administered as a jetting fluid is moderately effective against the chewing louse *B. ovis* (Rugg *et al.*, 1995) and other lice infestations (Thompson *et al.*, 1994).

FLY INFESTATIONS. Treatment of keds with MLs is the same as for lice. A long-acting formulation of ivermectin given subcutaneously at 0.3 mg kg^{-1} at shearing has been shown to exhibit effective control of *M. ovinus* (Roberts *et al.*, 1998).

Oestrosis can be treated effectively by injectable ivermectin, doramectin or moxidectin or by an oral drench of ivermectin (all administered at 0.2 mg kg^{-1}). These MLs are effective against all instar larval stages of the nasal bot. The prophylactic effect of injectable ivermectin was only moderate and the oral formulation poor (Dorchies *et al.*, 1997). The CR ivermectin bolus will also control and prevent *O. ovis* infestations (Rugg *et al.*, 1997).

Ivermectin has been shown to be effective for control of blowfly strike in experimental studies and large field trials when applied as a jetting fluid (Eagleson *et al.*, 1993; Thompson *et al.*, 1994). Ivermectin is effective prophylactically and will also kill larvae on strike lesions. Protection was demonstrated for about 3 months after jetting treatment. Recent trials with the ivermectin CR capsule (0.02–0.04 mg kg^{-1} day^{-1} for 100 days) showed that ivermectin was an effective aid for control of breech strike but was less efficacious against body strike (Rugg *et al.*, 1998).

Injectable ivermectin has been shown to be highly efficacious against the larvae of the warble fly *Przhevalskiana silenus* in goats (Khan *et al.*, 1994).

Conclusions

Although research is being conducted on new anthelmintics and ectoparasiticides, we simply cannot afford a cavalier attitude with regard to our current compounds since new antiparasitics tend to appear on the market relatively infrequently (Geary *et al.*, 1999). The pivotal role that MLs now play in the control of many of our most important parasitic diseases suggests that the conservation of efficacy of these compounds should be given some priority in the allocation of research and advisory funding by both the industry and academia.

References

Alvinerie, M., Sutra, J.F. and Galtier, P. (1993) Ivermectin in goat plasma and milk after subcutaneous injection. *Veterinary Research*, 24, 417–421.

Alvinerie, M., Lacoste, E., Sutra, J.F. and Chartier, C. (1999) Some pharmacokinetic parameters of eprinomectin in goats following pour-on administration. *Veterinary Research Communications* 23, 449–455.

Armour, J. (1980) The epidemiology of helminth disease in farm animals. *Veterinary Parasitology* 6, 7–46.

Bairden, K., Duncan, J.L. and Mudd, A.J. (1995) The persistent activity of moxidectin against some of the common gastrointestinal nematodes of sheep. *Proceedings of the Sheep Veterinary Society* 18, 153–156.

Banks, D.J.D., Singh, R., Barger, I.A., Pratap, B. and Le Jambre, L.F. (1990) Development and survival of infective larvae of *Haemonchus contortus* and *Trichostrongylus colubriformis* in a tropical environment. *International Journal for Parasitology* 20, 155–160.

Barger, I.A., Siale, K., Banks, D.J.D. and Le Jambre, L.F. (1994) Rotational grazing for control of gastrointestinal nematodes of goats in a wet tropical environment. *Veterinary Parasitology* 53, 109–116.

Bates, P. (1999) Chewing lice, sheep scab and systemic endectocides. *Veterinary Record* 144, 243.

Bates, P.G., Groves, B.A., Courtney, S.A. and Coles, G.C. (1995) Control of sheep scab (*Psoroptes ovis*) on artificially infested sheep with a single injection of doramectin. *Veterinary Record* 137, 491–492.

Bogan, J., Benoit, E. and Delatour, P. (1987) Pharmacokinetics of oxfendazole in goats: a comparison with sheep. *Journal of Veterinary Pharmacology and Therapeutics* 10, 305–309.

Bridi, A.A., Rehbein, S., Carvalho, L.A., Barth, D., Barrick, R.A. and Eagleson, J.S. (1998) Efficacy of ivermectin in the controlled release formulation against *Psoroptes ovis* (Hering, 1838), Gervais, 1841 (acari: psoroptidae) on sheep. *Veterinary Parasitology* 78, 215–221.

Corba, J., Varady, M., Praslicka, J. and Tomasovicova, O. (1995) Efficacy of injectable moxidectin against mixed (*Psoroptes ovis* and *Sarcoptes-scabiei* var *ovis*) mange infection in sheep. *Veterinary Parasitology* 56, 339–344.

Dorchies, P., Alzieu, J.P. and Cadiergues, M.C. (1997) Comparative curative and preventive efficacies of ivermectin and closantel on *Oestrus ovis* (Linne, 1758) in naturally infected sheep. *Veterinary Parasitology* 72, 179–184.

Eagleson, J.S., Thompson, D.R., Scott, P.G., Cramer, L.G. and Barrick, R.A. (1993) Field trials to confirm the efficacy of ivermectin jetting fluid for control of blowfly strike in sheep. *Veterinary Parasitology* 51, 107–122.

Eddi, C., Caracastantogolo, J., Moltedo, H., Derozier, C. and Cutulle, C. (1999) Persistent efficacy of doramectin injectable against *Psoroptes ovis* infestations in sheep. *Proceedings of the 18th International Conference of the World Association for the Advancement of Veterinary Parasitology (WAAVP)*, Copenhagen, Denmark (g.6.59).

FAOSTAT Agricultural database (2000) *Food and Agriculture Organisation, Livestock Statistics*. Available at: www.fao.org

Forbes, A.B., Pitt, S.R., Baggott, D.G., Rehbein, S., Barth, D., Bridi, A.A., Carvalho, L.A. and O'Brien, D.J. (1999) A review of the use of a controlled-release formulation of ivermectin in the treatment and prophlylaxis of *Psoroptes ovis* infestations in sheep. *Veterinary Parasitology* 83, 319–326.

Fthenakis, G.C., Papadopoulos, E., Himonas, C., Leontides, L., Kritas, S. and Papatsas, J. (2000) Efficacy of moxidectin against sarcoptic mange and effects on milk yield of ewes and growth of lambs. *Veterinary Parasitology* 87, 207–216.

Geary, T.G., Sims, S.M., Thomas, E.M., Vanover, L., Davis, J.P., Winterrowd, C.A., Klein, R.D., Ho, N.F.H. and Thompson, D.P. (1993) *Haemonchus contortus* – ivermectin-induced paralysis of the pharynx. *Experimental Parasitology* 77, 88–96.

Geary, T.G., Sangster, N.C. and Thompson, D.P. (1999) Frontiers in anthelmintic pharmacology. *Veterinary Parasitology* 84, 275–295.

Ghosh, S.K. and Nanda, S.K. (1997) Management of sarcoptic mange infestation in goats in Tripura – a clinical report. *Indian Veterinary Journal* 74, 248–249.

Gill, J.H., Kerr, C.A., Shoop, W.L. and Lacey, E. (1998) Evidence of multiple mechanisms of avermectin resistance in *Haemonchus contortus* – comparison of selection protocols. *International Journal for Parasitology* 28, 783–789.

Gomez Blanco, J.C., Clos, P., Casan, A.A., Margueritte, J.A., Elordi, L., Filippe, J., Lamberti, J.C., Barrios, E.A. and Bulman, G.M. (1998) First report of the efficacy of a new formulation of long acting 1% ivermectin (Bovifort-registerd trade mark, L.A. Biogenesis, S.A.) in a single subcutaneous dose against ovine mange (*Psoroptes ovis*) in Argentina. *Veterinaria Argentina* 15, 14–17.

Hennessy, D.R., Sangster, N.C., Steel, J.W. and Collins, G.H. (1993) Comparative pharmacokinetic behaviour of albendazole in sheep and goats. *International Journal for Parasitology* 23, 321–325.

Huntley, J.F., Patterson, D.M., McKellar, A., Jackson, F., Stevenson, L.M. and Coop, R.L. (1995) A comparison of the mast cell and eosinophil responses of sheep and goats to nematode infections. *Research in Veterinary Science* 58, 5–10.

Jackson, F. and Coop, R.L. (2000) The development of anthelmintic resistance in sheep nematodes. *Parasitology* 120, s95-s107.

Jallow, O.A., McGregor, B.A., Anderson, N. and Holmes, J.H.G. (1994) Intake of trichostrongylid larvae by goats and sheep grazing together. *Australian Veterinary Journal* 71, 361–364.

Kerboeuf, D., Hubert, J., Cardinaud, B. and Bold-Riou, F. (1995) The persistence of the efficacy of injectable or oral moxidectin against *Teladorsagia*, *Haemonchus* and *Trichostrongylus* species in experimentally infected sheep. *Veterinary Record* 137, 399–401.

Khan, M.Q., Akhtar, S. and Cheema, A.V. (1994) Efficacy of ivermectin against goat warbles (*Przhevalskiana silenus* Brauer) in Pakistan. *Veterinary Record* 135, 361.

Le Jambre, L.F. and Royal, W.M. (1976) A comparison of worm burdens in grazing Merino sheep and Angora goats. Australian *Veterinary Journal* 52, 181–183.

Manurung, J., Stevenson, P., Beriajaya and Knox, M.R. (1990) Use of ivermectin to control sarcoptic mange in goats in Indonesia. *Tropical Animal Health and Production* 22, 206–212.

McKellar, Q.A. (1994) Chemotherapy and delivery systems – helminths. *Veterinary Parasitology* 54, 249–258.

McKellar, Q.A., Bogan, J.A., Horspool, L. and Reid, K. (1988) Effect of ivermectin on the reproductive potential of *Cooperia curticei*. *Veterinary Record* 122, 444.

McKenzie, M.E. (1997) Efficacy of doramectin against field outbreaks of *Psoroptes ovis* in sheep. *Proceedings of the 17th International Conference of the World*

Association for the Advancement of Veterinary Parasitology (WAAVP), Sun City, South Africa, p. 21.

National Office of Animal Health Limited (NOAH) *Compendium of Data Sheets for Veterinary Products 1999–2000*.

Oakley, G.A. (1990) *Ivermectin: the Veterinary Handbook.* Anpar Books, Berkhampstead, UK, pp. 100–105, 129–137.

O'Brien, D.J., Gray, J. and O'Reilly, P.F. (1993) Control of sheep scab by subcutaneous injection of ivermectin. *Irish Veterinary Journal* 46, 99–101.

O'Brien, D.J., Gray, J.S. and O'Reilly, P.F. (1994) The use of moxidectin 1% injectable for the control of psoroptic mange in sheep. *Veterinary Parasitology* 52, 91–96.

O'Brien, D.J., Parker, L.D., Menton, C., Keaveny, C., McCollum, E. and O'Laoide, S. (1996) Treatment and control of psoroptic mange (sheep scab) with moxidectin. *Veterinary Record* 139, 437–439.

O'Brien, D.J., Forbes, A.B., Pitt, S.R. and Baggott, D.G. (1999) Treatment and prophylaxis of psoroptic mange (sheep scab) using an ivermectin intra-ruminal controlled-release bolus for sheep. *Veterinary Parasitology* 85, 79–85.

Ortega-Mora, L.M., Diez Banos, N., Martinez Gonzalez, B. and Rojo Vazquez, F.A. (1998) Controlled field efficacy of injectable moxidectin against naturally acquired psoroptic mange in sheep. *Small Ruminant Research* 29, 271–276.

Papadopoulos, E. and Fthenakis, G.C. (1999) Administration of moxidectin for treatment of sarcoptic mange in a flock of sheep. *Small Ruminant Research* 31, 165–168.

Parker, L.D., O'Brien, D.J. and Bates, P.G. (1999) The use of moxidectin for the prevention and treatment of psoroptic mange (scab) in sheep. *Veterinary Parasitology* 83, 301–308.

Peter, R.J., Boelema, E., Grove, J.T. and Rall, M. (1994) The residual anthelmintic efficacy of moxidectin against selected nematodes affecting sheep. *Journal of the South African Veterinary Association* 65, 167–169.

Rehbein, S., Batty, A.F., Barth, D., Visser, M., Timms, B.J., Barrick, R.A. and Eagleson, J.A. (1998) Efficacy of an ivermectin controlled-release capsule against nematode and arthropod endoparasites in sheep. *Veterinary Record* 142, 331–334.

Rehbein, S., Barth, D., Visser, M., Winter, R. and Langholff, W.K. (2000a) Efficacy of an ivermectin controlled-release capsule against some rarer nematode parasites of sheep. *Veterinary Parasitology* 88, 293–298.

Rehbein, S., Barth, D., Visser, M., Winter, R., Cramer, L.G. and Langholff, W.K. (2000b) Effects of *Psoroptes ovis* infection and its control with an ivermectin controlled-release capsule on growing sheep. 1. Evaluation of weight gain, feed consumption and carcass value. *Veterinary Parasitology* 91, 107–118.

Rehbein, S., Oertel, H., Barth, D., Visser, M., Winter, R., Cramer, L.G. and Langholff, W.K. (2000c) Effects of *Psoroptes ovis* infection and its control with an ivermectin controlled-release capsule on growing sheep. 2. Evaluation of wool production and leather value. *Veterinary Parasitology* 91, 119–128.

Roberts, G.R., Paramidani, M., Bulman, G.M., Lamberti, J.C., Elordi, L., Filippi, J. and Margueritte, J.A. (1998) Efficacy of a new formulation of 1% injectable ivermectin in a single subcutaneous dose against *Melophagus ovinus* (Linnaeus, 1758) in sheep in Patagonia (Argentina). *Veterinaria Argentina* 15, 91–95.

Rugg, D., Thompson, D.R., Boyle, R. and Eagleson, J.S. (1995) Field efficacy of an ivermectin jetting fluid for control of the sheep body louse *Bovicola (Damalinia) ovis* in New Zealand. *New Zealand Veterinary Journal* 43, 48–49.

Rugg, D., Gogolewski, R.P., Barrick, R.A. and Eagleson, J.S. (1997) Efficacy of ivermectin controlled-release capsules for the control and prevention of nasal bot infestations in sheep. *Australian Veterinary Journal* 75, 36–38.

Rugg, D., Thomson, D., Gogolewski, R.P., Allerton, G.R., Barrick, R.A. and Eagleson, J.S. (1998) Efficacy of ivermectin in a controlled-release capsule for the control of breech strike in sheep. *Australian Veterinary Journal* 76, 350–354.

Scott, E.W., Kinabo, L.D. and McKellar, Q.A. (1990) Pharmacokinetics of ivermectin after oral or percutaneous administration to adult milking goats. *Journal of Veterinary Pharmacology and Therapeutics* 13, 432–435.

Sengupta, P.P., Basu, A. and Pahari, T.K. (1997) Studies on comparative efficacy of three potential miticides against caprine sarcoptic mange in field conditions. *International Journal of Animal Sciences* 12, 245–248.

Shoop, W.L., Egerton, J.R., Eary, C.H. and Suhayda, D. (1990) Anthelmintic activity of the macrocyclic lactone F28249-alpha in sheep. *American Journal of Veterinary Research* 51, 1873–1874.

Shoop, W.L., Demontigny, P., Fink, D.W., Williams, J.B., Egerton, J.R., Mrozik, H., Fisher, M., Skelly, B.J. and Turner, M.J. (1996) Efficacy in sheep and pharmacokinetics in cattle that led to the selection of eprinomectin as a topical endectocide for cattle. *International Journal for Parasitology* 26, 1227–1235.

Soll, M.D. and Carmichael, I.H. (1988) Efficacy of injectable ivermectin against the itch mite (*Psorergates ovis*) of sheep. *Parasitology Research* 75, 81–82.

Soll, M.D., Carmichael, I.H., Swan, G.E. and Abrey, A. (1992) Treatment and control of sheep scab (*Psoroptes ovis*) with ivermectin under field conditions in South Africa. *Veterinary Record* 130, 572–574.

Sutherland, I.A., Leathwick, D.M. and Brown, A.E. (1999) Moxidectin: persistence and efficacy against drug-resistant *Ostertagia circumcincta*. *Journal of Veterinary Pharmacology and Therapeutics* 22, 2–5.

Taylor, S.M. (1999) Sheep scab – environmental considerations of treatment with doramectin. *Veterinary Parasitology* 83, 309–317.

Taylor, S.M., Kenny, J., Edgar, H.W., Ellison, S. and Ferguson, L. (1997) Efficacy of moxidectin, ivermectin and albendazole oral drenches for suppression of periparturient rise in ewe worm egg output and reduction of anthelmintic treatment for lambs. *Veterinary Record* 141, 357–360.

Thompson, D.R., Rugg, D., Scott, P.G., Cramer, L.G. and Barrick, R.A. (1994) Rainfall and breed effects on the efficacy of ivermectin jetting fluid for the prevention of fly strike and treatment of infestations of lice in long-wooled sheep. *Australian Veterinary Journal* 71, 161–164.

Vlassoff, A., Bissett, S.A. and McMurty, L.W. (1999) Faecal egg counts in Angora goats following natural or experimental challenge with nematode parasites: within flock variabilities and repeatabilities. *Veterinary Parasitology* 84, 113–123.

Waller, P.J. (1997) Anthelmintic resistance. *Veterinary Pathology* 72, 391–412.

Waller, P.J. and Thomas, R.J. (1978) Nematode parasitism in sheep in North-East England: the epidemiology of *Ostertagia* species. *International Journal for Parasitology* 8, 275–283.

Waller, P.J., Echevarria, F., Eddi, C., Maciel, S., Nari, A. and Hansen, J.W. (1996) The prevalence of anthelmintic resistance in nematode parasites of sheep in Southern Latin America: general overview. *Veterinary Parasitology* 62, 181–187.

Williams, H.G. and Parker, L.D. (1996) Control of sheep scab (*Psoroptes ovis*) by a single prophylactic injection of moxidectin. *Veterinary Record* 139, 598–599.

Chapter 8

The Use of Macrocyclic Lactones to Control Parasites of Horses

C.M. Monahan and T.R. Klei

Introduction

Effectiveness against the most economically important nematode parasites coupled with the lack of drug resistance to date has made the macrocyclic lactones (MLs) the industry standard for parasite control programmes for horses. Development and availability of MLs for use in equids are limited to ivermectin (IVM) and moxidectin (MOX). A review of the early development and efficacy of IVM was published in 1989 (Campbell *et al.*, 1989). Ivermectin is used in horses at a dose of 200 μg kg^{-1} body weight and is available in both a 1.87% oral paste formulation and a 1% liquid drench. The original formulation, a micellar intramuscular injectable formulation, was withdrawn from the market in 1984 due to adverse reactions (Campbell *et al.*, 1989). Some of these were associated with clostridial contamination of the injection site while others were anaphylactoid reactions. While potentially useful, other injectable products have not been developed for equids. MOX is available as a 2% oral gel. It is sold for use at a dose rate of 400 μg kg^{-1} body weight. With the recent general availability of IVM for pharmaceutical development, a mixture of IVM and praziquantel has been developed for use in equids. Marketed in a paste formulation, this product has the combined efficacy of ivermectin against nematodes and arthropods and praziquantel against tapeworms, *Anoplocephala* spp. Publications on its efficacy in peer-reviewed journals currently are not available. Although as yet not marketed in the USA or Europe, it is likely that these registrations will occur soon.

Efficacy of the MLs

Both IVM and MOX have a broad range of efficacies against adult and migrating larval stages of nematode and arthropod parasites of equids. Reported efficacies of the two products differ slightly (Table 8.1).

Nematodes

Prior to the development of MOX, the long-term use of IVM had apparently reduced the prevalence of a number of equine nematode parasites

Table 8.1. Efficacy[a] of ivermectin and moxidectin against major internal parasites of horses.

	Ivermectin	Moxidectin
Large strongyles		
Strongylus vulgaris, adult	100 (Campbell *et al*., 1989)	100 (Monahan *et al*., 1995a)
arterial larval stage	100 (Campbell *et al*., 1989)	100 (Monahan *et al*., 1995b)
S. edentatus, adult tissue	100 (Campbell *et al*., 1989)	99.9 (Monahan *et al*., 1995a)
stages	100 (Campbell *et al*., 1989)	100 (Monahan *et al*., 1995a)
S. equinus, adult	100 (Campbell *et al*., 1989)	NR
Triodontophorus spp., adult	100 (Campbell *et al*., 1989)	100 (Monahan *et al*., 1996)
Small strongyles		
Cyathastominae[b]	99 (Torbert *et al*., 1982)	99 (Monahan *et al*., 1995a)
Lumenal L_4	100 (Torbert *et al*., 1982)	99 (Monahan *et al*., 1995a)
Mucosal developing L_3 and L_4	0–77 (see Discussion)	55–84 (see Discussion)
Other nematodes		
Parascaris equorum, adult	100 (Campbell *et al*., 1989)	100 (Monahan *et al*., 1995a)
migrating larvae	100 (Campbell *et al*., 1989)	100 (Monahan *et al*., 1995a)
Oxyuris equi, adult larval	100 (Monahan *et al*., 1996)	96 (Monahan *et al*., 1995a)
stages	100 (Monahan *et al*., 1996)	99 (Monahan *et al*., 1995a)
Trichostrongylus axei, adult	100 (Torbert *et al*., 1982)	100 (Monahan *et al*., 1995a)
Harbronema muscae, adult	100 (Campbell *et al*., 1989)	100 (Monahan *et al*., 1995a)
cutaneous L_3 stage	100 (Campbell *et al*., 1989)	NR
Draschia spp., cutaneous L_3 stage	100 (Campbell *et al*., 1989)	NR
Onchocerca spp., microfilaria	100 (Campbell *et al*., 1989)	100 (Monahan *et al*., 1995c)
Dictyocaulus arnfieldi, adult and L_4 stage	100 (Campbell *et al*., 1989)	100 (Coles *et al*., 1998)
Strongyloides westeri, adult	100 (Campbell *et al*., 1989)	100 (Costa *et al*., 1998)
Bots		
Gasterophilus spp., second and third instars	100 (Campbell *et al*., 1989)	20–99 (see Discussion)

[a]Efficacy is reported in percentage reduction in selected controlled studies. In general, the highest percentage reduction reported is cited. In the case of ivermectin, many of the original papers are cited in the review by Campbell *et al.* (1989).
[b]Individual species of cyathostomes are not listed, see Discussion.
NR, data not reported.

and associated disease conditions in countries where the drug is used extensively. These include the spirurid stomach worms, *Draschia megastomum* and *Habronema* spp., *Onchocerca cervicalis* microfilariae and the large strongyles, *Strongylus vulgaris, S. edentatus* and *S. equinus*. It is likely that differences in reported efficacies between these two compounds have been affected by the availability of horses infected with a wide range of parasites for testing with MOX. In other cases, efficacies of MOX against some species such as *Stongyloides westeri* which only occurs in very young horses, and *Dictyocaulus arnfieldi* which is rare, have not been reported extensively. This is probably due to the difficulty of finding sufficient numbers of horses with these infections for experimental purposes. In other cases, some publications have shown efficacy of IVM against rarely studied parasites such as the adult stage of *Setaria equina* (Campbell *et al*., 1989). Differences in reports of efficacy against species of cyathostomes is probably due to the diversity of species recovered from horses in different trials and the different protocols used to recover and identify these parasites. Recently, it has been shown that the routine examination of 100–200 cyathostomes per large intestinal aliquot, which is the most common approach, only identifies 50–60% of the species present which are identified when large numbers (>1000) are examined from the same individual horses (Chapman *et al*., 1999a). A majority of adult cyathostome species have been identified in controlled efficacy trials with both IVM and MOX, suggesting that these compounds are equally effective against all luminal stages of cyathostomes.

It is important to note that label claims are controlled by governmental regulatory agencies, and thus may vary between countries due to legislation rather than actual differences in drug efficacy. Pharmaceutical firms will seek a label claim only for the most economically important parasites for a region; thus, label claims should not be misinterpreted as an extent of the parasite spectrum affected by a certain drug. In the case of IVM and MOX, the early success of IVM reduced the prevalence of certain equine parasites below economic importance, or below a level of sufficient numbers to perform viable drug trials with MOX against these species. The omission of some equine parasites on a label claim for MOX is most likely to represent this decline in prevalence or economic importance, rather than lack of efficacy. Thus, it is likely that, with a few exceptions, the efficacy of IVM (200 $\mu g\ kg^{-1}$ body weight) and MOX (400 $\mu g\ kg^{-1}$ body weight) against internal adult parasites are essentially the same. Significant differences in efficacy of IVM and MOX have been noted against the mucosal larval stages of the cyathostomes and larvae of *Gasterophilus* spp.

Activity against the mucosal larval stages has been difficult to assess and can be influenced by the post-treatment necropsy examination time point as well as by methods used to count these larvae (Chapman *et al*., 1999b; Eysker and Klei, 1999). It is important to note that these techniques are only now beginning to become minimally standardized. Stages found in the mucosa, and often designated as mucosal larvae or encysted larvae, include the early hypobiotic third stage (EL_3), the developing third stage (LL_3) and the developing fourth stage (DL_4) (Chapman *et al.*, 1994). The EL_3, LL_3 and DL_4 can be counted following digestion of the mucosa. Transmural illumination (TMI) of the intestine can be used to count but not differentiate the LL_3 and DL_4, but not count the EL_3. When these data are reported, the LL_3 and DL_4 are generally considered together as developing larvae (DL). Earlier work on MOX activity against DL reported some efficacy of the 400 $\mu g\ kg^{-1}$ dosage when evaluated at a 2-week post-treatment time point. These efficacies were 63% (Xiao *et al.*, 1994), 84% (Monahan *et al.*, 1995a) and 87% (Monahan *et al.*, 1996) against DL, as determined by TMI; and 86% (Monahan *et al.*, 1995a) and 50% (Monahan *et al.*, 1996) against DL as determined by digestion. Efficacy against EL_3 (0–37%) in these studies was not significant. Questions were raised with regards to the sensitivity of differentiating moribund larvae from viable larvae when necropsies were conducted at 2 weeks post-treatment. This 2-week post-treatment period was extended to 35 days to allow the dying larvae to be more recognizable as such (Eysker *et al.*, 1997). These studies demonstrated a 90% efficacy against large developing L_4 but no effect on the total DL population or the EL_3. Extending the post-treatment necropsy time point to 5 weeks, however, may also obscure the true activity of MOX. Removal of the LL_3 or DL_4 subsets by an effective drug treatment could then trigger the unaffected and larger population of EL_3 to resume development to LL_3, DL_4 or luminal L_4 stages during that 5-week period (Eysker *et al.*, 1997).

In comparison with MOX, IVM has not shown significant efficacy against the encysted cyathostomes at therapeutic dosages (Eysker *et al.*, 1992; Xiao *et al.*, 1994; Monahan *et al.*, 1996), nor at elevated dosages (Klei *et al.*, 1993). The single report of efficacy against mucosal larvae was measured only against LL_3 and DL_4 following an experimental infection (Love *et al.*, 1995). Although the latter data are interesting, further investigations on this phenomenon are necessary. The pharmacokinetic data (Perez *et al.*, 1999) may explain this difference in activities: the terminal half-life of IVM is 4.3 days compared with 23.1 days for MOX. This suggests that the plasma bioavailability of IVM has fallen below an effective concentration by the time encysted or hypobiotic cyathostomes receive signals to resume development, but the plasma persistence of MOX removes many of the reactivated larvae until it itself falls below that threshold. The suboptimal concentrations of MOX that persist in horses longer than day 80 (Perez *et al.*, 1999) could select for drug resistance among this pool of reactivated cyathostome larvae (Sangster, 1999).

Based on evidence from an experimental challenge to assess the residual activity of MOX, Vercruysse and co-workers estimated that the persistence of MOX in the circulation can provide horses with 2–3 weeks of protection from infective cyathostome larvae (Vercruysse *et al.*, 1998). This protection may contribute to the prolonged egg reappearance period (ERP) achieved following use of MOX versus IVM; however, this also highlights that past this 2–3 week protection period, suboptimal concentrations of MOX will be in circulation that do not provide protection from cyathostomes but which could select for resistance to the entire class of ML. Although similar persistence studies have not been reported for IVM, observations on experimental *S. vulgaris* infections have suggested a 2-week persistence of activity of IVM against the migrating stages of this nematode (Slocombe *et al.*, 1984).

Arthropods

External parasites of horses, such as mites and ticks, have received much less attention in controlled efficacy trials, and little is published in this area (Campbell *et al.*, 1989). None the less, based on these limited reports and efficacy in other species, IVM is recommended for the treatment of equine sarcoptic, psoroptic and chorioptic mange (Logas and Barbet, 1999). The impact of MOX in control of these infections is unclear.

Clear differences in IVM and MOX efficacy that have emerged are the levels of activity of the two drugs against *Gasterophilus* larvae. In almost all reports, IVM is very active (>99%) against all instars of *Gasterophilus intestinalis* (Campbell *et al.*, 1989). The activity of MOX, however, is variable and ranges between 20 and 100%, and activity has not been demonstrated against the first instars. This variability in the reported efficacy of MOX may relate to testing protocols used, particularly as this relates to treatment at times of the year when the third instars may not be actively feeding (Scholl *et al.*, 1998). For practical purposes, these differences are minor and the activity of MOX against bot fly larvae, shown in most reports to be greater than 90%, is more than acceptable for control purposes, particularly when one considers the absence of evidence that these parasites are of major significance (Blagburn *et al.*, 1991).

The Control of Nematodes and Egg Reappearance Periods (ERPs) of the MLs

Considerable debate has taken place in the equine literature over the strengths and weaknesses of IVM or MOX when used in strategic control programmes for horses. Anthelmintic programmes, once based on fixed intervals of anthelmintic administration during the grazing season (reviewed by Bennett, 1983), more recently have been based on strategic treatments delivered to all horses in a herd and targeted for control of both large and small strongyles (reviewed by Ewert *et al.*, 1991; Klei, 1997;

Hutchens *et al.*, 1999). These recent recommendations have addressed the rotation of drug classes that are utilized in order to target the entire spectrum of important equine parasites and to avoid the development of drug resistance by the small strongyles. Ideally, the fixed time interval would fall within the ERP, a hypothetical time period between a zero faecal egg count following an effective treatment and the return of strongyle eggs in a faecal sample. These intervals differ depending on the anthelmintic employed. Pyrantel salts provide the shortest ERP (4–6 weeks), benzimidazoles an intermediate ERP (up to 8 weeks), and MLs the longest ERP (>8–10 weeks). The goal of a strategic treatment programme is to reduce or eliminate the developing subadult nematodes before sexual maturity leads to egg production responsible for pasture contamination. By controlling pasture contamination, the strategic treatment approach prevents heavy infections, clinical parasitism and associated losses in all the animals of a herd. Under a strategic programme, all members of a herd are treated, rather than an individual within the herd. Anthelmintic treatment of an individual with clinical parasitism has been termed a 'therapeutic treatment' or 'salvage treatment'. Therapeutic treatment may remove the parasitic stages and temporarily alleviate the symptoms, but has very little effect on the level of free-living infective larvae on pasture. For this reason, strategic treatments at the end of the ERP are preferable for the health of all the occupants of the pasture. Compounds that extend the ERP are clinically advantageous because fewer treatments are needed during the course of a grazing season. Several reports have cited different ERPs for IVM and MOX, but some examination of the methodology used to determine the ERP is needed for valid comparisons to be made between IVM and MOX.

Intuitively, the ERP represents a theoretical time point at which eggs reappear in faecal samples after an effective anthelmintic treatment reduced their numbers to zero. Complete efficacy in strongyle egg elimination is not always achieved by any treatment, and eggs may be found at low levels in faecal samples within days or weeks following treatment. For this reason, Herd defined the ERP as the time from administration to the reappearance of an arithmetic mean value of 100 eggs per gram (EPG) in the test herd (Herd, 1990). Several authors have used this as the measure of the ERP (Boersema *et al.*, 1995, 1996), although this definition has not been adopted as a standard measure within the literature. Some authors cite the return to an arithmetic mean value of 200 EPG in 25% of the horses as marking the ERP (Uhlinger, 1990); others have used a return to an arithmetic mean EPG for the herd of 200 EPG with 50% of the animals demonstrating eggs in the faeces (Demeulenaere *et al.*, 1997). Still other authors cite the return to a geometric mean value of 200 EPG with 50% positive for strongyle eggs. Some authors do not define their criteria for the ERP in their materials and methods. At a recent parasitology conference, a non-scientific survey of several attending equine parasitologists provided as many definitions for the ERP as parasitologists questioned. A non-standard definition of the ERP can affect reports of the ERP after treatment with any compound, including IVM or MOX. A return to EPG levels of an arithmetic mean value of 200 in 25% of the herd may occur at a different time point than the geometric mean of 100 EPG. These differences are most important when comparing reports in the literature that tested either IVM or MOX independently.

Additional differences of the herd composition, climatic conditions or previous anthelmintic treatments during the previous grazing season will also affect reports of the ERP. The age composition of the sampled population will directly affect the length of the ERP. Herd and Gabel (1990) demonstrated that young horses develop higher EPG at earlier time points than older horses, findings that were supported by Uhlinger (1992), and Boersema and co-workers (Boersema *et al.*, 1996). The relative stratification of the sampled herd will influence either the mean EPG of the herd, or the time when a fixed percentage of the herd returns to a predetermined EPG level. This factor can complicate the interpretation of reported ERP for different compounds unless the data to be compared are collected from the same herd under the same conditions.

The anthelmintic regime employed previously will also affect the ERP during the subsequent grazing season (Uhlinger, 1992). During a multiyear study of six horse farms, treatment intervals based on ERP lengthened over the course of the study, possibly as a result of lower pasture contamination and lower challenge infections. The ERP for IVM during this study ranged from 6 to 18 weeks, and was influenced by the age stratification of the herd.

Studies using IVM or MOX that are performed using concurrent administrations can control for many, but not all, of the variables that can influence the ERP. Based on this premise, several reports of the difference in ERP following concurrent treatments with either IVM or MOX should be considered. Taylor and Kenny (1995) reported the ERP of IVM to be 8 weeks, whereas the ERP for MOX was 15 weeks. Jacobs and co-workers found the ERP for IVM to be 14 weeks and for MOX to be more than 24 weeks (Jacobs *et al.*, 1995). Demeulenaere and co-workers demonstrated an ERP for IVM ranging between 10 and 13 weeks, and that of MOX ranging between 22 and 24 weeks (Demeulenaere *et al.*, 1997). Differences in the experimental design, ages and management of the horses, as well as the definition of the ERP between these studies can account for some interstudy variation, but this does highlight that use of MOX prolongs the ERP and increases the treatment interval.

Differences in reported ERP as well as potential activity against mucosal stages of the cyathostomes of horses may be partially explained by a comparison of the pharmacokinetics of MOX and IVM (Marriner *et al.*, 1987; Afzal *et al.*, 1997; Perez *et al.*, 1999).

Pharmacokinetics of MLs in Horses

Both IVM and MOX have similar chemical structures and share a common mechanism of action involving disruption of nerve conduction of nematodes (Conder *et al.*, 1993; Shoop, 1993; McKellar and Benchaoui, 1996; see Chapter 3). Marriner *et al.* (1987) compared subcutaneous injection with oral administration of IVM in horses. The study of Perez *et al.* (1999) compared oral formulations of MOX and IVM concurrently, whereas the study of Afzal and co-workers compared oral administration and intravenous injection of MOX (Afzal *et al.*, 1997). Collectively, these studies revealed that IVM and MOX are absorbed following oral administration and both reach peak plasma concentrations between 6 and 8 h following oral administration (Perez *et al.*, 1999). IVM is slower to reach peak plasma concentrations following subcutaneous administration than by the oral route (Marriner *et al.*, 1987). This is also true when comparing oral with subcutaneous injections of IVM in sheep (Shoop *et al.*, 1997). In contrast, MOX is rapidly absorbed from the site of injection in cattle, and IVM has the slower time to peak plasma concentrations (Lanusse *et al.*, 1997). Following oral administration in horses, peak concentrations of MOX are higher than those of IVM, as would be expected from the higher dosage formulation commercially available (400 μg kg^{-1} for MOX versus 200 μg kg^{-1} for IVM). This adds to a greater area under the plasma concentration curve for MOX versus IVM via the oral route (Perez *et al.*, 1999). Following peak plasma concentrations, the distribution of MOX differs significantly from IVM, showing a two-compartment elimination consistent with the results found by Afzal and co-workers examining MOX alone (Afzal *et al.*, 1997). This distribution of MOX resulted in a fourfold difference in mean plasma residence time (18.4 days for MOX versus 4.8 days for IVM), and a longer terminal half-life (23.1 days for MOX versus 4.3 days for IVM) (Perez *et al.*, 1999). Additionally, the plasma concentration of MOX remained above 1 ng ml^{-1} for more than 80 days (the limit of the study), but IVM fell below this value in fewer than 20 days (Perez *et al.*, 1999). This multiphasic distribution of MOX in horses is consistent with previous findings, although the study of Afzal and co-workers terminated at 168 h after administration (Afzal *et al.*, 1997). A similar difference between IVM and MOX was found to exist in sheep following oral adminstration (Shoop *et al.*, 1997).

The longer plasma bioavailability of MOX in horses may account for reports of the longer ERP following administration of MOX compared with groups treated simultaneously with IVM (Jacobs *et al.*, 1995; Taylor and Kenny, 1995; Demeulenaere *et al.*, 1997). As encysted, hypobiotic cyathostomes receive signals to resume development following removal of luminal stages, higher concentrations of MOX persist in the plasma to target those activated larvae. Based on the work by Vercruysse and co-workers, this activity can last between 2 and 3 weeks, after which time the circulating levels may not be sufficient to affect reactivated larvae (Vercruysse *et al.*, 1998). The protocol of Vercruysse utilized experimental challenge of parasite-free foals, thus the protection afforded to the treatment groups may have been directed towards the exsheathed L_3s rather than reactivated larvae within tissues. Due to the complexity inherent in studying the cyathostome life cycle, this question may never be answered. None the less, the persistent levels of MOX appear to provide additional protection, but risk the selection for ML resistance during the prolonged period of suboptimal concentrations.

This prolonged plasma retention of MOX raises fears that avermectin-resistant cyathostomes will be selected during the period of suboptimal drug levels as voiced by Shoop and co-workers regarding trichostrongylid nematodes of sheep (Shoop *et al.*, 1997), and addressed by Sangster regarding cyathostomes of horses (Sangster, 1999). Plasma persistence of anthelmintics has been proposed as a major factor contributing to selection for drug resistance (Dobson *et al.*, 1996; Sangster and Gill, 1999). Use of IVM in horses since the early 1980s has not resulted in selection for drug-resistant cyathostomes, perhaps due to the short retention time and negligible effects on the mucosal stages. These important differences can be exploited when designing anthelmintic control programmes, and the potential for drug resistance can be reduced by judicious use of MOX, as will be discussed.

Resistance

Despite the intensive use of IVM on many horse farms since the advent of IVM in the early 1980s, there has never been a demonstration of drug-resistant small strongyles in horses. An explanation proposed for this phenomenon is that the encysted and hypobiotic EL_3, LL_3 and DL_4 stages of cyathostomes represent the majority of the cyathostome gene pool and these stages are not subjected to selection pressure due to the short terminal half-life of IVM following oral administration. These mucosal larvae represent a refugium of the cyathostome gene pool in a similar fashion to the population of infective larvae on pasture (Sangster, 1999). Due to this refugium of unselected, hypobiotic cyathostomes, any larvae reactivated for development will most probably be as susceptible to IVM as the adult population just removed by treatment. In contrast, the persistent plasma time of MOX could subject any reactivated larvae to selection pressure, theoretically resulting in an adult population less susceptible to treatment than the adults just removed by MOX treatment. This is the basis for concerns that widespread use of MOX could lead to drug resistance for the entire class of MLs.

Although the concern about drug selection for resistant populations is real, Eysker and co-workers have suggested that MOX exerts its principal effect against the DL_4 and not against the EL_3 when examined at 5-week post-treatment intervals (Eysker *et al.*, 1997). As the EL_3 constitute the majority of mucosal larvae, lack of MOX activity against EL_3 may translate into negligible selection pressure in spite of the prolonged bioavailability of MOX versus IVM. Additionally, the extended ERP following treatment with MOX could reduce the overall number of anthelmintic treatments, thus reducing the exposure of larvae to drug selection. Judicious use of MOX in a portion of the herd may be an alternative that further reduces the likelihood of drug resistance. By restricting the use of MOX to younger animals (1–6 years), with older animals in the herd receiving IVM treatments, the overall cyathostome genome in a horse herd would not receive identical selection pressure, thus the establishment of a resistance gene in those cyathostomes would be less possible. As the younger horses in a herd are more likely to benefit from some activity against the mucosal stages than older horses with an acquired resistance to infection, such restriction of MOX usage would target the most needy population. Additionally, younger horses have a shorter ERP following treatment than older horses (Herd and Gabel, 1990); thus, the use of MOX in younger horses while treating older horses with IVM could lead to a more uniform ERP in a herd with wide age stratification.

Use of MLs in Control Programmes

Internal parasite control programmes for horses in developed countries currently target cyathostomes because of their ubiquitous presence and potential for induction of acute and chronic disease. In recent years, most programmes emphasized the rotation of drug classes to avoid the development of drug resistance in the cyathostomes, as has been seen with benzimidazoles (reviewed by Bennett, 1983; Ewert *et al.*, 1991; Klei, 1997; Hutchens *et al.*, 1999). Rotation will expose the population to different selection pressures based on the mechanism of action for the drug class. Theoretically, this will reduce the chance of drug resistance developing to any single class of anthelmintic. Programmes can be classified as follows: interval treatment or fast rotation; annual rotation or slow rotation; non-rotational; selective treatment of infected individuals in a given herd; or strategic programmes which are designed to eliminate parasites and avoid pasture contamination. The pros and cons of these programmes and the drug classes available for use have been reviewed (Klei, 1997). Both IVM and MOX can be used effectively in a strategic control programme.

The exclusive use of IVM on horse farms has not been effective for controlling the entire spectrum of equine parasites because the MLs have no efficacy against cestodes, such as *Anoplocephala perfoliata*, and the proposed significance of this parasite in the induction of colic recently has gained support (Proudman and Trees, 1999). The use of higher dosages of pyrantel pamoate or prolonged administration of benzimidazoles have been recommended at some time point during the grazing season to target the equine tapeworms (French and Chapman, 1992). Praziquantel is also effective for removal of *A. perfoliata* (Lyons *et al.*, 1995, 1998), and approval for a combination product of IVM and praziquantel is available in some regions of the world. This combination would effectively target the spectrum of important adult equine helminths but would not remove the mucosal stages of cyathostomes.

Because the greatest susceptibility to heavy mucosal burdens is seen in younger horses that have not yet developed resistance to heavy infections (Klei, 1992; Klei and Chapman, 1999), the use of MOX in this age group may be advantageous but must be weighed against the risk of drug resistance. An advantage to the use of MOX in a strategic control programme is the prolonged ERP following treatment, which would reduce the number of treatments during the course of a grazing season. Treatment frequency has been cited as a major factor in the establishment of drug resistance within a nematode population (Sangster, 1999). To benefit from this ERP extension, a veterinary practitioner may need to monitor the faeces for the return of strongyle eggs rather than retreating at a defined interval.

For removal of the mucosal cyathostomes, a 5-day course of fenbendazole appears to be more effective than a single treatment with MOX (DiPietro *et al.*, 1997; Duncan *et al.*, 1998), because fenbendazole was efficacious against the EL_3 stages that MOX does not affect. This regimen of multiple, elevated dosages was effective against a population of cyathostomes that showed resistance to therapeutic doses of fenbendazole (DiPietro *et al.*, 1997). As benzimidazole resistance is widespread within the cyathostomes, these findings suggest that use of benzimidazoles could be restricted to a single annual course of treatment of the younger, more susceptible horses in the herd, and IVM could be used as the principal anthelmintic for the rest of the herd. Such a strategic programme would avoid the potential selection pressure posed by use of MOX while accomplishing more control than is possible by use of MOX alone. Larvicidal treatments of younger horses may also speed the development of acquired resistance in the younger members of the herd because these larvae would die *in situ*, theoretically supplying the immune system with antigen during their decay that could generate an anamnestic response during future infections.

As noted in previous reviews of anthelmintic control programmes (Bennett, 1983; Ewert *et al.*, 1991; Klei, 1997; Hutchens *et al.*, 1999), there is no single regimen applicable to every stable or farm. Veterinary practitioners must weigh the advantages of anthelmintic formulations against the disadvantages each may have in terms of the spectrum of activity or risk of drug resistance. The needs or risks will differ between the casual horse owner with a backyard pasture, or a breeding farm with a wide age stratification, or a training facility needing to maximize the athletic performance of the resident horses. Judicious use of anthelmintics coupled with a monitoring programme to follow the ERP would maximize the returns and minimize the potential for drug resistance. For any size horse operation, the expense of monitoring the programme is compensated by the savings in reduced numbers of anthelmintic treatments and the gains in the health status of the resident horses.

References

Afzal, J., Burke, A.B., Batten, P.L., DeLay, R.L. and Miller, P. (1997) Moxidectin: metabolic fate and blood pharmacokinetics of ^{14}C-labeled moxidectin in horses. *Journal of Agricultural and Food Chemistry* 45, 3627–3633.

Bennett, D.G. (1983) Drug resistance and the control of equine strongyles. *Compendium of Continuing Education for the Practicing Veterinarian*, Supplement 5, S343–S349.

Blagburn, B.L., Lindsay, D.S., Hendrix, C.M. and Schumacher, J. (1991) Pathogenesis, treatment and control of gastric parasites in horses. *Compendium of Continuing Education for the Practicing Veterinarian* 13, 850–857.

Boersema, J.H., Borgsteede, F.H.M., Eysker, M. and Saedt, I. (1995) The reappearance of eggs in faeces of horses after treatment with pyrantel embonate. *Veterinary Quarterly* 17, 18–20.

Boersema, J.H., Eysker, M., Maas, J. and van der Aar, W.M. (1996) Comparison of the reappearance of strongyle eggs in foals, yearlings and adult horses after treatment with ivermectin or pyrantel. *Veterinary Quarterly* 17, 18–20.

Campbell, W.C., Leaning, W.H.D. and Seward, R.L. (1989) Use of ivermectin in horses. In: Campbell, W.C. (ed.) *Ivermectin and Abomectin*. Springer-Verlag, New York, pp. 234–244.

Chapman, M.R., Hutchinson, G.W., Cenac, M.J. and Klei, T.R. (1994) *In vitro* culture of equine Strongylidae to the fourth larval state in a cell-free medium. *Journal of Parasitology* 80, 225–231.

Chapman, M.R., French, D.D. and Klei, T.R. (1999a) Intestinal helminths of ponies; a comparison of species prevalent in Louisiana pre- and post-ivermectin. *Proceedings of the American Association of Veterinary Parasitologists 44th Annual Meeting*, New Orleans, Louisiana, 10–13 July, p. 74.

Chapman, M.R., Kearney, M.T. and Klei, T.R. (1999b) An experimental evaluation of methods used to enumerate mucosal cyathostome larvae in ponies. *Veterinary Parasitology* 86, 191–202.

Coles, G.C., Hillyer, M.H., Taylor, F.G. and Parker, L.D. (1998) Activity of moxidectin against bots and lungworms in equids. *Veterinary Record* 6, 169–170.

Conder, G.A., Thompson, D.P. and Johnson, S.S. (1993) Demonstration of co-resistance of *Haemonchus contortus* to ivermectin and moxidectin. *Veterinary Record* 132, 651–652.

Costa, A.J., Barbosa, O.F., Moraes, F.R., Acuna, A.H., Rocha, U.F., Soares, V.E., Paullilo, A.C. and Sanches, A. (1998) Comparative efficacy of mosicectin gel and ivermectin paste against internal parasites of equines in Brazil. *Veterinary Parasitology* 80, 29–36.

Demeulenaere, D., Vercruysse, J., Dorny, P. and Claerebout, E. (1997) Comparative studies of ivermectin and moxidectin in the control of naturally acquired cyathostome infections in horses. *Veterinary Record* 141, 383–386.

DiPietro, J.A., Klei, T.R. and Reinemeyer, C.R. (1997) Efficacy of fenbendazole against encysted small strongyle larvae. *American Association of Equine Practitioners, Proceedings* 43, 343–344.

Dobson, R.J., LeJambre, L. and Gill, L. (1996) Management of anthelmintic resistance and selection with persistent drugs. *International Journal for Parasitology* 26, 993–1000.

Duncan, J.L., Bairden, K. and Abbott, E.M. (1998) Elimination of mucosal cyathostome larvae by five daily treatments with fenbendazole. *Veterinary Record* 142, 268–271.

Ewert, K.M., DiPietro, J.A., Foreman, J.H. and Todd, K.S. (1991) Control programs for endoparasites in horses. *Compendium of Continuing Education for the Practicing Veterinarian* 13, 1012–1018.

Eysker, M. and Klei, T.R. (1999) Mucosal larval recovery techniques of cyathostomes: can they be standardized? *Veterinary Parasitology* 85, 137–144.

Eysker, M., Boersema, J.H. and Kooyman, F.N.J. (1992) The effect of ivermectin treatment against inhibited early third stage, late third stage and fourth stage larvae and adult stages of the cyathostomes in Shetland ponies and spontaneous expulsion of these helminths. *Veterinary Parasitology* 42, 295–302.

Eysker, M., Boersema, J.H., Grinwis, G.C.M., Kooyman, F.N.J. and Poot, J. (1997) Controlled dose confirmation study of a 2% moxidectin equine gel against equine internal parasites in The Netherlands. *Veterinary Parasitology* 70, 165–173.

French, D.D. and Chapman, M.R. (1992) Tapeworms of the equine gastrointestinal tract. *Compendium of Continuing Education for the Practicing Veterinarian* 44, 655–662.

Herd, R.P. (1990) The changing world of worms: the rise of cyathostomes and the decline of *Strongylus vulgaris*. *Compendium of Continuing Education for the Practicing Veterinarian* 12, 732–736.

Herd, R.P. and Gabel, A.A. (1990) Reduced efficacy of anthelmintics in young compared with adult horses. *Equine Veterinary Journal* 22, 164–169.

Hutchens, D.E., Paul, A.J. and DiPietro, J.A. (1999) Treatment and control of gastrointestinal parasites. *Veterinary Clinics of North America: Equine Practice* 15, 561–573.

Jacobs, D.E., Hutchinson, M.J., Parker, L. and Gibbons, L.M. (1995) Equine cyathostome infection: suppression of faecal egg output with moxidectin. *Veterinary Record* 137, 545.

Klei, T.R. (1992) Recent observations on the epidemiology, pathogenesis, and immunology of equine helminth infections. In: Ploughwright, W., Rossdale, P.O. and Wade, J.F. (eds) *Equine Infectious Diseases, VI*. R and W Publications, Newmarket, UK, pp. 129–136.

Klei, T.R. (1997) Parasite control programs. In: Robinson, N.E. (ed.) *Current Therapy in Equine Medicine 4*. W.B. Saunders and Co., Philadelphia, Pennsylvania, pp. 709–713.

Klei, T.R. and Chapman, M.R. (1999) Immunity in equine cyathostome infections. *Veterinary Parasitology* 85, 123–133.

Klei, T.R., Chapman, M.R., French, D.D. and Taylor, H.W. (1993) Evaluation of ivermectin at an elevated dose against encysted equine cyathostome larvae. *Veterinary Parasitology* 47, 99–106.

Lanusse, C., Lifshitz, A., Virkel, G., Alvarez, L., Sanchez, S., Sutra, J.F., Galtier, P. and Alvinare, M. (1997) Comparative plasma disposition kinetics of ivermectin, moxidectin and doramectin in cattle. *Journal of Veterinary Pharmacology* 20, 91–99.

Logas, D.B. and Barbet, J.L. (1999) Diseases characterised by pruritus and hair loss. In: Colahan, P.T., Mayhew, I.G., Merritt, A.M. and Moore, J.N. (eds) *Equine Medicine and Surgery*, Vol. II, 5th edn. Mosby, New York, pp. 1901–1909.

Love, S., Duncan, J.L., Parry, J.M. and Grimshaw, W.T.R. (1995) Efficacy of oral ivermectin paste against mucosal stages of cyathostomes. *Veterinary Record* 136, 18–19.

Lyons, E.T., Tolliver, S.C., Stamper, S.S., Drudge, J.H., Granstrom, D.E. and Collins, S.S. (1995) Activity of praziquantel (0.5 mg/kg) against *Anoplocephala perfoliata* (Cestoda) in equids. *Veterinary Parasitology* 56, 255–257.

Lyons, E.T., Tolliver, S.C. and Ennis, L.E. (1998) Efficacy of praziquantel (0.25 mg/kg) on the cecal tapeworm (*Anoplocephala perfoliata*) in horses. *Veterinary Parasitology* 78, 287–289.

Marriner, S.E., McKinnon, I. and Bogan, J.A. (1987) The pharmacokinetics of ivermectin after oral and subcutaneous administration to sheep and horses. *Journal of Veterinary Pharmacology and Therapeutics* 10, 175–179.

McKellar, Q.A. and Benchaoui, H.A. (1996) Avermectins and milbemycins. *Journal of Veterinary Pharmacology and Therapeutics* 19, 331–351.

Monahan, C.M., Chapman, M.R., French, D.D., Taylor, H.W. and Klei, T.R. (1995a) Dose titration of moxidectin oral gel against gastrointestinal parasites of ponies. *Veterinary Parasitology* 59, 241–248

Monahan, C.M., Chapman, M.R., Taylor, H.W., French, D.D. and Klei, T.R. (1995b) Dose titration of moxidectin oral gel against migrating *Strongylus vulgaris* and *Parascaris equorum* larvae in pony foals. *Veterinary Parsitology* 60, 103–110.

Monahan, C.M., Chapman, M.R., French, D.D. and Klei, T.R. (1995c) Efficacy of moxidectin oral gel against *Onchocerca cervicalis* microfilariae. *Journal of Parasitology* 81, 117–118.

Monahan, C.M., Chapman, M.R., Taylor, H.W., French, D.D. and Klei, T.R. (1996) Comparison of moxidectin oral gel and vermectin oral paste against a spectrum of internal parasites of ponies with special attention to encysted cyathastome larvae. *Veterinary Parasitology* 63, 225–235.

Perez, R., Cabezas, I., Garcia, M., Rubilar, L., Sutra, J.F., Galtier, P. and Alvinerie, M. (1999) Comparison of the pharmacokinetics of moxidectin (Equest) and ivermectin (Equalan) in horses. *Journal of Veterinary Pharmacology and Therapeutics* 22, 174–180.

Proudnam, C.J. and Trees, A.J. (1999) Tapeworms as a cause of intestinal disease in horses. *Parasitology Today* 15, 156–159.

Sangster, N.C. (1999) Pharmacology of anthelmintic resistance in cyathostomes: will it occur with the avermectin/milbemycins? *Veterinary Parasitology* 85, 189–204.

Sangster, N.C. and Gill, J. (1999) Pharmacology of anthelmintic resistance. *Parasitology Today* 15, 141–146.

Scholl, P.J., Chapman, M.R., French, D.D. and Klei, T.R. (1998) Efficacy of moxidectin 2% oral gel against second and third instars of *Gasterophilus intestinalis* De Geer. *Journal of Parasitology* 84, 656–657.

Shoop, W.L. (1993) Ivermectin resistance. *Parasitology Today* 9, 154–159.

Shoop, W.L., Michael, B.F., Haines, H.W., Murphy, T.P., Faidley, T.D., Hajdu, R. and Thompson, D.R. (1997) Moxidectin and ivermectin in lambs: plasma depletion and efficacy against helminths. *Journal of Veterinary Pharmacology and Therapeutics*, Supplement 20, 10–19.

Slocombe, J.O.D., McGraw, B.W., Pennock, P.W. and Vasey, J. (1984) Ivermectin in the control of *Strongylus vulgaris* in the equine. *Proceedings of the MSD AGVET Symposium: Recent Developments in the Control of Animal Parasites, XXII World Veterinary Congress*, Perth, Australia, pp. 265–270.

Taylor, S.M. and Kenny, J. (1995) Comparison of moxidectin with ivermectin and pyrantel embonate for reduction of fecal egg counts in horses. *Veterinary Record* 137, 516–518.

Torbert, B.J., Kramer, B.S. and Klei, T.R. (1982) Efficacy of injectable and oral paste formulations of ivermectin against gastrointestinal parasites in ponies. *American Journal of Veterinary Research* 43, 1451–1453.

Uhlinger, C. (1990) Effects of three anthelmintic schedules on the incidence of colic in horses. *Equine Veterinary Journal* 4, 251–255.

Uhlinger, C. (1992) Preliminary studies into factors affecting the variability of egg reappearance period and anthelmintic treatment intervals in the control of equine cyathostomes. In: Ploughwright, W., Rossdale, P.O. and Wade, J.F. (eds) *Equine Infectious Diseases, IV*. R and W Publications, Newmarket, UK, pp. 157–161.

Vercruysse, J., Eysker, M., Deleulenaere, D., Smits, K. and Dorny, P. (1998) Persistence of the efficacy of a moxidectin gel on the establishment of cyatostominae in horses. *Veterinary Record* 143, 307–309.

Xiao, L., Herd, R.P. and Majewski, G.A. (1994) Comparative efficacy of moxidectin and ivermectin against hypobiotic and encysted cyathostomes and other equine parasites. *Veterinary Parasitology* 53, 83–90.

Chapter 9

The Use of Macrocyclic Lactones to Control Parasites of Pigs

J. Arends and J. Vercruysse

Introduction

The species of endoparasites found in swine depend upon the physical location of the animals and, more specifically, the rearing system utilized. In general, in the rearing systems currently in use, the number of endoparasite species infesting swine will be fewest in total confinement, with a larger variety of species found as one moves from total confinement to a more extensive management system. The total number of any specific parasite is dependent upon the age of the host and the exposure to either eggs or infective larvae, the flooring system used and the manure management system.

Ectoparasites of swine are restricted to two species, *Haematopinus suis*, the sucking louse, and sarcoptic mange, *Sarcoptes scabiei* var. *suis*. Other ectoparasites such as fleas and ticks are rarely found in modern production systems.

Ivermectin has been used in swine as an injectable formulation, with a subcutaneous injection site, and as an in-feed formulation for breeding and growing swine. Generic formulations (e.g. ivermectin) are entering the market place as the patent expirations allow. Doramectin has been used in breeding and growing swine as an injectable formulation, both subcutaneous and intramuscular. Both ivermectin and doramectin are marketed in their formulations on a worldwide basis. Moxidectin has been evaluated in swine as an injectable and as a pour-on, and is marketed as an injectable in 16 nations in Eastern Europe. The recommended dose for the injectable formulations is 0.3 mg kg^{-1}. The in-feed formulation of ivermectin is given over 7 consecutive days at 0.1 mg kg^{-1} day^{-1}.

Efficacy of the Macrocyclic Lactones (MLs) against Helminths

Helminths of pigs

In general, the MLs have shown good efficacy against helminths of swine. As a group, efficacy against *Trichuris* has been variable, and remains the dose-limiting helminth for swine. The MLs have been evaluated against: *Oesophagostomum* spp., *Ascaris suum, Strongyloides ransomi, Trichuris suis, Metastrongylus* spp., *Stephanurus dentatus* and *Hyostrongylus rubidus*. Studies evaluating the efficacy and duration of activity of the MLs were conducted on induced infections and naturally acquired infections.

Ascaris, Oesophagostomum spp, *Trichuris* and *Hyostrongylus* all have life cycles that are similar in nature. Infections are acquired by the consumption of embryonated eggs or larvae that are deposited in the faeces and the environment. Following development to adult stage, mating takes place and eggs are shed into the gut where they are passed in the faeces. *S. ransomi* has a distinctive life cycle and is primarily a parasite of neonatal piglets. Infective larvae can be transported across the placenta or in sows' milk, ensuring that piglets are infected at birth, or piglets can be infected from L_3 larvae in the piglets' environment, which will infect the animals via trans-dermal penetrations. *Metastrongylus* spp., the lungworms, require an intermediate host, an earthworm. The L_3 larvae are released after the host has consumed the infected worm. *S. dentatus*, the kidney worm, also requires an intermediate host, generally a beetle. Following the consumption of the infected intermediate host, the kidney worm can complete its development. The adult worms are found in the ureter or peri-renal fat, and eggs are released into the urinary tract and expelled in the urine.

Efficacy of MLs

The efficacy of all MLs has been studied extensively against several worm species, but the economically important and most commonly found parasites in modern production systems are *Ascaris, Oesophagostomum* spp. and, to a lesser extent, *Trichuris*. All other swine internal parasites, while still found, are not found in the production systems that produce virtually 99% of the swine.

Ivermectin injectable

Ivermectin given at a dose of 300 μg kg^{-1} via a subcutaneous injection has been shown to have efficacy against the majority of adult and larval forms of gastrointestinal parasites of swine. Efficacy against *A. suum* was more than 99%; *H. rubidus*, 98%; *Oesophagostomum* spp., 95%; *S. ransomi*, 99%; *Metastrongylus* spp., 100%; and variable efficacy against *T. suis*, 60–95%

(Benz *et al.*, 1989). Other parasites such as *Ascarops* and *Stephanurus* have been evaluated, where efficacy is greater than 99%. Stewart *et al.* (1991) evaluated efficacy of injectable ivermectin against pigs that were infected with round, nodular and thread worms. These pigs were held on solid concrete floors, orally dosed with *A. suum* and *Oesophagostomum* and dosed subcutaneously with *Strongyloides*. At day 34, they were treated with ivermectin injectable and were euthanized on day 75 and data taken: 7% of the pigs harboured round worms, 40% thread worms and 47% harboured nodular worm when necropsy was completed at the end of the grow-out period.

Efficacy against *Oesophagostomum* spp. has been variable. Varady *et al.* (1996) evaluated the efficacy of ivermectin at 150, 300 and 600 μg kg^{-1} against *O. dentatum* and *O. quadrispiculatum* and found that ivermectin was less effective against *O. quadrispiculatum* (53.2% reduction) compared with *O. dentatum* (96.9% reduction) at 300 μg kg^{-1}. They also found a higher efficacy of ivermectin against immature stages than adults, which agreed with observations reported by Stewart *et al.* (1981).

The effect that ivermectin would have on the embryonation and infectivity of helminths of swine has been evaluated for *Ascaris* and *Trichuris*. Boes *et al.* (1998) evaluated the impact of ivermectin on *A. suum* eggs by collecting expelled female worms (post-treatment), removing the eggs and then injecting the embryonated eggs into mice and evaluating the migrating larvae in the lungs for determination of infectivity. While albendazole showed ovicidal activity against *A. suum*, ivermectin had only a limited effect on either embryonation or infectivity of eggs.

Ivermectin in feed

The delivery of a parasiticide via an in-feed formulation is an accepted method for producers. The introduction of ivermectin in feed meant that this was the first endectocide to become available by this traditional delivery method. Efficacy for the in-feed delivery of ivermectin has been reported as equal to or in some species exceeding the injectable. Alva-Valdes *et al.* (1989) administered ivermectin at rates of 100 and 200 μg kg^{-1} day^{-1} to measure efficacy against naturally acquired infections. Efficacy of 97.7% was achieved for *A. suum*, 97.8% for *Metastrongylus*, greater than 99% for *Oesophagostomum* spp., 100% for *Macracanthorhynchus hirudinaceus* and 89.7% for *Ascarops strongylina* at 100 μg kg^{-1} day^{-1}. This concentration tested against induced fourth-stage larvae provided 100% control for *A. suum* and 96.9% for *Oesophagostomum* spp.

Evaluations using 200 μg kg^{-1} day^{-1} against naturally acquired infections showing efficacy of 100% was achieved for *A. suum*, *Metastrongylus* and *A. strongylina*, 99.6% for *Oesophagostomum* spp. and 85.9% for *M. hirudinaceus*. This concentration tested against induced fourth-stage larvae provided 100% control for *A. suum* and 95% for *Oesophagostomum* spp. In both of these studies, pigs were fed the in-feed ivermectin for a 7-day period and necropsy preformed 42–44 days post-treatment.

Primm *et al.* (1992) evaluated in-feed ivermectin in two trials, finding efficacies against all the commonly found swine parasites that were equal to or exceeded those published by Alva-Valdes *et al.* (1989). Drag *et al.* (1998) and Barth *et al.* (1996) reported efficacy of in-feed ivermectin against *S. ransomi*, when pregnant females were treated prior to parturition, of 100%, with no transmission of worms to any of the piglets.

Long acting ivermectin (Virbamec LA)

The safety of a long-acting formulation of ivermection was examined by Mercier *et al.* (2000c) who found no safety issues when used as an intramuscular injection in pigs at the recommended dose rate. Houffschmitt *et al.* (2000a) evaluated the efficacy of LA ivermectin against gastrointestinal nematodes in naturally infected pigs. Pigs infected with *Ascaris*, *Oesophagostomum* spp., *S. ransomi* and *T. suis* were utilized in the trial. Pigs were treated and returned to infected pens, mixing all pigs together. Efficacy was reported at 98% and the worm burden of animals in the LA ivermectin and doramectin groups was not significantly different.

Doramectin injectable

Doramectin injectable was evaluated in a number of studies at a rate of 300 μg kg^{-1} delivered intramuscularly. The intramuscular delivery of doramectin resulted in higher and more prolonged plasma concentrations than those observed with ivermectin utilizing subcutaneous injection (Friis and Bjorn, 1995). Logan *et al.* (1996) reported efficacy against experimentally induced infections of gastrointestinal parasites from 22 studies conducted with common protocols. Efficacy of 98% or greater was reported for fourth-stage larvae and adults of: *H. rubidus*, *A. suum*, *S. ransomi*, *O. dentatum*, *O. quadrispiculatum*, *Metastrongylus* spp. and *S. dentatus*. Efficacy for *T. suis* was reported as 87 and 79% for adult and fourth-stage larvae, respectively. Yazwinski *et al.* (1997) reported similar efficacies on the above parasites plus reported efficacies of 99.5 and 62.1% against *A. strongylina* and *M. hirudinaceus*, repectively. Stewart *et al.* (1997) evaluated doramectin against naturally acquired parasites and found efficacies against all parasites that agreed with those reported by Logan *et al.* (1996).

Moxidectin

Moxidectin, a milbemycin related to the avermectins, has been reported to have broad anthelmintic properties. Moxidectin is formulated as an injectable and as a pour-on for use in food animals. Gundlach *et al.*

(1994) reported that from research in Poland, moxidectin injectable at 300 μg kg^{-1} was 100% effective against *S. ransomi* and *T. suis*, but had lower efficacy against *O. dentatum* and *A. suum*. Further work by Gundlach *et al.* (1992), in which moxidectin was compared for efficacy when administered at an injectable or as a pour-on, showed that the pour-on at 500 μg kg^{-1} was more effective for helminths than the injectable at 400 μg kg^{-1}, but the pour-on at 750 μg kg^{-1} and the injectable at 400 μg kg^{-1} were equally active against mange. Stewart *et al.* (1999) evaluated a 5% pour-on formulation of moxidectin at rates of 0.75, 1.0, 1.25 and 1.5 mg kg^{-1} against artificially infected animals. Efficacies reported were: *A. suum*, 98.3% at 1.25; *Metastrongylus* spp., 100% at 0.75; *O. quadrispiculatum*, 100% at 1.5; and *T. suis*, 93.5% at 0.75 mg kg^{-1}.

Persistent efficacy of MLs against helminths

Persistent activity of doramectin and ivermectin is of great interest, as a long-lasting persistent material would allow for greater utility when using these materials in the production facility. Lichtensteiger *et al.* (1999) examined the persistent activity of doramectin and ivermectin against *A. suum*, the most common swine parasite. Animals were treated on day 0, −7, −14 or −21 and then challenged with 50,000 embryonated *A. suum* eggs on day 0 of the trial. The data indicated that the anthelmintic activity of ivermectin persisted for less than 7 days and the activity of doramectin persisted from more than 7 but less than 14 days. Houffschmitt *et al.* (2000a) reported that a generic LA ivermectin protected pigs from *Oesophagostomum* spp. for a minimum of 6 weeks.

Efficacy of the MLs Against Ectoparasites

Introduction

Sarcoptes scabiei var. *suis* is the mite responsible for causing sarcoptic mange in swine and affects the majority of swine herds throughout the world. Recent surveys conducted in North America and Western Europe estimate that as many as 50–95% of herds may be infested with *S. scabiei* mites (Garcia *et al.*, 1994). Another common ectoparasite of pigs is the sucking louse *H. suis*. Today, the MLs are well recognized as effective parasiticides with strong activity against mites and lice, and the high level of activity and the availability of different formulations to facilitate administration have made the MLs the 'gold standard' for controlling mange and lice in swine.

Efficacy of the MLs against mange mites

Ivermectin injectable

The high efficacy of ivermectin injectable, at a dose of 300 μg kg^{-1} once by subcutaneous injection, against *S. scabiei* var. *suis* can be explained, amongst others, by the fact that the drug penetrates very well in target tissues. Scott and McKellar (1992) studied the distribution of ivermectin administered subcutaneously at a dose rate of 300 μg kg^{-1}. They measured very high concentrations in the ear wax of the pigs (mean of 3359 ng g^{-1}). Ivermectin is a lipophilic drug and therefore could accumulate in wax or may be excreted in wax. Ivermectin could also be detected significantly in the skin (74 ng g^{-1}) and, to a lesser extent, in the ear tissues (10 ng g^{-1}). As a comparison, concentrations in other tissues (gut wall, lung, injection sites) were always lower than 20 ng g^{-1}. Higher concentrations, up to 210 ng ml^{-1}, are seen in bile and gut contents, suggesting that faeces is the main route of excretion of the drug. The high concentrations in skin and ear wax, a predilection site for mites, made this drug particularly useful against mites.

Martineau *et al.* (1984) studied the efficacy of ivermectin injectable against clinical mange. General improvement of lesions was noted as early as 8 days after treatment. Scratching stopped by day 21 in all treated animals, and on day 21 no mites could be detected. Yang and Jeng (1986) treated with injectable ivermectin 70 pigs of five different age classes who were all confirmed by scrapings to be infected with mites. Based on scrapings, the cure rate was 100% on day 21, and animals remained negative for up to 167 days. Soll and Smith (1987) studied the efficacy of injectable ivermectin on 20 pigs with confirmed infections (live mites present). Twelve pigs were treated and the other eight remained as untreated controls. Based on scrapings on day 28 and 42 post-treatment, the cure rate was 100%. Cramer *et al.* (1996) described the results of a series of eight studies designed to study the efficacy of ivermectin injectable against sarcoptic mange under various conditions worldwide. All trials were conducted as controlled studies and included a total of 42 untreated controls and 50 ivermectin-treated pigs. The infestations were naturally acquired in seven studies, and induced in one. Scrapings to detect live mites were performed pre-treatment and on days 14, 28 and 42. On day 14 post-treatment, mites were still present in three trial sites out of the six sampled; mites were still present in one trial on day 28 and in two trials on day 42.

Ivermectin in feed

Differences have been described in the bioavailability and half-life of ivermectin in pigs when administered by the oral or subcutaneous routes.

The concentration of ivermectin in the plasma attains its peak more quickly following the oral dose than the parenteral. These peaks appeared at 0.5 days and 2 days, respectively, indicating a faster absorption rate via oral administration than subcutaneous. The slower absorption associated with the parenteral route is possibly due to precipitation of the drug at the injection site (Lo *et al.*, 1985). The bioavailability as estimated by the area under the curve (AUC; 0–40 days) of ivermectin after oral administration of 300 $\mu g\ kg^{-1}$ was 41% of a comparable subcutaneous administration. Administration of ivermectin in the feed may be the preferable route of administration when treating large groups of swine. Because pigs redose themselves daily when ivermectin is incorporated in the ration, acaricidal drug concentration is maintained throughout the period from oviposition to hatching. Alva-Valdes *et al.* (1989) studied the dose of an in-feed ivermectin formulation against *S. scabiei*. Twenty-four animals were divided in three groups of eight animals each. The pigs had a severe infestation with the mange mite. One group was a control group and the other two groups received ivermectin in feed over 7 consecutive days calculated at a dosage of 100 and 200 $\mu g\ kg^{-1}\ day^{-1}$, respectively. A significant decrease in number of mites was seen after treatment. Primm *et al.* (1992) evaluated the efficacy of an in-feed preparation of ivermectin in 40 naturally infected pigs in two trials. In trial 1, examination of skin scrapings before treatment revealed eight of ten control animals to be scabies positive, and five pigs out of the ten pigs in the treatment group were infected. On post-treatment day 14, three non-treated control pigs remained scabies positive, whereas there were none in the treated group. In trial 2, pre-trial skin scraping from all pigs ($n = 20$, ten for each group) contained live mites. On post-treatment days 7, 21 and 35, live mites were found on eight, six and one control pigs, respectively. No mites were found in any of the treated pigs from 7 days post-treatment and throughout the remainder of the study. The study showed that ivermectin in feed completely cured scabies by day 7. Cramer *et al.* (1996) describes the results of a series of seven studies designed to assess the efficacy of ivermectin in-feed formulation against sarcoptes mange under various conditions worldwide. All trials were conducted as controlled studies which included a total of 82 untreated controls and 82 ivermectin-treated pigs. The infestations were naturally acquired in five studies, in one study infections were superimposed to natural infections and in one study infections were induced. Scrapings to detect live mites were performed pre-treatment and on days 14, 27/28 and 42. On day 14 and 42 post-treatment, all treated animals were negative for mites; on day 27/28 mites were present in one trial. Wallace *et al.* (1996) studied the efficacy of ivermectin pre-mix in 14 adult swine, naturally infected. Animals were allocated on the basis of day –3 mite counts to seven replicates to determine efficacy of ivermectin in-feed. On days –3, 14, 28 and 42, skin scrapings were taken for the detection of live mites. From day 14, all seven treated sows were negative for mites.

Roppa *et al.* (1996) described the efficacy of the in-feed formulation of ivermectin on 128 pigs naturally parasitized during the fattening period. A total of eight pens with eight animals each were kept as untreated controls; the pigs in the other eight pens were treated. On days 0 and 28, scrapings from both ears of four animals in each pen were examined for the presence of live mange mites. On day 0, one animal in the treated group was positive for live mites and on day 28 one control animal was positive; otherwise no live mites were detected on these occasions. On day 28 post-treatment, 99.3% of the control animals had a positive dermatitis score, in contrast to only 1.2% of the treated animals.

Recently, Mercier *et al.* (2000a) showed that an in-feed preparation of abamectin was 100% efficient against sarcoptic mites at 100 $\mu g\ kg^{-1}\ day^{-1}$ for 7 days.

LA ivermectin injectable

Two studies have been published on the efficacy of LA ivermectin against swine scabies. Mercier *et al.* (2000b) examined the efficacy of ivermectin injected subcutaneously at 300 $\mu g\ kg^{-1}$ and LA ivermectin injected intramuscularly at 300 $\mu g\ kg^{-1}$ in pigs with an induced mange infestation. Efficacy of 100% was found in animals treated 7 and 14 days prior to infestation. Houffschmitt *et al.* (2000b) evaluated the efficacy of LA ivermectin under a natural challenge model. Doramectin and LA ivermectin treatments were used in the study, and by day 14 all treated pigs were determined to be free of mites based upon skin scrapings and remained mite free until day 49, the conclusion of the study.

Doramectin injectable

The intramuscular administration of doramectin at 300 $\mu g\ kg^{-1}$ resulted in higher and more prolonged plasma concentrations than after subcutaneous administration of ivermectin at 300 $\mu g\ kg^{-1}$ (Friis and Bjorn, 1995). According to the authors, this may probably be attributed to higher bioavailability for doramectin.

Several studies were performed to measure acaricidal effectiveness of doramectin in swine naturally or experimentally infested with S. *scabiei*. Fujii *et al.* (1994) achieved a 99% reduction from day 0 to day 28 post-treatment. On the 28th day after treatment, 46 of the 52 doramectin-treated animals had no detectable mites and mite eggs. Scores of skin lesions and of pruritis in doramectin-treated animals were significantly reduced by day 14 and day 7 post-treatment, respectively, compared with a control group. Cargill *et al.* (1994) studied the efficacy of doramectin injectable in experimentally infested pigs, of which 16 were treated.

Assessments (scrapings and scratching) were made on days 7, 14, 21 and 28. Doramectin was 100% effective in eliminating *S. scabiei*. The intensity of clinical and post-mortem signs associated with hypersensitivity to mange mites was significantly reduced in all treated pigs compared with the majority of controls. Logan *et al.* (1996) observed in 63 naturally infected and treated pigs, by day 21, a 100% cure in mite counts. In a study designed to measure acaricidal effectiveness of doramectin in naturally infested pigs, a total abolition of live mite populations was observed for all doramectin-treated pigs at each post-treatment sampling date (7, 14, 21 and 28 days) (Yazwinski *et al.*, 1997).

Residual activity of MLs in pigs

Arends and Skogerboe (1995) demonstrated that pigs treated with ivermectin subcutaneously and challenged on days 3 and 6 did not develop lesions during the observation period following challenge, but animals challenged with mites 9 days post-treatment exhibited lesions 21 days post-challenge. Cargill *et al.* (1999) confirmed the persistent efficacy of ivermectin injectable, with no mites recovered for 6 weeks after pigs were challenged 12 days post-treatment.

Arends *et al.* (1999) performed two studies to compare the persistent efficacy of doramectin and ivermectin in swine experimentally infected with *S. scabiei*. Persistent efficacy of doramectin was determined to be between 18 and 21 days on the basis of mite recovery. This was approximately twice as long as the persistent efficacy of ivermectin, which was only 9–12 days also on the basis of mite recovery. In contrast, Cargill *et al.* (1999) found no difference between ivermectin injectable and doramectin when pigs were challenged more than 12 days post-treatment.

It must be stressed that since all active mite stages (larvae, nymphs, adults) are highly susceptible to ivermectin and doramectin, a product must, to break the mange mite life cycle, only outlast the mite egg incubation period, estimated to be 3–10 days.

Eradication of scabies

The eradication of mange from large pig herds on many farms in Denmark (Ebbesen and Henriksen, 1986; Ebbesen, 1998), The Netherlands (Rambage *et al.*, 1998), Belgium (Smets *et al.*, 1999), Australia (Cargill *et al.*, 1996), Germany (Koheler and Zabke, 1998) and the USA (Mohr, 1999) illustrates that, in practice, the efficacy and prolonged protection induced by ivermectin injection or in-feed effectively breaks the mange–mite cycle (Hogg, 1984; Yang and Jeng, 1986; Boraski and Brown, 1992). Eradication of swine scabies by using ivermectin twice was at first combined with treatment of the environment (Ebbesen and Henriksen, 1986; Hogg, 1989; Lambers, 1994). Practical experiences have shown later that spraying acaricides in the stables is not necessary (Jacobsson *et al.*, 1998). The medication programme most in use is two treatments with ivermectin injectable (0.2 mg kg^{-1}) with an interval of 14 days; any piglets born after day 0 and up to day 7 must receive an ivermectin injection on day 7. Researchers in Belgium used a combination of ivermectin injectable and in feed for a closed breeding herd. Sows, boars and piglets were treated by subcutaneous injection on days 0 and 14, while growers and finishers were given oral powder in feed for two periods of 7 days, with an interval of 1 week between them. Feed dosages are 2 ppm for 25–40 kg pigs and 2.4 ppm for those in the 40–100 kg range (Smets *et al.*, 1999). Feed medication (with ivermectin) for 14–21 consecutive days also seems to be very successful (T.J. Ebbesen, personal communication). Mange was eliminated successfully from two herds using a single dose of doramectin (Arnason *et al.*, 2000). Compared with two injections with ivermectin, this model is much less labour intensive and less endectocide is used. However, since only one treatment is given, great care should be taken that all animals are treated.

Efficacy of the MLs against lice

Only a limited number of studies have been published on the efficacy of MLs against lice. Efficacies were between 99 and 100% for ivermectin (Stewart *et al.*, 1981) and 100% for doramectin (Logan *et al.*, 1996).

Conclusions

The MLs have proven to be highly efficacious against ecto- and endoparasites of swine in modern production schemes. The broad choice of delivery systems, duration of activity and safety have made them the treatment of choice worldwide. The largest impact MLs have had on swine production has been their use in mange eradication programmes where sarcoptic mange has been eliminated from single herds as well as entire integrated production systems. The use of MLs for deworming swine in modern production systems has been much less than their use for swine mange. Production systems tend to have few worm problems that are of significant economic impact. In systems that are less intensive or utilize either a solid concrete flooring or dirt, helminths are an economic issue and the MLs have tremendous, yet unrealized potential to manage parasite infections in swine production.

References

Alva-Valdes, R., Wallace, D.H., Foster, A.G., Erricsson, G.F. and Wooden, J.W. (1989) Efficacy of an in-feed ivermectin formulation against gastrointestinal helminths, lungworms, and sarcoptic mites in swine. *American Journal of Veterinary Research* 50, 1392–1395.

Arends, J.J. and Skogerboe, T.L. (1995) Persistence of efficacy of doramectin and ivermectin against *Sarcoptes scabiei* var. *suis*. *15th International Conference of the World Association for Advances in Veterinary Parasitology*, Yokohama, 1–3 September, pp. 35–40.

Arends, J.J., Skogerboe, T.L. and Ritzhaupt, L.K. (1999) Persistent efficacy of doramectin and ivermectin against experimental infestations of *Sarcoptes scabiei* var. *suis* in swine. *Veterinary Parasitology* 82, 71–79.

Arnason, T., Nielsen, L.H., Jensen, J.C.E. and Cracknell, V. (2000) Elimination of mange mites (*Sarcoptes scabiei* var. *suis*) from two natural infested Danish sow herds using a one injection regime with doramectin. *Proceedings of the 16th Conference of the International Pig Veterinary Society*, Melbourne, 17–20 August, p. 273.

Barth, D., Rehbein, S., Reid, J.F.S. and Barrick, R.A. (1996) Efficacy of an in-feed formulaton of ivermectin against adult worms and somatic larvae of *Strongyloides ransomi*. *Veterinary Parasitology* 65, 89–97.

Benz, G.W., Roncalli, R.A. and Gross, S.J. (1989) Use of ivermectin in cattle, sheep, goats and swine. In: Campbell, W.C. (ed.) *Ivermectin and Abamectin*, Springer-Verlag, New York, pp. 215–229.

Boes, J., Eriksen, L. and Nansen, P. (1998) Embryonation and infectivity of *Ascaris suum* eggs isolated from worms expelled by pigs treated with albendazole, pyrantel pamoate, ivermectin or piperazine dihydrochloride. *Veterinary Parasitology* 75, 181–190.

Boraski, E.A. and Brown, J.R. (1992) A comprehensive approach to mange and lice elimination. *Proceedings of the 12th Conference of the International Veterinary Pig Society*, The Hague, 17–20 August, p. 377.

Cargill, C.F., Davies, P.R., Carmichael, L., Hooke, F. and Moore, M. (1994) Treatment of pigs with doramectin to control sarcoptic mange. *Proceedings of the 13th Conference of the International Pig Veterinary Society*, Bangkok, 26–30 June, p. 238.

Cargill, C.F., Pointon, A.M., Moore, M. and Garcia, R. (1996) A retrospective evaluation based on slaughter monitoring of using ivermectin to control and eradicate sarcoptic mange. *Proceedings of the 14th Conference of the International Pig Veterinary Society*, Bologna, 7–10 July, p. 356.

Cargill, C., Garcia, R. and Ryan, W. (1999) Evaluation of the duration of efficacy of doramectin and ivermectin in the prevention of *Sarcoptes scabiei* var.*suis* infestation in pigs. *Proceedings of the 30th Annual Meeting of the American Association of Swine Practitioners*, St Louis, Missouri, 27 Feb–2 March 1999, p. 243.

Cramer, L.G., Cox, J.L., Cifelli, S., Garcia, R. and Gross, S.J. (1996) Efficacy of ivermectin as a control agent for *Sarcoptes scabiei* in swine. *Proceedings of the 14th Conference of the International Pig Veterinary Society*, Bologna, 7–10 July, p. 363.

Drag, M.D., Green, S.E., Houser, R.A., Wallace, D.H., Cox, J.L. and Barrick, R.A. (1998) Efficacy of an in-feed formulation of ivermectin against somatic larvae of *Strongyloides ransomi* in pregnant swine. *American Journal of Veterinary Research* 59, 277–279.

Ebbesen, T.J. (1998) Eradication of sarcoptic mange in farrow-to finish herds with Ivomec vet premix and iomec vet injection. *Proceedings of the 15th Conference of the International Pig Veterinary Society*, Birmingham, 5–9 July, p. 120.

Ebbesen, T.J. and Henriksen, S.V. (1986) Eradication of sarcoptic mange in two sow herds with ivermectin. *Proceedings of the 9th Conference of the International Pig Veterinary Society*, Barcelona, 15–18 July, p. 364.

Friis, C. and Bjorn, H. (1995) Pharmacokinetics of doramectin and ivermectin in pigs. *Proceedings of the 15th International Conference of the World Association for Advances in Veterinary Parasitology*, Yokohama, 1–3 September, pp. 27–32.

Fujii, T., Furuya, T., Yamada, Y., Nakumara, Y. and Kagota, K. (1994) Field efficacy trials of doramectin against ectoparasites of swine in Japan. *Proceedings of the 13th Conference of the International Pig Veterinary Society*, Bangkok, 26–30 June, pp. 26–30.

Garcia, R., Piche, C., Davies, P. and Gross, S. (1994) Prevalence of sarcoptic mange mites and dermatitis in slaughter pigs in North America and Western Europe. *Proceedings of the 13th Conference of the International Pig Veterinary Society*, Bangkok, 26–30 June, p. 250.

Gundlach, J.L., Sadzizkowski, A.B., Tomczuk, K. and Uchacz, S. (1992) Effectiveness of drugs containing moxidectin in the control of parasites in pigs. *Medycyna Weterynaryjna* 48, 209–211 [Polish].

Gundlach, J.L., Sadszizkowski, A.B. and Tomczuk, K. (1994) Usefulness of moxidectin (Cydectin) in the elimination of internal parasites and scab mites in pigs under different housing conditions. *Medycyna Weterynaryjna* 50, 72–74 [Polish].

Hogg, A. (1984) Eradication of sarcoptic mange in swine with ivermectin. *Proceedings of the 8th Conference of the International Pig Veterinary Society*, Ghent, 27–30 June, p. 206.

Hogg, A. (1989) The control and eradication of sarcoptic mange in swine herds. *Agri-practice* 10, 8–10.

Houffschmitt, P., Frayssinet, L., Najbar, N. and Gundlach, J. (2000a) Efficacy of a novel long acting ivermectin against gastrointestinal nematodes in naturally infected pigs. *Proceedings of the International Pig Veterinary Society Congress*, Melbourne, Australia, 17–21 September.

Houffschmitt, P., Frayssinet, L., Mercier, P. and Farkas, R. (2000b) Efficacy of a novel long acting ivermectin against sarcoptic mange in pigs submitted to a high natural challenge. *Proceedings of the International Pig Veterinary Society Congress*, Melbourne, Australia, 17–21 September.

Jacobsson, M., Bornstein, S. and Wallgren, P. (1998) Experiences from eradication systems directed against *Sarcoptes scabiei*. *Proceedings of the 15th Conference of the International Pig Veterinary Society*, Birmingham, 5–9 July, p. 118.

Koheler, U. and Zabke, J. (1998) Mange eradication with ivomec premix and ivomec injection in a large swine unit. *Proceedings of the 15th Conference of the International Pig Veterinary Society*, Birmingham, 5–9 July, p. 255.

Lambers, J.H. (1994) Elimination of *Sarcoptes scabiei* in a Dutch pig breeding herd. *Proceedings of the 13th Conference of the International Pig Veterinary Society*, Bangkok, 26–30 June, p. 252.

Lichtensteiger, C.A., DiPietro, J.A., Paul, A.J., Neumann, E.J. and Thompson, L. (1999) Persistent activity of doramectin and ivermectin against *Ascaris suum* in experimentally infected pigs. *Veterinary Parasitology* 82, 235–241.

Lo, P.-K.A., Fink, D.W., Williams, J.B. and Blodinger, J. (1985) Pharmacokinetic studies of ivermectin: effects of formulation. *Veterinary Research Communications* 9, 251–268.

Logan, N.B., Weatherley, A.J. and Jones, R.M. (1996) Activity of doramectin against nematode and arthropod parasites of swine. *Veterinary Parasitology* 66, 87–94.

Martineau, G.P., Vaillencourt, J. and Fréchette, J.-L. (1984) Control of *Sarcoptes scabiei* infestation with ivermectin in a large intensive breeding piggery. *Canadian Veterinary Journal* 25, 235–238.

Mercier, P., White, C.R., Eddi, C. and Caracostantogolo, J. (2000a) Efficacy of an abamectin in-feed preparation against mites in pigs. *Veterinary Record* 147, 52.

Mercier, P., White, C.R., Eddi, C., Caracostantogolo, J. and Houffschmitt, P. (2000b) Persistent efficacy of two injectable endectocides against induced sarcoptic mange mites in Pigs. *Proceedings of the International Pig Veterinary Society Congress*, Melbourne, Australia, 17–21 September.

Mercier, P., Chick, B. and Houffschmitt, P. (2000c) Safety of a novel long acting formulation of ivermectin administered by intramuscular route in pigs. *Proceedings of the International Pig Veterinary Society Congress*, Melbourne, Australia, 17–21 September.

Mohr, M. (1999) *Sarcoptes scabiei* var. *suis* elimination with ivomec premix and ivomec injection. *Proceedings of the 30th Annual Meeting of the American Association Swine Practitioners*, St Louis, Missouri, 27 Feb–2 March 1999, pp. 245–247.

Primm, N.D., Hall, W.F., DiPietro, J.A. and Bane, D.P. (1992) Efficacy of an in-feed preparation of ivermectin against endoparasites and scabies mites in swine. *American Journal of Veterinary Research* 53, 508–512.

Rambage, P.G.M., Vesseur, P.C. and van der Heijden, H.M.J. (1998) Mange (*Sarcoptes scabiei* var. *suis*) eradication programme and possibilities for certification in Dutch pig farms. *Proceedings of the 15th Conference of the International Pig Veterinary Society*, Birmingham, 5–9 July, p. 253.

Roppa, L., Cruz, J.B., Maciel, A.E., Bordin, E.L. and Garcia, R. (1996) Evaluation of the efficacy of in-feed ivermectin and the impact on productivity in growing pigs. *Proceedings of the 14th Conference of the International Pig Veterinary Society*, Bologna, 7–10 July, p. 360.

Scott, E.W. and McKellar, Q.A. (1992) The distribution and some pharmacokinetic parameters of ivermectin in pigs. *Veterinary Research Communications* 16, 139–146.

Smets, K., Neirynck, W. and Vercruysse, J. (1999) Eradication of sarcoptic mange from a Belgian pig breeding farm with a combination of injectable and in-feed ivermectin. *Veterinary Record* 145, 721–724.

Soll, M.D. and Smith, C.J.Z. (1987) Efficacy of ivermectin against the pig mange mite *Sarcoptes scabiei* var. *suis*. *Journal of the South African Veterinary Association* 58, 29–30.

Stewart, T.B., Marti, O.G. and Hale, O.M. (1981) Efficacy of ivermectin against five genera of swine nematodes and the hog louse, *Haematopinus suis*. *American Journal of Veterinary Research* 42, 1425–1426.

Stewart, T.B., Leon, D.L., Fox, M.C., Southern, L.L. and Bodak-Koszalka, E. (1991) Performance of pigs with mixed nematode infections before and after ivermectin treatment. *Veterinary Parasitology* 39, 253–266.

Stewart, T.B., Fox, M.C. and Wiles, S.E. (1997) Doramectin efficacy against gastrointestinal nematodes in pigs. *Veterinary Parasitology* 66, 101–108.

Stewart, T.B., Wiles, S.E., Miller, J.E. and Rulli, R.D. (1999) Efficacy of moxidectin 0.5% pour-on against swine nematodes. *Veterinary Parasitology* 87, 39–44.

Varady, M., Petersen, M.B., Bjorn, H. and Nansen, P. (1996) The efficacy of ivermectin against nodular worms of pigs: the response to treatment using three different dose levels against *Oesophagostomum dentatum* and *Oesophagostomum quadrispiculatum*. *International Journal of Parasitology* 26, 369–374.

Wallace, D.H., Holste, J., Bierschwal, C.J., Cox, J.L. and Garcia, R. (1996) Efficacy and safety of Ivomec premix in adult swine. *Proceedings of the 14th Conference of the International Pig Veterinary Society*, Bologna, 7–10 July, p. 361.

Yang, D.P.C. and Jeng, C.R. (1986) The efficacy of ivermectin on the eradication program of swine scabies in a breeding herd. *Proceedings of the 9th Conference of the International Pig Veterinary Society*, Barcelona, 15–18 July, p. 363.

Yazwinski, T.A., Tucker, C., Fetherston, H., Johnson, Z. and Wood-Huels, N. (1997) Endectocidal efficacies of doramectin in naturally parasitized pigs. *Veterinary Parasitology* 70, 123–128.

Chapter 10

The Use of Macrocyclic Lactones in the Control and Prevention of Heartworm and Other Parasites in Dogs and Cats

J. Guerrero, J.W. McCall and C. Genchi

Introduction

Among the internal and external parasites of dogs and cats, *Dirofilaria immitis* is one of the most important to the health and well-being of those animals. Frequently, this parasite produces infections that can be fatal to the host. Dirofilariosis, however, is a preventable disease due to the availability of highly effective preventive drugs that are safe, effective, convenient and easy to administer. According to the recently published American Heartworm Society guidelines, all animals at risk for contracting the disease should routinely receive heartworm preventive medications (Knight, 1999).

Chemoprophylactic drugs for heartworm infection fall into two basic classes, the macrocyclic lactones (MLs) or macrolides and diethylcarbamazine (DEC). In heartworm-endemic areas, puppies and kittens born during the transmission season should be given their first dose of macrocyclic lactones between 6 and 8 weeks of age.

The present chapter will briefly review information on the epidemiology of *D. immitis* infections in dogs and cats and the use of MLs in the prevention of heartworm infections as well as their effect on other important ecto- and endoparasites of companion animals when used according to label instructions. No information on the 'off label' use of the MLs will be reviewed since most of the available data are based on the use of these compounds in exaggerated potentially dangerous doses.

Epidemiology

More than 70 species of mosquitoes have been shown to be capable of developing the L_1 stage (microfilariae) of *D. immitis* to the infective, third-stage larvae (L_3), but fewer than a dozen of these species are believed to be major vectors (Otto and Jachowski, 1981). Although the susceptibility of different geographical strains may vary, there is a high probability that at least one susceptible vector species is present in a geographic area that is conducive to the propagation of mosquitoes. Once the ubiquitous heartworm parasite is introduced into an area, its transmission is virtually ensured.

The prevalence and distribution of heartworm infections in dogs are better described than for other animal species, but gradually more information on the frequency of diagnosis of infection in cats is becoming available. It is now generally accepted that heartworm disease occurs in cats in any area where dogs are infected (Guerrero *et al.*, 1992a; McCall *et al.*, 1994), but the geographical distribution and level of infection are less predictable in cats than in dogs. In highly endemic areas with sufficient rainfall, essentially every unprotected dog becomes infected (McTier *et al.*, 1992b). In contrast, about 75% of cats can be infected experimentally with *D. immitis* L_3 (McCall *et al.*, 1992a). However, the prevalence of natural infections in cats is between 5 and 20% that for dogs in the same geographical area (Ryan and Newcomb, 1995).

Outdoor cats and strays who are seemingly exposed to high numbers of bites from infective mosquitoes may be able to mount an effective immune response that could be partially protective (Dillon *et al.*, 1996); however, studies to determine levels of susceptibility have not been performed. For indoor cats, it seems that even one encounter with an infective vector may lead to the development of a large proportion of transmitted larvae to the adult stage, causing severe illness (Genchi *et al.*, 1992). Recently, it has been reported that between 10 and 16% of apartment-dwelling cats with respiratory disease in the USA were seropositive for *D. immitis* (Miller *et al.*, 1998; Robertson-Plouch *et al.*, 1998). Therefore, chemoprophylactic treatment is a viable option for cats – even cats living more sheltered lives – residing in any area where heartworm is considered endemic in dogs. As suggested by Atkins (1997), it seems rational to recommend chemoprophylaxis for feline heartworm infection, given that the disease has a higher incidence than both feline leukaemia virus (FeLV) and feline immunodeficiency virus (FIV), infections for which vaccination protocols are increasingly advocated.

Prevalence

D. immitis infections in dogs and cats have been described in all continents. What was considered to be a regional problem of dogs in the USA is now a national problem that is considered to be endemic in 49 states of the Union, with no clearly documented cases in Alaska; however, focal transmission appears to be possible in all 50 US states. Growing prevalence has also been described in Mexico, Puerto Rico and the Caribbean (Genchi *et al.*, 2001). Heartworm infections in cats are now being reported by veterinarians with increasing frequency in the USA and southern Canada (Guerrero *et al.*, 1992c; Robertson-Plouch *et al.*, 1998). The prevalence for heartworm infections in dogs in South America has been reviewed previously by Guerrero *et al.* (1992b), while infections in cats have only been reported by Labarthe *et al.* (1997). Heartworm disease in dogs and cats is well established in Australia (Kendall *et al.*, 1991) and in Japan (Guerrero *et al.*, 1992a; Roncalli *et al.*, 1998). The prevalence and spread of heartworm infection in Europe has been reviewed comprehensively by Genchi *et al.* (1998). The disease is diagnosed mainly in the southern European countries of Spain, Italy, Portugal and France, with scattered reports from Greece, Turkey and some Eastern European countries.

Chemoprophylactic Treatment of Canine Heartworm Disease and its Effect on Internal Parasites

Since the discovery of ivermectin and the initial description of activity against developmental stages of *D. immitis*, a large number of publications have appeared reporting on their attributes. In essence, ivermectin at doses of 6 $\mu g\ kg^{-1}$ kills the tissue larval stages of *D. immitis* and interrupts the development of these larvae to the adult stage. As the first heartworm preventive able to be administered once a month, ivermectin changed the way heartworm prophylaxis was conducted in small animal clinics. The studies that demonstrated this exquisite efficacy were reviewed previously by Campbell (1989).

Presently, monthly oral administration of ivermectin at 6–12 $\mu g\ kg^{-1}$, milbemycin oxime at 500–999 $\mu g\ kg^{-1}$, moxidectin at 3–6 $\mu g\ kg^{-1}$ or topical selamectin at 6–12 $mg\ kg^{-1}$ are approved and provide effective protection against heartworm infections in dogs (Table 10.1).

An ivermectin/pyrantel pamoate chewable formulation has been available for veterinary use since 1993. This formulation is available in Europe and in the USA. This product expands the indications to include treatment and control of certain gastrointestinal parasites. Clark *et al.* (1992) described the bioequivalence of ivermectin (6 $\mu g\ kg^{-1}$) in the single as well as the combined formulation (ivermectin 6 $\mu g\ kg^{-1}$ and pyrantel pamoate at 5 $mg\ kg^{-1}$). The lack of interference of the pyrantel pamoate portion on the activity of ivermectin as a heartworm preventive was shown by Paul *et al.* (1992a). The authors describe 100% efficacy against *D. immitis* larvae in the ivermectin pyrantel pamoate combinations. Dzimianski *et al.* (1992) reported efficacy of the combination product of

Table 10.1. Macrocyclic lactones approved for use in companion animal medicine.

Macrocyclic lactone	Presentation	Frequency of use	Host	Indications
Ivermectin	Tablets	Monthly	Dogs	Di
Ivermectin	Chewables	Monthly	Dogs	Di
Ivermectin	Chewables	Monthly	Cats	Di, At, Ab
Ivermectin/ pyrantel pamoate	Chewables	Monthly	Dogs	Di, Tc, Tl, Ac, Ab, Us
Milbemycin oxime	Flavour tablets	Monthly	Dogs	Di, Tc, Tl, Ac, Tv
Milbemycin oxime	Flavour tablets	Monthly	Cats	Di, Tc, At
Milbemycin oxime/lufenuron	Flavour tablets	Monthly	Dogs	Di, Tc, Tl, Ac, Tv, Cf
Moxidectin	Tablets	Monthly	Dogs	Di
Moxidectin	Injectable	Every 6 months	Dogs	Di, Ac
Selamectin	Topical	Monthly	Dogs	Di, Tc, Cf, Ss, Oc, Dv
Selamectin	Topical	Monthly	Cats	Di, Tct, At, Cf, Oc

Di, *Dirofilaria immitis*; Tc, *Toxocara canis*; Tct, *Toxocara cati*; Tl, *Toxascaris leonina*; Ac, *Ancylostoma caninum*; Ab, *Ancylostoma braziliense*; At, *Ancylostoma tubaeforme*; Us, *Uncinaria stenocephala*; Tv, *Trichuris vulpis*; Cf, *Ctenocephalides felis*; Ss, *Sarcoptes scabiei*; Oc, *Otodectes cynotis*; Dv, *Dermacentor variabilis*.

100% against *D. immitis* larvae, adult *Toxascaris leonina* and adult *Ancylostoma caninum*; 99.3% against adult *Uncinaria stenocephala*; and 88.2% against adult *Toxocara canis*. These investigators also concluded that the efficacy of the combined formulation was as good or better than that of either of the two products given alone. Other investigators (Clark *et al.*, 1992; Daurio *et al.*, 1993) also demonstrated high equivalent levels of anthelmintic efficacy of the ivermectin/pyrantel pamoate chewable formulation against *T. canis, T. leonina, A. caninum* and *U. stenocephala*. Clark *et al.* (1992) reported 90.1% efficacy against *T. canis*, 99.2% against *T. leonina*, 98.5% against *A. caninum* and 98.7% against *U. stenocephala*. Efficacy against *A. braziliense* was later shown by Shoop *et al.* (1996) who reported a 100% efficacy of the combined formulation against adult parasites.

Milbemycin oxime, an analogue of milbemycin D, is highly efficacious against developmental stages of *D. immitis* in dogs when administered in monthly oral doses as low as 0.05 mg kg^{-1} (Grieve *et al.*, 1989) or 0.1 mg kg^{-1} (Bradley, 1989). The dosage defined for commercial presentation of milbemycin oxime is 0.5 mg kg^{-1} administered monthly (Bater, 1989; Grieve *et al.*, 1989), and at that dosage the efficacy against developmental stages of *D. immitis* is 100%. Against intestinal parasites, an early study by Bowman *et al.* (1988) described 100% efficacy against adult *T. canis* even at doses as low as 270–390 μg kg^{-1}. At the recommended dose for heartworm prophylaxis (500 μg kg^{-1}), milbemycin oxime is also effective against *T. canis*, *T. leonina*, *A. caninum* and *Trichuris vulpis*. Blagburn *et al.* (1989) reported a range of 89.6 to 95.4% efficacy of a tablet formulation of milbemycin oxime against *A. caninum*. In a second study, Blagburn *et al.* (1992) demonstrated a 95% efficacy against adult *Ancylostoma* spp. (mixture of *A. caninum* and *A. braziliense*) and 49–81% efficacy against larval stages of the same parasites. In this quoted study, the investigators utilized milbemycin oxime at the actual recommended dose. Against adult *T. vulpis*, these authors reported a 97% efficacy. A combination milbemycin oxime/lufenuron tablet is available in the USA for the control of *D. immitis* infections and *Ctenocephalides felis* infestations in dogs, given on a monthly schedule. The tablet given as recommended provides 0.5 mg milbemycin oxime kg^{-1} and 10 mg lufenuron kg^{-1}. Blagburn *et al.* (1998) reported that the combined formulation was 100% effective for preventing maturation of *D. immitis*, and demonstrated a weekly range of 79.7–100% in preventing maturation of *C. felis* eggs up to day 28 after treatment in one study, while the same authors report 100% of flea control for 3 months in a second study.

Moxidectin tested in dogs experimentally infected with *D. immitis* was shown to be effective in preventing the development of third-stage larvae when administered orally at doses of 1.5, 3 or 6 μg kg^{-1} 1 month after infection (McTier *et al.*, 1992a). McCall *et al.* (1992b) reported that oral moxidectin given monthly at 1 or 3 μg kg^{-1} was 100% effective in preventing natural infections with *D. immitis*. When given at 3 μg kg^{-1} every 2 months, moxidectin was still effective. Similar findings were reported by King *et al.* (1992) in tests of the 3 μg kg^{-1} oral dosage under clinical conditions. Lock *et al.* (2001) reported that the sustained-release formulation of moxidectin prevents the development of *D. immitis* even when infective larvae were injected 6 months after administration of moxidectin at doses of 170 or 500 μg kg^{-1}. The formulation was also found to be safe at these dosage levels. Recently, a sustained-release injectable formulation of moxidectin to be utilized solely by veterinarians in dogs for prevention of heartworm infections has been introduced into the market. The commercial formulation is approved in the USA and Europe for use in dogs 6 months of age and older. This product also treats infections with *A. caninum*. The recommended schedule for treatments is every 6 months.

A newly developed avermectin, selamectin, an avermectin monosaccharide (Bishop *et al.*, 2000) recently has been approved for use in Europe and in the USA. There are several important characteristics unique to selamectin: the commercial formulation is given as a topical formulation, thereby avoiding problems associated with compliance and oral administration. Selamectin is also a true endectocide since its activity encompasses internal and external parasites.

The primary use of selamectin for the prevention of heartworm infections in dogs and cats was described by McTier *et al.* (1998) who demonstrated that this compound administered topically to dogs at 8 mg kg^{-1} 30, 45 or 60 days after infection was completely effective in preventing the development of experimental infections of *D. immitis* L_3s. Recently, selamectin was found to be 100% efficacious when used as a single topical dose at 3 or 6 mg kg^{-1} administered to dogs either 30 or 45 days after infection (McTier *et al.*, 2000a). The same authors reported 100% efficacy when selamectin was used at 6 mg kg^{-1} 60 days after infection. According to McTier *et al.* (2000a), neither bathing nor shampooing performed between 2 and 96 h after treatment affected the efficacy of selamectin. The clinical efficacy of the selamectin administered monthly at 6 mg kg^{-1} was demonstrated by Boy *et al.* (2000) and by Clemence *et al.* (2000).

McTier *et al.* (2000e) studied the effect of selamectin against induced and natural infections of *T. canis*. Induced infections were reduced by 93.9–98.1% after a single treatment, 88.3–98.6% after two monthly applications, and 100% after three. With natural infections, the authors reported 84.6–97.9% efficacy after two monthly treatments and 91.1–97.6% utilizing the 'monthly plus' regimen of treatments. The authors also reported 93.3% efficacy against naturally acquired infections of *T. leonina* in dogs after two monthly doses of selamectin.

Campbell (1989) reviewed the attempts at controlling transmission of *T. canis* from infected bitches to their puppies utilizing elevated doses of ivermectin; however, no effect on *T. canis* larvae has ever been reported at the recommended dosage rate of 6 μg kg^{-1} administered monthly. Recently, Payne-Johnson *et al.* (2000) reported the results of a trial designed to determine the efficacy of selamectin administered topically to pregnant and lactating bitches for treatment and prevention of *T. canis* and infestations with *C. felis* in the dams and the pups. In this study, the dams received topical selamectin at 6 mg kg^{-1} approximately 40 and 10 days before whelping and 10 and 40 days after parturition. Results showed a 99.7% reduction in *T. canis* faecal egg counts for the dams and more than 96% for the pups up to day 34 after birth. The number of adult worms recovered from the gastrointestinal tract of the pups treated with selamectin was reduced by 98.2%, compared with controls. The population of fleas was reduced in 99.8% for the dams and 100% for the pups.

Chemoprophylactic Treatment of Feline Heartworm Disease and its Effect on Internal Parasites

Heartworm preventive treatment in cats follows the same regimen established for dogs, that is, monthly dosing should begin within 1 month from the start of the transmission season and the last dose should be given within 1 month from the end of the risk period. Ivermectin is marketed in the USA for use as a prophylactic agent in cats given monthly at the dose of 24 μg kg^{-1} (McTier *et al.*, 1992c; Paul *et al.*, 1992b). This oral dosage is also highly effective for treatment and control of *A. tubaeforme* (94.5%) and *A. braziliense* (98.5%) (Robertson *et al.*, 1992). Nolan *et al.* (1992) reported 92.8% efficacy against adult *A. braziliense* and 90.7% against adult *A. tubaeforme*. The effect of ivermectin on microfilaraemic cats was studied recently. Each of 20 cats was experimentally infected with four pairs of adult male and female heartworms. Ten cats were treated with ivermectin and ten remained untreated as controls. Nineteen of the 20 cats became microfilaraemic, and three of the ten control cats remained microfilaraemic for 4–9 months after treatment. All ten infected cats treated with ivermectin became amicrofilaraemic within 2 months after initiation of treatment, and there were no signs of adverse effects associated with the killing of microfilariae (J.W. McCall, Georgia, 2001, personal communication)

Milbemycin oxime is also known to be effective in cats for heartworm prophylaxis (Stewart *et al.*, 1992) and is commercially available for this purpose in the same flavour tablet formulation used in dogs. The recommended dosage for cats is 2000 μg kg^{-1}. At this dosage milbemycin oxime is effective in the prevention of *D. immitis* and the removal of adult *A. tubaeforme* and *T. cati*. Efficacy of moxidectin for heartworm prevention in cats has not been determined.

The exquisite sensitivity of the larvae of *D. immitis* to all MLs was also found to be true with selamectin. McTier *et al.* (1998) demonstrated that the topically applied single dose of selamectin at 6 mg kg^{-1} completely prevented the development of *D. immitis* in experimentally infected cats. More recently, McTier *et al.* (2000a) confirmed the efficacy of the commercial formulation of selamectin in cats experimentally infected with *D. immitis*. In this study, selamectin was 100% effective against heartworm infections when administered topically at 6 mg kg^{-1}. The authors also report that bathing with water or shampoo between 2 and 96 h after treatment did not affect the efficacy of selamectin.

Selamectin has also been found to be effective against the roundworm *T. cati* and the hookworm *A. tubaeforme* in cats (McTier *et al.*, 2000f). The authors reported 100% reduction in adult *T. cati* after a single application of selamectin. A single application reduced naturally occurring adult populations of *A. tubaeforme* in cats by 99.4%, while in experimentally infected cats, the reduction ranged from 84.7 to 99.7%. The same authors reported that two doses of selamectin given at an interval of one month provided 91.9% efficacy in the reduction of adult *A. tubaeforme* in experimentally infected cats (McTier *et al.*, 2000f). Under clinical conditions, Six *et al.* (2000b) reported virtually complete reduction of faecal ascarid and hookworm egg counts after one or two treatments of client-owned cats with selamectin.

Evans *et al.* (2001) conducted an experiment in cats similar to that performed in dogs by Payne-Johnson *et al.* (2000). Results show a similar effect of selamectin against *T. cati* infections and *C. felis* infestations in cats. In this study, the pregnant queens received the selamectin treatment 6 and 2 weeks before and 2 and 6 weeks after parturition. The faecal egg count reduction in the selamectin-treated queens was 100% on day 49 post-partum. Likewise, intestinal worm counts in the kittens born from selamectin-treated queens was reduced by 100% as compared with kittens born to placebo-treated queens. No fleas were detected in the queens treated with selamectin or in their kittens. These recently reported findings in dogs and cats represent a major development in the effort to control the source of such an important zoonosis as visceral and ocular larval migrans. The continuous monthly use of selamectin in dogs and cats offers the practitioner an invaluable tool in the fight against parasitic zoonosis.

Testing of cats for the presence of adult *D. immitis* should be considered after the first season of preventive treatment and is advisable at the beginning of each new transmission season before preventive therapy is to be initiated, unless the clinician has wisely chosen to utilize ivermectin, milbemycin oxime, moxidectin or selamectin chemoprophylaxis year-round. The relatively long life cycle of the parasite in the cat, as well as the difficulty in accurately diagnosing infection, increase the risk of inadvertently treating an infected animal; however, based on results reported by McCall *et al.* (J.W. McCall, Georgia, 2001, personal communication), monthly administration of ivermectin to cats infected with adult worms did not precipitate any negative reactions. Based on this limited set of data, as well as the actual in-field use of ivermectin for several years, the continuous use of this product in cats living in heartworm-endemic areas may be highly recommended.

Heartworm Clinical Prophylaxis in Dogs and Cats

All MLs are completely efficacious against *D. immitis* larvae, which allows them to be administered every 30 days. For instance, the oral chewable formulation of ivermectin has been found to be 100% effective in preventing development of *D. immitis* larvae in dogs and cats when administered 30 or 45 days after challenge with infective larvae (McTier *et al.*, 1992c; Paul *et al.*, 1992a). This characteristic provides a safeguard in the case of omission or delay of a monthly treatment, or when the chemoprophylactic history of the dog cannot be verified. Ivermectin and milbemycin oxime have both been found to provide a high degree of protection when administered on a regular basis, beginning 3 months after infection

(Atkins *et al.*, 1995). In fact, monthly treatment with ivermectin over a 1-year period has been shown to be greater than 95% effective in preventing development of *D. immitis* larvae that were 4 months old at the time of treatment; however, under the same conditions, milbemycin oxime was only 41.4–49.3% effective as a clinical prophylactic agent (McCall *et al.*, 1995). This retroactive or 'reachback' effect has been reported for orally administered moxidectin, even at a dosage as low as 0.5 $\mu g\ kg^{-1}$ (McTier *et al.*, 1992a), and products with this compound have a label claim for efficacy of 2 months' duration. This so-called 'reachback' effect of the MLs is very useful to compensate for missed or delayed treatments, but should not be considered as justification to modify the recommended monthly interval for treatment.

Recently published information has shown that ivermectin given at prophylactic doses (6 $\mu g\ kg^{-1}$) monthly for 16 months reduced the adult heartworm population by 56% in dogs with transplanted worms (McCall *et al.*, 1998). In a study reported by McCall *et al.* (forthcoming a), the ivermectin/pyrantel pamoate combination was 98.7 and 94.9% effective in dogs when treatment was initiated either 5 or 7 months, respectively, after receiving an experimental infection with 50 L_3 and treatment was given year-round for 31 and 29 consecutive months. Milbemycin oxime under the same conditions was ineffective against adult parasites. Recent work reported by Lock *et al.* (2001) on the injectable, sustained-release formulation of moxidectin (170 $\mu g\ kg^{-1}$) indicates that a single injection protects dogs against heartworm infection for 6 months. Its retroactive (i.e. reachback, clinical prophylactic) efficacy against 4-month-old infections has been reported as high (85.9%) when a single injection is given and even higher (97.2%) when it is followed by a second injection 6 months later (i.e. 10 months after infection) (McCall *et al.*, forthcoming b). This injectable formulation of moxidectin was essentially ineffective against 6-month-old infections, even when three injections were given at 6-month intervals.

Selamectin has been found to have a 98.5% efficacy on reduction of development of 3-month-old larvae of *D. immitis* (McCall *et al.*, forthcoming a). Dzimianski *et al.* (forthcoming) have found selamectin to have a partial effect on adult *D. immitis* (39.4% efficacy) when administered topically at the recommended dose for 18 consecutive months (Table 10.2).

In cats, a more recent study demonstrated that the recommended prophylactic dosage of ivermectin was 66.5% effective in clearing a 7-month-old *D. immitis* infection that had been transplanted from an infected dog. A dramatic decrease in circulating antigen levels also was detected in ivermectin-treated cats. These findings are highly noteworthy for the feline veterinarian since adulticide treatments are not considered a viable option for cats (J.W. McCall, Georgia, 2001, personal communication).

Table 10.2. Clinical prophylactic (reachback) and adulticidal activity of macrocylic lactones against *Dirofilaria immitis* in dogs.

Drug	Age of heartworms (months)	No. of treatments	% Efficacy	Appearance and/or motility of live heartworms	Reference
Ivermectin	3	13	97.7	Abnormal	McCall *et al.* (1996)
(6 μg kg^{-1}	4	14	97.8	Abnormal	McCall *et al.* (1995)
per os,	4	12	95.1	Abnormal	McCall *et al.* (1995)
monthly)	5	31	98.7	Abnormal	McCall *et al.* (forthcoming a)
	7	29	94.9	Abnormal	McCall *et al.* (forthcoming a)
	8	16	56.3	Abnormal	McCall *et al.* (1998)
Milbemycin	3	13	96.7	Normal	McCall *et al.* (1996)
(500 μg	4	14	49.3	Normal	McCall *et al.* (1995)
kg^{-1} *per os*,	4	12	41.4	Normal	McCall *et al.* (1996)
monthly)	8	16	0	Normal	McCall *et al.* (1998)
Selamectin (6 mg kg^{-1}	3	12	98.5	ND	McCall *et al.* (forthcoming a)
topical,	6	On-going			
monthly)	Adult	18	39.0	Abnormal	Dzimianski *et al.* (forthcoming)
Moxidectin (0.5 μg kg^{-1} *per os*)	2	1	100	NA	McTier *et al.* (1992a)
Moxidectin (0.17 mg	4	1	85.9	Abnormal	McCall *et al.* (forthcoming b)
kg^{-1} s.c., every 6	4 and 10	2	97.2	Abnormal	McCall *et al.* (forthcoming b)
months)	6	1	≤25	Abnormal	McCall *et al.* (forthcoming b)
	6, 12 and 18	3	≤25	Abnormal	McCall *et al.* (forthcoming b)

ND, not done; NA, not applicable.

Effect of Selamectin Against Ectoparasites of Dogs and Cats

At the dose recommended for heartworm prevention in dogs and cats (6 mg kg^{-1}), selamectin is also effective in preventing and controlling cat flea (*C. felis*) and dog flea infestations (*C. canis*) (Benchaoui *et al.*, 2000; Boy *et al.*, 2000; McTier *et al.*, 2000b,c,d; Shanks *et al.*, 2000a). Selamectin has demonstrated a larvicidal effect on fleas (McTier *et al.*, 2000d) and is also effective for treatment and control of ear mites (*Otodectes cynotis*) and sarcoptic mange in dogs (*Sarcoptes scabiei*) (Shanks *et al.*, 2000b,c; Six *et al.*,

2000a). In controlled trials using naturally infected dogs and cats, the investigators reported 100% efficacy against *O. cynotis* following a single treatment with selamectin (Shanks *et al.*, 2000c). Against *S. scabiei* and in clinical trials against *O. cynotis*, the investigators reported that two monthly treatments were needed to obtain clinical and parasitological cure (Shanks *et al.*, 2000b; Six *et al.*, 2000a). Complete efficacy for the recommended dose of selamectin against biting lice (*Trichodectes canis* and *Felicola subrostratus*) has also been described recently in dogs and cats (McTier *et al.*, 2001). For ticks in dogs, selamectin has been found to be efficacious against *Dermacentor variabilis* and *Rhipicephalus sanguineus* when used in a treatment every 2 weeks or the so-called 'monthly plus' regimen (addition of a treatment on day 14 to the standard monthly treatment regimen) (Jernigan *et al.*, 2000). Although the authors describe fair efficacy against these parasites 3 days after treatment, the peak effect is not found until day 5 after administration, a bit slow from the cosmetic as well as the sanitary benefits expected from a true tickicide. No information is available on the effect of selamectin on other important ticks of veterinary importance as *Ixodes scapularis*, *I. ricinus*, *I. pacificus*, *D. andersoni* and *Amblyomma americanum*.

Concluding Remarks

Since the mid-1980s, veterinarians, pet owners and producers around the world have counted on the MLs as great tools to control endo- and ectoparasites of domestic animals. The initial discovery and development of ivermectin followed by the other MLs gave all of us involved in parasite control a tremendous sense of security; however, it is up to us, the users of these magnificent molecules, to secure the responsible and judicious use of them to ensure their continuous efficacy for many years to come. We believe that a major revolution like the discovery and development of ivermectin may not happen again in our lifetime; however, we are encouraged by the innovations offered by moxidectin sustained-release formulation and the development of selamectin, the first true endectocide for use in small animal medicine.

References

Atkins, C.E. (1997) Feline vascular disease: therapeutic considerations. *Proceedings of the 13th American College of Veterinary Internal Medicine Forum*, pp. 150–151.

Atkins, C.E., Atwell, R.B., Dillon, R., Genchi, C., Hayasaki, M., Holmes, R., Knight, D.H., Lukof, D.K., McCall, J.W. and Slocombe, J.O. (1995) Guidelines for the diagnosis, treatment, and prevention of heartworm (*Dirofilaria immitis*) infection in cats. In: Soll, M.D. and Knight, D.H. (eds) *Proceedings of the Heartworm Symposium '95*. American Heartworm Society, Batavia, Illinois, pp. 309–312.

Bater, A.K. (1989) Efficacy of oral milbemycin against naturally acquired heartworm infections in dogs. In: Otto, G.F. (ed.) *Proceedings of the Heartworm Symposium '89*. American Heartworm Society, Batavia, IIllinois, pp. 107–108.

Benchaoui, H.A., Clemence, R.G., Clements, P.J.M., Jones, R.I., Watson, P., Shanks, D.J., Smith, D.G., Sture, G.H., Jernigan, A.D. and Rowan, T.G. (2000) Efficacy and safety of selamectin against fleas on dogs and cats presented as veterinary patients in Europe. *Veterinary Parasitology* 91, 223–232.

Bishop, B.F., Bruce, C.I., Evans, N.A., Goudie, A.C., Gration, K.A.F., Gibson, S.P., Pacey, M.S., Perry, D.A., Walshe, N.D.A. and Witty, M.J. (2000) Selamectin: a novel broad-spectrum endectocide for dogs and cats. *Veterinary Parasitology* 91, 163–176.

Blagburn, B.L., Hendrix, C.H., Lindsay, D.S., Vaughan, J.L., Mysinger, R.H. and Hepler, D.I. (1989) Milbemycin: efficacy and toxicity in beagle and collie dogs. In: Otto, G.F. (ed.) *Proceedings of the Heartworm Symposium '89*. American Heartworm Society, Batavia, Illinois, pp. 109–113.

Blagburn, B.L., Hendrix, C.M., Lindsay, D.S., Vaughan, J.L., Hepler, D.L. and Wright, J.C. (1992) Efficacy of milbemycin oxime against naturally acquired or experimentally induced *Ancylostoma* spp. and *Trichuris vulpis* infections in dogs. *American Journal of Veterinary Research* 53, 513–516.

Blagburn, B.L., Lindsay, D.S., Vaughan, J.L., Rippey, N.S., Butler, J.M. and Parks, S.C. (1998) Efficacies of a milbemycin oxime and lufenuron combination tablet for prevention of experimental heartworm infections and flea infestations in dogs. In: Seward, R.L. and Knight, D.H. (eds) *Proceedings of the Heartworm Symposium '98*. American Heartworm Society, Batavia, Illinois, pp. 201–207.

Bowman, D., Parsons, J., Grieve, R. and Hepler, D.I. (1988) Effects of milbemycin on adult *Toxocara canis* in dogs with experimentally induced infections. *American Journal of Veterinary Research* 49, 1986–1989.

Boy, M.G., Six, R.H., Thomas, C.A., Novotny, M.J., Smothers, C.D., Rowan, T.G. and Jernigan, A.D. (2000) Efficacy and safety of selamectin against fleas and heartworms in dogs and cats presented as veterinary patients in North America. *Veterinary Parasitology* 91, 233–250.

Bradley, R.E. (1989) Dose titration and efficacy of milbemycin oxime for prophylaxis against *Dirofilaria immitis* infection in dogs. In: Otto, G.F. (ed.) *Proceedings of the Heartworm Symposium '89*. American Heartworm Society, Batavia, Illinois, pp. 115–120.

Campbell, W.C. (1989) Use of ivermectin in dogs and cats. In: Campbell, W.C. (ed.) *Ivermectin and Abamectin*. Springer-Verlag, New York, pp. 245–259.

Clark, J.N., Daurio, C.P., Plue, R.E., Wallace, D.H. and Longhofer, S.L. (1992) Efficacy of ivermectin and pyrantel pamoate combined in a chewable formulation against heartworm, hookworm and ascaris infections in dogs. *American Journal of Veterinary Research* 53, 517–520.

Clemence, R.G., Sarasola, P., Genchi, C., Smith, D.G., Shanks, D.J., Jernigan, A. and Rowan, T.G. (2000) Efficacy of selamectin in the prevention of adult heartworm (*Dirofilaria immitis*) infection in dogs in northern Italy. *Veterinary Parasitology* 91, 251–258.

Daurio, C.P., Robertson, E.L. and Seward, R.L. (1993) Efficacy of ivermectin in a beef-based chewable formulation against *Ancylostoma caninum* and *Uncinaria stenocephala* in dogs. *Journal of Parasitology* 79, 768–770.

Dillon, R., Atkins, C., Knight, D., Clekis,T., McCall, J., Genchi, C. and Miller, M.W. (1996) Roundtable discussion: feline heartworm disease, Part 1. *Feline Practice* 24 (6), 12–16.

Dzimianski, M.T., Robertson, E.L., McTier, T., Pule, R. and McCall, J.W. (1992) Prevention of heartworm infection and treatment of ascarid and hookworm infections in dogs using a chewable formulation of ivermectin plus pyrantel pamoate. In: Soll, M.D. (ed.) *Proceedings of the Heartworm Symposium '92.* American Heartworm Society, Batavia, Illinois, pp. 201–204.

Dzimianski, M.T., McCall, J.W., Steffens, W.L., Supakorndej, N., Mansour, A.E., Ard, M.B., McCall, S.D. and Hack, R. (forthcoming) The safety of selamectin in heartworm infected dogs and its effect on adult worms and microfilariae. *Proceedings of the Heartworm Symposium 2001* (submitted for publication).

Evans, N.A., Payne-Johnson, M., Maitland, T.P., Cooke, D.J., Murphy, M.G., McLoughlin, D.J., Shanks, D.J., Sherington, J., Rowan, T.G. and Jernigan, A.D. (2001) The efficacy of selamectin administered to cats during pregnancy and lactation against *Toxocara cati* and *Ctenocephalides felis* in queens and their offspring. *Proceedings of the 46th Annual Meeting of the American Association of Veterinary Parasitologists*, p. 38.

Genchi, C., Di Sacco, B. and Cancrini, G. (1992) Epizootiology of canine and feline heartworm infection in Northern Italy: possible mosquito vectors. In: Soll, M.D. (ed.) *Proceedings of the Heartworm Symposium '92.* American Heartworm Society, Batavia, Illinois, pp. 39–46.

Genchi, C., Solari-Basano, F., Marrone, R.V. and Petruschke, G. (1998) Canine and feline heartworm infection in Europe with special emphasis on Italy. In: Seward, R.L. and Knight, D.H. (eds) *Proceedings of the Heartworm Symposium '98.* American Heartworm Society, Batavia, Illinois, pp. 75–82.

Genchi, C., Kramer, L. and Prieto, G. (2001) Epidemiology of canine and feline dirofilariosis. In: Simon, F. and Genchi, C. (eds) *Heartworm Infection in Humans and Animals.* Ediciones Universidad de Salamanca, Salamanca, Spain, pp. 121–133.

Grieve, R.F., Frank, G.R., Stewart, V.A., Parsons, J.C., Abraham, D., MacWilliams, P.S. and Hepler, D.I. (1989) Effect of dosage and dose timing on heartworm (*Dirofilaria immitis*) chemoprophylaxis with milbemycin. In: Otto, G.F. (ed.) *Proceedings of the Heartworm Symposium '89.* American Heartworm Society, Batavia, Illinois, pp. 121–124.

Guerrero, J., Newcomb, K.M. and Fukase, T. (1992a) Prevalence of *Dirofilaria immitis* in dogs and cats in the United States and Japan. *Proceedings of the XVII WSAVA World Congress*, Rome, Vol. I, pp. 211–217.

Guerrero, J., Ducos de la Hitte, J., Genchi, C., Rojo, F., Gomez-Bautista, M., Carvalho Valera, M., Labarthe, N., Bordin, E., Gonzales, G., Mancebo, O., Patino, F., Uribe, L.F. and Samano, R. (1992b) Update on the distribution of *Dirofilaria immitis* in dogs from southern Europe and Latin America. In: Soll, M.D. (ed.) *Proceedings of the Heartworm Symposium '92.* American Heartworm Society, Batavia, Illinois, pp. 31–37.

Guerrero, J., McCall, J.W., Dzimianski, M.T., McTier, T.L., Holmes, R.A. and Newcomb, K.M. (1992c) Prevalence of *D. immitis* infection in cats from the southeastern United States. In: Soll, M.D. (ed.) *Proceedings of the Heartworm Symposium '92.* American Heartworm Society, Batavia, Illinois, pp. 91–95.

Jernigan, A.D., McTier, T.L., Chieffo, C., Thomas, C.A., Krautmann, M.J., Hair, J.A., Young, D.R., Wang, C. and Rowan, T.G. (2000) Efficacy of selamectin against experimentally induced tick (*Rhipicephalus sanguineus* and *Dermacentor variabilis*) infestations in dogs. *Veterinary Parasitology* 91, 359–375.

Kendall, K., Collins, G.H. and Pope, S.E. (1991) *Dirofilaria immitis* in cats from inner Sydney. *Australian Veterinary Journal* 68, 356–357.

King, R.R., Courtney, C.H. and Aguilar, R. (1992) Heartworm prophylaxis with moxidectin: field trial results from a hyperenzootic area. In: Soll, M.D. (ed.) *Proceedings of the Heartworm Symposium '92*. American Heartworm Society, Batavia, Illinois, pp. 179–182.

Knight, D.H. (1999) Guidelines for the diagnosis, prevention, and management of heartworm (*Dirofilaria immitis*) infection in dogs. In: Seward, R.L. and Knight, D.H. (eds) *Proceedings of the Heartworm Symposium '98*. American Heartworm Society, Batavia, Illinois, pp. 257–264.

Labarthe, N., Ferreira, A.M.R., Guerrero, J., Newcomb, K. and Paes-de-Almeida, E. (1997) Survey of *Dirofilaria immitis* (Leidy, 1856) in random source cats in metropolitan Rio de Janeiro, Brazil, with descriptions of lesions. *Veterinary Parasitology* 71, 301–306.

Lock, J.B., Knight, D.H., Wang, G.T., Doscher, M.E., Nolan, T.J., Hendrick, M.J., Steber, W. and Heaney, K. (2001) Activity of an injectable, sustained-release formulation of moxidectin administered prophylactically to mixed breed dogs to prevent infection with *Dirofilaria immitis*. *American Journal of Veterinary Research* 62, 1721–1726.

McCall, J.W., Dzimianski, M.T., McTier, T.L., Jernigan, A.D., Jun, J.J., Mansour, A.E., Supakorndej, P., Plue, R.E., Clark, J.N., Wallace, D.H. and Lewis, R.E. (1992a) Biology of experimental heartworm infections in cats. In: Soll, M.D. (ed.) *Proceedings of the Heartworm Symposium '92*. American Heartworm Society, Batavia, Illinois, pp. 71–79.

McCall, J.W., McTier, T.L., Holmes, R.A., Greene, T., Strickland, J. and Aguilar, R. (1992b) Prevention of naturally acquired heartworm infections in heartworm-naive beagles by oral administration of moxidectin at an interval of either one or two months In: Soll, M.D. (ed.) *Proceedings of the Heartworm Symposium '92*. American Heartworm Society, Batavia, Illinois, pp. 169–178.

McCall, J.W., Calvert, C.A. and Rawlings, C.A. (1994) Heartworm infection in cats: a life-threatening disease. *Veterinary Medicine* 89, 639–648.

McCall, J.W., McTier, T.L., Supakorndej, N., Ricketts, R. (1995) Clinical prophylactic activity of macrolides on young adult heartworms. In: Soll, M.D. and Knight, D.H. (eds) *Proceedings of the Heartworm Symposium '95*. American Heartworm Society, Batavia, Illinois, pp. 187–195.

McCall, J.W., McTier, T.L., Ryan, W.G., Gross, S.J. and Soll, M.D. (1996) Evaluation of ivermectin and milbemycin oxime efficacy against *Dirofilaria immitis* infections of three and four months' duration in dogs. *American Journal of Veterinary Research* 57, 1189–1192.

McCall, J.W., Ryan, W.G., Roberts, R.E. and Dzimianski, M.T. (1998) Heartworm adulticidal activity of prophylactic doses of ivermectin (6 μg/kg) plus pyrantel administered monthly to dogs. In: Seward, R.L. and Knight, D.H. (eds) *Proceedings of the Heartworm Symposium '98*. American Heartworm Society, Batavia, Illinois, pp. 209–215.

McCall, J.W., Hack, R., McCall, S.D., Mansour, A.E., Supakorndej, N., Supakorndej, P. and Steffens, W.L. (forthcoming a) Evaluation of repeated monthly dosing of selamectin against *Dirofilaria immitis* beginning three months after experimental inoculation of heartworm larvae in dogs. *Proceedings of the Heartworm Symposium 2001* (submitted for publication).

McCall, J.W., Supakorndej, P. and Dzimianski, M.T. (forthcoming b) Evaluation of retroactive and adulticidal activity of moxidectin canine SR (sustained-release) injectable formulation against *Dirofilaria immitis* infection in dogs. *Proceedings of the Heartworm Symposium 2001* (submitted for publication).

McTier, T.L., McCall, J.W., Dzimianski, M.T., Aguilar, R. and Wood, I. (1992a) Prevention of experimental heartworm infection in dogs with single oral doses of moxidectin. In: Soll, M.D. (ed.) *Proceedings of the Heartworm Symposium '92*. American Heartworm Society, Batavia, Illinois, pp. 165–168.

McTier, T.L., McCall, J.W., Dzimianski, M.T., Raynaud, J.P., Holmes, R.A. and Keister, D.M. (1992b) Epidemiology of heartworm infection in beagles naturally exposed to infection in three southeastern states. In: Soll, M.D. (ed.) *Proceedings of the Heartworm Symposium '92*. American Heartworm Society, Batavia, Illinois, pp. 47–57.

McTier, T.L., McCall, J.W., Dzimianski, M.T., Mansour, A.E., Jernigan, A., Clark, J.N., Plue, R.E. and Daurio, C.P. (1992c) Prevention of heartworm infection in cats by treatment with ivermectin at one month post-infection. In: Soll, M.D. (ed.) *Proceedings of the Heartworm Symposium '92*. American Heartworm Society, Batavia, Illinois, pp. 111–116.

McTier, T.L., McCall, J.W., Jernigan, A.D., Rowan, T.G., Giles, C.J., Bishop, B.F., Evans, N.A. and Bruce, C.I. (1998) Efficacy of UK-124,114, a novel avermectin, for the prevention of heartworm in dogs and cats. In: Seward, R.L. and Knight, D.H. (eds) *Proceedings of the Heartworm Symposium '98*. American Heartworm Society, Tampa, Florida, pp. 187–192.

McTier, T.L., Shanks, D.J., Watson, P., McCall, J.W., Genchi, C., Six, R.H., Thomas, C.A., Dickin, S.K., Pengo, G., Rowan, T.G. and Jernigan, A.D. (2000a) Prevention of experimentally induced heartworm (*Dirofilaria immitis*) infections in dogs and cats with a single topical application of selamectin. *Veterinary Parasitology* 91, 259–268.

McTier, T.L., Jernigan, A.D., Rowan, T.G., Holbert, M.S., Smothers, C.D., Bishop, B.F., Evans, N.A., Gration, K.A.F. and Giles, C.J. (2000b) Dose selection of selamectin for efficacy against adult fleas (*Ctenocephalides felis felis*) on dogs and cats. *Veterinary Parasitology* 91, 177–185.

McTier, T.L., Jones, R.L., Holbert, M.S., Murphy, M.G., Watson, P., Sun, F., Smith, D.G., Rowan, T.G. and Jernigan, A.D. (2000c) Efficacy of selamectin against adult flea infestations (*Ctenocephalides felis felis* and *Ctenocephalides canis*) on dogs and cats. *Veterinary Parasitology* 91, 187–199.

McTier, T.L., Shanks, D.J., Jernigan, A.D., Rowan, T.G., Murphy, M.G., Wang, C., Smith, D.G., Holbert, M.S. and Blagburn, B.L. (2000d) Evaluation of the effects of selamectin against adult and immature stages of fleas (*Ctenocephalides felis felis*) on dogs and cats. *Veterinary Parasitology* 91, 201–212.

McTier, T.L., Siedek, E.M., Clemence, R.G., Wren, J.A., Bowman, D.D., Hellman, K., Holbert, M.S., Murphy, M.G., Young, D.R., Cruthers, L.R., Smith, D.G., Shanks, D.J., Rowan, T.G. and Jernigan, A.D. (2000e) Efficacy of selamectin against experimentally induced and naturally acquired ascarid (*Toxocara canis* and *Toxascaris leonina*) infections in dogs. *Veterinary Parasitology* 91, 333–345.

McTier, T.L., Shanks, D.J., Wren, J.A., Six, R.H., Bowman, D.D., McCall, J.W., Pengo, G., Genchi, C., Smothers, C.D., Rowan, T.G. and Jernigan, A.D. (2000f) Efficacy of selamectin against experimentally induced and naturally acquired infections of *Toxocara cati* and *Ancylostoma tubaeforme* in cats. *Veterinary Parasitology* 91, 311–319.

McTier, T.L., Gautier, P., Pengo, G., Shanks, D.J., Evans, N.A., Rowan, T.G. and Jernigan, A.D. (2001) The efficacy of selamectin against biting lice on dogs and cats. *Proceedings of the AAVP 46th Annual Meeting*, p. 43.

Miller, M.W., Atkins, C.E., Stemme, K., Robertson-Plouch, C. and Guerrero, J. (1998) Prevalence of exposure to *Dirofilaria immitis* in cats in multiple areas of the United States. In: Seward, R.L. and Knight, D.H. (eds) *Proceedings of the Heartworm Symposium '98*. American Heartworm Society, Batavia, Illinois, pp. 161–166.

Nolan, T.J., Niamatali, S., Bhopale, V., Longhofer, S.L. and Schad, G.A. (1992) Efficacy of a chewable formulation of ivermectin against a mixed infection of *Ancylostoma braziliense* and *Ancylostoma tubaeforme* in cats. *American Journal of Veterinary Research* 53, 1411–1413.

Otto, G.F. and Jachowski, L.A. Jr (1981) Mosquitoes and canine heartworm disease. In: Otto, G.F. (ed.) *Proceedings of the Heartworm Symposium '80*. Veterinary Medicine Publishing Co., Edwardsville, Kansas, pp. 17–32.

Paul, A.J., Todd, K.S., Wallace, D.H., Wallig, M.A. and Daurio, C.P. (1992a) Efficacy of a chewable formulation containing ivermectin and pyrantel pamoate against the development of *Dirofilaria immitis* in dogs 30 days post-infection. In: Soll, M.D. (ed.) *Proceedings of the Heartworm Symposium '92*. American Heartworm Society, Batavia, Illinois, pp. 197–199.

Paul, A.J., Acre, K.E., Todd, K.S., Wallace, D.H., Jernigan, A.D. and Wallig, M.A. (1992b) Efficacy of ivermectin against *Dirofilaria immitis* in cats 30 and 45 days post-infection. In: Soll, M.D. (ed.) *Proceedings of the Heartworm Symposium '92*. American Heartworm Society, Batavia, Illinois, pp. 117–120.

Payne-Johnson, M., Maitland, T.P., Sherington, J., Shanks, D.J., Clements, P.J.M., Murphy, M.G., McLoughlin, A., Jernigan, A.D. and Rowan, T.G. (2000) Efficacy of selamectin administered topically to pregnant and lactating female dogs in the treatment and prevention of adult roundworm (*Toxocara canis*) infections and flea (*Ctenocephalides felis felis*) infestations in the dams and their pups. *Veterinary Parasitology* 91, 333–345.

Robertson, E.D., Schad, G.A., Ambrose, D.L., Nolan, T.J., Jernigan, A.D., Longhofer, S.L., Clark, J.N., Plue, R.E. and Daurio, C.P. (1992) Efficacy of ivermectin against hookworm infections in cats. In: Soll, M.D. (ed.) *Proceedings of the Heartworm Symposium '92*. American Heartworm Society, Batavia, Illinois, pp. 121–126

Robertson-Plouch, C.K., Dillon, A.R., Brawner, W.R. and Guerrero, J. (1998) Prevalence of feline heartworm infections among cats with respiratory and gastrointestinal signs: results of a multicenter study. In: Seward, R.L. and Knight, D.H. (eds) *Proceedings of the Heartworm Symposium '98*. American Heartworm Society, Batavia, Illinois, pp. 57–62.

Roncalli, R.A., Yamane, Y. and Nagata, T. (1998) Prevalence of *Dirofilaria immitis* in cats in Japan. *Veterinary Parasitology* 75, 81–89.

Ryan, W.G. and Newcomb, K.M. (1995) Prevalence of feline heartworm disease: a global review. In: Soll, M.D. and Knight, D.H. (eds) *Proceedings of the Heartworm Symposium '95.* American Heartworm Society, Batavia, Illinois, pp. 79–86.

Shanks, D.J., Rowan, T.G., Jones, R.L., Watson, P., Murphy, M.G., Smith, D.G. and Jernigan, A.D. (2000a) Efficacy of selamectin in the treatment and prevention of flea (*Ctenocephalides felis felis*) infestations on dogs and cats housed in simulated home environments. *Veterinary Parasitology* 91, 213–222.

Shanks, D.J., McTier, T.L., Behan, S., Pengo, G., Genchi, C., Bowman, D.D., Holbert, M.S., Smith, D.G., Jernigan, A.D. and Rowan, T.G. (2000b) The efficacy of selamectin in the treatment of naturally acquired infestations of *Sarcoptes scabiei* on dogs. *Veterinary Parasitology* 91, 269–281.

Shanks, D.J., McTier, T.L., Rowan, T.G., Watson, P., Thomas, C.A., Bowman, D.D., Hair, J.A., Pengo, G., Genchi, C., Smothers, C.D., Smith, D.G. and Jernigan, A.D. (2000c) The efficacy of selamectin in the treatment of naturally acquired aural infestations of *Otodectes cynotis* on dogs and cats. *Veterinary Parasitology* 91, 283–290.

Shoop, W.L., Michael, B.F., Soll, M.D. and Clark, J.N. (1996) Efficacy of an ivermectin and pyrantel pamoate combination against adult *Ancylostoma braziliense* in dogs. *Australian Veterinary Journal* 73, 84–85.

Six, R.H., Clemence, R.G., Thomas, C.A., Behan, S., Boy, M.G., Watson, P., Benchaoui, H.A., Clements, P.J.M., Rowan, T.G. and Jernigan, A.D. (2000a) Efficacy and safety of selamectin against *Sarcoptes scabiei* on dogs and *Otodectes cynotis* on dogs and cats presented as veterinary patients. *Veterinary Parasitology* 91, 291–309

Six, R.H., Sture, G.H., Thomas, C.A., Clemence, R.G.S., Benchaoui, H.A., Boy, M.G., Watson, P., Smith, D.G., Jernigan, A.D. and Rowan, T.G. (2000b) Efficacy and safety of selamectin against gastrointestinal nematodes in cats presented as veterinary patients. *Veterinary Parasitology* 91, 321–331.

Stewart, V.A., Hepler, D.I. and Grieve, R.B. (1992) Efficacy of milbemycin oxime in chemoprophylaxis of dirofilariasis in cats. *American Journal of Veterinary Research* 53, 2274–2277.

Chapter 11

The Use of Macrocyclic Lactones to Control Parasites of Domesticated Wild Ruminants

S.E. Marley and G.A. Conder

Introduction

This chapter will attempt to present available information on the efficacy and safety of macrocyclic lactones (MLs) in domesticated wild ruminants, those species that are managed as production and/or working animals. Species covered in other chapters or for which literature information is not available will not be addressed herein. Given the economic importance of domestic wild ruminants, the availability of cost-effective, efficacious and safe antiparasitic drugs is an essential element in managing these species. The introduction of the MLs as endectocides provided useful control agents for both nematode and arthropod parasites across a wide range of hosts. The first drug of this class, ivermectin, has been used extensively and effectively in domesticated wild ruminants, and it has been joined in recent years by a number of related drugs, that is, abamectin, doramectin, eprinomectin and moxidectin.

Water Buffalo (*Bubalus bubalis*)

Nematodes

Efficacy of ivermectin against gastrointestinal nematodes of water buffalo was investigated by Gill *et al.* (1989). Water buffalo naturally infected with gastrointestinal nematodes were treated subcutaneously with ivermectin at 200 μg kg^{-1} or left untreated. Efficacy was evaluated based on faecal egg counts. Nematodes expelled in the faeces were identified to genus. No gastrointestinal nematode eggs (*Toxocara* (syn. *Neoascaris*) *vitulorum*, *Strongyloides papillosus*, *Haemonchus* spp., *Trichostrongylus* spp., *Ostertagia ostertagi*, *Oesophagostomum* spp. or *Bunostomum phlebotomum*) were observed in ivermectin-treated animals by day 28 post-treatment, with the exception of a few *Trichuris* spp. eggs, while eggs were seen in undiminished numbers in non-treated animals throughout the study. The 200 $\mu g\ kg^{-1}$ subcutaneous dose of ivermectin was also found by Waghmare *et al.* (1991) to be 100% effective against *S. papillosus*, based on egg counts done at 15 days post-treatment. Suphalucksana and Ching (1991) examined subcutaneously administered ivermectin at 200 $\mu g\ kg^{-1}$ against a mixed nematode population; in their study, the drug appeared to be only marginally effective, although their results are difficult to interpret. Orally administered ivermectin at 200 $\mu g\ kg^{-1}$ was 100% effective against *Trichuris discolor*, based on faecal egg counts done at 7, 14 and 21 days after treatment (Garg *et al.*, 1998). Waghmare *et al.* (1991), Sindhu *et al.* (1996) and Rao *et al.* (2000) found subcutaneously administered ivermectin at 200 $\mu g\ kg^{-1}$ to be effective (99.3–100%) against *T. vitulorum* in buffalo calves based on egg counts done approximately 2 weeks post-treatment, while Shastri (1989) observed similar results based on egg counts done 4–8 days post-treatment. In contrast, Maqbool *et al.* (1997) saw only an 82% reduction in *T. vitulorum* eggs at 18 days post-treatment with the 200 $\mu g\ kg^{-1}$ subcutaneous dose of ivermectin. Subcutaneously administered ivermectin given at 200 $\mu g\ kg^{-1}$ was effective in clearing *Thelazia rhodesii* from the eye of a buffalo (Udupa *et al.*, 1995).

Gill *et al.* (1991) evaluated the effectiveness of ivermectin for the treatment of earsore (*Stephanofilaria zaheeri*) in water buffalo under continuous challenge in the field. Animals exhibiting clinical signs of earsore were treated with ivermectin subcutaneously at 200 $\mu g\ kg^{-1}$. Based on parasite counts of skin scrapings, ivermectin eliminated microfilariae in 14 out of 17 animals by day 28 post-treatment, while mean numbers of microfilariae in non-treated control animals increased over the same period. No appreciable effect on adult worm counts was observed in ivermectin-treated animals. Mean severity scores for earsore lesions were reduced in ivermectin-treated animals but increased in the control water buffalo during the study. No adverse reactions to treatment were observed. An 85% cure rate for *S. zaheeri* was observed in buffalo treated twice subcutaneously at a 28-day interval with 200 $\mu g\ kg^{-1}$ of ivermectin, as judged by disappearance of lesions and evidence of scar tissue formation (Gopal *et al.*, 1992). Singh *et al.* (1999) found that a single 200 $\mu g\ kg^{-1}$ dose of ivermectin administered subcutaneously reduced microfilariae of *Setaria* spp. to zero in buffalo, while S.P. Sharma (1991) using this treatment observed that nine of 13 animals became amicrofilaraemic by 36 h post-treatment, with the remaining buffalo becoming amicrofilaraemic by

15 days post-treatment following two additional treatments at 36–40 h intervals.

Arthropods

Gill *et al.* (1989) evaluated ivermectin administered subcutaneously at 200 μg kg^{-1} against mange mites (*Sarcoptes scabiei* and/or *Psoroptes ovis*) on water buffalo. Within 2 weeks of treatment, mites were not found in the skin scrapings of most ivermectin-treated animals, and pre-existing skin lesions had improved. A second dose given 28 days after the initial treatment was required to affect a cure in a few buffalo (four of 20) with an extensive amount of body surface affected by mange. All non-treated water buffalo had mites throughout the study and severe progressive lesions. Fahmy *et al.* (1996) found ivermectin topically applied once at 500 μg kg^{-1} was 100% effective against *Psoroptes natalensis* by 28 days post-treatment. Similar cures were obtained following one subcutaneous dose of ivermectin at 200 μg kg^{-1} for sarcoptic mange (Hayat *et al.*, 1996; Qudoos *et al.*, 1996; Ahmad *et al.*, 1997; Purohit *et al.*, 1997), after two doses given at a 2-week interval to treat psoroptic (Zaitoun *et al.*, 1998; Singh *et al.*, 1999) or sarcoptic (Singh *et al.*, 1999) mange, or after two doses at a 5-week interval for sarcoptic mange (Saha *et al.*, 1996). In contrast, Kumar and Suryanarayana (1995) found that 200 μg kg^{-1} of ivermectin administered subcutaneously on three occasions at 7 day intervals cured sarcoptic mange of only 80% of treated buffalo.

Efficacy of ivermectin was evaluated against *Haematopinus tuberculatus* by Lau and Singh, as reported by Soll (1989). Although louse numbers were reduced by 85 or 100% in water buffalo treated subcutaneously with ivermectin at 200 or 400 μg kg^{-1}, respectively, efficacy diminished to ≤50% by day 33 post-treatment. Similarly, Suphalucksana and Ching (1991) observed an 84% reduction in *Haematopinus* spp. numbers to 28 days post-treatment following a 200 μg kg^{-1} subcutaneous dose of ivermectin. In contrast, Shastri (1991) found a single dose of ivermectin administered subcutaneously at 200 μg kg^{-1} to be 100% effective against *H. tuberculatus*. Fahmy *et al.* (1996) found ivermectin topically applied once at 500 μg kg^{-1} was 100% effective against *Haematopinus eurysternus* by 14 days post-treatment and through to the end of the study (56 days post-treatment).

Larvae of *Chrysomya bezziana* were cleared from buffalo within 24–48 h following subcutaneous injection of ivermectin at 200 μg kg^{-1} (Reddy and Krishna, 1995; Senthilvel and Raman, 1999). Kumar and Joshi (1995) found that ivermectin administered subcutaneously at 200 μg kg^{-1} was effective in clearing cutaneous myiasis (caused primarily by *Calliphora* spp.) within 120 h post-treatment, and the drug applied topically in aqueous dilutions at 20 to 200 μg kg^{-1} cleared the myiasis within 24 h post-treatment.

Leech

Mahato and Thakur (as reported by Mahato, 1989) observed that nasal leeches (*Dinobdella ferox*) could be removed from buffalo effectively by first sprinkling water on the nose of the buffalo, which makes the leech protrude from the nose, and subsequently soaking the leech in a 10 μg kg^{-1} solution of ivermectin. Leeches gradually become flaccid and drop out of the nose within 3 h. Alternatively, the ivermectin can be administered as nasal drops, but subcutaneously or intramuscularly applied ivermectin is ineffective.

American Bison (*Bison bison*)

Nematode

Efficacy of ivermectin pour-on against *O. ostertagi* in bison has been reported (Marley *et al.*, 1995). Bison inoculated with third-stage larvae of *O. ostertagi* were treated 42 days post-inoculation with ivermectin pour-on at 500 μg kg^{-1} or an equal volume of carrier. Based on worm counts, ivermectin was 100% effective against *O. ostertagi*.

Arthropod

Soll (1989), reporting on the work of Schillhorn van Veen, Sikarskie and Braselton, noted that bison treated with ivermectin subcutaneously at 200 μg kg^{-1} in the autumn for *Hypoderma bovis* had no grub lesions in the ivermectin-treated animals the following spring compared with a mean of 11.3 grub lesions for control animals.

Pharmacokinetics, safety and residues

Determination of ivermectin residues in bison has been reported (Marley *et al.*, 1995). For ivermectin pour-on at 500 μg kg^{-1}, residue levels of 22,23 dihydroavermectin B_{1a} (ivermectin marker residue) in liver (6.5–54.8 ppb; mean 32.1 ± 5.4 ppb) and adipose tissue (1.5–16.7 ppb; mean 6.8 ± 1.7 ppb) at 18 days post-treatment, as measured by high-performance liquid chromatography, are similar to those seen in cattle. Soll (1989) reporting on the work of Schillhorn van Veen, Sikarskie and Braselton, demonstrated that the drug was safe at 200 or 1000 μg kg^{-1} in bison. The United States Food and Drug Administration (FDA) has determined a 56-day withdrawal period for ivermectin administered subcutaneously at 200 μg kg^{-1} in bison.

European Bison (*Bison bonasus*)

Nematodes

Demiaszkiewicz *et al.* (1997), using ivermectin in a medicated feed that provided a single dose of 300 μg kg^{-1}, found the drug to be ≥91.6% effective against *Dictyocaulus viviparus*, Trichostrongylidae and *Trichuris* (syn. *Trichocephalus*) *ovis* in European bison.

Reindeer (*Rangifer tarandus*)

Nematodes

Based on worm count data 50 days after treatment, Nordkvist *et al.* (1983) found that ivermectin given subcutaneously at 200 μg kg^{-1} was 94% effective against *Elaphostrongylus rangiferi* larvae and was 100% effective in clearing trichostrongylids and *D. viviparus* from reindeer, although some evidence of viable larvae of *D. viviparus* was found microscopically in the lungs of some animals. Nordkvist *et al.* (1984) saw similar results for worm counts at approximately 150 days following treatment for *D. viviparus* and *E. rangiferi*, but trichostrongylid numbers in ivermectin-treated reindeer were comparable with those of non-treated control animals, probably as a result of reinfection. Oksanen and Nieminen (1998) showed that both ivermectin and moxidectin administered subcutaneously to reindeer at 200 μg kg^{-1} reduced faecal egg counts for trichostrongylids to zero following treatment. They also demonstrated an apparent persistent activity for at least 2 months following treatment for both drugs against trichostrongylids, as faecal egg counts were still near zero 3 months but not 4 months following treatment (i.e. recently acquired worms at the 3-month sampling would not have had time to develop to maturity and produce eggs, while these worms would be producing eggs at the 4-month sampling). The timing of the study was during the winter months, so the level of challenge during the early months of the study (persistent period) may have been extremely light. A similar period of persistence against trichostrongylids was observed by Oksanen *et al.* (1992, 1993) for subcutaneously but not orally or, in the latter study, topically administered ivermectin at 200 μg kg^{-1}; again the studies were conducted during the winter months. Oksanen *et al.* (1992) showed efficacy against *Capillaria* spp. for ivermectin at 200 μg kg^{-1} by both oral and subcutaneous routes.

Arthropods

The effectiveness of several MLs has been evaluated against natural infestations of larval warble fly, *Hypoderma* (syn. *Oedemagena*) *tarandi*, and throat bot, *Cephenemyia trompe*, in reindeer. Ivermectin administered at 200 μg kg^{-1} as an oral formulation (paste) (Soveri *et al.*, 1990; Oksanen *et al.*, 1992, 1993; Oksanen, 1996) or as an injectable formulation (subcutaneously) (Nordkvist *et al.*, 1983, 1984; Soveri *et al.*, 1990; Oksanen *et al.*, 1992, 1993; Oksanen and Nieminen, 1998) and doramectin as an injectable formulation (subcutaneously) (Oksanen and Nieminen, 1996) or ivermectin administered at 500 μg kg^{-1} as a topical formulation (Oksanen *et al.*, 1993) were 100% effective against *H. tarandi*, while moxidectin as an injectable formulation (subcutaneously) (Oksanen and Nieminen, 1998) was 92.8% effective. Dieterich and Craigmill (1990) did observe low numbers of warbles in reindeer treated subcutaneously with ivermectin at 200 μg kg^{-1}, although all larvae in animals that were subjected to necropsy were dead. Against *C. trompe*, ivermectin (Nordkvist *et al.*, 1983, 1984; Haugerud *et al.*, 1993; Oksanen and Nieminen, 1998) and doramectin (Oksanen and Nieminen, 1996) administered subcutaneously or ivermectin administered orally (Oksanen, 1996) at 200 μg kg^{-1} to reindeer were 100% effective, while a similar dose of moxidectin subcutaneously provided a clearance of 70.8% (Oksanen and Nieminen, 1998).

Ivermectin given subcutaneously at 200–250 μg kg^{-1} to reindeer calves was effective against a pentastomid, adults of the reindeer sinus worm (*Linguatula artica*) (Haugerud *et al.*, 1993). Although no actual percentage efficacy could be calculated due to the experimental procedures, only two sinus worms were recovered from a single treated animal, while 68.4% of non-treated animals examined were infected with a mean of 10.6 worms.

Pharmacokinetics, safety and residues

Dieterich and Craigmill (1990) examined safety and residues in reindeer treated subcutaneously with ivermectin at 1 and 2 mg kg^{-1} (five and ten times the 200 μg kg^{-1} level). No drug-related adverse reactions were observed in any animal (fawn or adult) and, based on residue data comparable with those from cattle, the FDA established a withdrawal period of 56 days in reindeer. Oksanen *et al.* (1995) demonstrated that peak plasma concentrations in reindeer for subcutaneously administered ivermectin (200 μg kg^{-1}) were comparable with those seen in cattle and significantly greater than those resulting from orally (200 μg kg^{-1}) or topically (500 μg kg^{-1}) administered ivermectin in reindeer.

White-tailed Deer (*Odocoileus virginianus*)

Nematodes

Garris *et al.* (1991) demonstrated that feeding ivermectin-treated maize once at 500 g (~100 μg kg^{-1}) controlled trichostrongyles in white-tailed deer, based on an egg count of zero in the faeces on day 3 post-treatment in previously positive animals.

Olsen (1982) and Kocan (1985) investigated the effect of ivermectin against *Parelaphostrongylus tenuis* in white-tailed deer exposed prior to treatment to infective larvae of the nematode. Drug was administered subcutaneously at 200 or 400 μg kg^{-1} or 100 μg kg^{-1}, respectively in the two studies. Treatment protected the animals from infection when given 24 h following exposure to the parasite in both studies, but failed to prevent infection when administered in the latter study at 10 or 30 days after exposure, at which time the parasite resides in the spinal cord. These authors also showed that ivermectin, at the doses noted above, was ineffective against established, adult populations of the parasite in artificially infected white-tailed deer, although Kocan (1985) did observe a temporary reduction in larvae shed from the treated animals. For the related species, *Parelaphostrongylus andersoni*, Samuel and Gray (1988) evaluated the efficacy of ivermectin by treating white-tailed deer (held indoors on cement to prevent reinfection) subcutaneously at 200 or 400 μg kg^{-1}. The number of first-stage larvae in faeces dropped to zero at 17–18 days post-treatment. Reappearance of larvae in the faeces occurred 1.5–6 weeks later in six of seven deer. A second 200 μg kg^{-1} treatment was administered to four deer approximately 9 weeks following the initial treatment and, in 12–18 days, first-stage larvae again dropped to zero in the faeces and remained so for 14–49 days. These results and those of Kocan (1985) suggest that larval production by the adult female *Parelaphostrongylus* spp. may be suppressed and/or first-stage larvae are killed in the lungs for a period of time following ivermectin treatment.

Arthropods

White-tailed deer, naturally infested with *Psoroptes cuniculi*, were fed ivermectin-treated (technical grade) maize at a rate of 1000 g day^{-1} (~200 μg kg^{-1} day^{-1}) twice for 3 days at approximately a 4-week interval in an attempt to eliminate the ear mites (Garris *et al.*, 1991). This regime cleared 11 out of 12 deer of ear mites by 17 days after the last treatment. Several additional days of treatment at the 1000 g day^{-1} level were effective in eradicating the parasite in all deer, except one doe that exhibited all stages of the mite and, based on social behaviour within the herd and an abscessed jaw, was believed to not have eaten a sufficient amount of the treated maize. One intramuscular injection of ivermectin at 400 μg kg^{-1} ultimately was effective in clearing the mite infestation in this doe.

Miller *et al.* (1989) reported that daily oral doses for 48 days of 35 or 50 μg kg^{-1} of ivermectin in milk formula was 100% effective against adult and approximately 90% effective against nymphal lone star ticks (*Amblyomma americanum*) when ticks were artificially applied to the deer 7 days after ivermectin treatment was initiated and again 25 days later. When the authors gave a single 50 μg kg^{-1} dose orally, ivermectin was more than 90% effective against attached adult and nymphal ticks, but protection was dramatically less against ticks placed on deer 3 days after treatment. Ivermectin-treated, whole-kernel maize (10 mg ivermectin 0.45 kg^{-1} maize) fed to white-tailed deer at a target rate of 0.45 kg maize per deer day^{-1} to provide approximately 30 ng of ivermectin ml^{-1} of serum was evaluated for efficacy against *A. americanum* (Pound *et al.*, 1996) and *Ixodes scapularis* (Rand *et al.*, 2000) under field conditions. In the former study, feed was provided daily and split into an early morning and late afternoon feeding from February to September in both 1992 and 1993 and in both medicated and non-medicated pastures. This treatment resulted in 83.4, 92.4 and 100% fewer adults, nymphs and larvae of *A. americanum*, respectively, in the treated pasture compared with the non-treated pasture. In the second study (Rand *et al.*, 2000), treated maize was fed during five consecutive periods (1994–1996) corresponding to the presence of questing adult *I. scapularis*. Deer that achieved a serum concentration of ivermectin of ≥15 ng ml^{-1} exhibited reductions of 91, 95 and 94% in female tick infestation, oviposition and larval eclosion, respectively, for *I. scapularis* relative to deer with serum levels less than 15 ng ml^{-1} or no detectable ivermectin.

Treatment of community-housed white-tailed deer and mule deer fawns with ivermectin intramuscularly at 200 μg kg^{-1} twice at a 21-day interval reduced but did not eliminate the sucking louse, *Linognathus africanus* (Foreyt *et al.*, 1986), although only animals exhibiting lice were treated.

Ivermectin administered subcutaneously at 200 μg kg^{-1} was effective in treating *Cephenemyia* spp. larvae in white-tailed deer, resolving clinical signs by 10 days post-treatment (Weber, 1992).

Red Deer (*Cervus elaphus*)

Nematodes

Andrews and Lancaster (1988) and Andrews *et al.* (1993) reported, on the basis of field observations, that a single dose of injectable ivermectin at 200 μg kg^{-1} as recommended for cattle, sheep and horses appeared to be insufficient to control nematode infections in red deer, as is the case of goats treated at the sheep dose (reviewed by Conder and Campbell, 1995). Based on egg/larval count (Andrews *et al.*, 1993) or egg/larval/worm count (Mackintosh *et al.*, 1993) data, 400 μg kg^{-1} of ivermectin did appear to be slightly more efficient than 200 μg kg^{-1} when administered subcutaneously against *D. viviparus* and gastrointestinal worms; the former dose is widely used now in red deer (Connan, 1996). Consistent with the above data for injectable ivermectin, Connan (1991) reported that a group of red deer, treated orally with ivermectin at 200 μg kg^{-1} at housing (December) and again 4 weeks later, exhibited poor condition and appreciable worm burdens (predominantly ostertagids) at necropsy (April). The stage of development of the worms at necropsy, the presence of umbilicated nodules consistent with those of ostertagiasis and the lack of exposure to infection during housing suggested an inhibited population that had not been controlled adequately by ivermectin treatment. Further, Connan (1997) found that 400 μg kg^{-1} of ivermectin administered subcutaneously was 100% effective against adults and developing larvae and 95% effective against hypobiotic larvae of ostertagids, based on worm counts. *D. viviparus* may not require the higher dose, since a subcutaneous dose (Mackintosh and Mason, 1985; Mackintosh *et al.*, 1985) or an oral dose (Mackintosh *et al.*, 1990) of ivermectin at 200 μg kg^{-1} was found to be 100% effective against this nematode, based on worm counts or faecal larval counts, respectively. Kutzer (1990), using two oral doses of ivermectin administered at 200 μg kg^{-1} and given at a 4-week interval, found the treatment regime to be 100% effective against gastrointestinal nematodes, *D. viviparus*, and *Varestrongylus sagittatus*, but not effective against *Elaphostrongylus cervi*, based on egg and larval counts. Two 300 μg kg^{-1} doses of ivermectin given on consecutive days in a granulated food (2.5 kg ivermectin pre-mix (0.6% ivermectin) in 500 kg of granulate) provided 89.3, 91.2, 90.3, 95.3 and 99.6% reductions in Trichostrongylidae, *Trichuris* spp., *E. cervi*, *V. sagittatus* and *Dictyocaulus noerneri*, respectively, 18 days post-treatment compared with pre-treatment levels, as determined by flotation and Baermann methods (Malczewski *et al.*, 1998). Kutzer (2000) also reported that use of an ivermectin pre-mix (0.6%) administered at 400 μg kg^{-1} twice at an interval of 1 week was effective in controlling intestinal and lung nematodes in red deer. To assess persistence, red deer artificially infected with *D. viviparus* were treated topically with ivermectin pour-on at 500 μg kg^{-1} (Rehbein and Visser, 1997). Ivermectin was more than 99% effective against infection with *D. viviparus* for a mininum of 28 days following treatment. Mackintosh *et al.* (1990) and Mason *et al.* (1990) observed similar results based on faecal larval counts, taking an approximately 21-day pre-patent period for the parasite into consideration.

Moxidectin in a pour-on formulation administered at 500 μg kg^{-1} was evaluated in red and wapiti × red hybrid deer against natural infections of lungworms (*D. viviparus*) and abomasal nematodes (*Ostertagia* type)

(Waldrup *et al.*, 1998). In red deer, moxidectin was 100% effective against all stages of the lungworm and, in wapiti × red hybrids, the treatment was 100 and 99.7% effective against adult and immature stages, respectively. Against the *Ostertagia*-type nematodes (genera assumed to be *Ostertagia, Spiculopteragia, Skrjabinagia* and *Apteragia*) in both red and wapiti × red hybrid deer, moxidectin was ≥99.9% effective against all stages. Similar results in red deer had been reported previously by Mackintosh *et al.* (1993) and Middelberg (1994); they also found moxidectin to be effective against *Haemonchus, Trichostrongylus* and *Oesophagostomum* spp. Mackintosh *et al.* (1997) found that based on worm counts, moxidectin in the pour-on formulation administered at 500 μg kg^{-1} provided protection for up to 42 days following treatment against artificial challenge with a mixed nematode population, including *D. viviparus, Spiculopteragia asymmetrica, S. spiculoptera, Ostertagia leptospicularis, Skrjabinagia kolchida, Trichostrongylus axei, Cooperia* spp., *Oesophagostomum* spp. and *Chabertia* spp.

Eprinomectin pour-on applied at 500 μg kg^{-1} was reported to be safe and effective against adult and fourth-stage larvae of *D. viviparus, Ostertagia* spp., *Trichostrongylus* spp. and adult *Oesophagostomum* spp. in red deer (Gogolewski *et al.*, 1997a,b).

Arthropods

As an anecdote, Fletcher (1984) reported resolution of fly strike in a red deer stag treated 3 days previously with 4 ml of 1% ivermectin. Although no details were provided, Rafferty (1982) reported that ivermectin was effective in controlling warble fly larvae in red deer, and Soll (1989) reported that Petrov and colleagues used ivermectin-medicated salt licks for treatment and prevention of warble fly infection. Kutzer (1988, 1990) using two oral doses of ivermectin administered at 200 μg kg^{-1} and given on 2 successive days or at a 4-week interval found these treatment regimes to be at or near 100% effective against *Pharyngomyia picta* and *Hypoderma actaeon*. Kutzer (2000) also demonstrated that use of an ivermectin pre-mix (0.6%) administered at 400 μg kg^{-1} twice at an interval of 1 week was at or near 100% effective in controlling *P. picta, Cephenemyia stimulator* and *H. actaeon* in red deer.

Pharmacokinetics, safety and residues

Although the data are not directly comparable, having been collected in different studies, it does appear that the C_{max} resulting from a 200 μg kg^{-1} dose of ivermectin is dramatically reduced in red deer (15.3 ng ml^{-1} (Andrews *et al.*, 1993); 15.8 ng ml^{-1} (Mackintosh *et al.*, 1985)) relative to cattle (44 ng ml^{-1} (Fink and Porras, 1989)). A 400 μg kg^{-1} dose in red deer provides a higher C_{max} (28.3 ng ml^{-1} (Andrews *et al.*, 1993)), but one still dramatically less than that for cattle. Gogolewski *et al.* (1997a) found eprinomectin administered topically at five times the recommended dose of 500 μg kg^{-1} to be safe in red deer.

Roe Deer (*Capreolus capreolus*)

Nematodes

Moxidectin or ivermectin administered subcutaneously at 200 μg kg^{-1} were shown to be more than 96% effective against *D. viviparus* and *E. cervi* in roe deer as judged by larval shedding 21–49 days after treatment (Sugár, 1995). Kutzer (2000) reported that use of an ivermectin pre-mix (0.6%) administered at 400 μg kg^{-1} twice at an interval of 1 week was effective in controlling intestinal and lung nematodes in roe deer.

Arthropods

Lamka *et al.* (1996) investigated whether ivermectin would control larval stages of warble fly (*Hypoderma diana*) in roe deer. The study was conducted for 3 years. During winter feeding, ivermectin was administered orally for 2 days at a daily dose of 300 μg kg^{-1}. Subsequently, any deer that was killed during the hunting season was examined for larvae. An appreciable decrease in prevalence and intensity of infection was observed. In addition, direct checks of treated, tame deer showed that oral ivermectin was highly effective against the larval stage of the warble fly. Using the same winter treatment dose and regime, Lamka *et al.* (1997) found that ivermectin was effective in markedly reducing the prevalence and intensity of infection with larval *C. stimulator* in roe deer during the spring and summer. Kutzer (1988) found 200 μg kg^{-1} of ivermectin given orally on 2 successive days when added to feed was highly effective against *P. picta* and *C. stimulator* in roe deer on hunting grounds. Kutzer (2000) also demonstrated that use of an ivermectin pre-mix (0.6%) administered at 400 μg kg^{-1} twice at an interval of 1 week was at or near 100% effective in controlling *C. stimulator*.

Fallow Deer (*Dama dama*)

Nematodes

Ivermectin administered orally at 200 μg kg^{-1} to fallow deer was ineffective against mixed nematode infections based on faecal egg counts 13 days following treatment (Mylrea *et al.*, 1991). A second treatment of ivermectin (1 ml per deer) given subcutaneously 13 days after the original treatment resulted in egg counts of zero at 40 and 47 days after the second treatment in at least two deer; however, at necropsy at the latter time, *S. asymmetrica* was present in both and *O. ostertagi* was found in one of the two deer. These results could be explained by either reinfection or suppression of egg production following the second treatment. Two 300 μg kg^{-1} doses of ivermectin given on consecutive days in a granulated food (2.5 kg ivermectin pre-mix (0.6% ivermectin) in 500 kg of granulate) provided 95.5, 100 and 100% reductions in Trichostrongylidae, *Trichuris* spp. and *D. noerneri*, respectively, 18 days post-treatment compared with pre-treatment levels, as determined by flotation and Baermann methods (Malczewski *et al.*, 1998).

One-humped Camel (*Camelus dromedarius*)

Nematodes

Robin *et al.* (1989) reported that based on worm counts, ivermectin given subcutaneously at 200 μg kg^{-1} effectively controls (clearance >90%) gastrointestinal nematodes, including adults of *Haemonchus contortus, H. longistipes, Ostertagia* spp., *Trichostrongylus axei, T. colubriformis, T. probolurus, T. vitrinus, Camelostrongylus mentulatus* (also L_4), *Impalaia tuberculata, Oesophagostomum columbianum, O. venulosum, Chabertia ovina* and, although in small numbers, *Cooperia* spp. Mixed results were obtained by these authors between studies for *Nematodirus spathiger* and *Trichuris globulosa.* Similar results, based on worm counts, for ivermectin using the same dose and route against *H. longistipes* and *I. tuberculata* were reported by Tager-Kagan and Robin (1986). Dafalla *et al.* (1987), Raisinghani *et al.* (1989), Makkar *et al.* (1991) and L.K. Sharma (1991) found that faecal egg counts went down to zero by 4–15 days following subcutaneous administration of ivermectin at 200 μg kg^{-1} and remained at this level for a minimum of 30 days following treatment. Reductions of approximately 97 and 99%, based on faecal counts of nematode eggs, were reported by Kumar and Yadav (1991) and Partani *et al.* (1995) 21 days following a 200 μg kg^{-1} subcutaneous dose of ivermectin, with the residual infection being attributed respectively to *Strongyloides* spp. or mixed *Strongyloides, Nematodirella* and *Trichuris* spp. Njanja (1991) saw egg reductions of ≥93% for a mixed population of nematodes at 1 week following subcutaneous treatment with ivermectin at 200 μg kg^{-1}. Against *H. longistipes*, Maqbool *et al.* (1994) observed a 95% reduction in faecal egg counts 18 days after subcutaneous treatment with ivermectin at 200 μg kg^{-1}. Based on faecal egg counts, Olaho-Mukani and Kimani (1999) observed reductions relative to pre-treatment counts of 71, 86 and 78% on consecutive monthly examinations following subcutaneous treatment at 200 μg kg^{-1} for a mixed strongyle population, and the authors suggested that resistance may be developing to ivermectin. Boyce *et al.* (1984) showed that orally administered ivermectin at 200 μg kg^{-1} was ≥96.7% effective against *Haemonchus*, *Trichostrongylus* and *Cooperia* spp. for up to 4 weeks post-treatment, as judged by mean reductions in faecal egg counts and identification following larval culture, while subcutaneous treatment at 200 μg kg^{-1} provided more modest mean reductions of ≥74.1% over the same period. Further, Boyce *et al.* (1984) demonstrated mean egg count reductions of 85.2% for *Trichuris* spp. at 4 weeks following oral ivermectin treatment and no reduction by the subcutaneous route.

Elamin *et al.* (1993) found that a single subcutaneous dose of ivermectin at 200 μg kg^{-1} was effective in eliminating microfilariae of an unidentified filarial species (morphology and measurements consistent with *Dipetalonema evansi*) within 2–5 days and resulted in marked clinical improvement by 1–2 weeks following treatment. Treated camels remained amicrofilaraemic through to the end of the study (133 days post-treatment). Two camels treated at a time when microfilarial counts were high died, presumably due to an anaphylactic-type response resulting from a rapid and massive release of microfilarial antigens, as has been reported previously for microfilaricidal drugs (e.g. Greene *et al.*, 1989). The 200 μg kg^{-1} dose of subcutaneously administered ivermectin was also found to be effective in treating *D. evansi* in camels by Agag *et al.* (1993).

Arthropods

Camels with mange due to *S. scabiei* were treated subcutaneously with ivermectin at 200 μg kg^{-1} twice at a weekly interval (Njanja, 1991; Nayee *et al.*, 1994), twice (Opferman, 1985; Hashim and Wasfi, 1986; Raisinghani *et al.*, 1989; Makkar *et al.*, 1991) or three times (Maqbool *et al.*, 1996) at approximately a 2-week interval, or three times (Radwan *et al.*, 1987; Chellappa *et al.*, 1989) at weekly intervals. The infestations resolved and no recurrence was evident during the course of the study, 12 months following treatment in one report. Hayat *et al.* (1997) found that a single 200 μg kg^{-1} subcutaneous dose of ivermectin reduced *S. scabiei* numbers by 99% at 45 days post-treatment in camels; Njanja (1991) found the same presentation and dose of ivermectin to be 100% effective against mild infestations by 3 weeks post-treatment, while Hassan *et al.* (1989) observed that a single subcutaneous injection of ivermectin at either 100 or 200 μg kg^{-1} was effective in eliminating *S. scabiei* from infested camels by 60 days post-treatment, although the lower dose was not examined against severe infestations. Dafalla *et al.* (1987) cleared sarcoptic mange from six of seven camels with a single subcutaneous treatment of ivermectin at 200 μg kg^{-1}, but the last animal required a second treatment, which was administered 35 days after the first. In contrast, Abu-Samra (1999) reported that poor results (40% cure) were obtained against *S. scabiei* following two treatments with ivermectin at 200 μg kg^{-1} at an interval of 5 weeks, probably as a result of eggs surviving treatment and re-establishing infestation, suggesting that a second treatment is required before eggs have the opportunity to produce patent adults.

Ivermectin administered subcutaneously at 200 μg kg^{-1} was ineffective against *Hyalomma* spp. ticks on camels (Tager-Kagan and Robin, 1986; van Straten and Jongejan, 1993) and larvae of the fly *Cephalopina titillator* (Tager-Kagan and Robin, 1986; Robin *et al.*, 1989). In contrast, Raisinghani *et al.* (1989) found the same dose and route to be effective in removing *Hyalomma dromedarii* by 5 days after treatment, and Sharma (1992) found this treatment to be effective against *C. titillator*.

Pharmacokinetics, residues and safety

Pharmacokinetic evaluation of subcutaneously administered (200 μg kg^{-1}) ivermectin or moxidectin levels following treatment in plasma and milk of lactating camels (Oukessou *et al.*, 1999) and for ivermectin in plasma of non-lactating camels (Oukessou *et al.*, 1996) yielded area under the curve (AUC), C_{max} and T_{max} values that generally suggested dramatically reduced bioavailability of the two drugs in camels relative to published literature on cattle, but more comparable bioavailability relative to published information on sheep and goats for the respective compound. The plasma AUC (ng ml^{-1} day^{-1}), C_{max} (ng ml^{-1}) and T_{max} (days) values, respectively, observed for ivermectin were approximately 30, 2 and 12 (lactating camels) and 66, 3 and 6 (non-lactating camels), and for moxidectin were 71, 9 and 1 (non-lactating camels). Respective values observed in milk of lactating camels for ivermectin were 38, 3 and 17, and for moxidectin were 291, 29 and 4. In general, the data from lactating camels suggest a greater bioavailability for moxidectin than for ivermectin in this species.

A 28-day withdrawal period has been established by the FDA for ivermectin administered subcutaneously at 200 μg kg^{-1} in camels. Transitory discomfort has been observed in some camels following subcutaneous administration, and a low incidence of soft tissue swelling, which resolved without treatment, has been seen at the injection site (Hashim and Wasfi, 1986; Robin *et al.*, 1989; Maqbool *et al.*, 1994). Radwan *et al.* (1987) found no adverse effects on pregnancy or resulting offspring following monthly subcutaneous treatments with ivermectin at 600 μg kg^{-1} during the second and third trimesters in camels.

Two-humped Camel (*Camelus bactrianus*)

Arthropod

Kuntze and Kuntze (1991) reported that subcutaneously administered ivermectin given once at 200 μg kg^{-1} was effective against sarcoptic mange in the two-humped camel.

Llama (*Lama glama*)

Nematodes

For gastrointestinal nematodes and lungworms in llamas, Cheney and Allen (1989) and Fowler (1989) state that ivermectin given subcutaneously or orally at 200 μg kg^{-1} is safe and effective. Johnson *et al.* (1992) observed a reduction in the number of animals exhibiting strongyle eggs following topical treatment with ivermectin at 500 μg kg^{-1} in each of three trials; in two of the trials, no animals shed strongyle eggs after treatment. Mixed results were obtained by these authors for *Nematodirus* and *Trichuris* spp. Windsor (1997) reported that despite treatment with ivermectin (dose and route not specified), three llamas from various locations with type II ostertagiasis died due to the parasite.

Mixed results have been obtained therapeutically and prophylactically for ivermectin against *P. tenuis* in llamas as reviewed by Pugh *et al.* (1995). Results may vary due to treatment timing relative to that of infection and subsequent development/location of the parasite, as discussed above for ivermectin treatment of *P. tenuis* in white-tailed deer.

Arthropods

Foreyt *et al.* (1992) found 200 μg kg^{-1} of ivermectin administered subcutaneously plus two drops of ivermectin diluted in saline (no specifics provided) administered topically in each ear was an effective treatment for *Psoroptes* spp. in llamas based on the absence of mites at 3 weeks post-treatment. Against sarcoptic mange, 200 μg kg^{-1} of ivermectin given subcutaneously once (Kuntze and Kuntze, 1991) or twice at a 10–14 day interval was effective (Kress, 1983). Ivermectin in the injectable formulation was not effective against *Chorioptes* spp. in llamas (Long, as reported by Rickard, 1994), and Johnson (1994) suggested that 400 μg kg^{-1} of ivermectin administered subcutaneously may be needed to elicit an effective result for chorioptic mange.

Larvae and nymphs of the spinose ear tick (*Otobius megnini*) can be treated effectively with ivermectin, administered subcutaneously at

200 μg kg^{-1} (Fowler, 1989). This same treatment was found to be effective against sucking lice (*Microthoracius* spp.) but not biting lice (*Damalinia breviceps*) by Fowler (1986).

Fowler and Paul-Murphy (1985) observed resolution of clinical signs associated with a nasopharygeal bot of the genus *Cephenemyia* following subcutaneous treatment with 200 μg kg^{-1} of ivermectin.

Alpaca (*Lama pacos*)

Nematodes

For gastrointestinal nematodes and lungworms in alpacas, Fowler (1989) states that ivermectin given subcutaneously or orally at 200 μg kg^{-1} is safe and effective. Windsor *et al.* (1992) reported that ivermectin administered subcutaneously at 200 μg kg^{-1} reduced but did not eliminate mixed nematode infections (including *Lamanema chavezi* and *Nematodirus* spp.) from alpacas at 1 month after treatment, based on faecal egg counts.

Arthropods

Ivermectin given subcutaneously at 200 μg kg^{-1} reduced, but did not eliminate, mange (*S. scabiei*) and lice (*Microthoracius praelongiceps*) infections in alpacas at 1 month following treatment, based on skin scrapings or five wool partings (Windsor *et al.*, 1992). Two treatments of moxidectin given subcutaneously at 200 μg kg^{-1} at a 7–10 day interval is reportedly effective against mange (psoroptic and sarcoptic) and lice (*Microthoracius mazzai*) (Cicchino *et al.*, 1998). Three injections of ivermectin at 200 μg kg^{-1} given at 10 day intervals in alpacas resulted in a regression of lesions due to chorioptic mange and, although skin scrapings were negative during treatment, the effect was considered to be transitory (Petrikowski, 1998). An unspecified mixture of ivermectin and dimethylsulphoxide (DMSO) applied topically was effective in eliminating chorioptic mange from an alpaca (Rickard, 1994).

Llama (*Lama guanicoe*)

Arthropod

Against sarcoptic mange, 200 μg kg^{-1} of ivermectin given subcutaneously once was effective in the guanaco (Kuntze and Kuntze, 1991).

References

Abu-Samra, M.T. (1999) The efficacy of Sebacil E.C. 50%, Gamatox and Ivomec in the treatment of sarcoptic mange in camel (*Camelus dromedarius*). *Journal of Camel Practice and Research* 6, 61–67.

Agag, B.I., Nasser, M.H., Abu-El-Magd, M.M. and Hafez, I.A. (1993) Clinical and biochemical studies on microfilaria and trypanosoma infected camels. *Assiut Veterinary Medical Journal* 29, 125–134.

Ahmad, M., Khan, I.Q., Gill, G.M., Maqbool, A. and Ali, S. (1997) Prevalence and chemotherapy of sarcoptic mange in buffaloes. *Veterinary Medical Journal, Giza* 45, 397–401.

Andrews, S.J. and Lancaster, M.B. (1988) Use of ivermectin in deer. *Veterinary Record* 123, 354.

Andrews, S.J., Ferrari, M.M., Pow, J.D.E. and Lancaster, M.B. (1993) Nematode egg output and plasma concentration of ivermectin after its administration to red deer (*Cervus elaphus elaphus*). *Veterinary Record* 132, 161–163.

Boyce, W., Kollias, G., Courtney, C.H., Allen, J. and Chalmers, E. (1984) Efficacy of ivermectin against gastrointestinal nematodes in dromedary camels. *Journal of the American Veterinary Medical Association* 185, 1307–1308.

Chellappa, E., Thiruthallinathan, R. and Ravishankar, M. (1989) Efficacy of 'ivermectin' against mange in camels. *Indian Veterinary Journal* 66, 451–452.

Cheney, J.M. and Allen, G.T. (1989) Parasitism in llamas. *Veterinary Clinics of North America: Food Animal Practice* 5, 217–225.

Cicchino, A.C., Muñoz Cobeñas, M.E., Bulman, G.M., Diaz, J.C. and Laos, A. (1998) Identification of *Microthoracius mazzai* (Phthiraptera: Anoplura) as an economically important parasite of alpacas. *Journal of Medical Entomology* 35, 922–930.

Conder, G.A. and Campbell, W.C. (1995) Chemotherapy of nematode infections of veterinary importance with special reference to drug resistance. In: Baker, J.R., Muller, R. and Rollinson, D. (eds) *Advances in Parasitology*, Vol. 35. Academic Press, San Diego, California, pp. 1–84.

Connan, R.M. (1991) Type II ostertagiosis in farmed red deer. *Veterinary Record* 128, 233–235.

Connan, R.M. (1996) Observations on the epidemiology of gastrointestinal nematodes of farmed red deer in central southern England. *Veterinary Record* 139, 228–232.

Connan, R.M. (1997) Hypobiosis in the ostertagids of red deer and the efficacy of ivermectin and fenbendazole against them. *Veterinary Record* 140, 203–205.

Dafalla, E.I., El Gezuli, A.Y. and Saad, M.B. (1987) A preliminary study on the efficacy of ivermectin (Ivomec) against natural infections with sarcoptic mange and gastrointestinal nematodes of the one-humped camel (*Camelus dromedarius*) in the Sudan. *Sudan Journal of Veterinary Science and Animal Husbandry* 26, 39–45.

Demiaszkiewicz, A.W., Krasiński, Z., Dróżdż, J. and Lachowicz, J. (1997) Skuteczność preparatu Ivomec premix w zwalczaniu helmintoz żubrów w rezerwatach Bialowiesskiego Parku Narodowego. *Magazyn Weterynaryjna* 6, 366–367.

Dieterich, R.A. and Craigmill, A.L. (1990) Safety, efficacy, and tissue residues of ivermectin in reindeer. *Rangifer* 10, 53–56.

Elamin, E.A., Mohamed, G.E., Fadl, M., Elias, S., Saleem, M.S. and Elbasher, M.O.A. (1993) An outbreak of cameline filariasis in the Sudan. *British Veterinary Journal* 149, 195–200.

Fahmy, M.M., Mousa, W.M. and El-Ghaysh, A. (1996) Application of a new formulation of ivermectin (Ivomec Pour-On) to control external parasites of farm animals in Egypt. *Veterinary Medical Journal, Giza* 44, 339–347.

Fink, D.W. and Porras, A.G. (1989) Pharmacokinetics of ivermectin in animals and humans. In: Campbell, W.C. (ed.) *Ivermectin and Abamectin*. Springer-Verlag, New York, pp. 113–130.

Fletcher, J. (1984) Ivermectin in fly strike. *Veterinary Record* 115, 475.

Foreyt, W.J., Rice, D.H. and Kim, K.C. (1986) Pediculosis of mule deer and white-tailed deer fawns in captivity. *Journal of the American Veterinary Medical Association* 189, 1172–1173.

Foreyt, W.J., Rickard, L.G. and Boyce, W. (1992) *Psoroptes* sp. in two llamas (*Lama glama*) in Washington. *Journal of Parasitology* 78, 153–155.

Fowler, M.E. (1986) Lice in llamas. *Avian/Exotic Practice* 3, 22–25.

Fowler, M.E. (1989) Parasites. In: *Medicine and Surgery of South American Camelids*. Iowa State University Press, Ames, Iowa, pp. 132–165.

Fowler, M.E. and Paul-Murphy, J. (1985) *Cephenemyia* sp. infestation in the llama. *California Veterinarian* 39, 10–12.

Garg, S.K., Chauhan, P.P.S., Bhushan, C. and Katoch, R. (1998) Efficacy of ivermectin and closantel against whipworm infection in buffalo calves. *Indian Veterinary Journal* 75, 82–83.

Garris, G.I., Prullage, J.B., Prullage, J.L., Wright, F.C. and Miller, J.A. (1991) Control of *Psoroptes cuniculi* in captive white-tailed deer with ivermectin-treated corn. *Journal of Wildlife Diseases* 27, 254–257.

Gill, B.S., Singh, J., Gill, B.S., Singh, A., Khehra, S.S., Rai, A. and Hussain, O. (1989) Efficacy of ivermectin against mange and gastrointestinal nematodes of buffalo (*Bubalus bubalis*). *Veterinary Parasitology* 31, 141–147.

Gill, B.S., Balakrishnan, P., Lumsden, G.G. and Jones, P.G.H. (1991) Treatment of stephanofilariosis ('earsore') with ivermectin. *Veterinary Parasitology* 40, 159–163.

Gogolewski, R.P., Familton, A.S., Slacek, B., Allerton, G.R., Alexander, A., Langholff, W. and Eagleson, J.S. (1997a) The safety and anthelmintic efficacy of a topical formulation of eprinomectin in red deer. *World Association for the Advancement of Veterinary Parasitology*, Abstract 122.

Gogolewski, R.P., Plue, R.E., Rugg, D., Allerton, G.R., Familton, A.S., Langholff, W. and Eagleson, J.S. (1997b) Field trials in New Zealand demonstrating the anthelmintic efficacy of a topical formulation of eprinomectin in red deer. *World Association for the Advancement of Veterinary Parasitology*, Abstract 123.

Gopal, K.V., Rao, R. and Gaffar, A.A. (1992) Therapy of stephanofilarial ear-sore in buffaloes. *Indian Veterinary Journal* 69, 165–166.

Greene, B.M., Brown, E.R. and Taylor, H.R. (1989) Use of ivermectin in humans. In: Campbell, W.C. (ed.) *Ivermectin and Abamectin*. Springer-Verlag, New York, pp. 311–323.

Hashim, N.H. and Wasfi, I.A. (1986) Ivermectin treatment of camels naturally infected with sarcoptic mange. *World Animal Review* 57, 26–29.

Hassan, A.B., El-Atrash Kawther, S.M. and El-Hady, A. (1989) Treatment of sarcoptic mange in dromedary camels with ivermectin. *Indian Veterinary Journal* 66, 1164–1167.

Haugerud, R.E., Nilssen, A.C. and Rognmo, A. (1993) On the efficacy of ivermectin against the reindeer sinus worm *Linguatula artica* (Pentastomida), with a review on ivermectin treatment in reindeer. *Rangifer* 13, 157–162.

Hayat, B., Rahim, M.A., Hayat, C.S., Khan, M.N. and Qudoos, A. (1997) Evaluation of the efficacy of different acaricides against *Sarcoptes scabiei* var *cameli* of dromedary camel in Pakistan. *Indian Veterinary Journal* 74, 164–166.

Hayat, C.S., Maqbool, A., Amin, K.K. and Hayat, B. (1996) Clinical epidemiology and chemotherapy of sarcoptic mange in buffaloes. *Pakistan Journal of Zoology* 28, 123–126.

Johnson, L.W. (1994) Llama herd health. *Veterinary Clinics of North America: Food Animal Practice* 10, 248–258.

Johnson, L.W., Cheney, J.M., Taton-Allen, G. and Getzy, D.M. (1992) Use of ivermectin (Ivomec®) topically in llamas. *Proceedings of the 1992 Symposium on the Health and Disease of Small Ruminants*, pp. 36–38.

Kocan, A.A. (1985) The use of ivermectin in the treatment and prevention of infection with *Parelaphostrongylus tenuis* (Dougherty) (Nematoda: Metastrongyloidea) in white-tailed deer (*Odocoileus virginianus* Zimmermann). *Journal of Wildlife Diseases* 21, 454–455.

Kress, P.J. (1983) Use of ivermectin for treatment of scabies in llamas. *Llama World* 1, 18–19.

Kumar, A. and Joshi, B.P. (1995) Studies on the clinical efficacy of ivermectin against cutaneous myiasis in domestic animals. *Indian Veterinary Journal* 72, 277–280.

Kumar, G.S. and Suryanarayana, C. (1995) Clinico-biochemical and therapeutic studies on mange in buffalo calves. *Indian Veterinary Journal* 72, 77–79.

Kumar, S. and Yadav, C.L. (1991) Comparative efficacy of ivermectin, fenbendazole and levamisole against gastrointestinal nematodiasis in camels. *Indian Veterinary Journal* 68, 787–789.

Kuntze, A. and Kuntze, O. (1991) Ivermectin in exotic animals: scabies in tylopodes (*Camelus bactrianus, Lama guanicoe, L. glama*) and scabies and ascariasis in bears (*Thalarctos maritimus, Ursus arctos*). *Berliner und Münchener Tierärzliche Wochenschrift* 104, 46–48.

Kutzer, E. (1988) Treatment of ectoparasites by ivermectin (Ivomec®) in red deer, roe deer and wild boar. *Mitteilungen der Deutschen Gesellschaft für Allgemeine und Angewandte Entomologie* 6, 217–222.

Kutzer, E. (1990) Zum Einsatz von ivermectin (Ivomec®) bei rotwild (*Cervus elaphus hippelaphus*) unter besonderer berücksichtigung der lungenwurmbehandlung. *Wiener Tierärztliche Monatsschrift* 77, 309–312.

Kutzer, E. (2000) The treatment of oestrinosis and hypodermosis in red deer (*Cervus elaphus hippelaphus*) and roe deer (*Capreolus* c. *capreolus*) by means of ivermectin (Ivomec®). *Berliner und Münchener Tierärztliche Wochenschrift* 113, 149–151.

Lamka, J., Suchý, J. and Štaud, F. (1996) Efficacy of perorally administered ivermectin against larval stages of warble fly (*Hypoderma diana* B.) in roe deer. *Veterinarni Medicina* 41, 251–254.

Lamka, J., Suchý, J. and Štaud, F. (1997) Efficacy of orally administered ivermectin against larval stages of bot fly (*Cephenemyia stimulator* C.) in roe deer. *Acta Veterinaria Brno* 66, 51–55.

Mackintosh, C.G. and Mason, P.C. (1985) Anthelmintics and lungworm in red deer. *The Royal Society of New Zealand, Bulletin* 22, 131–133.

Mackintosh, C.G., Mason, P.C., Manley, T., Baker, K. and Littlejohn, R. (1985) Efficacy and pharmacokinetics of febantel and ivermectin in red deer. *New Zealand Veterinary Journal* 33, 127–131.

Mackintosh, C.G., Mason, P.C. and Taylor, M. (1990) A pilot study of the efficacy of topical ivermectin against lungworm in young red deer (*Cervus elaphus*). *New Zealand Veterinary Journal* 38, 112–113.

Mackintosh, C.G., Waldrup, K., Labes, R. and Taylor, M. (1993) Efficacy of ivermectin injection and moxidectin pour-on formulations in young red deer, *Cervus elaphus*. *Deer Branch of the New Zealand Veterinary Association*, Course No. 10, pp. 143–150.

Mackintosh, C.G., Qureshi, T., Waldrup, K., Labes, R.E., Taylor, M., Murphy, A. and Johnstone, P. (1997) Persistence of moxidectin activity against nematodes in red deer. *Deer Branch of the New Zealand Veterinary Association* 14, 149–154.

Mahato, S.N. (1989) Leech infestation of livestock. *Veterinary Record* 124, 641–642.

Makkar, M.S., Patel, P.R., Suthar, B.H., Momin, R.R., Mody, S.K., Tripathi, R.M. and Siddiqui, G.M. (1991) Efficacy of ivermectin against sarcoptic mange of camels. *Indian Journal of Veterinary Medicine* 11, 76.

Malczewski, A., Dróżdż, J., Demiaszkiewicz, A.W., Lachowicz, J. and Dmuchowski, B. (1998) Efficacy of Ivomec against helminths in red deer and fallow deer. *Medycyna Weterynaryjna* 54, 46–48.

Maqbool, A., Abdel Rahim, M., Iftikhar, M., Khan, M.N. and Butt, A.A. (1994) Haemonchosis in camels and its treatment with ivermectin. *Assiut Veterinary Medical Journal* 31, 130–134.

Maqbool, A., Athar, M. and Shakoor, A. (1996) Efficacy of ivermectin against sarcoptic mange in camel. *Pakistan Journal of Scientific and Industrial Research* 39, 32–33.

Maqbool, A., Hayat, C.S., Raza, A., Badar, N. and Ahmad, M. (1997) Prevalence and chemotherapy of ascariasis in buffaloes. *Journal Veterinar Malaysia* 9, 47–50.

Marley, S.E., Knapp, S.E., Rognlie, M.C., Thompson, J.R., Stoppa, T.M., Button, S.M., Wetzlich, S., Arndt, T. and Craigmill, A. (1995) Efficacy of ivermectin pour-on against *Ostertagia ostertagi* infection and residues in the American bison, *Bison bison*. *Journal of Wildlife Diseases* 31, 62–65.

Mason, P.C., Mackintosh, C.G., Alexander, A., Taylor, M.J. and Booth, D. (1990) Activity of ivomec pour-on against lungworm in red deer. A pilot study. *Bulletin de la Societe France de Parasitologie* 8 (Supplement 2), 1101–1106.

Middelberg, A. (1994) Efficacy of moxidectin pour-on in young red deer. *Deer Branch of the New Zealand Veterinary Association*, Course No. 11, pp. 203–205.

Miller, J.A., Garris, G.I., George, J.E. and Oehler, D.D. (1989) Control of lone star ticks (Acari: Ixodidae) on Spanish goats and white-tailed deer with orally administered ivermectin. *Journal of Economic Entomology* 82, 1650–1656.

Mylrea, G.E., Mulley, R.C. and English, A.W. (1991) Gastrointestinal helminthosis in fallow deer (*Dama dama*) and their response to treatment with anthelmintics. *Australian Veterinary Journal* 68, 74–75.

Nayee, A.S., Avastthi, B.L., Kathiria, L.G. and Gill, B.S. (1994) Efficacy of ivermectin against *Sarcoptes scabiei* infection in camels. *Indian Journal of Animal Sciences* 64, 270–272.

Njanja, J.C. (1991) Therapeutic use of ivermectin against sarcoptic mange in camels. *Bulletin of Animal Health Production of Africa* 39, 275–279.

Nordkvist, M., Rehbinder, C., Christensson, D. and Rönnbäck, C. (1983) A comparative study on the effects of four anthelmintics on some important reindeer parasites. *Rangifer* 3, 19–38.

Nordkvist, M., Christensson, D. and Rehbinder, C. (1984) A deworming field trial with ivermectin (MSD) in reindeer. *Rangifer* 4, 10–15.

Oksanen, A. (1996) Influence of timing of endectocide antiparasitic treatment on its efficacy in overwintering reindeer. *Rangifer* 16, 147–150.

Oksanen, A. and Nieminen, M. (1996) Larvicidal effectiveness of doramectin against natural warble (*Hypoderma tarandi*) and throat bot (*Cephenemyia trompe*) infections in reindeer. *Medical and Veterinary Entomology* 10, 395–396.

Oksanen, A. and Nieminen, M. (1998) Moxidectin as an endectocide in reindeer. *Acta Veterinaria Scandinavica* 39, 483–489.

Oksanen, A., Nieminen, M., Soveri, T. and Kumpula, K. (1992) Oral and parenteral administration of ivermectin to reindeer. *Veterinary Parasitology* 41, 241–247.

Oksanen, A., Nieminen, M. and Soveri, T. (1993) A comparison of topical, subcutaneous and oral administrations of ivermectin to reindeer. *Veterinary Record* 133, 312–314.

Oksanen, A., Norberg, H., Nieminen, M. and Bernstad, S. (1995) Influence of route of administration on the plasma concentrations of ivermectin in reindeer. *Research in Veterinary Science* 58, 286–287.

Olaho-Mukani, W. and Kimani, J.K. (1999) Efficacy of parenteral formulation of levamisole and ivermectin against strongylosis in dromedary camels. *Journal of Camel Practice and Research* 6, 73–75.

Olsen, S. (1982) The efficacy of ivermectin (MK-933) for treatment and prevention of infection of *Parelaphostrongylus tenuis* (Metastrongyloidea) in cervids. MS Thesis, Oklahoma State University, Stillwater, Oklahoma.

Opferman, R.R. (1985) Treatment of sarcoptic mange in a dromedary camel. *Journal of the American Veterinary Medical Association* 187, 1240–1241.

Oukessou, M., Badri, M., Sutra, J.F., Galtier, P. and Alvinerie, M. (1996) Pharmacokinetics of ivermectin in the camel (*Camelus dromedarius*). *Veterinary Record* 139, 424–425.

Oukessou, M., Berrag, B. and Alvinerie, M. (1999) A comparative kinetic study of ivermectin and moxidectin in lactating camels (*Camelus dromedarius*). *Veterinary Parasitology* 83, 151–159.

Partani, A.K., Kumar, D., Manohar, G.S. and Kumar, R. (1995) Comparative efficacy of some anthelmintics against gastrointestinal nematodes in camel. *Journal of Camel Practice and Research* 2, 97–99.

Petrikowski, M. (1998) Chorioptic mange in an alpaca herd. *Advances in Veterinary Dermatology* 3, 450–451.

Pound, J.M., Miller, J.A., George, J.E., Oehler, D.D. and Harmel, D.E. (1996) Systemic treatment of white-tailed deer with ivermectin-medicated bait to control free-living populations of lone star ticks (Acari: Ixodidae). *Journal of Medical Entomology* 33, 385–394.

Pugh, D.G., Causey, M.K., Blagburn, B.L. and Wolfe, D.F. (1995) Clinical parelaphostrongylosis in llamas. *Compendium for Continuing Education for the Practicing Veterinarian – Food Animal* 17, 600–605.

Purohit, G.N., Vyas, S.K. and Yadav, S.B.S. (1997) Efficacy of ivermectin treatment for sarcoptic mange in Surti buffalo calves. *Indian Veterinary Journal* 74, 689–690.

Qudoos, A., Khan, M.N. and Hayat, C.S. (1996) Evaluation of the efficacy of different acaricides against *Sarcoptes scabiei* in naturally infested buffaloes in Pakistan. *Cheiron* 25, 58–62.

Radwan, Y.A., Abdou, O.M.A., Hamid, S.A. and Arab, R.M.H. (1987) Efficacy and safety of Ivomec against camel mange. *Veterinary Medical Journal* 35, 83–94.

Rafferty, G.C. (1982) Red deer endoparasites. *Veterinary Record* 111, 565.

Raisinghani, P.M., Kumar, D. and Rathore, M.S. (1989) Efficacy of ivermectin against *Sarcoptes scabiei* var *cameli* infestation in Indian camel (*Camelus dromedarius*). *Indian Veterinary Journal* 66, 1160–1163.

Rand, P.W., Lacombe, E.H., Holman, M.S., Lubelczyk, C. and Smith, R.P. Jr (2000) Attempt to control ticks (Acari: Ixodidae) on deer on an isolated island using ivermectin-treated corn. *Journal of Medical Entomology* 37, 126–133.

Rao, R.S., Kumar, M.N. and Rao, R. (2000) Therapeutic trial of drugs in ascariasis in buffalo calves. *Indian Veterinary Journal* 77, 57–59.

Reddy, P.M.T. and Murali Krishna, B.V. (1995) Ivermectin in cutaneous myiasis of buffaloes. *Buffalo Bulletin* 14, 75.

Rehbein, R. and Visser, M. (1997) Persistent anthelmintic activity of topically administered ivermectin in red deer (*Cervus elaphus* L.) against lungworms (*Dictyocaulus viviparus*). *New Zealand Veterinary Journal* 45, 85–87.

Rickard, L.G. (1994) Parasites. *Veterinary Clinics of North America: Food Animal Practice* 10, 239–247.

Robin, B., König, K. and Anstey, M.D. (1989) Efficacy of ivermectin against internal parasites in dromedaries (*Camelus dromedarius*). *Revue Scientifique et Technique/Office International des Épizooties* 8, 155–161.

Saha, G.R., Mitra, M., Sur, S.K., Sasmal, N.K. and Das, A.K. (1996) A clinical report on ivermectin against sarcoptic mange in camel. *Indian Veterinary Medical Journal* 20, 69–70.

Samuel, W.M. and Gray, J.B. (1988) Efficacy of ivermectin against *Parelaphostrongylus andersoni* (Nematoda, Metastrongyloidea) in white-tailed deer (*Odocoileus virginianus*). *Journal of Wildlife Diseases* 24, 491–495.

Senthilvel, K. and Raman, M. (1999) Efficacy of ivermectin against cutaneous myiasis in buffalo. *Indian Journal of Veterinary Medicine* 19, 82.

Sharma, L.K. (1991) Efficacy of some anthelmintics against gastro-intestinal nematodes in camels (*Camelus dromedarius*). *Indian Veterinary Journal* 68, 1069–1072.

Sharma, L.K. (1992) Efficacy of ivermectin against nasal bots in camels. *Indian Veterinary Journal* 69, 835–836.

Sharma, S.P. (1991) Treatment of clinical microfilariasis in buffaloes with ivermectin. *Indian Veterinary Journal* 68, 972–974.

Shastri, U.V. (1989) Efficacy of ivermectin (MSD) against *Toxocara vitulorum*. *Journal of Veterinary Parasitology* 3, 153–154.

Shastri, U.V. (1991) Efficacy of ivermectin (MSD) against lice infestation in cattle, buffaloes, goats and dogs. *Indian Veterinary Journal* 68, 191.

Sindhu, S.A., Ashraf, M., Ahmad, R., Ullah, K. and Pervaiz, K. (1996) Anthelmintic efficacy of Caesalpinia crista, ivermactin, levamisole and oxfendazole against *Toxocara vitulorum* in buffalo-calves. *Pakistan Journal of Scientific Research* 48, 108–111.

Singh, O.V., Singh, J.L. and Dabas, Y.P.S. (1999) Efficacy of ivermectin in concurrent infestation of *Setaria* and mange in buffalo bulls. *Indian Journal of Veterinary Medicine* 19, 68–69.

Soll, M.D. (1989) Use of ivermectin in laboratory and exotic mammals, and in birds, fish, and reptiles. In: Campbell, W.C. (ed.) *Ivermectin and Abamectin.* Springer-Verlag, New York, pp. 260–286.

Soveri, T., Nikander, S. and Nieminen, M. (1990) Efficiency of parenteral and oral ivermectin treatment on parasites in reindeer. *Rangifer* Special Issue 4, 64.

Sugár, L. (1995) Importance and control of the lungworm infection in cervidae. *Magyar Állatorvosok Lapja* 50, 161–164.

Suphalucksana, W. and Ching, F.A. (1991) A field trial on the efficacy of ivermectin, albendazole and coumaphos against parasites in buffaloes. *Kasetsart Journal, Natural Sciences* 25, 256–260.

Tager-Kagan, P. and Robin, B. (1986) Resultats de l'expérimentation de l'ivermectine (Ivomec®) sur les parasites du dromadaire au Niger. *Revue de l'Elevage et Medecine Véterinaire des Pays Tropicaux* 39, 333–340.

Udupa, K.G., Reddy, P.M.T. and Prakash, N. (1995) Treatment of thelaziasis in a buffalo – a case report. *Buffalo Bulletin* 14, 68.

van Straten, M. and Jongejan, F. (1993) Ticks (Acari: Ixodidae) infesting the Arabian camel (*Camelus dromedarius*) in the Sinai, Egypt with a note on the acaricidal efficacy of ivermectin. *Experimental and Applied Acarology* 17, 605–616.

Waghmare, S.P., Rode, A.M., Sapre, V.A. and Sarode, D.B. (1991) Anthelmintic efficacy of ivermectin and albendazole against gastrointestinal helminth infection in buffalo calves. *Indian Journal of Veterinary Medicine* 11, 70–71.

Waldrup, K.A., Mackintosh, C.G., Duffy, M.S., Labes, R.E., Johnstone, P.D., Taylor, M.J. and Murphy, A.W. (1998) The efficacy of a pour-on formulation of moxidectin in young red and wapiti-hybrid deer. *New Zealand Veterinary Journal* 46, 182–185.

Weber, M. (1992) Valoración clínica del efecto de la ivermectina contra *Cephenemyia* spp en venados cola blanca. *Veterinaria Mexico* 23, 239–242.

Windsor, R.H.S., Teran, M. and Windsor, R.S. (1992) Effects of parasitic infestations on the productivity of alpacas (*Lama pacos*). *Tropical Animal Health and Production* 24, 57–62.

Windsor, R.S. (1997) Type II ostertagiasis in llamas. *Veterinary Record* 141, 608.

Zaitoun, A.M., Ali, H.S., Ahmed, L.S. and Abu-Zeid, A.S.I. (1998) Field observations on buffaloe's mange in Assiut Governorate – Egypt. *Assiut Veterinary Medicine Journal* 39, 245–256.

Chapter 12

The Use of Macrocyclic Lactones to Control Parasites of Exotic Pets

S.E. Little, C.B. Greenacre and R.M. Kaplan

Introduction

Ivermectin and other avermectin/milbemycin compounds (MLs) have been used extensively as antiparasitic agents in a wide variety of exotic pets including ferrets, rabbits, rodents and other 'pocket pets', birds and reptiles. Although the activity of these compounds was first established in laboratory animal parasite systems, their use in rodents and other exotic pets is extralabel, and treatment protocols are often established through empirical clinical experience rather than controlled studies. Our discussion is limited in focus to exotic pets commonly seen by veterinarians. However, these compounds have also been used extensively as endectocides in laboratory animals, wildlife, exotic livestock and animals held in zoological and aquaria collections.

Mammals other than Dogs and Cats

Ferrets

Ivermectin has been used in ferrets as both a heartworm preventative and a mange mite treatment. Domestic ferrets are very susceptible to natural and experimental infection with *Dirofilaria immitis*. Although not labelled for this use, monthly administration of ivermectin *per os* at 3 or 6 μg kg^{-1} prevents development of adult heartworms in ferrets (Supakorndej *et al.*, 1992). Ivermectin at 400 μg kg^{-1} applied topically (dose divided between each ear) and systemically has been used in ferrets to treat ear mite infestations with *Otodectes cynotis* (Patterson and Kirchain, 1999); topical application reportedly is more efficacious (Patterson and Kirchain, 1999).

Rabbits

Ivermectin alone or in combination with other products has been used to treat a variety of mange mites in rabbits. The efficacy of avermectin analogues in the treatment of the rabbit ear mange mite *Psoroptes cuniculi* is well established (Wilkins *et al.*, 1980). The dose of ivermectin effective in the treatment of *P. cuniculi* is 200–400 μg kg^{-1} subcutaneously (Pandey, 1989; Bowman *et al.*, 1992). Treatment is more effective and lesions resolve more quickly when the higher dose is used (Wright and Riner, 1985; Pandey, 1989). Similar treatment protocols applied large scale to commercial rabbitries resulted in the elimination of *P. cuniculi* infestations (Curtis and Brooks, 1990; Curtis *et al.*, 1990). A combination of subcutaneous ivermectin and topical fipronil successfully treated a generalized *P. cuniculi* infestation in a pet rabbit (Cutler, 1998). *Sarcoptes scabiei* infestations in rabbits have also been eliminated using ivermectin (400 μg kg^{-1} subcutaneously; Nfi, 1992). A single subcutaneous injection (400 μg kg^{-1}) of ivermectin successfully eliminated *Notoedres cati* var. *cuniculi* infestation in 15 rabbits, while the infestation persisted in five untreated controls (Isingla *et al.*, 1996).

Pet Rodents

Although the efficacy of MLs is often established in laboratory animal screens, these compounds are not labelled for clinical use in rodents. None the less, extralabel use in rodents is common, and many protocols have been reported that describe the use of the avermectins/milbemycins as endectocides in rodents.

Eradication of pinworms (*Syphacia* spp.) has been achieved in both mice and rats through strategic use of MLs. Although single oral treatments with ivermectin (2000 μg kg^{-1}) were not effective in eliminating *Syphacia obvelata* from naturally infected mice, feeding a diet containing 0.0005% ivermectin for 6 consecutive days successfully eradicated the parasite (Ostlind *et al.*, 1985). Other protocols proposed for eliminating *S. obvelata* from mouse colonies include a high dose of oral ivermectin (2000 μg kg^{-1}, repeat in 11 days) (Flynn *et al.*, 1989) and a weekly treatment initiated with piperazine (2.1 mg ml^{-1} in tap water for 24 h on day 0 and day 7) followed by ivermectin (0.007 mg ml^{-1} in tap water on day 14 and day 21) (Lipman *et al.*, 1994). Topical misting of mice and caging with a dilute ivermectin solution (0.9–1.8 mg per cage) have also been proposed to help control pinworms (Le Blanc *et al.*, 1993), but others have reported toxicity problems in young mice using this protocol (Skopets *et al.*, 1996). In rats, *S. muris* was effectively treated by administering ivermectin (200 μg kg^{-1} day^{-1}) via gastric intubation for 5 consecutive days (Battles *et al.*, 1987). *S. muris* was eliminated from a rat colony with three doses of ivermectin orally at 2000 μg kg^{-1} at 7–10 day intervals; a two-dose regimen was ineffective at eradicating the parasite (Huerkamp, 1993). Doramectin (200 μg kg^{-1} day^{-1} for 4 days) was also reported to be effective in eliminating *S. muris* infections in rats, while moxidectin (200 μg kg^{-1} day^{-1} for 4 days) was apparently ineffective (Oge *et al.*, 2000).

Other nematodes of mice for which ivermectin has been reported to be effective include *Aspicularis tetraptera* (1.0–1.6 mg kg^{-1} in drinking water for 24 h; Hasslinger and Wiethe, 1987), *Nematospiroides dubius* (300 μg kg^{-1} orally; Rajasekariah *et al.*, 1986), *Strongyloides ratti* (two treatments at 300 μg kg^{-1}; Rajasekariah *et al.*, 1986) and *Strongyloides stercoralis* (single treatment at 100 μg kg^{-1}; Grove, 1983). However, attempted treatment of *Trichuris muris* in mice using repeated doses of ivermectin at 10 mg kg^{-1} was not effective (Rajasekariah *et al.*, 1986). In rats, ivermectin has been used to treat larvae of *Parastrongylus malaysiensis* (400 and 800 μg kg^{-1} 14 days post-infection), but was ineffective in clearing the adults (Ambu *et al.*, 1992).

Ivermectin has also been used to treat mange mites of rodents, including *Myobia muscili* in mice (two 200 μg kg^{-1} doses, subcutaneously, 1 week apart; Wing *et al.*, 1985); *Mycoptes musculinus* in mice (two 200 μg kg^{-1} doses, subcutaneously, 1 week apart; Wing *et al.*, 1985); *Trixacarus caviae* in guinea pigs (three doses at 200 μg kg^{-1} subcutaneously at 7-day intervals (Harvey, 1987), or a single dose at 500 μg kg^{-1} subcutaneously (McKellar *et al.*, 1992)); and *Chirodiscoides caviae* in guinea pigs (spraying with diluted ivermectin at 200 μg ml^{-1} followed by application of 10,000 μg ml^{-1} directly on the fur, re-treated in 2 weeks; Hirsjarvi and Phyala, 1995). However, ivermectin (500 μg kg^{-1} subcutaneous injection, retreated in 2 weeks) was ineffective in treating *Acarus farris* infestation in gerbils (Jacklin, 1997).

Toxicity in outbred, adult rodents due to use of MLs appears rare. However, ivermectin toxicity has been reported in young mice (Skopets *et al.*, 1996) and some mutant strains (Lipman *et al.*, 1994). A study on the effect of ivermectin on the behaviour of mice showed a small, significant effect on some sensitive behaviours; thus, caution should be used when treating animals involved in behavioural experiments (Davis *et al.*, 1999).

Birds

The use of MLs has not been well studied in birds other than poultry, and documentation of the efficacy of these compounds in pet birds is limited. None the less, ivermectin is used routinely as a nematocide and ectoparasiticide in pet birds at a dose of 200 μg kg^{-1} (Greiner and Ritchie, 1994; Ritchie and Harrison, 1994; Tully, 1997; Bowman, 1999; Clyde and Patton, 2000; Samour, 2000). Ivermectin has been used to treat pigeons for *Capillaria* spp. (1.5 mg per pigeon *per os* or 300 μg per pigeon intramuscularly, single dose) and *Ascaridia columbae* (1.5 mg per pigeon *per os* or 300 μg per pigeon intramuscularly, two doses 3 weeks apart). Toxicity, including death, was reported in pigeons using ivermectin at a dose of 7.5–15 mg/pigeon *per os* (Schepkens *et al.*, 1985). The standard 200 μg kg^{-1} dose of ivermectin was ineffective in treating a yellow-collared macaw (*Ara aurocollis*) with a subcutaneous nodule containing 35 filarids (*Pelictus* spp.) (Allen, 1985). This result was not surprising as the MLs have limited efficacy against adult filarids when administered as a single treatment.

Ivermectin has also been used at 200 μg kg^{-1} orally, intramuscularly, subcutaneously, and topically to treat scaly leg and face mites of birds. A single dose of ivermectin (200 μg kg^{-1} intramuscularly) given to a wild caught great horned owl (*Bubo virginianus*) was effective in treating the only documented case of *Knemidokoptes mutans* in an owl (Schulz *et al.*, 1989). A group of migratory American robins (*Turdus migratorius*) infected with *Knemidokoptes jamacensis* showed variable response to one or two doses of ivermectin at 200 μg kg^{-1} *per os* (Pence *et al.*, 1999). Ivermectin treatment is generally considered more effective in treating the superficial scaly face mite, *Knemidokoptes pilae*, of budgerigars and less effective in treating the deeper scaly leg mite, *K. jamacensis*, of canaries. Apparently, it is difficult for ivermectin to kill mites in poorly vascularized areas such as the dense hyperkeratotic skin lesions associated with *K. jamacensis* or air sac mites found in finches (Greiner and Ritchie, 1994).

Adverse reactions have been associated with the use of MLs in birds. Doses greater than 400 μg kg^{-1} are thought to be toxic in bullfinches (Ritchie and Harrison, 1994). Intramuscular administration of ivermectin to small birds such as budgerigars, waxbills and bullfinches can cause death. Although the exact mechanism for this reaction is unknown, affected birds develop respiratory difficulty prior to death; it has been suggested that a shock-like reaction to the propylene glycol base is responsible for this response. Topical or oral administration of ivermectin appears to be safer and as efficacious as administration by the subcutaneous or intramuscular route (Ritchie and Harrison, 1994). However, dilution of ivermectin in a propylene glycol base with water or saline prior to administration has not been evaluated in birds and may cause precipitation of the drug and subsequent adverse consequences (Ritchie and Harrison, 1994; Clyde and Patton, 2000).

Reptiles

Snakes

Ivermectin is used frequently as a nematocide in snakes at a dose of 200 μg kg^{-1} administered either orally or by injection. This dose is considered to be safe in most species of snakes, and corn snakes (*Elapha guttata*) have tolerated doses of 1000 μg kg^{-1} without apparent adverse effects (Teare and Bush, 1983). However, there are anecdotal reports of ivermectin toxicity in some Australian snakes (Jacobson, 1993). Ivermectin also has been used successfully as an acaricide in many species of snakes. When administered subcutaneously or intramuscularly at a dose of 200–400 μg kg^{-1}, ivermectin was both safe and effective at eliminating mite (*Ophionyssus* spp.) and tick infestations in snakes (Lawrence, 1984; Clyde, 1996). When treating *Ophionyssus* spp., it is recommended that the ivermectin dose be repeated in 14 days, that all snakes and lizards in a collection be treated simultaneously, and that measures be taken to cleanse the environment (Clyde, 1996).

Lizards

Ivermectin is frequently used orally and subcutaneously as a nematocide and acaricide in pet lizards at a dose of 200 μg kg^{-1} (Barten, 1993). Although ivermectin is not known to kill pentastomes (common parasites of the reptilian respiratory system) in snakes (Jacobson, 1993), ivermectin treatment (200 μg kg^{-1} subcutaneously, repeated in 10 days) of a Bosc's monitor lizard suffering from pentastomiasis resulted in a clinical cure (Flach *et al.*, 2000).

Chelonians

Ivermectin toxicity has been demonstrated at subtherapeutic doses in chelonians. Of four species of chelonians treated with ivermectin (leopard tortoise (*Geochelone pardalis*), red-footed tortoise (*Geochelone carbonaria*), red-eared turtle (*Chrysemys scripta*) and eastern box tortoise (*Terrapene carolina*)), the leopard tortoise was the most sensitive, showing mild signs of toxicosis at a dosage of 25 μg kg^{-1} and moderate to severe signs at 50 μg kg^{-1} (Teare and Bush, 1983). Box tortoises (*T. carolina*) were more resistant, demonstrating only mild signs of toxicosis at 100 μg kg^{-1}, but a dose of 300 μg kg^{-1} caused a flaccid paralysis that resulted in the death of the tortoise. Based on these data, ivermectin is not recommended for use in chelonians (Conboy *et al.*, 1993).

In contrast to the toxicity reported for ivermectin, milbemycin (A_3–A_4 oxime) appears to be safe and effective in chelonians (Bodri *et al.*, 1993). In these studies, milbemycin technical grade material or powdered tablets of Interceptor (Novartis AH Inc.) were dissolved in propylene glycol and administered by the oral or subcutaneous routes. Red-eared sliders (*Pseudemys scripta elegans*) were treated orally with 500 or 1000 μg kg^{-1}, and Gulf Coast box turtles (*Terrapene carolina major*), ornate box turtles

(*Terrapene carolina ornata*) and red-eared sliders were treated subcutaneously with 250 or 500 μg kg^{-1}. None of the turtles demonstrated any deleterious effects from milbemycin treatment. Prior to treatment with milbemycin, faecal examinations were performed on all turtles used in the study to confirm the presence of nematode eggs. Following treatment, most turtles tested negative for nematode ova and a few required a second treatment before faecal examination became negative. However, necropsy examination to identify nematodes was not performed on any of these turtles; thus, actual anthelmintic efficacy was not determined (Bodri *et al.*, 1993).

Amphibians

Ivermectin diluted in propylene glycol and administered to leopard frogs (*Rana pipiens*) at doses of 200 and 400 μg kg^{-1} by intramuscular injection or 2.0 and 20 mg kg^{-1} percutaneously did not induce any signs of toxicosis (Letcher and Glade, 1992). In the same report, ten of ten wild-caught leopard frogs testing positive for nematode eggs or larvae by faecal examination and treated with ivermectin (2.0 mg kg^{-1} percutaneous over the dorsal thorax) became faecal negative by 1 week post-treatment and remained negative at 2, 3 and 4 weeks post-treatment. Three untreated control frogs were examined at necropsy and found to be infected with *Strongyloides* spp., *Oswaldocruzia* spp. and *Capillaria* spp. (Letcher and Glade, 1992).

Ivermectin administered orally or by injection into the dorsal lymph sac at 200 μg kg^{-1} has also been used to to treat cutaneous capillariasis successfully in South African clawed frogs (*Xenopus laevis*) (Cromeens *et al.*, 1987; Dawson *et al.*, 1992; Iglauer *et al.*, 1997). Due to the small size and excitable nature of frogs and toads, these animals are difficult to medicate by the oral and parenteral routes (Letcher and Glade, 1992). Since the percutaneous route appears to be an effective means of anthelmintic delivery in frogs, several of the newer ML pour-on products may become useful therapeutics in these species once safety and efficacy studies are performed.

References

Allen, J.L., Kollias, G.V., Griener, E.C. and Boyce, W. (1985) Subcutaneous filarisis (*Pelecitus* sp.) in a yellow collared macaw. *Avian Diseases* 29, 891–894.

Ambu, S., Mak, J.W. and Ng, C.S. (1992) Efficacy of ivermectin against *Parastrongylus malayensis* infection in rats. *Journal of Helminthology* 66, 293–296.

Barten, S. (1993) The medical care of iguanas and other common pet lizards. *Veterinary Clinics of North America: Small Animal Practice* 23, 6.

Battles, A.H., Adams, S.W., Courtney, C.H. and Mladinich, C.R.T. (1987) Efficacy of ivermectin against natural infection of *Syphacia muris* in rats. *Laboratory Animal Science* 37, 791–792.

Bodri, M.S., Nolan, T.J. and Skeeba, S.J. (1993) Safety of milbemycin (a(3)–a(4) oxime) in chelonians. *Journal of Zoo and Wildlife Medicine* 24, 171–174.

Bowman, D.D. (1999) Arthropods. In: *Georgis' Parasitology for Veterinarians*, 7th edn. W.B. Saunders, Philadelphia, Pennsylvania, pp. 1–78.

Bowman, D.D., Fogelson, M.L. and Carbone, L.G. (1992) Effect of ivermectin on the control of ear mites (*Psoroptes cuniculi*) in naturally infested rabbits. *American Journal of Veterinary Research* 53, 105–109.

Clyde, V.L. (1996) Practical treatment and control of common ectoparasites in exotic pets. *Veterinary Medicine* 91, 632–637.

Clyde, V.L. and Patton, S. (2000) Parasitism in caged birds. In: Olsen, G.H. and Orosz, S.E. (eds) *Manual of Avian Medicine*. Mosby, Philadelphia, Pennsylvania, pp. 424–448.

Conboy, G.A., Laursen, J.R., Averbeck, G.A. and Stromberg, B.E. (1993) Diagnostic guide to some of the helminth parasites of aquatic turtles. *Compendium of Continuing Education for the Practicing Veterinarian* 15, 1217–1222.

Cromeens, D., Robbins, V. and Stephens, C. (1987) Diagnostic exercise: cutaneous lesions in frogs. *Laboratory Animal Science* 37, 58–59.

Curtis, S.K. and Brooks, D.L. (1990) Eradication of ear mites from naturally infested conventional research rabbits using ivermectin. *Laboratory Animal Science* 40, 406–408.

Curtis, S.K., Housley, R. and Brooks, D.L. (1990) Use of ivermectin for treatment of ear mite infestation in rabbits. *Journal of the American Veterinary Medicine Association* 196, 1139–1140.

Cutler, S.L. (1998) Ectopic *Psoroptes cuniculi* infestation in a pet rabbit. *Journal of Small Animal Practice* 39, 86–87.

Davis, J.A., Paylor, R., McDonald, M.P., Libbey, M., Ligler, A., Bryant, K. and Crawley, J.N. (1999) Behavioral effects of ivermectin in mice. *Laboratory Animal Science* 49, 288–296.

Dawson, D., Schultz, T. and Shroeder, E. (1992) Laboratory care and breeding of the African clawed frog. *Laboratory Animals* 21, 31–36.

Flach, E.J., Riley, J., Mutlow, A.G. and McCandlish, I.A.P. (2000) Pentastomiasis in Bosc's monitor lizards (*Varanus exanthematicus*) caused by an undescribed Sambonia species. *Journal of Zoo and Wildlife Medicine* 31, 91–95.

Flynn, B.M., Brown, P.A., Eckstein, J.M. and Strong, D. (1989) Treatment of *Syphacia obvelata* in mice using ivermectin. *Laboratory Animal Science* 39, 461–463.

Greiner, E.C. and Ritchie, B.W. (1994) *Parasites*. In: Ritchie, B.W., Harrison, G.J. and Harrison, L.R. (eds) *Avian Medicine: Principles and Application*. Wingers Publishing, Lake Worth, Florida, pp. 1007–1029.

Grove, D.I. (1983) The effects of 22,23 dihydroavermectin B1 on *Strongyloides ratti* and *S. stercoralis* infections in mice. *Annals of Tropical Medicine and Parasitology* 77, 405–410.

Harvey, R.G. (1987) Use of ivermectin for guinea pig mange. *Veterinary Record* 120, 351.

Hasslinger, M.A. and Wiethe, T. (1987) Oxyurids of small laboratory animals and their control with ivermectin. *Tierarztliche Prax* 15, 93–97.

Hirsjarvi, P and Phyala, L. (1995) Ivermectin treatment of a colony of guinea pigs infested with fur mite (*Chirodiscoides caviae*). *Laboratory Animals* 29, 200–203.

Huerkamp, M.J. (1993) Ivermectin eradication of pinworms from rats kept in ventilated cages. *Laboratory Animal Science* 43, 86–90.

Iglauer, F., Willmann, F., Hilken, G., Huisinga, E. and Dimigen, J. (1997) Anthelmintic treatment to eradicate cutaneous capillariasis in a colony of South African clawed frogs (*Xenopus laevis*). *Laboratory Animal Science* 47, 477–482.

Isingla, L.D., Juyal, P.D. and Gupta, P.P. (1996) Therapeutic trial of ivermectin against *Notoedres cati* var. *cuniculi* infection in rabbits. *Parasite* 3, 87–89.

Jacklin, M.R. (1997) Dermatosis associated with *Acarus farris* in gerbils. *Journal of Small Animal Practice* 38, 410–411.

Jacobson, E. (1993) Snakes. *Veterinary Clinics of North America: Small Animal Practice* 23, 1179–1212.

Lawrence, K. (1984) Ivermectin as an ectoparasiticide in snakes. *Veterinary Record* 115, 441–442.

Le Blanc, S.A., Faith, R.E. and Montgomery, C.A. (1993) Use of topical ivermectin treatment for *Syphacia obvelata* in mice. *Laboratory Animal Science* 43, 526–528.

Letcher, J. and Glade, M. (1992) Efficacy of ivermectin as an anthelmintic in leopard frogs. *Journal of the American Veterinary Medicine Association* 200, 537–538.

Lipman, N.S., Dalton, S.D., Stuart, A.R. and Arruda, K. (1994) Eradication of pinworms (*Syphacia obvelata*) from a large mouse breeding colony by combination oral anthelmintic therapy. *Laboratory Animal Science* 44, 517–520.

McKellar, Q.A., Midgley, D.M., Galbraith, E.A., Scott, E.W. and Bradley, A. (1992) Clinical and pharmacological properties of ivermectin in rabbits and guinea pigs. *Veterinary Record* 130, 71–73.

Nfi, A.N. (1992) Ivomec, a treatment against rabbit mange. *Revue de l'Elevage et de Médecine Vétérinaire des Pays Tropicaux* 45, 39–41.

Oge, H., Ayaz, E., Ide, T. and Dalgic, S. (2000) The effect of doramectin, moxidectin, and netobimin against natural infections of *Syphacia muris* in rats. *Veterinary Parasitology* 88(3–4), 299–303.

Ostlind, D.A., Nartowicz, M.A. and Mickle, W.G. (1985) Efficacy of ivermectin against *Syphacia obvelata* in mice. *Journal of Helminthology* 59, 257–261.

Pandey, V.S. (1989) Effect of ivermectin on the ear mange mite, *Psoroptes cuniculi*, of rabbits. *British Veterinary Journal* 145, 54–56.

Patterson, M.M. and Kirchain, S.M. (1999) Comparison of three treatments for control of ear mites in ferrets. *Laboratory Animal Science* 49, 655–657.

Pence, D.B., Cole, R.A., Brugger, K.E. and Fischer, J.R. (1999) Epizootic podoknemidokoptiasis in American robins. *Journal of Wildlife Diseases* 35, 1–7.

Rajasekariah, G.R., Deb, B.N., Dhage, K.R. and Bose, S. (1986) Response of laboratory-adapted human hookworms and other nematodes to ivermectin. *Annals of Tropical Medicine and Parasitology* 80, 615–621.

Ritchie, B.W. and Harrison, G.J. (1994) Formulary. In: Ritchie, B.W., Harrison, G.J. and Harrison, L.R. (eds) *Avian Medicine: Principles and Application*. Wingers Publishing, Lake Worth, Florida, pp. 457–481.

Samour, J. (2000) *Avian Medicine*. Mosby, St Louis, Missouri.

Schepkens, E., Duchatel, J.P. and Vindevogel, H. (1985) Ivermectin treatment of ascaridiasis and capillariosis in pigeons. *Annales de Medicine Veterinarie* 129, 775–785.

Schulz, T.A., Stewart, J.S. and Fowler, M.E. (1989) *Knemidokoptes mutans* (Acari: Knemidokoptidae) in a great-horned owl (*Bubo virginianus*). *Journal of Wildlife Diseases* 25, 430–432.

Skopets, B., Wilson, R.P., Griffith, J.W. and Lang, C.M. (1996) Ivermectin toxicity in young mice. *Laboratory Animal Science* 46, 111–112.

Supakorndej, P., McCall, J.W., Lewis, R.E., Rowan, S.J., Mansour, A.E. and Holmes, R.A. (1992) Biology, diagnosis, and prevention of heartworm infection in ferrets. In: Soll, M.D. (ed.) *Proceedings of the Heartworm Symposium '92*. American Heartworm Society, Batavia, Illinois, pp. 56–59.

Teare, J.A. and Bush, M. (1983) Toxicity and efficacy of ivermectin in chelonians. *Journal of the American Veterinary Medicine Association* 183, 1195–1197.

Tully, T. (1997) Formulary. In: Altman, R.B., Clubb, S.L., Dorrestein, G.M. and Quesenberry, K.E. (eds) *Avian Medicine and Surgery*. W.B. Saunders, Philadelphia, Pennsylvania, pp. 671–678.

Wilkins, C.A., Conroy, J.A., Ho, P., O'Shanny, W.J. and Malatesta, P.F. (1980) Treatment of psoroptic mange with avermectins. *American Journal of Veterinary Research* 41, 2112–2114.

Wing, S.R., Courtney, C.H. and Young, M.D. (1985) Effect of ivermectin on murine mites. *Journal of the American Veterinary Medical Association* 187, 1191–1192.

Wright, F.C. and Riner, J.C. (1985) Comparative efficacy of injection routes and doses of ivermectin against *Psoroptes* in rabbits. *American Journal of Veterinary Research* 46, 752–754.

Chapter 13

The Use of Macrocyclic Lactones to Control Parasites of Humans

K.R. Brown

Macrocyclic lactones (MLs) exhibit *selective toxicity*, which defines an ideal anti-infective agent as one that has a chemical target in the microorganism causing infection, but either has no target or has no access to a target in the infected host. Of the MLs described in this book, only ivermectin has been studied in humans. Should another of these compounds be active against most of the gastrointestinal helminths of humans in a single dose and have a good safety profile (including being negative in the Ames assay), it would be a welcomed product addition. No currently available anti-parasitic drug has all of these properties. If one of the MLs could be shown to be an effective macrofilaricide, this would make it a very attractive development candidate, not necessarily as a commercially profitable product but one that would be welcomed by the public sector. This chapter will review briefly the use of ivermectin for various human tissue and gastrointestinal helminths (Greene *et al.*, 1989) and its use in mass distribution programmes (Foege, 1998). It will then cover some potential uses against less well known parasites.

Onchocerciasis

After the efficacy of ivermectin against *Onchocerca cervicalis* was shown in horses, Dr William Campbell at Merck suggested its potential utility in humans infected with *O. volvulus*. Since so much experience in food and companion animals had accrued, the process of introduction into humans was expected to be straightforward. However, some rough-coated collie dogs and one strain of mice (CF-1) developed neurotoxicity when given

 Macrocyclic Lactones in Antiparasitic Therapy (eds J. Vercruysse and R.S. Rew)

ivermectin, while other species including most food and companion animals showed no serious toxicity. As a precaution, very low doses were explored beginning at 5 $\mu g\ kg^{-1}$. Given the absence of toxicity, studies progressed rapidly through ascending doses to a level of 150–200 $\mu g\ kg^{-1}$ with no serious safety problems.

Ivermectin was efficacious in reducing the levels of skin microfilariae in individuals at a moderate rate. By the third day after treatment, skin levels of microfilariae fell almost to undetectable levels. This fall was gradual, not precipitous, as was the case following the use of diethylcarbamazine (DEC); the frequency and severity of Mazzotti-like reactions following ivermectin both decreased. The optimum time interval between doses was 6 or 12 months (depending on the mean community microfilarial load); the optimum dose was 150–200 $\mu g\ kg^{-1}$ (Brown and Neu, 1990). Furthermore, the killing of microfilariae in the eye was slower than that by DEC; it provoked no adverse response, and halted progression of eye disease in many patients. Studies of children as small as 17 kg were included since they share the risk of developing disease as well as infection (Brown, 1998). Extensive studies with or without a placebo control or a comparator drug confirmed these favourable findings. Relief of itching and passage of roundworms encouraged the patients to return for repeat visits. A halt to the progression of eye damage after the treatment with ivermectin stimulated broad interest in a long-term programme. Ivermectin was not a macrofilaricide nor could it reverse many of the changes associated with the chronic skin disease. Finally, while ivermectin could not be shown to be a chemosterilant, it had a distinct damaging effect on the production and release of microfilariae *in utero* (Schulz-Key *et al.*, 1986), thus suggesting an additional benefit in control of transmission.

Before broad use of ivermectin in the hands of minimally trained personnel could be allowed, there was the need for extensive field experience. With the use of field staff from the Onchocerciasis Control Program, studies that would carefully monitor the recipients of ivermectin for 72 h after taking it were initiated to enrol approximately 100,000 persons with onchocerciasis (Anonymous, 1989). In 1987, Merck and Co., Inc., the manufacturer, announced that it would donate MECTIZAN™ (ivermectin formulated for human onchocerciasis). An independent committee was formed to review and approve applications for ivermectin to be used in mass treatment programmes with a target of approximately 18,000,000 persons needing treatment every year; it would take 10 years before enabling that number of treatments (Foege, 1998). Programmes of ivermectin distribution for onchocerciasis, most on an annual basis, have been initiated in virtually all of the onchocerciasis endemic countries in Africa and Latin America.

Lymphatic Filariasis (LF)

As is the case for DEC in onchocerciasis, the tolerance of DEC in patients with LF is poor, and many patients are reluctant to use the drug as scheduled (Ottesen, 1987). Furthermore, patients with LF and coincident onchocerciasis run the risk of the Mazotti reaction, which occurs because DEC kills the microfilariae of *O. volvulus* so rapidly. On the other hand, DEC has the advantage of some macrofilaricidal activity (Ottesen, 1985). Because DEC was not an ideal therapy, ivermectin was considered as a possible treatment for LF (Diallo *et al.*, 1987). These authors showed that 100 μg kg^{-1} of ivermectin caused a greater reduction on the blood microfilarial counts of patients infected with *Wuchereria bancrofti* than did a dose of 50 μg kg^{-1}. This study encouraged a series of more detailed and broad-ranging efforts to assess the possible benefits of ivermectin in LF, whether caused by *W. bancrofti* or by *Brugia malayi*. Those studies are summarized in a recent extensive review of the literature (Brown *et al.*, 2000). The emphases in these studies were on dose ranging, comparisons with placebo or with DEC, efficacy without a comparator drug, evaluation of a very low dose and/or safety of ivermectin. Study data were acquired in Asia for the most part and also in Africa and Latin America. As a result of these studies, the following conclusions suggested the potential for ivermectin as a microfilaricide (Brown *et al.*, 2000).

1. Ivermectin, by lowering microfilaraemia, would be useful in the control of transmission of LF caused by either *W. bancrofti* or *B. malayi*.
2. Ivermectin did not have macrofilaricidal activity.
3. For consistent activity in lowering microfilaraemia, doses of 150–200 μg kg^{-1} would be considered as the minimum.
4. A higher dose (400 μg kg^{-1}) was even more effective than 200 μg kg^{-1} in terms of lowering of microfilaria levels and duration of effect (statistically significant, clinical importance not determined).
5. If the lower dose (150–200 μg kg^{-1}) were to be used, the preferred frequency of dosing would probably be every 6 months rather than every 12 months.
6. The concept of a very low dose (20 μg kg^{-1}) used to 'clear' microfilariae in order to diminish the potential for adverse experiences was not shown to be useful.
7. When used as a microfilaricide in LF patients, ivermectin was very safe.
8. Adverse experiences reported following ivermectin use were more closely related to the level of microfilaraemia in the patient than they were to the dose of ivermectin.
9. Ivermectin plus albendazole produced a more profound and more sustained suppression of microfilaraemia than did either drug alone.

It is important to understand the different goals in the use of ivermectin for onchocerciasis and for LF. In the former case, treatment is given to provide symptomatic relief of skin and eye disease as well as a possible interruption of transmission. For LF, ivermectin has no direct clinical benefits but its use is to help lower the counts of microfilariae in the blood in order to block transmission. In order to obtain the maximum effect on transmission of LF, it may be necessary to use both ivermectin and albendazole. Finally, there are insufficient data on the beneficial effects of the use of ivermectin in patients with *Mansonella perstans*, *M. ozzardi* or *M. streptocerca* to recommend it over DEC. For patients with loiasis, the risks may outweigh the benefits (Boussinesque *et al.*, 1998).

Human Gastrointestinal Parasites

The broad spectrum of activity of ivermectin against many gastrointestinal and tissue nematodes and ectoparasites in animals created high hopes that a similar spectrum of activity would be shown against the parasites of humans. This hope was not fully borne out in that it was demonstrated that the human hookworms were only minimally susceptible in the very first studies using ivermectin in the treatment of human enteric nematodes. Thus the emphasis in clinical testing was directed to the remaining gastrointestinal nematodes expected to be susceptible.

While there are many available safe and effective agents against *Ascaris* and pinworm (including ivermectin), the choices for the threadworm and the whipworm are much more limited if one considers the importance of one-time dosing. Studies of ivermectin against the enteric roundworms in humans were encouraging, especially for *Strongyloides* in which case activity was demonstrated at a dose as low as 25 $\mu g\ kg^{-1}$, with progressive activity of rising doses levelling off at about 200 $\mu g\ kg^{-1}$ (Naquira *et al.*, 1989). A high degree of efficacy after a single dose has been confirmed by others (Scaglia *et al.*, 1990; Marti *et al.*, 1996). Similarly, the activity of single dose ivermectin against *Trichuris* was in the range of 85% efficacy, as was the coverage of *Ascaris* and pinworms (Beach *et al.*, 1999). Another study (Ismail and Jayakody, 1999) suggested that the combination of ivermectin and albendazole was superior to albendazole alone in that the combination had both a high cure rate (79.3%) and a high rate (93.8%) of egg reduction in the stool.

Human Ectoparasites and Other Potential Targets for Ivermectin

Animal lice and mites are often treated successfully with ivermectin (see Chapters 6 and 7). The theoretical utility of ivermectin against human lice was suggested in one study from Israel in which human body lice feeding on rabbits were killed in the first 3 days after the rabbits were treated with a single dose of 200 μg kg^{-1} ivermectin (Mumcuoglu *et al.*, 1990).

Glaziou *et al.* (1994) failed to effect a complete cure of head lice with a single dose of ivermectin in 26 patients, using 200 μg kg^{-1}, but noted that on the 14th day, 20 patients were improved while six were cured of all signs and symptoms. Using patient recognition of a cure, Bell (1998) showed that a single dose of orally administered ivermectin cured 73% of the patients with head lice; interpretation of that study was complicated because some patients used the LiceMeister™ comb as an adjunctive therapy. In most lice treatment studies, re-exposure and reinfestation confound the interpretation of efficacy. Describing the treatment of a less well known but common infestation, Burkhart and Burkhart (2000) also showed the utility of ivermectin against the pubic or crab louse infecting the eyelashes (phthiriasis).

In contrast to most human studies in which ivermectin is given by mouth, Youssef *et al.* (1995) used topical application of ivermectin (15–25 ml of a 0.8% solution over the whole body) on 75 patients with either head lice or scabies. They claimed a complete clinical cure, as well as elimination of the lice after a single dose; in the case of scabies, there was a symptomatic relief after a single dose, but 50% of the patients required a second dose 5 days later. Studies have concluded that oral ivermectin is efficacious against human mites in uncomplicated cases and in Norwegian or crusted scabies. Both in an open study of over 1100 patients (Nnoruka and Agu, 2001) and in a controlled study (Leppard and Nebari, 2000) of 58 patients with scabies, oral ivermectin treatment showed a rate of clinical cure/improvement as high as 95% and with a follow-up as long as 8 weeks. Including both adults as well as children as young as 5 years in these two studies, there was a wide margin of safety. However, in the case of Norwegian or crusted scabies, there was sometimes a need for a second dose of oral ivermectin or concomitant or follow-up treatment with a topical agent such as lindane in adults or permethrins in children. Because of its safety and efficacy, ivermectin may become the treatment of choice for scabies. Its successful use in an infestation with an unspeciated demodex causing a rosaceiform rash about the nose in an immunocompromised patient has been described (Forstinger *et al.*, 1999). Studies of the efficacy of ivermectin in the therapy of creeping eruption – larva migrans caused by *Ancylostoma brasiliensis* (Caumes *et al.*, 1992; Van den Enden *et al.*, 1998) and larva currens caused by *Strongyloides stercoralis* (Caumes *et al.*, 1994) – have provided an alternative to topical use of thiabendazole or the oral use of albendazole; both of these drugs require more than 1 day's use for larva migrans. Because of the demonstrated broad spectrum, some have suggested that ivermectin could become the single agent dermatologists seek for human ectoparasites (del Giudice and Marty, 1999).

Safety Considerations

Ivermectin has been shown to be very safe in humans, as in other species. The specific issues that necessitated a very conservative start to the development programme in humans included concerns about use in pregnancy and possible central nervous system toxicity. Such concerns have not been validated in extensive clinical experience. One study of pregnant women in Ecuador showed that there was a lower rate of miscarriages in ivermectin-treated women than in untreated women, possibly related to the lowering of microfilarial loads (Guderian *et al.*, 1997). The most serious clinical adverse experiences attributed to ivermectin have been the cases of encephalopathy in patients with concomitant onchocerciasis and loiasis, with high levels of *Loa loa* microfilariae in the blood and sometimes in the cerebrospinal fluid (Boussinesq *et al.*, 1998). The presumed explanation for the high degree of safety in humans is that P-glycoprotein in the placenta (MacFarland *et al.*, 1994) and in the blood–brain barrier serves to exclude ivermectin from critical receptor access, as is the case for animals (Lankas *et al.*, 1997). The extensive programmes in humans have demonstrated the high degree of safety to the point where the manufacturer, while *not recommending ivermectin use in pregnancy*, does not withhold the drug from mass distribution programmes that choose to treat pregnant women. Furthermore, its paediatric use has been extended to include children who are as small as 15 kg, regardless of age (Brown, 1998).

Summary

Ivermectin, the first ML to be used in humans, has a broad range of antiparasitic activity.[1] As an orally administered agent, it has become the drug of choice for the mass treatment and control of onchocerciasis and a potentially useful tool in the attempts to control lymphatic filariasis. In addition, ivermectin is highly effective in the single-dose treatment of strongyloidiasis (whether gastrointestinal or as larva currents), and relatively efficacious as well for larva migrans and scabies. It may find utility in the treatment of head lice. As a single dose in combination with albendazole, it is very effective against trichuriasis and highly suppressive of the microfilaraemia of lymphatic filariasis.

[1] Ivermectin is approved in the USA and Australia only for the treatment of onchocerciasis, lymphatic filariasis and strongyloidiasis. In France, it is also approved for the treatment of scabies. See the respective package circulars for the approved indications, which may vary from the descriptions in this chapter.

References

Anonymous (1989) *Report of a Meeting of the TDR/OCP/OCT Subcommittee for Monitoring of Community Trials of Ivermectin.* World Health Organization, Geneva, pp. 2–3.

Beach, M.J., Streit, T.G., Addiss, D.G., Prospere, R., Roberts, J.M. and Lammie, P.J. (1999) Assessment of combined ivermectin and albendazole for treatment of intestinal helminths and *Wuchereria bancrofti* infections in Haitian schoolchildren. *American Journal of Tropical Medicine and Hygiene* 60, 487–486.

Bell, T.A. (1998) Treatment of *Pediculus humanus* var. *capitis* infestation in Cowlitz County, Washington with ivermectin and the Licemeister™ comb. *Pediatric Infectious Disease Journal* 17, 923–924.

Boussinesq, M., Gardon, J., Gardon-Wendel, N., Kamgno, J., Ngoumou, P. and Chippaux, J. (1998) Three probable cases of *Loa loa* encephalopathy following ivermectin treatment for onchocerciasis. *American Journal of Tropical Medicine and Hygiene* 58, 461–469.

Brown, K.R. (1998) Changes in the use profile of Mectizan: 1987–1997. *Annals of Tropical Medicine and Parasitology* 92 (Suppl. 1), S62–S64.

Brown, K.R. and Neu, D.C. (1990) Ivermectin – clinical trials and treatment schedules in onchocerciasis. *Acta Leidensia* 59, 169–175.

Brown, K.R., Ricci, F.M. and Ottesen, E.A. (2000) Ivermectin: effectiveness in lymphatic filariasis. *Parasitology* 121, S133–S146.

Burkhart, C.N. and Burkhart, C.G. (2000) Oral ivermectin for phthiriasis palpebrum. *Archives of Ophthalmology* 118, 134–135.

Caumes, E., Datry, A., Paris, L., Danis, M., Gentilini, M. and Gaxotte, P. (1992) Efficacy of ivermectin in the therapy of cutaneous larva migrans. *Archives of Dermatology* 128, 994–995.

Caumes, E., Datry, A., Mayorga, R., Gaxotte, P., Danis, M. and Gentilini, M. (1994) Efficacy of ivermectin in the therapy of larva currens. *Archives of Dermatology* 190, 132.

del Giudice, P. and Marty, P. (1999) Ivermectin: a new therapeutic weapon in dermatology. *Archives of Dermatology* 135, 705–706.

Diallo, S., Aziz, M.A., N'dir, O., Badiane, S., Bah, I.B. and Gaye, O. (1987) Dose-ranging study of ivermectin in treatment of filariasis due to *Wuchereria bancrofti*. *Lancet* 1, 1030.

Foege, W.H. (1998) 10 Years of MECTIZAN. *Annals of Tropical Medicine and Parasitology* 92 (Supplement 1), S7-S10.

Forstinger, C., Kittler, H. and Binder, M. (1999) Treatment of rosacea-like demodicidosis with oral ivermectin and topical permethrin cream. *Journal of the American Academy of Dermatology* 41, 775–777.

Glaziou, P., Nguyen, L.N., Moulia-Pelat, J.P., Cartel, J.L. and Martin, P.M. (1994) Efficacy of ivermectin for the treatment of head lice (pediculosis capitis). *Tropical Medicine and Parasitology* 45, 253–254.

Greene, B.M., Brown, K.R. and Taylor, H.R. (1989) Use of ivermectin in humans. In: Campbell, W.C. (ed.) *Ivermectin and Abamectin*, Springer-Verlag, New York, pp. 311–323.

Guderian, R.H., Lovato, R., Anselmi, M., Mancero, T. and Cooper, P.J. (1997) Onchocerciasis and reproductive health in Ecuador. *Transactions of the Royal Society of Tropical Medicine and Hygiene* 91, 315–317.

Ismail, M.M. and Jayakody, R.R. (1999) Efficacy of albendazole and its combinations with ivermectin or diethylcarbamazine (DEC) in the treatment of *Trichuris trichiura* infections in Sri Lanka. *Annals of Tropical Medicine and Parasitology* 93, 501–504.

Lankas, G.R., Cartwright, M.E. and Umbenhauer, D. (1997) P-glycoprotein deficiency in a subpopulation of CF-1 mice enhances avermectin-induced neurotoxicity. *Toxicology and Applied Pharmacology* 143, 357–365.

Leppard, B. and Nebari, A.E. (2000) The use of ivermectin in controlling an outbreak of scabies in a prison. *British Journal of Dermatology* 143, 520–523.

MacFarland, A., Abramovich, D.R., Ewen, S.W. and Pearson, C.K. (1994) Stage-specific distribution of P-glycoprotein in first-trimester and full-term human placenta. *Histochemical Journal* 26, 417–423.

Marti, H., Haji, H.J., Savioli, L., Chwaya, H.M., Mjeni, A.F., Ameir, J.S. and Hatz, C. (1996) A comparative trial of single-dose ivermectin versus three days of albendazole for the treatment of *Strongyloides stercoralis* and other soil-transmitted helminths infections in children. *American Journal of Tropical Medicine and Hygiene* 55, 477–481.

Mumcuoglu, K.Y., Miller, J., Rosen, L.J. and Galun, R. (1990) Systemic activity of ivermectin on the human body louse (Anoplura: Pediculidae). *Journal of Medical Entomology* 27, 72–75.

Naquira, C., Jimenez, G., Guerra, J.G., Bernal, R., Nalin, D.R., Neu, D. and Aziz, M.A. (1989) Ivermectin for human strongyloidiasis and other intestinal helminths. *American Journal of Tropical Medicine and Hygiene* 40, 304–309.

Nnoruka, E.N. and Agu, C.E. (2001) Successful treatment of scabies with oral ivermectin in Nigeria. *Tropical Doctor* 31, 15–18.

Ottesen, E.A. (1985) Efficacy of diethylcarbamazine in eradicating infection with lymphatic-dwelling filariae in humans. *Reviews of Infectious Diseases* 7, 341–356.

Ottesen, E.A. (1987) Description, mechanism and control of reactions to treatment in the human filariases. In: Evered, D. and Clark, S. (eds) *Filariasis. Ciba Foundation Symposium 127.* John Wiley & Sons, London, pp. 265–283.

Scaglia, M., Marchi, L., Bruno, A., Chichino, G., Gatti, S. and Gaxotte, P. (1990) Effectiveness of ivermectin in human strongyloidiasis: a pilot study. *Therapy of Infectious Diseases* 4, 159–164.

Schulz-Key, H., Greene, B.M., Awadzie, K., Lariviere, M., Klager, S. and Dadzie, Y. (1986) Efficacy of ivermectin on the reproductivity of *Onchocerca volvulus*. *Annals of Tropical Medicine and Parasitology* 37, 89.

Van den Enden, E., Stevens, A. and Van Gompel, A. (1998) Treatment of cutaneous larva migrans. *New England Journal of Medicine* 39, 1246–1247.

Youssef, M.Y., Sadaka, H.A., Eissa, M.M. and el-Ariny, A.F. (1995) Topical application of ivermectin for human ectoparasites. *American Journal of Tropical Medicine and Hygiene* 53, 652–653.

Chapter 14

Macrocyclic Lactones as Antiparasitic Agents in the Future

T.G. Geary

Two distinct questions come to mind when we ask how macrocyclic lactone antiparasitic agents (MLs) will be used in the future. First, what general role will anthelmintics and endectocides play in the future? Secondly, and in that context, how will MLs be used? Consideration of these broad questions can be focused by separately addressing future challenges to the use of chemotherapy for metazoan parasites (including MLs), and the opportunities for new or expanding applications of MLs, in contrast to new management, new vaccines, new drugs, etc., that may diminish ML markets.

Challenges

Cultural and political trends, economic forces and scientific/technical factors will interact to change the use of anthelmintics in animal agriculture. Obviously, the influence of these factors will be moderated by largely unpredictable shifts in economic and political climates around the world. None the less, they deserve serious consideration because they could markedly alter how anthelmintics are used in the future.

It is important for this discussion to distinguish (rather arbitrarily) between the economies of two types of countries or regions based on economic profiles. The first is composed of regions that have capital/service/information-based economies (CSI countries); these consist mostly of what typically have been called 'developed' nations. The second includes those with economies based primarily on labour and natural resources (LNR countries). These countries typically derive much of their gross national product (GNP) from export of natural resources and possess an excess of cheap labour; they are usually termed 'developing' regions.

Cultural/political trends

An already strong and still emerging social trend in CSI countries is the demand for 'organic' food products. Premiums are now paid for products derived from animals raised in 'chemical-free' operations. Despite higher cost to the consumer, these products are met with enthusiasm in the marketplace. The lower productivity associated with raising animals in the absence of purposefully introduced chemicals (especially antibiotics, hormones and anthelmintics, though there is less pressure on the last) can be tolerated because higher prices can be charged to consumers motivated by non-economic principles.

How this trend will influence anthelmintic use depends upon the abundance of philosophically motivated consumers in target markets with enough disposable income to afford higher prices for food that is undetectably different from lower priced products. As long as the current period of economic prosperity lasts, this market will flourish. When the next recession occurs, it may significantly diminish. Abandoning modern methods of animal agriculture to pursue this upscale market segment would harm the larger fraction of the population, even in CSI countries, for which price is a major determinant of food purchases.

Another cultural/emotional trend leading away from anthelmintic use is the increasing profile of animal rights and vegetarianism in the CSI world (neither movement has yet had much influence in LNR countries). The trend toward reduction in meat and dairy product consumption based on health concerns can be partially countered by advertising campaigns or the development of lower fat items, at least in the short term. However, vegetarianism based on moral or emotional conversion, and promulgated by celebrities, could have a significant and increasing impact on the demand for meat and dairy products in the future. Stagnant or reduced market sizes in CSI countries will decrease the price of meat and dairy products as supply outstrips demand, and will result in decreased anthelmintic usage. Conversely, animal production may increase in LNR countries (see below). As these countries tend to cluster in warmer climates, where helminths are generally more problematic, global anthelmintic demand could actually increase.

The 'animal rights' movement has the goal of eliminating animal agriculture entirely, which would eliminate the major anthelmintic market. In the shorter term, however, proponents of animal rights aim to reverse the trend toward factory-style production of meat products. Of special concern are total indoor operations; the animal rights movement pushes for operations in which animals have ready access to the outdoors.

'Free range' chicken commands a premium in the market, and one can predict that free-range pork and beef will follow this trend. An interesting complication of this philosophy is that the incidence of helminthoses is almost certain to rise in such operations (Nansen and Roepstorff, 1999). The need for anthelmintics, for therapy if not prophylaxis, will increase as this trend develops.

Economic forces

Three major economic/political developments will alter the anthelmintic market. The trend toward 'corporatization' of agriculture in the USA can be expected to infiltrate animal production to an ever-greater extent. Individual beef and dairy producers may eventually be forced out of business, as has happened in the poultry sector and is happening in pork. This trend means that livestock producers will lose loyalty to traditional locales (in exchange for loyalty to profit), just as we have seen in the manufacturing industry. Secondly, a re-evaluation of the role of government financial support for agriculture, including livestock, should be expected as 'government minimalization' policies mature; this may result in a reduction in price supports. Finally, the international movement toward a global economy has increased pressure to reduce or eliminate tariffs and trade barriers that currently have dramatic effects on the stability of local animal production. It remains to be seen how this scenario will play out in CSI countries with different historical trajectories (the USA, Japan and Europe, for instance).

The combination of globalization and corporatization will shift animal production to the lowest cost/maximum return regions. Just as manufacturing jobs have fled CSI countries, we will see an increasing share of the livestock industry relocate to LNR countries. Tariff reduction will introduce even stiffer price competition into animal production operations in CSI countries. Operational costs (land and labour) and regulatory/social pressures typically are much lower in LNR areas than in (for example) the USA or Europe. Land use restrictions are also usually less onerous. As exports of beef and dairy increasingly flow from LNR to CSI countries, profit margins in the traditional animal health target countries will decline. All costs, including those for anthelmintics, will be carefully scrutinized. Conversely, the market for anthelmintics will increase in new producer countries, which, as noted above, typically have a greater need for helminth control. As wealth creation begins to lift personal income in LNR regions (or so we hope), consumption of animal protein (both dairy and meat) will increase. This shift will create new markets for anthelmintics, especially the MLs (due to their unparalleled spectrum and persistence).

Increased market penetration by generic MLs will accompany the expansion of the global market for cattle (and swine) anthelmintics. As cost competition intensifies, it will be difficult to support the higher price of branded versus generic drug. The availability of lower cost MLs will undercut profit margins in the anthelmintic field. It will also tend to displace the benzimidazole- and levamisole-type anthelmintics from the marketplace, to an even greater extent than has already occurred, as the price gap between MLs and these alternative anthelmintics narrows. Simply put, the persistence and spectrum of MLs are not equalled by any other class of anthelmintic. Cheap MLs will be heavily used, which will accelerate the development and spread of resistance in cattle parasites; the combination of a shift in use patterns to greater intensity of infection in more tropical climates, coupled with more frequent dosing, will exacerbate this trend.

Scientific/technical factors

Residues of introduced chemicals in foodstuffs have become a serious consumer concern, which has co-evolved with the 'Green' movement and the fascination with organic food products. To a neutral observer, the intense distrust of synthetic chemicals among a vocal sector of the CSI population appears to be almost a communal paranoia. Consumers readily ingest significant quantities of tens of thousands of chemicals of unknown structure and unknown risk in 'herbal' and other so-called natural products, without complaint. Simultaneously, they react with fear and loathing to trace amounts of synthetic molecules, usually derived from 'natural' sources as starting points, which suffer only from the fact that enzymes did not create all their chemical bonds (see Ames *et al.*, 1990). Coupled with the continuous development of increasingly sophisticated technology for detection of chemical residues, this concern eventually could lead to an extension of withdrawal times for MLs. While the scientific evidence used to calculate no-effect levels argues against such a move, it must be recognized that food quality decisions have become increasingly politicized. If consumers want zero residue levels, instead of assurances that only safe levels are permitted, pressure will mount to achieve undetectable levels. This argument is of most concern for the longest acting drugs, which are precisely those that provide the greatest economic benefit for the producer (fewer doses per growing season). Just as social pressures counteract the benefits of transgenic plants, the pressure for residue-free meat and dairy products may undermine the current use pattern of anthelmintics in animal agriculture as ever-more sensitive detection technologies come on-line. This pressure is not faced in the companion animal (dog, cat, horse) sector in CSI regions; this lack of regulatory concern vis-à-vis residues in companion animals is another factor turning the animal health industry in this direction. While residues are not yet a major concern in most LNR countries (compared with the overwhelming need to feed people), the evolving export market from these countries could depend upon certification of residue-free status.

Is this threat real? We have already seen a ban on bovine somatotropin and hormone-treated meat products in some CSI countries, despite a lack of scientific evidence that these products constitute a health risk to the consumer. It would be a mistake to ignore the issue.

Deleterious environmental impact of agricultural chemicals constitutes another simmering concern. It has surfaced most prominently for pesticide and herbicide use, in which relatively large amounts of chemicals are introduced into the environment deliberately. The issue gained significance in the anthelmintic venue with the introduction of the MLs. Their extraordinary potency against a broad variety of invertebrates means that the small amounts of drug (or active metabolites) excreted into the environment from treated animals could, in theory, affect invertebrate species living in the area of the deposit. However, this issue has not markedly altered use patterns of the MLs and should not be a major factor in limiting their use in the future, unless a novel competitor with no action outside the animal is marketed aggressively in a 'Green' sense.

Resistance to anthelmintics, including the MLs, is a major problem in small ruminant operations throughout the world (Conder and Campbell, 1995; Waller, 1997; Sangster, 1999). It is foolish to believe that a similar scenario will never develop in cattle or companion animals (especially horses), the most lucrative segments of the market. Indeed, reports of ML resistance in cattle populations of *Cooperia* species are now available from CSI and LNR countries (Vermunt *et al.*, 1995; Coles *et al.*, 1998; Echevarria and Pinheiro, 1999; Anziani *et al.*, 2001; Fiel *et al.*, 2001). How long will it take before ML resistance becomes an economic problem in cattle and horses? Guessing about the time frame is not useful, as we lack the scientific tools and the will to acquire the data needed for worthwhile estimates. However, it is crucial to realize that no class of chemotherapeutic drugs has ever evaded the selection of resistant pathogen populations, and there is no scientific reason to imagine that MLs will be the exception to this rule (Geary *et al.*, 1999a). Erosion of the economic benefits of MLs due to resistance among key nematode parasites would certainly have dramatic consequences for their continued use. One result could be the development of combination drugs in which an ML is given along with a narrower spectrum anthelmintic that specifically eliminates the ML-resistant species.

It is increasingly popular to advocate non-anthelmintic worm control strategies as an alternative to chemotherapy. Techniques such as pasture management or paddock rotation, the use of fungi that prey on larval stages of parasitic nematodes, the development by targeted breeding of parasite-resistant host animals and vaccination have garnered a great deal of attention in academic and industrial research laboratories. Techniques that reduce chemical use meet with approval of organic food and 'Green' movements, but not animal rights groups (which could set up an interesting and divisive conflict). However, it has yet to be shown that any of these strategies can completely replace broad-spectrum anthelmintics in a price-competitive manner, and there is reason to doubt that any will reach that stage (see Vercruysse and Dorny, 1999). None the less, adoption of some of these tactics as part of organic production may erode ML use, especially in the upscale sector of the meat industry in CSI countries.

Opportunities

Although serious challenges to anthelmintic use in animal agriculture are evident, there are also opportunities to protect and perhaps enlarge the markets for these drugs. As for the Challenges section, this discussion will be separated into cultural, economic and scientific factors that represent emergent opportunities for anthelmintics, including new compounds and the MLs.

Cultural and political trends

Although it may be extreme, it is not inaccurate to assert that companion animals have become surrogate children for many people in CSI countries. Animal health companies have recognized lately that pet owners will spend a great deal of money on veterinary medicines, including anthelmintics, for dogs, cats and horses. The emotional investment in companion animals drives consumers to pay a premium for value-added products. Currently, spectrum of action, coupled with persistence, is the gold standard set by the MLs for anthelmintics/endectocides. However, one can envisage the development of an upscale veterinary medical market based on perceived (due to advertisement) value of high-cost products, one that is less driven by efficacy than by social status (as has developed in, for instance, children's clothing or dog and cat food).

This trend is threatened by its dependence on the abundance of disposable income in CSI economies. Is the cultural trend to view pets as children strong enough to withstand a recession? If not, an anthelmintic marketing strategy based on constant growth in the companion animal sector will be held hostage to the vagaries of economic conditions in CSI countries. Similar trends are likely to develop in the rest of the world. As living standards and disposable income rise in some LNR countries, spending on pets will increase. These emerging markets initially will be driven by cost as a primary determinant, so generic MLs should dominate over branded versions.

The combination of affluence and a litigious culture may open new anthelmintic markets in CSI economies. One can envisage a demand for parasite eradication campaigns based on the high perceived value of pets; a ready example is canine heartworm disease. It is not impossible to imagine an interventionist government mandating heartworm prophylaxis for all registered dogs (along with, for instance, rabies vaccinations), or to the conception and implementation of a medicated bait programme, employing MLs, designed to interrupt transmission from wild or stray canid populations. Although rabies vaccination is designed to protect humans against the disease, the increasing tendency to view pets as children may facilitate a movement to protect them from infectious disease, just as polio vaccination campaigns were motivated.

The future may witness the growth of a fledgling veterinary public health market. The twin trends of the movement of human residential areas to formerly wild regions and a dramatic decline in hunting and trapping will inevitably increase wildlife contact with humans and their pets. Of particular concern among helminths will be zoonoses due to infection with *Baylisascaris procyonis* from raccoons, *Echinococcus* spp. from wild and domestic canids, and larval migrans conditions from various ascarids of wild and domestic origin. Concern regarding arthropods includes zoonoses such as Lyme's disease or Rocky Mountain spotted fever that are transmitted by ticks from residential deer populations. Even a few cases, if sufficiently publicized, could lead to a strong public demand for programmes to treat wildlife for helminths or disease-transmitting ticks. It is a repeated observation in CSI societies that affluence cultivates the demand for guaranteed safety, a trend reinforced by the abundance of attorneys in some CSI countries. Development of formulations and/or delivery strategies suitable for mass treatment of wildlife could drive this scenario into existence.

Economic changes

Generic anthelmintics will eventually dominate the market, as noted above, unless a novel, proprietary product appears. While challenging the market share of branded products (and thus the bottom lines of research-driven animal health firms), this situation provides a growth opportunity for companies that produce generic versions of veterinary medicines. The availability of ML products priced competitively with benzimidazoles and the nicotinic agonists (levamisoles, etc.) will lead to increases in the number of ML doses used per year.

As long as anthelmintic treatment can be shown to have production benefits, there will be a role for MLs in CSI economies for livestock and dairy operations. However, cost constraints that will accompany the impending market fission (between upscale, organic, high profit, high cost operations, and mass-production-oriented, chemical-dependent, cost-constrained, lower margin operations) will drive business toward generic MLs.

For veterinary anthelmintics, the only current unmet medical needs are encountered in the small ruminant market, in which multiple anthelmintic resistance has made control of parasitic nematodes almost unattainable in some regions (Waller, 1997). The scope and profit margin of small ruminant operations are generally considered too small to warrant a major discovery research effort for new sheep-specific anthelmintics. However, a new product introduced into this market would achieve rapid penetration and reasonably high profit if able to control nematodes resistant to commercially available compounds. The widespread presence of ML resistance suggests that an opportunity for new sales will not be realized by these drugs or by the other classes of anthelmintics currently available, unless a strategy for specifically combating resistant strains emerges (see below).

Scientific vistas

Political and cultural pressures to limit residues of anthelmintics can be partially offset by a better understanding of the pharmacology of the regulated molecules. Fortunately, we have a good preliminary understanding of the mechanism of action of anthelmintics, including the MLs (Cully *et al.*, 1996; Martin, 1997). However, our view of the mechanism of action of MLs at the molecular level is based largely on work done in the free-living nematode, *Caenorhabditis elegans* (Cully *et al.*, 1996). It remains an important objective to extend these observations to parasitic species, and to determine the extent to which different groups of parasites resemble *C. elegans* and each other in this regard.

Defining the mechanism(s) of anthelmintic resistance is a more pressing and more profitable task. As anthelmintic (particularly ML) pressure increases on cattle parasites, selection and expansion of resistant populations inevitably will occur. Currently available methods for tracking the spread of anthelmintic resistance alleles in populations of these pathogens are woefully inadequate, with the single exception of benzimidazole resistance (which is discussed below). Efforts to contain the spread of resistance will be fatally hampered if we cannot detect resistant parasites until they constitute a considerable proportion of the population. Research on the mechanism of anthelmintic resistance will have several benefits. First, knowledge of the mechanisms of resistance should allow the development of DNA-based techniques to detect and track the evolution and spread of resistance alleles in parasite populations. Some progress in this regard has been made for benzimidazoles (Elard *et al.*, 1999), although the story may be more complicated (Prichard,

2001). We may be able to limit the eventual impact of ML resistance by careful management practices if an accurate diagnostic can be found in time. Unfortunately, current research efforts in this direction are woefully inadequate.

A second benefit of basic research on anthelmintic resistance is that a strategy for specifically attacking resistant parasites may be revealed. For example, it appears that selection of an allele of β-tubulin with a tyrosine-for-phenylalanine substitution at position 200 is associated with benzimidazole resistance (Roos *et al.*, 1993). Identification of a new drug that selectively attacks β-tubulin with this substitution would permit the development of a powerful new anthelmintic combination with a generic benzimidazole. The combination should be effective against both benzimidazole-susceptible and -resistant parasites. Similarly, identification of the molecular mechanism of ML resistance could lead to a screening programme designed to discover new compounds that selectively target ML-resistant parasites.

Finally, additional research in this direction will reveal whether multiple mechanisms of ML resistance exist in a single species, as has been proposed for *Haemonchus contortus* (Prichard, 2001), or between different species. Work done in *C. elegans* shows that resistance to MLs is achieved by a combination of loss-of-function mutations in genes that encode subunits of glutamate-gated chloride channels (Dent *et al.*, 2000). How well is ML resistance in parasitic species predicted by studies done in this model organism?

Until (unless) ML resistance in cattle parasites becomes so widespread as to render the drugs essentially useless, they will dominate the anthelmintic market for a very long time. The quality of MLs is so high (spectrum, persistence, potency) that it is difficult to envisage the discovery of a competitive molecule in a novel structural class. Consequently, there is a common misperception that there is little incentive for animal health companies to sustain an (increasingly expensive) anthelmintic discovery campaign (see Geary *et al.*, 1999b). It is difficult to justify an anthelmintic discovery and marketing programme on the basis of the small ruminant market alone, which is ill-served by currently available drugs. Resistance in cattle and horses (if not other companion animals) will eventually demonstrate the need for new anthelmintics in these profitable market sectors, but it is impossible to predict accurately when this will be evident to consumers. However, the company that maintains an anthelmintic discovery programme, and so has an acceptable replacement for (or supplement to) the MLs when resistance becomes undeniable, will be well rewarded for its foresight.

The current lack of research investment devoted to anthelmintics is simply replaying the antibiotic scenario. Only a few years ago, the accepted dogma was that so many antibiotics were available that new compounds were really not needed. The consequent reduction in academic research on antibiotics and infectious diseases, and in industrial investment on antibiotic discovery, left us ill prepared to deal with the recent onslaught of multiply resistant strains of pathogenic bacteria. It does not appear that this lesson has been learned by those who target veterinary parasites for control.

Conclusions

Anthelmintics have a valuable role in most areas of animal agriculture and for companion animal medicine. The ML class of anthelmintics is clearly superior to other classes, offering an unprecedented combination of spectrum, safety, persistence and potency. Expiration of patent protection will shift the market from branded to generic products, reducing profit margins and providing a limited incentive for investment into the discovery of novel, proprietary anthelmintics. The twin forces of corporatization and globalization (profit maximization coupled with reduction/elimination of trade barriers) will shift animal production toward LNR countries, away from CSI regions. This shift will increase demand for anthelmintics, which will largely be met by generic MLs. The market share held by non-MLs will diminish even further. Increasing frequency of ML dosing in the face of increased parasite pressure will exacerbate the development and spread of ML resistance in cattle parasites. We are poorly positioned to deal with the situation, as prospects for the discovery and development of novel anthelmintics in the foreseeable future are dim.

References

Ames, B.N., Profet, M. and Gold, L.S. (1990) Nature's chemicals and synthetic chemicals: comparative toxicology. *Proceedings of the National Academy of Sciences of the USA* 87, 7782–7786.

Anziani, O.S., Zimmerman, G., Guglielmone, A.A., Vazquez, R. and Suarez, V. (2001) Avermectin resistance in *Cooperia pectinata* in cattle in Argentina. *Veterinary Record* 149, 58–59.

Coles, G.C., Stafford, K.A. and MacKay, P.H. (1998) Ivermectin-resistant *Cooperia* species from calves on a farm in Somerset. *Veterinary Record* 142, 255–256.

Conder, G.A. and Campbell, W.C. (1995) Chemotherapy of nematode infections of veterinary importance, with special reference to drug resistance. *Advances in Parasitology* 35, 1–84.

Cully, D.F., Wilkinson, H., Vassilatis, D.K., Etter, A. and Arena, J.P. (1996) Molecular biology and electrophysiology of glutamate-gated chloride channels of invertebrates. *Parasitology* 113, S191-S200.

Dent, J.A., Smith, McH.M., Vassilatis, D.K. and Avery, L. (2000) The genetics of ivermectin resistance in *Caenorhabditis elegans*. *Proceedings of the National Academy of Sciences of the USA* 97, 2674–2679.

Echevarria, F. and Pinheiro, A. (1999) Eficiência de anti-helminticos em bovinos. *Abstracts of the XI Seminário Brasiliero de Parasitologia Veterinária, 24–28 October, 1999*, Abstract TL-HB-274.

Elard, L., Cabaret, J. and Humbert, J.F. (1999) PCR diagnosis of benzimidazole-susceptibility or -resistance in natural populations of the small ruminant parasite, *Teladorsagia circumcincta*. *Veterinary Parasitology* 80, 231–237.

Fiel, C.A., Saumell, C.A., Steffan, P.E. and Rodriguez, E.M. (2001) Resistance to *Cooperia* to ivermectin treatments in grazing cattle of the Humid Pampa, Argentina. *Veterinary Parasitology* 97, 211–217.

Geary, T.G., Sangster, N.C. and Thompson, D.P. (1999a) Frontiers in anthelmintic pharmacology. *Veterinary Parasitology* 84, 275–295.

Geary, T.G., Thompson, D.P. and Klein, R.D. (1999b) Mechanism-based screening: discovery of the next generation of anthelmintics depends upon more basic research. *International Journal for Parasitology* 29, 105–112.

Martin, R.J. (1997) Modes of action of anthelmintic drugs. *Veterinary Journal* 154, 11–34.

Nansen, P. and Roepstorff, A. (1999) Parasitic helminths in the pig: factors influencing transmission and infection levels. *International Journal for Parasitology* 29, 877–891.

Prichard, R.K. (2001) Genetic variability following selection of *Haemonchus contortus* with anthelmintics. *Trends in Parasitology* 17, 445–453.

Roos, M.H., Kwa, M.S.G., Veenstra, J.G., Kooyman, F.H.J. and Boersema, J.H. (1993) Molecular aspects of drug resistance in parasitic helminths. *Pharmacology and Therapeutics* 60, 331–336.

Sangster, N.C. (1999) Anthelmintic resistance: past, present and future. *International Journal for Parasitology* 29, 115–124.

Vercruysse, J. and Dorny, P. (1999) Integrated control of nematode infections in cattle: a reality? A need? A future? *International Journal for Parasitology* 29, 165–175.

Vermunt, J.J., West, D.M. and Pomroy, W.E. (1995) Multiple resistance to ivermectin and albendazole in *Cooperia* species of cattle. *Veterinary Record* 137, 43–45.

Waller, P.J. (1997) Anthelmintic resistance. *Veterinary Parasitology* 72, 391–412.

Index

Page numbers in *italics* refer to tables. ML, macrocyclic lactones. Trade names are listed in upper case.